11-7-01

D0076830

OXFORD STUDIES
IN
NUCLEAR PHYSICS

GENERAL EDITOR
P. E. HODGSON

✓

OXFORD STUDIES IN NUCLEAR PHYSICS

Series Editor   Peter E. Hodgson   University of Oxford

1   J. L. Emmerson: *Symmetry principles in particle physics*

2   J. M. Irvine: *Heavy nuclei, superheavy nuclei, and neutron stars*

3   I. S. Towner: *A shell-model description of light nuclei*

4   P. E. Hodgson: *Nuclear heavy ion reactions*

5   R. D. Lawson: *Theory of nuclear shell model*

6   W. E. Frahn: *Diffractive purposes in nuclear physics*

7   S. S. M. Wong: *Nuclear statistical spectroscopy*

8   N. Ullah: *Matrix ensembles in the many-nucleon problem*

9   A. N. Antonov, P. E. Hodgson, I. Zh. Petkov: *Nucleon momentum and density distributions in nuclei*

10   D. Bonatsos: *Interacting boson models of nuclear structure*

11   H. Ejiri, M. J. A. deVoigt: *Gamma-ray and electron spectroscopy in nuclear physics*

12   B. Castel, I. S. Towner: *Modern theories of nuclear physics moments*

13   R. F. Casten: *Nuclear structure from a simple perspective*

14   I. Zh. Petkov, M. V. Stoitsov: *Nuclear density function theory*

15   I. Gadioli, P. E. Hodgson: *Pre-equilibrium nuclear reactions*

16   J. D. Walecka: *Theoretical nuclear & subnuclear physic*

17   D. N. Poenaru, W. Greiner: *Handbook of nuclear properties*

18   P. Fröbrich, R. Lipperheide: *Theory of nuclear reactions*

19   F. Stancu: *Group theory in subnuclear physics*

20   S. Boffi, C. Giusti, F. D. Pacati, M. Radici: *Electromagnetic response of atomic nuclei*

21   G. Ripka: *Quarks bound by chiral fields*

22   H. Ejiri, H. Toki (eds): *Nucleon hadron many body systems*

23   R. F. Casten: *Nuclear structure from a simple perspective, 2/e*

NUCLEAR STRUCTURE FROM A SIMPLE PERSPECTIVE

SECOND EDITION

R. F. Casten

Professor of Physics at Yale University
and
Director of the Wright Nuclear Structure Laboratory

OXFORD
UNIVERSITY PRESS

OXFORD
UNIVERSITY PRESS

Great Clarendon Street, Oxford OX2 6DP

Oxford University Press is a department of the University of Oxford.
It furthers the University's objective of excellence in research, scholarship,
and education by publishing worldwide in

Oxford New York

Athens Auckland Bangkok Bogotá Buenos Aires Calcutta
Cape Town Chennai Dar es Salaam Delhi Florence Hong Kong Istanbul
Karachi Kuala Lumpur Madrid Melbourne Mexico City Mumbai
Nairobi Paris São Paulo Singapore Taipei Tokyo Toronto Warsaw

with associated companies in Berlin Ibadan

Oxford is a registered trade mark of Oxford University Press
in the UK and in certain other countries

Published in the United States
by Oxford University Press Inc., New York

© R. F. Casten 1990, 2000

The moral rights of the author have been asserted

Database right Oxford University Press (maker)

First published 1990
Second edition 2000

A catalogue record for this book is available from the British Library

Library of Congress Cataloging in Publication Data
(Data available)

ISBN 0 19 850724 0

Typeset by Devi Information Systems, Chennai, India
Printed in Great Britain
on acid-free paper by
Biddles Ltd, Guildford & King's Lynn

PREFACE TO THE SECOND EDITION

This textbook represents a rather different approach than normally encountered. It emphasizes the conceptual basis of nuclear structure, and an intuitive (but hopefully neither sloppy nor handwaving) approach to nuclear models, rather than a regurgitation of formal elements that can be found elsewhere. I urge new readers to scan the Preface to the first edition for more information and explanation of the philosophy and content of the book.

I have always enjoyed teaching students from this point of view and hoped that they and others would welcome a text based on such an approach. As it happened, this was the case and I have been gratified by the enthusiastic reception accorded the first edition. There are, however, several motivations for a new edition, which I briefly outline.

First, it allows me to correct an astronomically large number of errors and misprints in the first edition – a number that seems barely exceeded by Avogadro's. Many readers have pointed out such errors, and I have discovered others in the process of teaching from this text at Yale. I am grateful to all who have informed me of mistakes. These include, but are not limited to, Kris Heyde, Ani Aprahamian, Jolie Cizewski, Victor Zamfir, Christian Guenther, Charles Barton, Mark Caprio, and, above all, Jean Kern who went systematically through the first edition and ferreted out the majority of errors I have corrected. His care and thoroughness are greatly appreciated.

The second motivation for the new edition is that, contrary to what some anticipated at the end of the 1980s, the field of nuclear structure has grown and prospered in the interim and is now enjoying a flowering renaissance. This is due in large part to the development of new technology (e.g., radioactive beams, advanced detectors, and the computers to handle the data) that now allows long unanswered questions to be resolved or re-attacked, and new horizons to be approached. To this end I have expanded the treatments of multi-phonon states, and of horizontal approaches to the evolution of structure, and have added a new chapter on radioactive beams (physics opportunities and experimental approaches). Astonishingly, and reflecting the rapidity of changes in the field, neither radioactive beams nor halo nuclei were even mentioned in the first edition.

I have also written new sections on isospin in Chapter 3 and the Geometric Collective Model in Chapter 6 that should have been in the first edition. I have added short discussions of numerous topics, such as the Fermi gas model, properties of the deuteron, implications of the Uncertainty principle, the relation between the 2-body nucleon–nucleon interaction and the treatment of nuclear structure in terms of a 1-body potential well, the basic properties of nuclear wave functions in such wells, the liquid drop model and semi-empirical mass formulas, dynamic and kinematic moments of inertia, and superdeformation. All these should also have been in the first edition. A large part of the discussion of the single particle shell model has been rewritten as have the discussions of γ and β vibrations and band mixing in Chapter 6. To keep the length within reasonable bounds, the material on exotic nuclei replaces Chapter 10 which never fit very well anyway.

In the Preface to the first edition I recommended several alternate textbooks which I particularly like and/or which I thought complemented the treatment here. I would now like to add three additional ones, those by Cohen, by Yang and Hamilton, and a second text by Heyde (1994): all three, like this book, use a number of physical and intuitive arguments to explain the results of various nuclear models. In some cases new material here was borrowed from, or inspired by, their treatments.

In the first edition of the book, I made the point that no attempt was made to reference the literature as this is a textbook, not a review article. In the second edition, I have basically kept to this philosophy except that, in a few cases where I explicitly update a topic with new data, I felt guilty about the anonymity of those whose recent research is being cited and hence in a few cases have made appropriate citations in the text.

In writing this new edition, I would like to add some acknowledgments to those from the first edition. I am especially grateful to Victor Zamfir for many exciting collaborations in the last 8 years, for helpful suggestions for this edition, and for creating most of the new figures that appear throughout the book. Many of these, especially in the section on Correlations of Collective Observables, summarize collaborative research with him. His ideas and insights, his amazing capacity for work, and his unselfish help in preparing this edition are very much appreciated.

I am particularly grateful to my graduate and undergraduate students at Yale who have taken courses based on this material and especially to Teha Kordich, Adam Hecht, Deseree Meyer and Mark Caprio for many insightful questions. Among the fruits of those is the new section on isospin. Mark Caprio is also responsible for the specific parametrization at the end of the discussion of the Geometric Collective Model in Chapter 6.

I am also grateful to Witek Nazarewicz for much inspiration, for his profound ideas on exotic nuclei, and for many collaborative experiences in the exciting new field of radioactive beams, of which he is one of the worldwide leaders.

My gratitude to Peter von Brentano, already noted in the Preface to the first edition, should be reiterated and expanded. I have been collaborating with him now for over 15 years and the experience has been both exciting and a joy. His approach to nuclear physics is unique, profound, and fruitful.

I would also like to thank Con Beausang and Reiner Krücken for many enlightening discussions of collective nuclei and of the γ ray spectroscopy which illuminates them. I am also grateful to them, and to Victor Zamfir, Norbert Pietralla, Mark Caprio, Alex Wolf, Lissa Zyromski, and Charles Barton for help in proofreading the galleys.

I am grateful to Yale University, its Physics Department, and the staff of the Wright Nuclear Structure Laboratory for help in many aspects of developing this new edition. In particular, I would like to express my deep appreciation to Jennifer Tenedine who did an absolutely marvelous job in preparing the new manuscript material out of illegible and convoluted chicken scratches, written on the backs of scraps of paper, in illegible faxes from faraway lands, and on paper napkins, and in persevering through a gazillion revisions. Amazingly, she never complained. This had two effects, both good. First, I didn't feel as guilty. Secondly, I wasn't as wary of still more revisions with the net result that (I hope) the final product is better for her efforts than it would have been.

New Haven, Connecticut   R. F. C
March 1999

PREFACE TO THE FIRST EDITION

This is a very personal book, reflecting the way I see and understand nuclear physics and my belief that if something cannot be explained simply it is not really understood, nor will it be a fertile inspiration for new ideas. It is quite different from existing texts. It is in no way intended to replace them, but aims instead to complement the standard approaches to nuclear structure.

The idea for the book and much of its approach arose gradually out of both formal and informal courses in nuclear structure I have given at the Institute Laue–Langevin in Grenoble, France, the Institut für Kernphysik in Köln, Germany, and at Drexel University, Philadelphia, as well as from a series of very informal tutorial-like sessions with several of my graduate students. The book represents an attempt to cut through the often heavy mathematical formalism of nuclear structure and to present the underlying physics of some pivotal models in a simple way that frequently emphasizes semi-classical pictures of nuclear and nucleonic motion and repeatedly exploits a few fundamental ideas. Such an approach has worked for me. I can only hope others will find it useful.

The emphasis in this book always centers on seeking the essence of the physics: rigor is therefore often sacrificed. Of course, rigor is absolutely essential to a proper development of nuclear models and to precise calculations. Yet it can also be terrifying with page upon page of complex formalism, Racah algebra, tensor expansions, and the like. Necessary as this is, especially to those who will become practitioners of particular models, many readers can become either discouraged or buried in the formalism. Unfortunately, in either case, the beauty, elegance, and conceptual economy of nuclear structure theory is often missed. The complexity necessary in formal treatments at times obscures rather than illuminates the simple physics at work. Moreover, rigor can be found elsewhere, in many excellent texts: there is no need for another book to repeat it. (Several of the best of these texts, such as those by de Shalit and Talmi, Bohr and Mottelson, Brussaard and Glaudemans, deShalit and Feshbach, or the recent work by Heyde, are cited in the reference list at the end. They are indispensable). What cannot be found so easily, though most individual parts of it probably exist scattered throughout the literature, is a systematic attempt to convey a more physically intuitive way of thinking about nuclear structure and of extracting the essential physics behind the derivations and models.

I honestly do not know whether this attempt will work. I feel it is worth a try since it is, in fact, a way of thinking used every day by practicing scientists but that is seldom presented in formal texts. If successful, it can deepen real understanding. As T. D. Lee once said (BNL Colloquium, May 1983): "That a thing is elementary does not mean it cannot be deep." Moreover, since physics research is fundamentally a creative act depending on insight and imagination (backed, of course, by hard science—experimental or theoretical) an intuitive sense of nuclear structure can foster new inspirations and remove much of the mystery surrounding formal or calculational complexity. Finally, an approach of this type can be of considerable practical use. My hope in fact is that the reader will come to appreciate how far one can go in obtaining many results of detailed calculations almost instantly, essentially "by inspection." It can help, for example, in anticipating the potential usefulness of a given model, in spotting errors in calculations, or in estimating the effects of particular parameter changes. One of the best examples is the famous Nilsson diagram: nearly all its features, and even the semiquantitative structure of Nilsson wave functions, can be deduced without calculation. The same applies to much of the study of residual interactions in the shell model, to collective models, to the structure of RPA vibrational wave functions, Coriolis coupling, or the IBA. To facilitate the development of this "sense" of the physics, and to provide contact between models and real data, concrete examples are almost always discussed in some detail and compared to exact calculations.

The book is no way intended as a complete treatise on nuclear structure, either in overall coverage or within each topic. Other texts are more comprehensive. For example, many active areas of modern nuclear physics (e.g., relativistic heavy ion physics and quark–gluon plasmas, mesonic and quark degrees of freedom, or baryon excitations such as delta resonances) are totally ignored. Nevertheless, the book relates to all of them in that it discusses the basic models of nuclear structure, which successfully describe virtually all low energy nuclear phenomena, and which subnucleonic approaches must eventually reproduce. Just as it is difficult to discuss relativistic effects in nuclei without first knowing how far a non-relativistic theory can go, it is necessary to understand how far traditional nuclear structure theory can go if one is to isolate the effects of quarks, mesons, nucleonic excitations, and the like.

Even within traditional nuclear structure, many areas are bypassed. One reason for this selectivity is that it allows deeper and more detailed discussions of the subjects that are treated, discussions that would normally be found only in specialized monographs. A more practical and honest reason is that the book reflects that small part of the subject with which I am somewhat familiar and where I felt I could attempt something that was more than just parroting existing texts.

Finally, in addition to selectivity in topics there is a selectivity in treatment of each subject. For many areas, there exist several equivalent approaches. For example, there are works on residual interactions in the shell model that barely mention the concept of seniority and others that stress it throughout. Likewise, high spin phenomena in deformed nuclei can be discussed in terms of the Nilsson model with Coriolis coupling or with the formalism of the cranked shell model. In each case, I have used the approach (in these examples, seniority and the Nilsson–Coriolis concepts, respectively) that makes the essential physics clearest to me (and, I hope, to the reader) and with which I feel most comfortable.

Upton, New York   R. C.

ACKNOWLEDGMENTS

This book, while my responsibility, owes much to many people, both to those who molded my own education or who were important in the development of the approach I use here, and to those who played specific roles in the book itself. I cannot mention all but would particularly like to single out, for special thanks and gratitude, the following:

Above all, my father, Daniel F. Casten, for everything.

Ed Kennedy, who gave me my initiation into the excitement of nuclear physics.

D. A. Bromley, who more than fulfilled those early expectations and who has continually encouraged me since.

D. D. Warner, who never lets me get away with anything, for his creativity and insight, and for years of exciting work together at a perfect time. The section of Chapter 6 on the IBA is only slightly modified from portions of a review article co-authored with him. A particular piece of Chapter 7 (so noted at the appropriate point) is also largely due to him.

F. Iachello and A. Arima whose work and inspiration forever altered the course of my own work.

Igal Talmi, for innumerable discussions and for making occasional friendly disagreements so interesting, and enlightening.

J. A. Cizewski, who started the whole business of this book by asking and caring deeply about nuclear structure and who seemed to enjoy my efforts to answer.

A. Aprahamian, whose excitement at learning and discovery is an inspiration, for our collaborations, friendship, and discussions.

Jolie and Ani were my principal "guinea pigs" in all this and I am very grateful for their enthusiasm, which inspired my own, both for research and for trying to understand what nuclear structure is all about. Without them, this book would not exist.

K. Heyde, for many clarifications of my muddled thinking and collaborations that taught me much. Part of Section 4.4 is based largely on discussions with him, and Figures 4.14a and b are based on ones in his book.

O. Hansen, S. Pittel, A. ("Aleee") Frank, D. S. Brenner, and Zhang, Jing-ye for many productive and enlightening collaborations and discussions that broadened my horizons.

J. Weneser, M. Goldhaber, W. R. Kane, and H. Feshbach for discussions, encouragement, and moral support.

H. Börner and K. Schreckenbach for a series of productive collaborations, for demonstrating to me the power of precision, and for their efforts to create such an attractive ambiance for nuclear physics at the ILL.

The students at the ILL in 1975–1976, those (particularly K. Schiffer) in Köln in 1984–1985, and those at Drexel University in 1987 who also bore the burden of listening to these ideas in embryonic form.

P. von Brentano and A. Gelberg for countless thought-provoking discussions and for creating the stimulating atmosphere in Köln where much of this book was written.

Gritti's Café, Dürenerstrasse, Köln, and its staff, where most of that writing actually took place, for its atmosphere, its music (by Achille), and its cappuccino conductive to physics and physics poetry.

The mountains near Wengen, Switzerland, for much of the same, at an especially crucial stage.

My wife, Jo Ann, for her patience during thousands of hours when I was trying to write this book and for submerging her feelings that nuclei are, after all, a silly sort of thing to get all excited about.

Bob and Marian Haight and Don, Jo Ann and Kim Reisler, and Tom and Ahng Suarez for years of friendship, talk, hiking, skiing, tennis, and general support.

Jacki Hartt, whose patience is outdone only by her tennis.

Judy Otto, Irma Reilly, Walter Palais, and Whitey Caiazza of the Illustration group at Brookhaven National Laboratory for their very professional, and enthusiastic, help with the figures. I am also grateful to the Photography Group for their excellent work and for putting up with so many "emergencies."

Jackie Mooney, whose skills, including cryptology, are absolutely amazing and without whose work, friendship, and dedication this book would have been sheer fantasy. I promise never to inflict such agony again.

I would also specifically like to thank K. Heyde, M. Buescher, W. Krips, R. Schuhmann, and Zhang, Jing-ye for very careful reading of many parts of the manuscript and for pointing out numerous errors.

I am very happy to acknowledge Brookhaven National Laboratory for permission to pursue this project and for the infrastructure that made it possible.

Finally, I am grateful to Oxford University Press for asking me to write this book, for forbearance during several delays and lapsed deadlines, and for making the whole process as pleasant as it has been.

NOTATION

In a work such as this, the question of notation always presents a dilemma—whether to adopt a rigorously unique set of symbols or to use those commonly found in the literature. With one significant exception (so noted at the appropriate place in the text), I have followed the latter course since it facilitates further study by the reader and because, in practice, there should be little ambiguity of meaning. Generally, different uses of a given symbol are widely separated in context and location (with one awful exception in Chapter 4 concerning the "A" dependence of the interaction strength). To further help avoid confusion it may be useful to list a few of the notational choices.

ψ refers to a physical wave function while ϕ usually refers to a basis state or an unperturbed wave function.

E is used for excitation energies but also for quasiparticle energies.

ε is used for single particle shell model energies and for one of the quadrupole deformation parameters.

δ refers to the "contact like" residual interaction and also to a quadrupole deformation parameter.

Generally p, n are used for proton and neutron (although occasionally π and ν are substituted). However, p also refers to particle as in p–p or p–h (particle–particle or particle–hole excitation).

N refers alternately to neutron number, boson number in the IBA, and to oscillator shell number.

A is the mass number of a nucleus but also the angular part of a residual interaction (especially in Chapters 4 and 5).

α is an amplitude of a mixed wave function and also the alignment quantum number in the rotation alignment scheme.

The literature commonly uses two symbols for the angular momentum of a state – J and I. I have chosen to use J, and reserve I for the moment of inertia (sometimes a script I is used in the literature, or a θ). Lower case j is used for the angular momentum of an individual nucleon, whose orbital angular momentum is denoted by l. Quantum mechanically, the actual total angular momentum is $[J(J+1)]^{1/2}\hbar$. In almost all cases, these angular momenta are given in units of $\hbar$ without writing the $\hbar$: thus $J = 60\hbar$ is written $J = 60$. Finally, note that in nuclear physics the total angular momentum of a state is (erroneously but traditionally) called its "spin". A 6^+ state is said to have "spin 6".

States in a nucleus are labeled by their angular momentum and parity (+ or −) and by a subscript giving their relative position in the level scheme. For example, the ground state of an even–even nucleus is the 0_1^+ state. The first and second excited states with angular momentum 2 and positive parity are denoted 2_1^+ and 2_2^+. Two notations are used more or less interchangeably (somewhat according to whim) for the energies of such states, namely $E(2_1^+)$ and $E_{2_1^+}$.

The symbol Q, for quadrupole moment, is sometimes taken in the literature and in textbooks to have units of ecm^2 and sometimes cm^2 (or barns). In the latter case, expressions for E2 matrix elements, for example, need to include the quantity eQ. We follow this practice, with Q in cm^2 (or equivalent areal units).

V is used for the potential, the potential energy, and 2-body interactions. Sometimes the potential is denoted by *U*.

Units: A few quantities that are useful throughout are: $c \sim 3 \times 10^{10}$ cm/s $= 30$ cm/ns; $\hbar = 6.582 \times 10^{-22}$ MeV·s $= 197.327$ MeV·fm/c; electric charge $e = 1.602 \times 10^{-19}$ C; proton mass 938.272 MeV; neutron mass 939.565 MeV; Atomic mass unit (AMU) 931.494 MeV. Also note the definitions: 1 fm $= 10^{-13}$ cm, 1 barn (b) $= 10^{-24}$ cm^2, and for E2 transitions, 1 W.u. $= (5.94 \times 10^{-6})A^{4/3}$ e^{2}b^2. Near A $\sim$ 150, 1 W.u. ~ 0.005 e^{2}b$^2 = 50$ e^{2}fm^4.

Finally, quantities functioning explicitly as operators, matrices, and vectors are usually given in bold face especially where it is useful to distinguish them from normal quantities. The same symbol (e.g., Q, n_d) in normal face usually stands for the eigenvalue of that operator.

CONTENTS

I INTRODUCTION

1 Introduction 3
  1.1 Introduction 3
  1.2 The nuclear force 6
  1.3 Pauli principle and antisymmetrization 16
  1.4 Two-state mixing 18
  1.5 Multistate mixing 24
  1.6 Two-state mixing and transition rates 27

2 The nuclear landscape 30

II SHELL MODEL AND RESIDUAL INTERACTIONS

3 The independent particle model 49
  3.1 Fermions in a potential—general properties 54
  3.2 Nuclear potentials 64
  3.3 Predictions of the independent particle model 73
  3.4 Mass dependence of single-particle energies 79
  3.5 Isospin 84

4 The shell model: two-particle configurations 98
  4.1 Residual interactions: the δ-function 100
  4.2 Geometrical interpretation 116
  4.3 Pairing interaction 123
  4.4 Multipole decomposition of residual interactions 125
  4.5 Implications 134

5 Multiparticle configurations 141
  5.1 J values in multiparticle configurations: the m scheme 141
  5.2 Coefficients of fractional parentage 144
  5.3 Multiparticle configurations j^n: the seniority scheme 146
  5.4 Some examples 156
  5.5 Pairing correlations 163

III COLLECTIVITY, PHASE TRANSITIONS, DEFORMATION

6 Collective excitations in even–even nuclei: vibrational and rotational motion 173
  6.1 An introduction to collectivity 173
  6.2 Collective excitations in spherical even–even nuclei 177
    6.2.1 Quadrupole vibrations 179

6.3 Deformed nuclei: shapes 197
6.4 Rotations and vibrations of axially symmetric deformed nuclei 202
    6.4.1 Bandmixing and rotation–vibration coupling* 221
6.5 Axially asymmetric nuclei 240
6.6 The interacting boson model 252
6.7 Geometric collective model (GCM) 288

7 Evolution of nuclear structure 297
7.1 Overview of structural evolution 297
7.2 Valence correlation schemes: the $N_p N_n$ scheme 300
7.3 Correlations of collective observables 314
7.4 Phase transitions in finite nuclei 321

8 The deformed shell model or Nilsson model 331
8.1 The Nilsson model 332
8.2 Examples 344
8.3 Prolate and oblate shapes 346
8.4 Interplay of Nilsson structure and rotational motion 348

9 Nilsson model: applications and refinements 356
9.1 Single nucleon transfer reactions 356
9.2 The Coriolis interaction in deformed nuclei 364
9.3 Coriolis mixing and single nucleon transfer cross sections 375
    9.3.1 Unique parity states 377
    9.3.2 Hexadecapole deformations and unique parity states 381
9.4 Coriolis effects at higher spins 385
    9.4.1 Rotation aligned coupling 385

10 Microscopic treatment of collective vibrations 398
10.1 Structure of collective vibrations 398
10.2 Examples: vibrations in deformed nuclei 408

11 Exotic nuclei and radioactive beams 418
11.1 Methods of producing RNBs 422
11.2 Nuclear structure physics opportunities in exotic nuclei 425
11.3 Facets of structure in exotic nuclei 431
11.4 Some simple signatures of structure 442

References 453

Index 462

PART I

INTRODUCTION

1

INTRODUCTION

1.1 Introduction

The atomic nucleus is not a single object but a collection of species ranging from hydrogen to the actinides, and displaying an unbelievably rich and fascinating variety of phenomena. The nucleus is extremely small, namely about 10^{-12} to 10^{-13} cm in diameter, and can contain up to a couple of hundred individual protons and neutrons that orbit relative to one another and interact primarily via the nuclear and Coulomb forces. This system may seem so complex that little could ever be learned of its detailed structure. Indeed, many of us involved in research into nuclear structure proclaim loudly and strenuously that we have barely scratched the surface (both literally and figuratively, as we shall see) in our understanding of nuclear structure. From another perspective, however, we have an immense number of facts about nuclei and we understand an enormous amount, often in great detail, concerning what the individual nucleons do in atomic nuclei, how this leads to the observed nuclear phenomena, how and why these phenomena change from nucleus to nucleus, and how certain nucleons interact with each other in the nuclear medium. We have basic models—the shell model and collective models—both geometric and algebraic, that provide a framework for our understanding and that are extremely simple, and yet subtle and refined. It is only after these models and framework are appreciated that the limitations in our knowledge become focused and identifiable; the identification of the problems that persist is a prerequisite to further advancement. In this book we emphasize the known and understood as a background, map, and guide to the unknown.

We hope the reader perusing this book will come to appreciate two principal facts: namely, the beautiful richness and variety of nuclear physics and the extent to which we can understand nuclear data and models by invoking a few extremely basic ideas and drawing upon arguments that are physically transparent and intuitive. We will see that it is possible, in many if not most cases, to understand the detailed results of complex calculations with an absolute minimum of formalism and often by inspection. As an example, even such seemingly complicated results of nuclear models such as the famous Nilsson diagram and the detailed structure of Nilsson wave functions, or of the microscopic RPA wave functions of collective vibrations, can be derived, nearly quantitatively, without any calculation whatsoever.

We emphasize that one must make a careful distinction between this approach and what is commonly called *handwaving*. The latter, to this author's mind, is what one does when one does not really know an answer or explanation and tries to account for some piece of nuclear data or the result of some calculation by an offhand, qualitative, "explanation" that is often little more than a slogan or a repetition of key words or venerated phrases. We have all encountered examples of such handwaving: supposedly forbidden γ-ray transitions glibly explained as "due to mixing," extra or unexpected excited states dismissed as "due to neglected degrees of freedom," unexplained peaks in transfer reaction spectra ascribed to "higher order processes in the reaction mechanism," or explanations of model calculations as "resulting from the symmetry properties," or "from an energy minimization" (of course, but why, how?). Indeed, in many cases, such statements are true and are reasonably accurate descriptions. Otherwise they would not have become catch phrases. But abuses abound to such an extent that they often represent a watered down substitute for real understanding that is to be discouraged.

The approach here, in contrast, attempts to *extract* the *basic physics ideas* that emerge either from an inspection of nuclear data or that lie behind the results of some model or calculation, and to do this by applying a minimum of key physical and geometrical ideas about the nucleus. When attempting such a program, there is always the danger of losing sight of important subtleties and of ignoring the importance of formal rigor that is so necessary in detailed and realistic model calculations. Undeniably, there are certain results of such calculations and certain model predictions that can only be understood by carrying out the fully detailed, often tedious, calculations. However, the author has always felt, and hopes that the reader will be convinced, that it is remarkable how far one can go in understanding complex nuclear structure phenomena by careful but simple physical arguments. Of course, this approach has the lurking danger of itself slipping into handwaving. If such sins are kept to the minimum here, there is a chance that the reader may emerge with an appealing physical understanding and intuition into nuclear structure that many working physicists have attained but that is seldom spelled out in textbooks because of a hesitancy or reluctance to commit to writing the nonrigorous and intuitive arguments that all of us use and, mentally, rely upon. If anything, the philosophy of this book is that such ideas and such an approach should not be a skeleton hidden in a closet for fear of ridicule but rather an important aid that is a constant and continual complement to necessary formal and rigorous calculations. As frequently as possible, qualitative physical explanations or "derivations" will not be left to stand alone; rather, the physical intuition behind them will be tested by, confirmed, and confronted with the results of actual calculations or with the data on atomic nuclei themselves.

We have already stated that many of the arguments here will rely on "a few basic simple ideas". In fact there are three of these that are of absolute and essential importance. They are:

- The generally attractive and short-range nature of the nuclear force.
- The effects of the Pauli principle.
- An understanding of two-state mixing—that is, the effects on energies and wave functions when two nuclear states mix due to some residual interaction.

These ideas, plus a constant reference to a kind of geometrical picture of the nucleus, form the basic ingredients behind many of the arguments to be presented.

A general outline of the book is as follows: After the discussion of these three basic ingredients in Chapter 1, Chapter 2 departs from the usual way of presenting nuclear structure by "surveying the nuclear landscape"; that is, by collecting a number of examples of nuclear data, level schemes, transition rates, systematics, and so on. In a normal text that relies on a systematic, step-by-step progression of ideas, such a chapter would be out of place since it utilizes terminology and concepts that will be formally introduced later. It is inserted here so that the beautiful and elegant consequences of nuclear models discussed in subsequent chapters will not be presented in a contextual vacuum. Often such results, when first encountered, seem highly abstract and of little practical importance and it is, unfortunately, often only years later, when the practicing nuclear physicist has gained a deeper familiarity with nuclear data, that their significance is finally understood.

One reason we feel justified in this approach is that this book is aimed not only at beginning nuclear physics graduate students, to whom much of the jargon in Chapter 2 will be unfamiliar, but also to practicing and experienced nuclear physicists who may be interested in the kind of alternative and complementary approach emphasized here.

Chapters 3, 4, and 5 will deal with the shell structure of nuclei, as is traditional, starting with the independent particle model and going on to the shell model for multiparticle configurations. The formalism and mathematical development of the shell model is one of the most remarkable creations of nuclear physics and allows one to account for many empirically observed features of atomic nuclei with an absolute minimum of physical input (e.g., many detailed predictions can be made without ever specifying the nature of the central shell model potential or the detailed structure of a residual interaction). Unfortunately, the shell model formalism, and derivations based on it, are often complex and, quite honestly, terrifying. This has the unfortunate consequence that this subject is often skimmed or glossed over by students. These chapters attempt to highlight and give plausibility arguments for many shell model results while at the same time avoiding, as much as possible, such formal treatments. Simpler derivations are sometimes possible and are given in appropriate cases.

The next section of the book (Chapter 6) deals with collective models for even–even nuclei, starting with macroscopic models of vibrational and rotational motion. This material emphasizes the profound importance of the residual proton–neutron interaction (especially the $T = 0$ component) and its role in the onset of collectivity, configuration mixing, and deformation in nuclei, in inducing nuclear phase transitions, and the assistance its understanding can provide in simplifying the systematics of nuclear data. Following this, a treatment of algebraic models, principally the interacting boson approximation (IBA) model is given. The chapter ends with a quantitative discussion of the Geometric Collective Model (GCM).

Chapter 7 is new. Part of it replaces the end of Chapter 6 from the first edition and the rest of it expands on this type of "horizontal" approach to nuclear structure, namely the study of the *evolution* of structure with nucleon number. Such an approach has received much study in recent years, especially with the advent of radioactive beams and access to new regions of nuclei. It embodies two ideas, that of Valence Correlation Schemes and of Correlations of Collective Observables, both of which give a remarkably simple and compact overview of structural evolution. Implications and applications of these correlations schemes will be given.

In Chapters 8 and 9, the discussion turns to odd-mass deformed nuclei, and an extensive discussion of the Nilsson model and its consequences, extensions, and testing via single nucleon transfer reactions.

Most of the collective models discussed up to this point will have been phenomeno-logical or macroscopic. We will not discuss detailed microscopic approaches at length but will introduce such approaches in Chapter 10 since they provide both the micro-scopic justification of macroscopic models and a simple physical picture of collective excitations (especially vibrations) that will allow the reader to anticipate their detailed structure without calculation.

Finally, in the new Chapter 11, we offer a glimpse of the new field of exotic nuclei and radioactive beams, discussing both the opportunities provided by an expansion of our horizons and the experimental hurdles en route to exploiting them.

With this discussion of the philosophy and outline of the material to follow, we turn now to the three "cornerstones" mentioned earlier that are of such central importance to everything that follows. Many readers know that the nuclear force is attractive, that the Pauli principle is important, and understand that residual interactions can mix neighbor-ing states. They might be tempted to skip over these sections and of course that is their prerogative; indeed, these pages contain nothing that is new or not to be found elsewhere. However, they do present a somewhat different perspective and provide a ground and background for what follows.

1.2   The nuclear force

Nuclei exist and are composed of neutrons and positively charged protons. Were the nuclear environment dominated by the repulsive Coulomb force, this could not be the case: one therefore deduces immediately that there must be a strong, attractive, interaction that can overcome the repulsive Coulomb force and bind nucleons together. The nuclear force is at first glance a mysterious one since it has few if any recognizable consequences in macroscopic matter (i.e., everyday phenomena). And, in fact, the exact nature of this force is still largely unknown. Nevertheless, it is remarkable how much we can learn about it from a few simple empirical facts. We have already stated that the very existence of nuclei implies a new force—the strong interaction that can overcome the Coulomb repulsion between protons. Beyond this many experiments point to two basic facts:

- Nuclei are small, on the order of 10^{-12} to 10^{-13} cm in diameter.
- For all practical purposes, the nuclear force can be neglected when considering atomic and molecular phenomena.

These two facts tell us that the nuclear force must be short range. A few further empirical observations allow us to refine this considerably.

- Nuclear binding energies, per nucleon, at first increase rapidly with A, until about $A \sim 10$ to 20, where they level off at approximately 8.5 MeV and remain roughly constant thereafter. These binding energies per nucleon are shown in Figure 1.1.
- The masses of mirror nuclei, which are defined as pairs of nuclei with interchanged numbers of protons and neutrons, $(Z, N)_1 = (N, Z)_2$, are nearly identical, after correcting for the different strengths of the Coulomb interaction in the two nuclei.

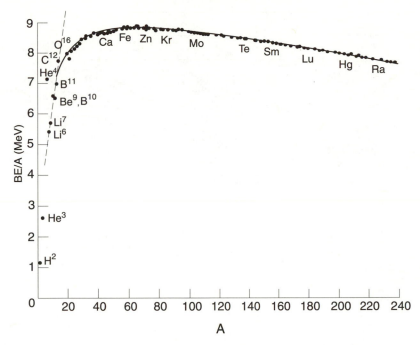

FIG. 1.1. Binding energies per nucleon (based on Eisberg, 1974). The solid curve is the result of a typical semi-empirical mass formula that includes corrections for surface effects, Coulomb repulsion, the Pauli principle, and pairing effects. The dashed line shows the extrapolated trend in binding energies if each nucleon interacted with all $A-1$ others.

- The sequencing, spin parity (J^{π}) values, and excitation energies of excited states in mirror nuclei are also nearly identical.

- Proton and neutron separation energies, denoted S(p) and S(n), are defined as the energies required to remove the last proton or neutron to infinity, and have a characteristic behavior with changing proton number and neutron number. Typical examples of such separation energies are shown in Figs. 1.2 and 1.3 from which it is evident that S(p) decreases with increasing Z and increases with increasing N while S(n) decreases with increasing N and increases with increasing Z. That is, each *decreases* with an increasing number of the *same* type of nucleon and *increases* with an increasing number of the *other* type. Although not exactly germane to the present discussion, we note for later use that the separation energies also show particularly large and sudden drops at certain special numbers of protons and neutrons, called *magic numbers*, namely N or $Z = 2, 8, 20, 28, 40, 50, 82, 126$. Those at 82 are evident in Figs. 1.2 and 1.3.

- Measurements of electron scattering provide abundant evidence of a nearly constant nuclear density independent of the number of nucleons A. This, in turn,

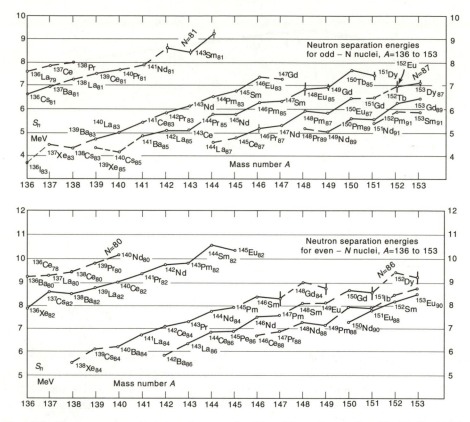

FIG. 1.2. Neutron separation energies near the $N = 82$ magic number (de Shalit, 1974).

implies that the nuclear volume must increase linearly with A. Neither of these facts may seem particularly surprising at first but it should be recalled that such is not the case with atomic systems whose sizes are nearly independent of Z. Note that if the nuclear volume $V \propto A$, then, assuming a roughly spherical nucleus, the nuclear radius scales as $A^{1/3}$. Innumerable studies have shown that a good approximation to the nuclear radius is $R = R_0 A^{1/3}$ where $R_0 \sim 1.2$ fm.

- There is only one bound state of the deuteron, the simplest nuclear system, with one proton and one neutron.

- This bound state has total angular momentum $J = 1$; that is, in the deuteron, the intrinsic spins (1/2) of the neutron and proton are aligned parallel to each other, not antiparallel. (This result assumes that the two nucleons have no relative orbital angular momentum, a result that will be justified in Chapter 3.)

- The deuteron has a nonzero quadrupole moment; that is, it has, on average, a preference for a nonspherical shape.

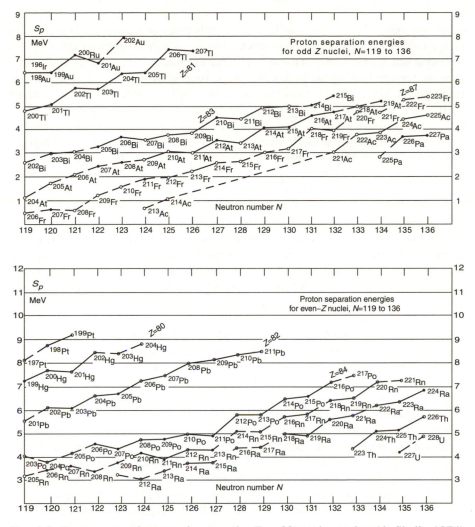

FIG. 1.3. Proton separation energies near the $Z = 82$ magic number (de Shalit, 1974).

Let us consider each of these facts in turn and see what we can learn about the nuclear force. Note that, except for the listing of the magic numbers, the preceding list contains virtually no specific nuclear structure information, although what we will learn from these facts about the nuclear force has many consequences for nuclear structure throughout the periodic table.

We have already deduced that the nuclear force is predominantly attractive and short range. (We neglect the short range repulsive core component.) The binding energy results, Fig. 1.1, and the fact that the nuclear density is approximately constant, allow us to go much further. The nuclear size, corresponding to a radius of $\sim 10^{-12}$ to 10^{-13} cm, tells us that the range of the force must be roughly this order of magnitude. However, the density and binding energy data tell us that the force is actually of a considerably smaller range than that of most nuclei. If the nuclear force extended more or less equally to all A nucleons, then the binding energy would increase roughly as $A(A-1)/2$, or the binding energy per nucleon (B.E./A) would increase with A and therefore so would the nuclear density. (See dashed line in Fig. 1.1.) That such is not the case, at least for $A \geq 10$, shows that the nuclear force saturates. The contrary fact that the binding energy per nucleon *does* increase for very light nuclei allows us to quantify this saturation and to make at least a crude estimate of the number of nucleons with which each other nucleon interacts.

To do this, we assume that all such interactions are equal and count the number of interactions assuming that each nucleon interacts with various numbers of other nucleons. Figure 1.4 gives a pictorial illustration of the connections and shows a plot of B.E./A deduced under different assumptions for the numbers of connections per nucleon. If one works through this figure, it becomes obvious how saturation arises when each nucleon interacts only with a small number, n_i, of others. For example, if $n_i = 3$, the number of interactions of each nucleon (and hence its binding energy) has already reached its maximum value when $A = 4$. We can use this approach to set some rough limits on n_i. Clearly $n_i = 2$ is unacceptable since it leads to immediate saturation at $A = 3$. Similarly $n_i = 3$ leads to premature saturation, but somewhere on the order of 6 to 10 mutual interactions leads to B.E./A values that approach saturation roughly where the data do. Thus, the empirically observed saturation in binding energy per nucleon data suggests that the range of the nuclear force is on the order of the size of nuclei such as Li or Be (i.e., approximately 2 to 4 fm). Crude as this analysis is, the idea behind it is qualitatively valid and the conclusion is more quantitatively correct than one might expect.

The properties of mirror nuclei also tell us much about the nuclear force. The data for three $A = 27$ nuclei are shown in Fig. 1.5 (note that ^{27}Mg is not mirror to the other two; it is shown for comparison and contrast). At the right, the figure gives the relative binding energies or masses of the three nuclei and on the left the low-lying excited states with their J^π values. The remarkable feature is the nearly identical spectra for the two mirror nuclei. The interactions between two nucleons can be divided into three categories: p–p, n–n, and p–n interactions. The data from mirror nuclei suggest that the nuclear force is "charge symmetric" (i.e., the p–p and n–n, interactions are equal). The fact that the absolute binding energy of ^{27}Al is greater than ^{27}Si does not reflect a breakdown of this idea, but rather the influence of the Coulomb interaction: ^{27}Si has more protons than ^{27}Al and, therefore, has a greater total repulsive Coulomb interaction that leads to lower total binding.

A more general characteristic of the nuclear force is charge *independence*, which means that the p–p, n–n, *and* p–n forces are equal. To examine this question, consider an isobar triplet such as $^{26}_{12}$Mg$_{14}$, $^{26}_{13}$Al$_{13}$, and $^{26}_{14}$Si$_{12}$. The low-lying levels of ^{26}Mg and ^{26}Si are similar as expected: ^{26}Mg has excited states $2^+(1.81$ MeV$)$, $2^+(2.94$ MeV$)$, $0^+(3.59$ MeV$)$, and so on, while ^{26}Si has $2^+(1.80$ MeV$)$, $2^+(2.78$ MeV$)$, $0^+(3.33$ MeV$)$, and so on. At first glance, the nucleus ^{26}Al appears quite different, but careful inspection of its level scheme shows a subset of excited states with similar binding energies as its isobars. Specifically, at energies *relative* to the lowest 0^+ state there are $2^+(1.84$ MeV$)$,

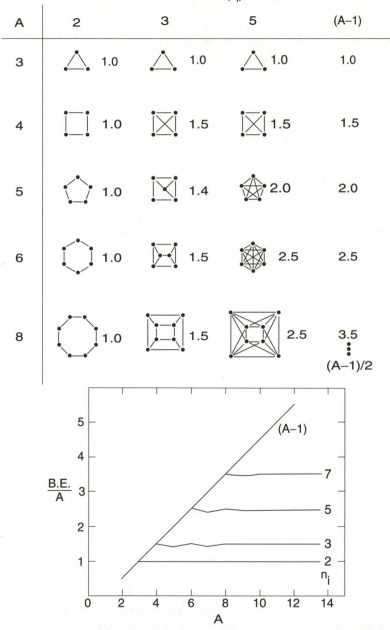

FIG. 1.4. Highly schematic calculation of the binding energy per nucleon under different "saturation" assumptions concerning nuclear forces. The number of connections indicated, n_i, is the number of nucleons with which each other nucleon is assumed to interact. All such interactions are considered to be of equal strength. The lower part shows a plot of the resulting binding energies per nucleon.

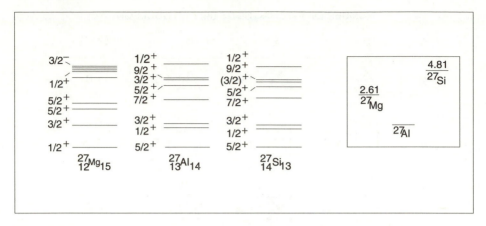

FIG. 1.5. Level schemes and binding energies (inset box at right with binding energies in MeV relative to ^{27}Al) of the mirror nuclei ^{27}Al and ^{27}Si, as well as the isobar ^{27}Mg.

Table 1.1 *Nucleon–nucleon interactions in A = 27 nuclei*

	$^{27}_{12}$Mg$_{15}$	$^{27}_{13}$Al$_{14}$	$^{27}_{14}$Si$_{13}$
p–p	66	78	91
n–n	105	91	78
p–n	180	182	182
Total	351	351	351

Based on deShalit, 1974.

2^+(2.93 MeV), 0^+(3.52 MeV) levels, and so on. This would seem to suggest that charge independence has approximate validity. However, there are other states that have no analogues in ^{26}Mg and ^{26}Si. We also note in Fig. 1.5 that ^{27}Mg is quite different than ^{27}Al and ^{27}Si, even though Table 1.1 shows that all three $A = 27$ isobars have the same total number of interactions (351).

   These results focus on a crucial aspect of the p–n system: it can exist in two different configurations. The concept of the proton and neutron as merely two states of the same particle, the *nucleon*, leads to the concept of *isospin*, which is formally (but not physically) similar to intrinsic spin. We will discuss isospin at some length at the end of Chapter 3. For now we just note the following basic ideas. In analogy with intrinsic spin, each nucleon is assigned an isospin $T = 1/2$: protons and neutrons are distinguished by the projection of this isospin on an imaginary isospin z-axis. This projection T_z is $-1/2$ for protons and $+1/2$ for neutrons. For multi-nucleon systems, the T_z projections add algebraically. $T_z = +1$ for a neutron–neutron system, -1 for a proton–proton system, and 0 for a proton–neutron system. Thus, since T cannot be less than its projection, a proton–proton or a neutron–neutron system must have $T = 1$. However, a proton–neutron system with T_z components of $-1/2$ and $+1/2$ has total $T_z = 0$, and T can be either 1 or 0. By charge independence, the interaction in the $T = 1$ p–n system must be identical to that in the p–p and n–n systems. There is, however, no reason why the p–n interaction has to be identical in the $T = 1$ and 0 couplings and, in fact, it is not: the $T = 0$ interaction is stronger.

Thus, there is no a priori reason to expect that ^{27}Mg, with fewer p–n interactions, should have the same sequence of energy levels as ^{27}Al or ^{27}Si and, indeed, it does not. Furthermore, ^{27}Mg is less bound than ^{27}Al, even though it has fewer protons and might, therefore, be expected to be more tightly bound. The reason is that it also has fewer $T = 0$ p–n interactions (see Table 1.1). This is already one piece of evidence suggesting the $T = 0$ interaction is stronger than the $T = 1$. The concept of the different and stronger, p–n interaction in the $T = 0$ channel will be of enormous importance later. For example, it determines why the excitation spectra of odd–odd nuclei differ so substantially from those of even–even nuclei. Moreover, its effects are intimately connected with those of the Pauli principle since $T = 1$ corresponds to a symmetric alignment of the two isospins in the p–n system, while $T = 0$ corresponds to an antisymmetric alignment.

Nucleon separation energies provide crucial information about the outermost nucleons and therefore about certain subtle aspects of the nuclear force in the "valence" region. As we will see, the most important nuclear model, the shell model, treats nuclei in terms of individual nucleons that orbit as independent particles in a central potential. Each orbit carries certain quantum numbers and a specific wave function. This is an excellent approximation of the actual motion except that there are important "residual interactions" beyond those encompassed by the central potential that must be considered when dealing with nuclei containing several particles outside closed or magic configurations. This will be a major topic of discussion in Chapters 4 and 5. We showed examples of separation energy data in Figs. 1.2 and 1.3 earlier and summarized the trends, which are valid for all mass regions: that S(p) decreases with increasing Z, that S(n) decreases with increasing N, and that each increases with increasing number of particles of the other type. Superimposed on this general behavior is a fine structure in that S(p) and S(n) display odd–even oscillation in Z and N such that nuclei with even numbers of either protons or neutrons have larger separation energies (i.e., are more bound). Though these separation energy data are widely familiar, it is seldom appreciated how much they tell us about the nuclear force.

The separation energies refer to the ground states of their respective nuclei: in nuclei with even numbers of protons and neutrons the ground state always has spin and parity $J^\pi = 0^+$. Invariably, this state is much lower in energy than any intrinsic excitation. The fact that S(p) and S(n) are larger when Z and N are even thus implies that there is a special attractive interaction in pairs of protons or neutrons coupled to $J^\pi = 0^+$. Later, we shall see that that is a property of short-range interactions resulting from the Pauli principle. The separation energy data also shows that the p–n interaction is strong and attractive since S(p) increases with increasing N and S(n) increases with increasing Z. In contrast, there is a decrease of each separation energy with increasing numbers of nucleons of the *same* type. (One needs to add that the decrease is larger than one would expect from the fact that the extra neutrons occupy higher energy, less bound orbits.) This gives the fundamentally critical result that, *aside from the pairing interaction*, the *residual* interaction between *like* nucleons is repulsive. This fact, pointed out in the early 1960s by Talmi, is seldom recognized or remembered; however, its consequences, are profound. For example, anticipating some concepts and jargon from upcoming chapters, it is one reason why singly magic nuclei do not become deformed and why the accumulation of proton neutron interaction strength is essential for the onset of collectivity and deformation in nuclei.

The properties of the simplest bound nuclear system, the *deuteron*, tell us still more about the nuclear force. The essential features, summarized earlier, are that there is only one bound state, that it has $J^\pi = 1^+$, and that the deuteron has a finite quadrupole moment. The fact that there is only one bound state and, moreover, that it is only weakly bound (the deuteron binding energy is 2.23 MeV) serves to emphasize the essential *weakness* of the so-called strong nuclear force. By *weak* here we mean weak in comparison to the kinetic energy of relative motion of the two nucleons. For example, this implies that the relative kinetic energy of two nucleons cannot be changed substantially by the strong interaction. This will be important in Chapter 3, when we discuss the reason why essentially independent particle motion is possible in a densely packed, strongly interacting nuclear medium.

It will also be shown in Chapter 3 that for a rather general central potential, the lowest energy state corresponds to zero orbital angular momentum (an S state). Thus, both the proton and the neutron in the deuteron must be in S orbital angular momentum states and the total angular momentum in the ground state can arise only from the proton and neutron intrinsic spins, $1/2\,\hbar$ (henceforth, in this book, we shall generally omit the units $\hbar$ in referring to angular momentum and intrinsic spin). There are two possible ways of coupling these two spins: to a total spin $S = 0$ or 1. The fact that the deuteron chooses the latter highlights an essential point: even though the nuclear force may have no explicit spin dependence, there can be large energy differences between states of different spins in multiparticle configurations. We shall discuss this point extensively in Chapter 4 where we shall see that the *implicit* spin dependence of nuclear forces is a reflection of the Pauli principle and that this has critical nuclear structure consequences.

We can learn more about the structure of the deuteron by considering its magnetic and quadrupole moments. Consider a circular loop of wire with current i, area A, and radius r. The magnetic moment is $\mu = iA$. If we view the magnetic moment as due to the motion of a charge e with mass m, velocity v, and orbital frequency ω, we have $\mu = e\omega\pi r^2$. But $\omega = v/2\pi r$ and the angular momentum $l = mvr$. So

$$\mu = \frac{evr}{2} = \frac{el}{2m}$$

Quantum mechanically, we can replace l by $l\hbar$, so

$$\mu = \frac{e\hbar}{2m}l$$

This is the magnetic moment of a quantal system where the charge and mass distributions coincide. To allow for cases where they do not, one introduces a g factor:

$$\mu = g\frac{e\hbar}{2m}l$$

For elementary particles, like the electron, the charge and mass distribution do coincide and one obtains from relativistic quantum theory that $\mu = e\hbar/2m$. Since $l = (1/2)\hbar$, this corresponds to $g = 2$. The proton and neutron have measured intrinsic magnetic moments $\mu_p = +2.79\ \mu_N$ and $\mu_n = -1.91\ \mu_N$ where $\mu_N = e\hbar/2m$ is the nuclear magneton. Again $l = (1/2)\hbar$, so $g_p = 5.58$ and $g_n = -3.82$. Since the neutron does not carry a net charge, the magnetic moment cannot arise from any orbital angular momentum. Hence, its finite magnetic moment implies that there must be a charge distribution inside the neutron. Clearly, then, from this alone we know that the neutron (and also analogously the proton) cannot be point particles.

Now, consider the deuteron. From the independent particle model single particle levels (see Chapter 3) both the proton and neutron are in $l = 0$ ($1s_{1/2}$) orbits. So the total magnetic moment is contributed solely by the intrinsic spin magnetic moments. If the spin angular momenta are parallel, the proton and neutron magnetic moments should add, giving

$$\mu_d = 2.79 - 1.91 = 0.88\ \mu_N$$

which is remarkably close to the measured value of $0.857\mu_N$. This simple result confirms three ideas: the additivity of the magnetic moments, the simple applicability of the shell model (see Chapter 3), and the $S = 1$ coupling of the proton and neutron spins in the deuteron.

We can easily further refine our understanding of the deuteron by reference to the experimental value of the quadrupole moment. The measured value is $Q_d = 2.7 \times 10^{-27}$ cm^2. In contrast, with a radius of $r_d \sim 1.4 \times 10^{-13}$ cm, the cross sectional area of the deuteron, πr_d^2, is much larger than Q_d. Thus, the deviation from sphericity must be very small. If the deuteron were composed solely of nucleons in $l = 0$ orbits, the quadrupole moment would vanish. [The quadrupole moment is given by $\langle\psi_d|\mathbf{Q}|\psi_d\rangle$ whose angular part for $l = 0$ orbits goes as (using very simplified notation) $\int Y_{00}Y_{20}Y_{00}\ dr$ which vanishes by the triangular properties of the spherical harmonics.] To conserve parity, the wave function of the deuteron can only contain either all even or all odd angular momentum terms. Since we know that there is a dominant S-wave component the quadrupole moment must be generated by a D-wave amplitude. Detailed analysis gives a D-wave probability of about $P_D \sim 0.05$. Finally, the non-zero quadrupole moment of the deuteron is our first indication of the tendency of the proton–neutron interaction to lead to non-spherical nuclear shapes.

It is worthwhile at this point to reiterate what we have learned about the nuclear force, and to emphasize that this rather detailed knowledge stems from some of the simplest empirical facts concerning nuclei. The essential characteristics of the nuclear force are:

- It is predominantly attractive
- It is short range
- It saturates
- It is charge independent (excluding, of course, the Coulomb part)

Moreover, we have learned that the residual interaction (the internucleon force not contained within an overall central potential) has the following properties:

- It exhibits the pairing property that favors the coupling of the angular momenta of like nucleons to 0^+.
- Aside from the pairing interaction, the like-nucleon residual interaction is, on average, repulsive.
- The $T = 0$ *component* of the p–n interaction, on the other hand, is predominantly attractive.
- The supposedly "strong" nuclear force is strong only in comparison with other forces: in the nuclear context, it is barely strong enough to overcome the relative kinetic energies of two nucleons in low energy orbits.
- On balance, as evidenced by the deuteron, the p–n interaction favors the coupling of the proton and neutron intrinsic spins to $S = 1$ rather than $S = 0$. In Chapter 4, we shall see that this is in striking contrast to the like-nucleon residual interaction that favors $S = 0$. Both of these are intimately connected with the effects of the Pauli principle.
- The proton–neutron system has a tendency to produce nonspherical shapes and provides evidence for spin-dependent tensor forces.

Before we end this discussion of the nuclear force, there is one other interesting point concerning its range. The short range, $\sim 10^{-12}$ to 10^{-13} cm, is not at all accidental, but may actually be derived by a simple consideration of its source. It is now generally accepted that all forces in nature result from the exchange of specific kinds of particles between the interacting entities. Between the time one of these entities emits such a "virtual" particle and the other absorbs it, there is a nonconservation of energy. Therefore, by the Heisenberg Uncertainty principle, $\Delta t \, \Delta E \geq \hbar$, there is only a finite amount of time during which the exchange can occur. Clearly, there is a relation between the *mass* of the exchanged particle and the possible *range* of the force: lighter (low E) virtual particles induce smaller violations of energy conservation and therefore can exist for longer periods of time, thus permitting longer-range forces. The outstanding example of this is the Coulomb interaction, which is mediated by massless virtual photons and is therefore of extremely long range. In the nuclear case, the mediation is carried by virtual mesons of which the lightest are the pions with mass ~ 140 MeV. Assuming that they travel at the speed of light, we immediately obtain an upper limit on their "lifetime:"

$$\Delta t \sim \hbar/140 \, \text{MeV} = \hbar/mc^2$$

where m is the pion mass. The distance they can travel in this period is

$$r = c\Delta t = \hbar/mc = 1.4 \times 10^{-13} \, \text{cm}$$

which is remarkably close to the typical range of the nuclear interaction.

1.3 Pauli principle and antisymmetrization

The Pauli principle is of fundamental importance to nuclear structure. For example, we will see in later chapters that it is essential in determining which nuclei are stable, that it provides a justification for the idea of independent particle motion in a dense nucleus, that it is the determining factor in the energy shifts that occur with various residual interactions in the shell model and, perhaps most importantly, that it is the principal reason why single nucleon configuration mixing depends on the valence proton–neutron interaction. In fact, these last two points may seem like structural details, but they explain in one stroke why all even–even nuclei have 0^+ ground states, why the low-lying states of these nuclei increase in energy with spin, why most low-lying negative parity states have odd spin, and, remarkably, the entire systematics of where collectivity, phase transitions, and deformation occur in nuclei.

The Pauli principle, in its simplest form, embodies the notion that no two identical nucleons can occupy the same place at the same time. More formally, no two nucleons can have identical quantum numbers. In this second form it plays an important role in proton–neutron systems where the two nucleons can be treated as two states of the same nucleon. Many applications of the Pauli principle, however, are best expressed in terms of a generalized mathematical formulation of it that the nuclear wave function must be totally antisymmetric—*totally* meaning antisymmetric in all coordinates, spatial, spin, and isospin (i.e., that the wave function must reverse its sign if all these coordinates are interchanged). To see the relation of this requirement of antisymmetry to the Pauli principle, consider a wave function of two identical particles, $\psi_{ab}(r_{12})$, where the orbits occupied by the particles are labeled a and b and where r_{12} is the distance between the two particles. Clearly, the Pauli principle requires that the wave function must vanish when $r_{12} = 0$; that is, when the particles are at the same point in space. A wave function such as $\psi_a(r_1)\psi_b(r_2)$ need not vanish at $r_{12} = 0$, and thus is not an acceptable two-particle state. However, consider the wave function

$$\psi_{ab}(r_{12}) = \psi_a(r_1)\psi_b(r_2) - \psi_a(r_2)\psi_b(r_1)$$

Obviously, $\psi_{ab}(r_{12}) = 0$ for $r_1 = r_2$ and thus satisfies the Pauli principle. But, for any r_{12}, it also follows that

$$\psi_{ab}(r_{12}) = [\psi_a(r_1)\psi_b(r_2) - \psi_a(r_2)\psi_b(r_1)]$$
$$= -[\psi_a(r_2)\psi_b(r_1) - \psi_a(r_1)\psi_b(r_2)] = -\psi_{ba}(r_{12}) \qquad (1.1)$$

So, the Pauli principle can be formulated mathematically by the statement that a two-particle nuclear wave function Ψ must be antisymmetric with respect to the interchange of the two partners. For multiparticle states, the antisymmetry must extend to interchanges of any pair of particles.

Although the present argument was phrased in terms of spatial coordinates, it can be extended to other spaces leading to the generalized antisymmetrization condition given earlier.

It is impossible to overemphasize the importance of the Pauli principle in nuclear physics. It has obvious and direct consequences as well as subtle, indirect, but no less real, effects. We shall encounter it continually.

1.4 Two-state mixing

In solving any quantum mechanical problem, the wave functions may be expressed in terms of any complete set of basis states spanning a Hilbert space. However, enormous calculational simplification and greater physical insight almost always result if a basis is chosen in terms of which the wave functions are simple. Of course, the limiting case occurs when the Hamiltonian and the basis mesh precisely (e.g., a Hamiltonian with harmonic oscillator potential $V = \frac{1}{2}kx^2$ and harmonic oscillator basis wave functions). In such a case the wave function for each state consists of only a single term, and the quantum numbers characterizing that basis state also characterize the eigenfunction of the Hamiltonian.

It is, of course, rare that this happens in practice. In realistic calculations, pure configurations are seldom encountered. Frequently, the actual nuclear states are complex admixtures of many components; an accurate treatment must involve the diagonalization of a large Hamiltonian matrix. Then, certain quantum numbers [e.g., the seniority (see Chapter 5) or the K quantum number for deformed nuclei (see Chapters 6, 8, 9)] are only approximate as well.

In dealing with complex situations, one frequently tries to decompose the Hamiltonian into one part, H_0, whose eigenfunctions are the basis states being used, and another part (hopefully only a relatively small perturbation), H_1, which mixes the basis states. Thus, one has

$$H = H_0 + H_1 = (T + V_0) + V_1$$

with the Schrödinger equation

$$H\psi = H_0\psi + H_1\psi = E\psi$$

To solve this equation, we expand ψ in terms of basic states φ_i which are eigenfunctions of H_0 with eigenvalues E_i. The Hamiltonian equation then becomes

$$\begin{bmatrix} E_1 & V_{12} & V_{13} & \ldots & V_{1n} \\ V_{21} & E_2 & V_{23} & \ldots & \\ \vdots & & & & \vdots \\ V_{n1} & & & & E_n \end{bmatrix} \begin{bmatrix} \\ \varphi \\ \\ \end{bmatrix} = \begin{bmatrix} \\ E \\ \\ \end{bmatrix} \begin{bmatrix} \\ \varphi \\ \\ \end{bmatrix}$$

where φ and E are column vectors whose components are φ_i and E_i. The solutions are obtained by diagonalizing the Hamiltonian H in the basis φ. The resulting wave functions (eigenfunctions of the full Hamiltonian H) will be given by

$$\psi_k = \sum_i \alpha_i^k \varphi_i$$

where the α_I^k are the expansion coefficients for the kth state obtained from the diagonalization.

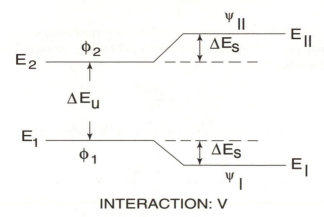

FIG. 1.6. Two-state mixing: definitions and notation.

Although this procedure is simple and straightforward in principle, it is often tedious in practice and the complexity of the solutions often makes it difficult, if not nearly impossible, to keep sight of the basic physics. In many, if not most cases, however, it is possible to regain a feeling for the underlying physics and at least a semiquantitative calculation by a simple two-state mixing calculation. In many cases, one can simulate the full diagonalization reasonably well using sequences of a few two-state mixing calculations. Two-state mixing is completely trivial. We will present the results in a slightly different form than normally encountered so that we will obtain universal analytic expressions. It is of the utmost importance to understand and to have an intuitive grasp of the relationships between the initial energy spacings and the mixing matrix element, on the one hand, and the final separations and admixed wave functions on the other. These ideas are exploited throughout this book. This section outlines the basic ideas and formulas, presents the universal mixing curve, and discusses some useful limiting cases. In addition, a few sample schematic multistate mixing calculations will be described in the next section.

Consider the situation illustrated in Fig. 1.6, in which two initial levels with energies E_1 and E_2 have wave functions ϕ_1 and ϕ_2. For an arbitrary interaction, V, the mixing matrix element is $\langle \phi_1 | V | \phi_2 \rangle$, which we denote simply by V. Note that V is normally negative (nuclear interactions are generally attractive). However, for simplicity, in the rest of this chapter, the notation V denotes $|V|$. Hence, for example, the quantities R and R' to be defined below are to be taken as positive. The final energies and wave functions are obtained by diagonalizing the 2×2 matrix

$$\begin{pmatrix} E_1 & V \\ V & E_2 \end{pmatrix}$$

The final wave functions are denoted by Roman numerals, Ψ_I and Ψ_{II} and have energies E_I and E_{II}. In general, the mixing depends both on the initial separation and on the matrix element. A large spacing reduces the effect of a given matrix element. Conversely, even a small matrix element may induce large mixing if the unperturbed states are close in energy. In order to present the results so that this two-parameter aspect is circumvented, yielding a single universal mixing expression valid for *any* interaction and *any* initial spacing, we define the ratio

$$R = \frac{\Delta E_u}{V} \tag{1.2}$$

of the unperturbed energy spacing to the strength of the matrix element. Then the perturbed energies are

$$\begin{aligned}
E_{\text{I,II}} &= \frac{1}{2}(E_1 + E_2) \pm \frac{1}{2}\sqrt{(E_2 - E_1)^2 + 4V^2} \\
&= \frac{1}{2}(E_1 + E_2) \pm \frac{\Delta E_u}{2}\sqrt{1 + \frac{4V^2}{\Delta E_u^2}} \\
&= \frac{1}{2}(E_1 + E_2) \pm \frac{\Delta E_u}{2}\sqrt{1 + \frac{4}{R^2}}
\end{aligned} \tag{1.3}$$

where the $+$ sign is for E_{II} and the $-$ sign for E_{I}. It follows that the final perturbed energy difference is

$$\Delta E_p = E_{\text{II}} - E_{\text{I}} = \Delta E_u\sqrt{1 + \frac{4}{R^2}} \tag{1.4}$$

or, in units of the unperturbed splitting ΔE_u, the final separation is given by the simple result

$$\frac{\Delta E_p}{\Delta E_u} = \frac{E_{\text{II}} - E_{\text{I}}}{\Delta E_u} = \sqrt{1 + \frac{4}{R^2}} \tag{1.5}$$

A more useful result is the amount, $|\Delta E_s|$, by which *each* energy is *shifted* by the interaction. $|\Delta E_s|$ is given by

$$|\Delta E_s| = |E_{\text{II}} - E_2| = |E_{\text{I}} - E_1| = \frac{\Delta E_u}{2}\left[\sqrt{1 + \frac{4}{R^2}} - 1\right]$$

or, again, in units of ΔE_u, one obtains a result *independent of the initial spacing*:

$$\frac{|\Delta E_s|}{\Delta E_u} = \frac{|E_{\text{II}} - E_2|}{\Delta E_u} = \frac{|E_{\text{I}} - E_1|}{\Delta E_u} = \frac{1}{2}\left[\sqrt{1 + \frac{4}{R^2}} - 1\right] \tag{1.6}$$

The mixed wave functions are

$$\psi_{\text{I}} = \alpha\phi_1 + \beta\phi_2$$
$$\alpha^2 + \beta^2 = 1 \tag{1.7}$$
$$\psi_{\text{II}} = -\beta\phi_1 + \alpha\phi_2$$

Table 1.2 *Examples of two-state mixing energy shifts and mixing amplitudes (from Eqs. 1.6 and 1.8).* $R = \Delta E_u / V$

| R^* | $\Delta E_s / \Delta E_u$ | β | Specific case: $\Delta E_u = 100$ keV | |
			V (keV)	ΔE_s (keV)
0.2	4.52	0.67	500	452
0.5	1.56	0.61	200	156
1	0.62	0.53	100	62
2	0.207	0.38	50	20.7
3	0.101	0.29	33.3	10.1
5	0.0385	0.19	20	3.85
10	0.0099	0.099	10	0.99
20	0.0025	0.050	5	0.25

*For $R = 0$, $\beta = 0.707$, and $\Delta E_s = V$.

where the smaller amplitude β is given by

$$\beta = \frac{1}{\left\{ 1 + \left[R/2 + \sqrt{1 + R^2/4} \right]^2 \right\}^{\frac{1}{2}}} \tag{1.8}$$

The essential point of Eqs. 1.6 and 1.8 is that both the final energy difference (in units of ΔE_u) and β are functions *only* of R, the ratio of the unperturbed energy splitting to the mixing matrix element. Equations 1.3–1.6 and 1.8 are *universal expressions* completely independent of the nature of the interaction or the initial splitting. The same ratio, $R = \Delta E_u / V$ always gives the same final wave functions, energies, and energy shifts (in units of ΔE_u).

These results are so important, and will be referred to, either quantitatively or qualitatively, so frequently that it is useful to dwell on them. Equations 1.6 and 1.8 are plotted in Fig. 1.7. To illustrate the results and get a feeling for the numbers involved, let us consider a couple of examples. Suppose two initial states are separated by 100 keV and admixed with a matrix element of -50 keV (a not uncommon situation, for example, in Coriolis mixing). Then $R = 2$, and we find that the mixing amplitude $\beta = 0.38$ and that each state is shifted by an amount 0.207 times the initial separation or, in this case, by 20.7 keV. Clearly, the final separation is 141.4 keV. Another common situation is that of rather weak mixing. Taking two states initially an MeV apart that mix with a -10 keV matrix element ($R = 100$), then Fig. 1.7 or Eqs. 1.6 and 1.8 instantly show that the mixing is negligible and the energy shift is virtually nil. Table 1.2 gives examples of ΔE_s and β for a range of R values.

Of course, in using these expressions in practical situations one normally knows the final perturbed energies not the initial separations. One often wants to extract the amount

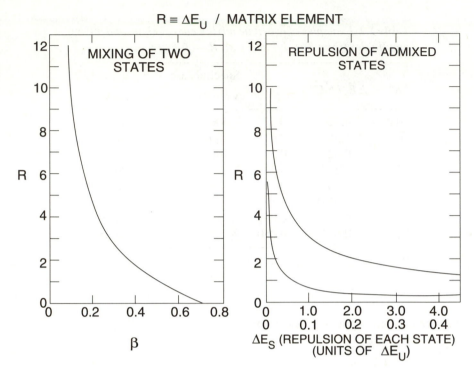

FIG. 1.7. Universal two-state mixing curves. The one on the left gives the smaller of the two mixing amplitudes, β, while the curves on the right give the energy shift of each level in units of the unperturbed energy separation. Here the lower curve goes with the upper abscissa scale, while the upper curve goes with the lower scale.

of mixing (β) or to deduce the interaction strength from some experimental measure of the mixing (e.g., the ratio of two transition strengths from the mixed state, one of which is allowed, the other forbidden in the unmixed limit: the branching ratio is then directly related to β^2). In principle it is then necessary to work the equations backward to solve first for V or for ΔE_u, and then for β. In practice, however, the mixing is often small and ΔE_s is a small fraction of ΔE_u so that an accurate approximation is obtained by taking $R \approx \Delta E_{\text{final}}/V = (E_{\text{II}} - E_{\text{I}})/V$. For example, for $R \geq 5$, the initial and final separations differ by less than 10%.

Alternately, a derivation similar to that leading to Eq. 1.4 can be carried out in terms of the quantity $R' \equiv \Delta E_p/V$, giving

$$\Delta E_u = \Delta E_p \sqrt{1 - \frac{4}{R'^2}} \tag{1.4a}$$

and

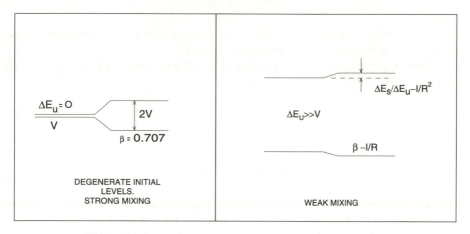

FIG. 1.8. The two limiting cases of strong and weak mixing.

$$\frac{\Delta E_s}{\Delta E_p} = \frac{1}{2}\left[1 - \sqrt{1 - \frac{4}{R'^2}}\right] \tag{1.5a}$$

Having dealt with some examples and these practical comments, we now consider two extremely important limiting cases where Eqs. 1.6 and 1.8 simplify: the situations of infinitely strong and relatively weak mixing illustrated in Fig. 1.8. The results in both cases have many useful and even profound implications and, for the latter case, the limiting situation has very wide applicability.

1. Consider first, then, the strong mixing limit. Suppose the two initial states are degenerate ($\Delta E_u = 0$), as in Fig. 1.8 (left). Of course, then, Eq. 1.6 cannot be used, but the initial expression we started with for $E_{I,II}$ tells us that

$$E_{I,II} = \frac{1}{2}\left[(E_1 + E_2) \pm 2V\right] = E_0 \pm V \tag{1.9}$$

where E_0 is the (common) initial energy. Thus each state is shifted by the *mixing matrix element*. This illustrates the important result that, for *any isolated* two-state system, the final separation can *never be closer than twice the mixing matrix element*. As trivial as this sounds, it is often forgotten. For example, it was one of the early arguments used to demonstrate that Coriolis matrix elements had to be attenuated: examples of isolated pairs of $13/2^+$ states were found that were closer than twice the calculated Coriolis mixing matrix elements.

In the case of degenerate orbits, it is clear that $\beta = 1/\sqrt{2} = 0.707$. Thus α is also 0.707 and the two states are completely mixed. This is conceptually obvious since the matrix element is "infinitely" stronger than the initial separation (i.e., $1/R \to \infty$). This seemingly trivial result also has profound consequences. For example, it means that the mixed wave functions for two initially degenerate states are *independent* of the strength of the interaction between them. (This argument will be used in Chapter 6 to show why the wave functions in the limiting symmetries of the IBA are independent of the coefficients—parameters—of the Hamiltonian as long as the *structure* of that Hamiltonian corresponds to the symmetry involved.)

2. The weak mixing limit corresponds to $R \gg 1$ (see Fig. 1.8, right). Equation 1.8 becomes

$$\beta \approx \frac{1}{R} \tag{1.10}$$

Hence,

$$V \approx \beta \Delta E_u \approx \beta \Delta E_{\text{final}} \tag{1.11}$$

since ΔE_s is small. Frequently (for example, from measured γ-ray branching ratios) one has empirical information on β and therefore Eq. 1.11 (or the exact Eq. 1.8) can be used to deduce V from the data. Similarly, for $R \gg 1$, Eq. 1.6 becomes

$$\frac{|\Delta E_s|}{\Delta E_u} \approx \frac{1}{2} \left[1 + \frac{2}{R^2} - 1 \right] = \frac{1}{R^2} \sim \beta^2 \tag{1.12}$$

An example is useful. Suppose $R = 10$. Equations 1.10 and 1.12 then give

$$\beta = 0.1 \quad \text{and} \quad \Delta E_s / \Delta E_u = 0.01$$

The exact results are $\beta = 0.0985$ and $\Delta E_s / \Delta E_u = 0.0099$. In fact, even for $R = 4$, Eqs. 1.10 and 1.12 are already quite satisfactory: β is correct to better than 10% and ΔE_s to 6%. Except in the case of rather strong mixing, Eqs. 1.10 and 1.12 thus provide quite accurate (instantaneous) results for two-state mixing.

There is one other important aspect of two-state mixing. Suppose we consider two states, 1 and 2, whose energies depend on some nuclear structure parameter x (as illustrated schematically in Fig. 1.9) along the horizontal direction. For example, x could be the deformation and the states might be two Nilsson orbits. For some x value, x_{crit}, the orbits would cross. Now suppose that the two levels mix. They can now never cross since they repel, and can never be closer than twice the mixing matrix element after mixing. Thus the actual behavior of the mixed states, labeled I and II, is as sketched by the solid lines in Figure 1.9. The energies have an *inflection point*. However, for $x > x_{\text{crit}}$, the wave function of state I will have a larger amplitude for unperturbed state 2 than for its own "parent" and vice versa. Such behavior is very common in structure calculations and is nearly always an indication of strong mixing. The point of closest approach of the two curves corresponds to the point where the mixed wave functions contain equal admixtures of each of the unperturbed states. In fact, from the separation at this point the mixing matrix element can be derived by inspection, as one-half the separation. This is another illustration of the usefulness of the limiting case of Eq. 1.9.

1.5 Multistate mixing

In general, a multistate mixing situation must be handled by explicit diagonalization. As noted earlier, this can often be simulated by a sequence of two-state mixing calculations.

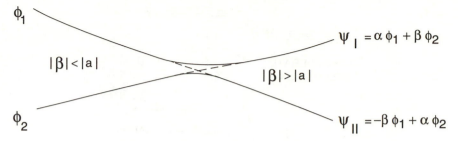

FIG. 1.9. Illustration of the noncrossing of two admixed levels.

In addition, a couple of idealized situations are particularly simple, often useful, and provide physical insight into the often misunderstood results of complex calculations (e.g., RPA calculations).

First, let us consider the case of N degenerate initial states and allow each of these levels to mix with each of the others with *equal attractive* matrix elements (i.e., between all pairs). This idea is illustrated in Fig. 1.10. It is then easy to show by explicit diagonalization that one state is lowered by $(N-1)V$ and each of the other states is raised by one unit in V. The wave function for the lowered state is totally mixed:

$$\psi_I = \frac{1}{\sqrt{N}}\phi_1 + \frac{1}{\sqrt{N}}\phi_2 + \frac{1}{\sqrt{N}}\phi_3 + \dots \frac{1}{\sqrt{N}}\phi_N \qquad (1.13)$$

Although this is a clear case of (optimum) multistate mixing, the result for the lowest eigenvalue is exactly what would result from applying a sequence of two-state mixing calculations: mixing with *each* of the other $N-1$ degenerate states lowers this state by V, giving a total lowering of $(N-1)V$.

This feature of one state emerging with special character, low energy, and a highly coherent wave function, is ultimately the microscopic basis for the physical idea behind the development of collectivity. Collective states result from many interactions of simpler (e.g., single particle or two quasi-particle) entities, and appear at low energies. As we shall see the RPA approach to the microscopic generation of collective vibrations is a prime illustration of this effect. So also is the effect of pairing among 0^+ states that leads to the well-known energy gap in even–even nuclei.

A second case is analogous, except that we lift the initial degeneracy and consider a set of N equally spaced levels. This situation is depicted in Fig. 1.10 for the case of $N = 6$. As before, one state is considerably lowered. Of course, the wave functions are now more complex, and are not of particular interest here. What is interesting is that the *ratio* of the lowering of the lowest level in the nondegenerate (ND) case to the lowering in the degenerate (D) case just considered

$$L_N(R) = \frac{\Delta E_{ND}}{\Delta E_D} = \frac{\Delta E_{ND}}{(N-1)V} \qquad (1.14)$$

is nearly independent of N. For $R = 1$, $L(R)$ for $N = 2, 4, 8$, and 12, respectively, is found, by diagonalization, to be 0.62, 0.60, 0.59, 0.59: that is the lowest state is lowered

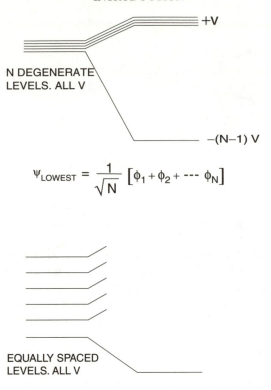

$$\Psi_{\text{LOWEST}} = \frac{1}{\sqrt{N}} \left[\phi_1 + \phi_2 + \cdots \phi_N \right]$$

FIG. 1.10. Illustration of two multistate mixing situations: (Top) N degenerate levels, all of which mix by equal attractive matrix elements V (i.e., $V < 0$); (Bottom) The same, except the initial levels are equally spaced.

by about 60% of what it would be if the initial states had been degenerate. The near-independence of N means that one can estimate the lowering, without calculation, simply by taking the two-state mixing result for the appropriate R value. As a test, suppose the (equal) spacings are all twice the matrix element V. Then, from Table 1.2, ΔE_s (two-state) is 0.414 V. For the degenerate case, it is of course $(2 - 1)V = V$. So $L_2(R = 2)$ is 0.414, which should now be approximately applicable to multistate mixing. The value for $N = 8$, obtained by diagonalization, is $L_8(R = 2) = 0.35$.

A third idealized case again concerns N degenerate levels, except that each level mixes with only the "adjacent" level (as shown in Fig. 1.11). This statement, however, is meaningless for degenerate levels, but it is clear that we can circumvent it by introducing an infinitesimal spacing, and therefore an "order" to the unperturbed levels, $1, 2, 3 \ldots N$. This limit, in fact, is not so far from the realistic situation of Coriolis mixing among a series of bands with $K = K_i, K_i + 1 \ldots K_f$, which frequently occurs in heavy nuclei. Again, one level is lowered, but now the mixed levels are symmetrically distributed with respect to the initial energy and the lowest state is not lowered nearly as much. One can write $\Delta E_{\text{lowest}} = f(N)V$ where the function $f(N)$ has the rough dependence sketched

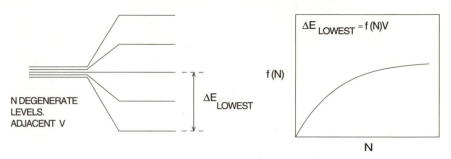

FIG. 1.11. Multistate mixing: N degenerate levels in which only "adjacent" levels are mixed with (equal) matrix elements V.

in Fig. 1.11.

Finally, note that in all the multistate mixing cases considered, *all* of the components of the lowest lying wave function have the *same* sign. Though this result depends on the phase conventions chosen, if consistent conventions are used for both wave functions and operators, then matrix elements (observables) will contain coherent, in-phase sums, and can be extremely large. The wave function has *coherence*, and such multistate mixing can lead to *collectivity* as reflected in enhanced transition rates, cross sections, and the like. Also, note that the sum of the initial and final energies is the same, as, of course, it must be. Since these energies appear on the diagonal of the matrix to be diagonalized, this is equivalent to the formal statement that the trace is conserved.

The importance and usefulness of the results in this section cannot be overemphasized. With them, and an understanding of the basically attractive nature of the nuclear force, and of the effects of the Pauli principle and of antisymmetrization, it is possible to understand nearly all of the detailed results of most nuclear model calculations in an extremely simple, intuitive way that illustrates the underlying physics that is often lost in complex formalisms and computations.

1.6 Two-state mixing and transition rates

One application of the concept of two-state mixing that is worth discussing, even though it invokes concepts and excitation modes that will be introduced later, is the effect of certain types of mixing on transition rates. Consider the simple level scheme in Fig. 1.12 with 2^+ levels from different intrinsic excitations (say, belonging to two bands of a deformed nucleus). Suppose that, according to some model, one 2^+ level has an allowed (A) ground state transition and the other has forbidden (F) transitions to both 0_1^+ and 2_1^+ states. One occasionally encounters statements of the following kind: "While the $2_2^+ \rightarrow 0_1^+$ transition is normally forbidden, its strength results from mixing the two 2^+ states: a similar argument accounts for the $2_2^+ \rightarrow 2_1^+$ transition." At first, this certainly sounds plausible: if the two 2^+ states mix, some of the strength of the allowed transition should be "distributed" to the forbidden one. Moreover, the two 2^+ states now share some of the same character and should be interconnected. Let us calculate the actual E2 matrix elements for the mixed states to see if the preceding conclusions are warranted.

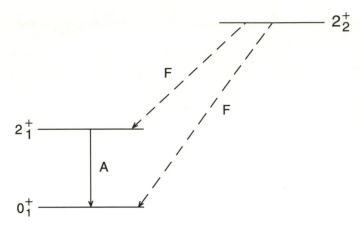

FIG. 1.12. Effect of mixing on allowed (A) and forbidden (F) γ-ray transitions.

Using the notation of Fig. 1.6 (Roman subscripts for the perturbed 2^+ wave functions, arabic for the unperturbed), we have

$$\langle 2_{II}^+|\mathbf{E2}|0_1^+\rangle = \langle(-\beta 2_1^+ + \alpha 2_2^+)|\mathbf{E2}|0_1^+\rangle$$
$$= -\beta\langle 2_1^+|\mathbf{E2}|0_1^+\rangle + \alpha\langle 2_2^+|\mathbf{E2}|0_1^+\rangle$$
$$= -\beta\langle 2_1^+|\mathbf{E2}|0_1^+\rangle \tag{1.15}$$

since the unperturbed $2_2^+ \rightarrow 0_1^+$ matrix element is forbidden. Thus the $2_{II}^+ \rightarrow 0_1^+$ transition is now finite and arises solely from the mixing with the 2_1^+ level as claimed. For the $2_{II}^+ \rightarrow 2_I^+$ transition we have

$$\langle 2_{II}^+|\mathbf{E2}|2_I^+\rangle = \langle(-\beta 2_1^+ + \alpha 2_2^+)|\mathbf{E2}|\alpha 2_1^+ + \beta 2_2^+\rangle$$
$$= -\alpha\beta\langle 2_1^+|\mathbf{E2}|2_1^+\rangle + \alpha\beta\langle 2_2^+|\mathbf{E2}|2_2^+\rangle$$
$$- \beta^2\langle 2_1^+|\mathbf{E2}|2_2^+\rangle + \alpha^2\langle 2_2^+|\mathbf{E2}|2_1^+\rangle \tag{1.16}$$

Since the unperturbed $2_2^+ \rightarrow 2_1^+$ transition is assumed forbidden, the last two terms vanish and

$$\langle 2_{II}^+|\mathbf{E2}|2_I^+\rangle = \alpha\beta\left[\langle 2_2^+|\mathbf{E2}|2_2^+\rangle - \langle 2_1^+|\mathbf{E2}|2_1^+\rangle\right] \tag{1.17}$$

The $2_{II}^+ \rightarrow 2_I^+$ transition vanishes in the limit of no mixing ($\beta = 0$). However, it is by no means clear that mixing will produce a strong transition. The resulting matrix element is proportional to the *difference* in the quadrupole moments of the two states. If the low-lying levels have nearly the same deformation, as is likely in a well deformed nucleus, this difference will be very small, and thus the second conclusion is at best risky. We have worked out this example explicitly because the error just discussed is widespread and partly because the derivation just given will be useful later in understanding the microscopic structure of the β vibration. The point can be summarized as follows: Consider two states (of the same spin) one of which has an allowed transition to some other level while the decay for the other is forbidden. The forbidden transition becomes finite if the two initial states mix, and its matrix element is proportional to the mixing. However, a forbidden transition *between* the two unperturbed levels becomes finite only if the states mix and if the intrinsic structure of the two unperturbed states differs in the moment corresponding to the operator for this transition.

With the background provided in this chapter on the properties of the nuclear force, the comments on the Pauli principle and our discussion of two and multistate mixing, we can now begin to develop an understanding of the rich diversity and unity of nuclear phenomena. We start with a cursory survey of some empirical features and then develop at some length the foundation models through which we try to understand these data and out of which new models and extensions arise.

2

THE NUCLEAR LANDSCAPE

One of the difficulties often faced by the student trying to understand nuclear models is that he/she cannot fully appreciate many of the truly simple and beautiful results that emerge from these models because there is no reservoir of familiar nuclear data to call upon. Therefore, when one derives the spin sequence for a δ-function interaction between two identical nucleons in the same orbit, the results are only of mathematical interest if he or she does not see that this instantly explains the low lying levels of literally dozens of near-closed shell even–even nuclei. The entire seniority scheme seems nothing but a labyrinth of Racah algebra when one does not understand how many well-known facets of nuclear structure are thereby trivially explained. Similarly, the simplicity and intuitiveness of many of the results of the Nilsson model may fall on barren ground unless one realizes the vast number of deformed heavy nuclei that display exactly these properties.

The main purpose of this chapter is to survey the nuclear landscape to display a few (definitely not all) typical patterns of nuclear spectra as well as some of the systematic changes in these patterns over sequences of nuclei, so that the reader will understand the motivation for each model and will benefit from an empirical context for their characteristic predictions. We will refer to the figures in this chapter frequently.

While this approach necessitates some repetition later, it allows us to see exactly *what* we are trying to explain with these models beforehand, and what kinds of data characterize atomic nuclei and are the most useful as tests of various models.

In principle, at this stage the data should be shown "blindly," without commentary on its meaning or implications. However, the purpose of this book is not to develop nuclear physics ab initio and, indeed, most readers will already be familiar with many of the major concepts and terminology. A "purist" approach here would be needlessly tedious and artificial. In the pages that follow, we will use many words and concepts freely that will be introduced formally later on. Those to whom these concepts are unfamiliar should concentrate simply on absorbing the data with the idea of using this base as a touchstone later.

When we speak of nuclear data, we are referring to a vast, varied, and rich reservoir of information about atomic nuclei—from the deuteron to the actinides—obtained by a most diverse array of techniques. The simplest information is the mass of atomic nuclei. A more useful form for these is nuclear binding energies, which focus on the *interactions* between nucleons in the nucleus when the masses of the individual nucleons are subtracted. Still more useful (in many cases) are nucleon separation energies, or the energy required to remove the last, outermost nucleons from the nucleus. (We have discussed these already in Chapter 1.) The nucleon separation energies give important data on the surface regions of nuclei. Later, we will show that the individual nucleons tend to orbit the nuclear center of mass in discrete shells and that, for many applications, it is possible to neglect all the underlying shells that are filled. Therefore, the outermost nucleons are frequently most crucial to understanding the observed properties of nuclear level schemes.

More detailed nuclear data consist of nuclear level schemes: the energies, and angular momenta, and parity values (J^π), of the ground state and low-lying excited states. The mirror nuclei shown in Fig. 1.5 were our first encounter with such schemes. Careful measurements of the γ-rays emitted when excited levels de-excite to lower lying ones are fundamental to both understanding and constructing nuclear level schemes. The crucial information here is, of course, the γ-ray energies, which help define their placements between nuclear levels and their absolute and relative intensities, which give direct measures of nuclear transition matrix elements.

A large amount of data has resulted from the study of scattering processes of one nucleus on another and direct reaction processes in which two interacting nuclei exchange one or more individual nucleons. Scattering experiments often use low energy projectiles with long wavelengths comparable in size to the nucleus itself. In such cases, they provide information on the overall nuclear shape and on the macroscopic, or collective, excitations of the nucleus as a whole (e.g., rotations and vibrations). Nuclear reactions, examples of which are single nucleon transfer reactions such as (d, p) or two-nucleon transfer reactions such as (p, t), can proceed by a direct process in which individual nucleons are inserted into or removed from specific orbits. These reactions provide detailed and microscopic information on the semi-independent particle motion characterizing atomic nuclei.

Heavy ion fusion reactions, which can bring in enormous amounts of energy and angular momentum and access neutron-deficient nuclei, or β decay experiments on fission product nuclei, which are extremely neutron rich, provide valuable sources of information, especially on unstable nuclei.

Of course, this listing of techniques barely touches the surface and highlights only a few that are most useful for studying low-energy nuclear structure. We now turn to the picture of the nuclear landscape they have provided us.

The most basic data for nuclei, of course, is a listing of which nuclei exist. Such a list is usually presented in an $N - Z$ plot as in Fig. 2.1, where the hatched area approximately outlines the stable nuclei. This is the so-called valley of stability (i.e., the valley of energy *vs.* Z and N: because nature prefers to minimize energies, the stable nuclei have the lowest energies). The general features are that $N \approx Z$ for light nuclei, and a preference for a neutron excess in heavier nuclei. We shall see that this pattern is easily explained by combining the concepts of independent particle motion, the Pauli principle, and the Coulomb force.

We have already seen some other examples of significant nuclear data, namely binding energies as a function of A, separation energies S(n) and S(p) and a couple of examples of level schemes for so-called mirror nuclei. These data were shown early on because they provide basic information on the nuclear force itself, which was already discussed in Chapter 1. Here we are concerned with a somewhat more detailed and much more extensive survey of nuclear excitations.

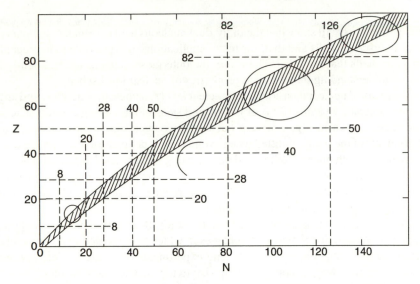

FIG. 2.1. The nuclear chart showing the path of stable nuclei (cross hatched), as well as the magic numbers and midshell deformed regions (circles or circular segments).

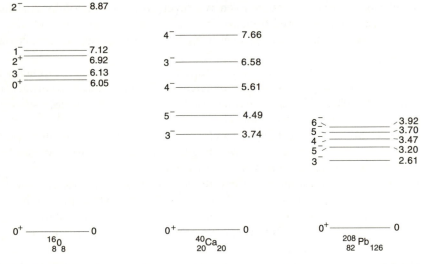

FIG. 2.2. Low-lying levels of three doubly magic nuclei. Energies on the right of each line are in MeV.

As noted in Chapter 1, there are large gaps in the nucleon separation energies that define so-called magic numbers, and point to the existence of nuclear shells, analogous to those in atomic physics. That is, certain *magic* numbers of nucleons of a given type correspond to the filling of a set of orbits constituting a "shell." Additional nucleons

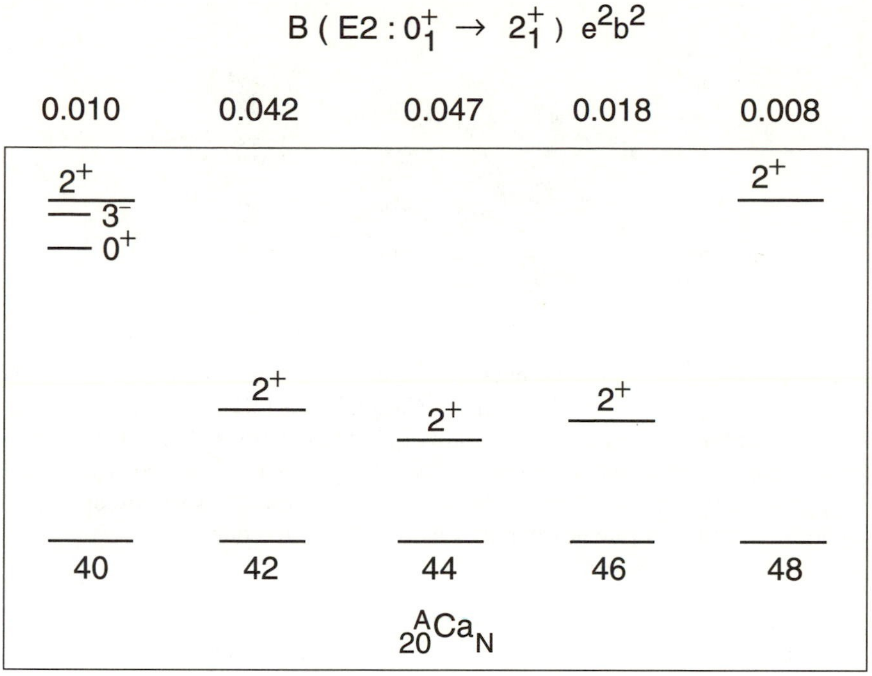

$$B (E2 : 0_1^+ \rightarrow 2_1^+) \; e^2b^2$$

0.010	0.042	0.047	0.018	0.008

2^+

— 3^-

— 0^+

2^+ (right)

2^+ (42) 2^+ (44) 2^+ (46)

| 40 | 42 | 44 | 46 | 48 |

$$_{20}^{A}Ca_N$$

FIG. 2.3. Low-lying levels and B(E2) values for the even–even Ca nuclei. ^{40}Ca and ^{48}Ca are doubly magic.

must then fill the next higher shell and are considerably less bound. In Figs. 1.2 and 1.3 we saw such data for the magic number 82. The complete set of the most important, empirically observed, magic numbers is 2, 8, 20, 28, 50, 82, and 126. There are also gaps or subshell gaps at 40 and 64, especially for protons, which exist for certain neutron numbers only (see further discussions in Chapters 3 and 7). Figure 2.2 shows examples of nuclei where *both* the proton and neutron numbers are magic. Note the extremely high energy of the first excited state, and the predominance of negative parity states. Figure 2.3 shows the Ca isotopes, two of which, ^{40,48}Ca, are "doubly" magic, while the others are singly magic. The abrupt change of 2^+ energies in $^{42-46}$Ca compared to ^{40,48}Ca is dramatic evidence of the difference in their magic structure.

One of the great successes of the independent particle model (Chapter 3) is the prediction of level sequences for nuclei near closed shells, in particular odd mass nuclei where, as we shall see, the total angular momenta J (this will often be colloquially referred to later as the "spin" of the level, although this nomenclature is clearly inaccurate) of the ground and low-lying excited states are given by the j values of the orbits into which the last odd nucleon can be placed. To illustrate this, we show the level schemes of two nuclei (^{41}Ca, ^{209}Pb) with one particle beyond a doubly magic nucleus in Fig. 2.4. The specific order and energies of these levels will be easy to understand and predict after we discuss the independent particle model in the next chapter. At this point, these

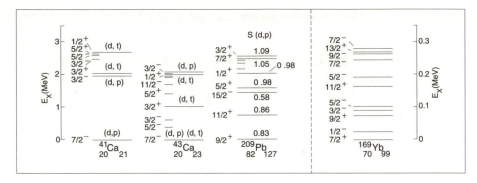

FIG. 2.4. Low-lying levels of two nuclei, ^{41}Ca and ^{209}Pb, with one nucleon beyond a doubly magic core, and two other nuclei, ^{43}Ca and ^{169}Yb, for comparison. Note that the energy scale for Yb is expanded by a factor of 10. The notations (d, t) and (d, p) indicate which levels are primarily populated in these single nucleon transfer reactions. The S(d, p) values given for ^{209}Pb are the single neutron transfer spectroscopic factors; values near unity indicate nearly pure single-particle structure.

sequences appear as unintelligible jumbles.

Of course, most nuclei have more than one valence nucleon. Two examples are included in the figure. One is $^{43}_{20}$Ca$_{23}$ with three valence neutrons. It is similar in many ways to ^{41}Ca but, with three valence nucleons, a proper treatment requires the study of multinucleon configurations and of the "residual interactions" occurring among nucleons in the *valence* shell.

The other is ^{169}Yb, which is far from magic in either protons or neutrons. This scheme, both in terms of its complexity (only hinted at in the figure) and its compressed energy scale, sets it completely apart from the other nuclei in the figure. We will see another example of this type of nucleus in a moment.

In treating multivalence particle nuclei such as these, a number of different approaches are used. Close to closed shells (e.g., ^{43}Ca), an extension of the independent particle model that includes residual interactions among the valence nucleons, has been enormously successful. Further from closed shells (e.g., ^{169}Yb), nonspherical shapes appear, and a deformed shell model (the Nilsson model) becomes appropriate. In the case of the shell model, it is often appropriate to consider a coupling scheme in which each nucleon has a given total angular momentum j. The coupling of these individual j values leads to the final J for the state in question. The energy of such a state clearly depends on the residual interactions among the nucleons in these orbits. It is certainly one of the triumphs of the shell model that one can easily derive expressions for these energies, often without a detailed knowledge of the residual interaction itself, that account reasonably well for a large body of data in both odd and even nuclei.

Figures 2.5–2.7 show the level spectra of some even mass nuclei in the general vicinity of closed shells. Typical of such nuclei, all have 0^+ ground states, first excited levels with $J^\pi = 2^+$, and mostly even-parity, low-lying excitations.

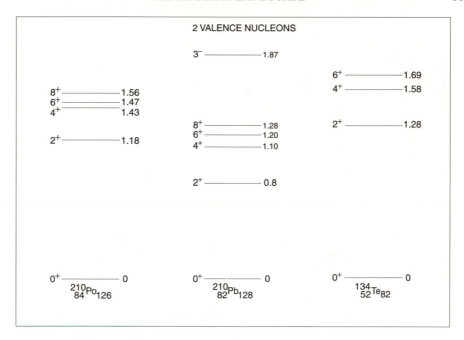

FIG. 2.5. Low-lying levels of three nuclei with two valence nucleons.

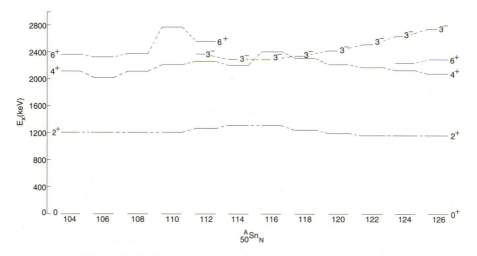

FIG. 2.6. Systematics of the lowest levels of the Sn nuclei.

All of these features will emerge later from very general considerations of the dependence of the like-nucleon interaction on total angular momentum and of the effects of the Pauli principle. Figures 2.5 and 2.6 deal with singly magic nuclei. They show two interesting features: relatively high-lying first excited states and a compression of

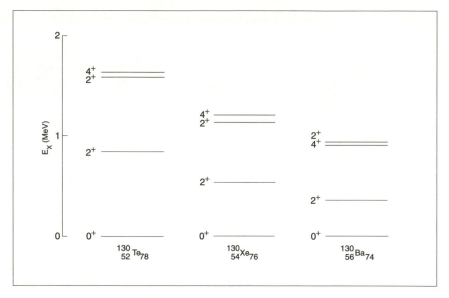

FIG. 2.7. Low-lying levels of the $A = 130$ isobars that show the effect of adding valence nucleons (in this case proton particles and neutron holes relative to $N = 82$).

positive-parity energy levels as J increases. Both features contrast sharply with "collective nuclei." Note that these features persist in the Sn isotopes even when there are many valence neutrons. It is only when there are both valence protons and neutrons that the excitation patterns change rapidly. Figure 2.7 shows this for three $A = 130$ nuclei. Note the systematic change as the total number of valence nucleons increases (here we count proton particles plus neutron *holes* relative to the nearest closed shells). Figure 2.8 illustrates this even more systematically and dramatically. Here, there is a sharp drop in $E_{2_1^+}$ from magic Sn to nonmagic nuclei and the lowering is greater for more and more valence neutrons. There is an additional drop when going from two valence protons (Te, Cd) to four (Xe).

In nuclei far from closed shells where the shell model is either intractable or unreliable, one normally takes recourse in other theoretical frameworks. One of the significant and most fruitful of these approaches can be called *geometrical* or *collective models,* which bypass the shell model by taking a more macroscopic approach of assigning a specific *shape* to the nucleus and examining the rotations and vibrations of such a (generally nonspherical) shape. Of course, a critical issue is whether or not such structures can in fact be derived microscopically from the shell model, and this will be a topic of some importance in a later chapter. Be that as it may, it is an undeniable *empirical* feature of many heavy nuclei that they exhibit properties that *seem* at variance with the concept of a shell model, and show evidence of "collective" behavior.

Individual nuclei exhibit several easily discernible types of collective behavior. Figure 2.9 shows typical *vibrational* nuclei, especially ^{118}Cd, where the first excited state is a

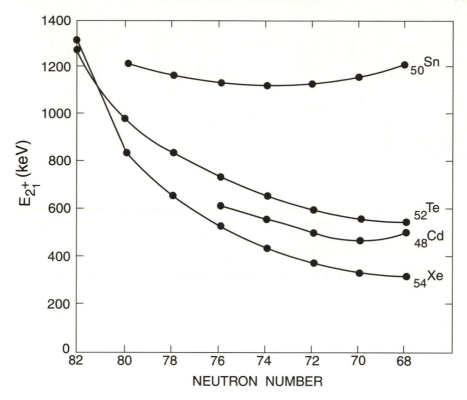

FIG. 2.8. $E_{2_1^+}$ values for nuclei in the Sn region.

quadrupole phonon excitation of a basically spherical shape. At about twice and three times this energy there are groups of states that can be described as two- and three-phonon excitations of the basic spherical structure.

Figure 2.10 shows three even–even deformed nuclei in the rare earth region. The lowest levels of spin $J = 0, 2, 4, 6, \ldots$ form a rotational structure whose energies closely follow the $J(J + 1)$ law for a rotating symmetric top. Above these are groups of levels, some of which we will interpret in Chapter 6 in terms of intrinsic excitations called β, γ, and octupole vibrations, each with rotational bands superimposed. Finally, Fig. 2.11 shows a typical odd-mass deformed nucleus. Here, the levels, which look hopelessly complicated on the left, are arranged in sequences of *single particle* intrinsic (Nilsson) states, each with a rotational band built on top of it on the right. Another example of a nucleus that can be classified in similar manner is ^{169}Yb, which we looked at in Fig. 2.4.

We note that the rotational bands in ^{161}Dy range in character from several ex-amples (labeled $5/2^-[523]$, $3/2^-[532]$, $5/2^-[512]$) with regular spacings that increase smoothly with J, to some with highly irregular sequences [e.g., $5/2^+[642]$, $1/2^-[521]$ and $1/2^-[530]$ (with missing $1/2^-$ level)]. An acceptable model for odd-mass deformed

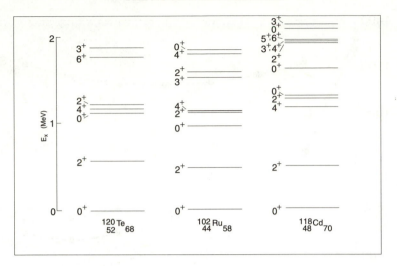

FIG. 2.9. Low-lying levels of some typical, near harmonic vibrational nuclei.

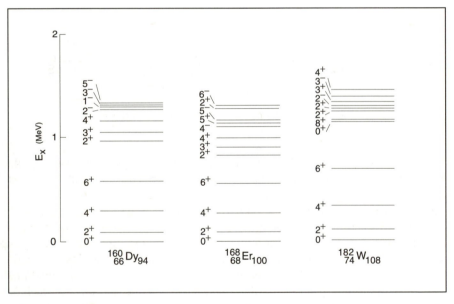

FIG. 2.10. Low-lying rotational, vibrational and particle excitations of some typical deformed rare earth nuclei.

nuclei must be able to account for both types of behavior. We will discuss these intrinsic excitations and their connection to rotational motion in detail in Chapters 8 and 9.

A dramatic way to illustrate both the collective behavior of nuclei far from closed shells and the evolution of structure is to examine a particular property over extended

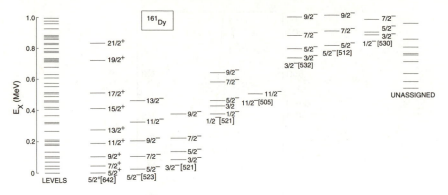

FIG. 2.11. Level scheme of ^{161}Dy. (Left) All levels. (Right) Levels arranged into rotational bands with Nilsson assignments.

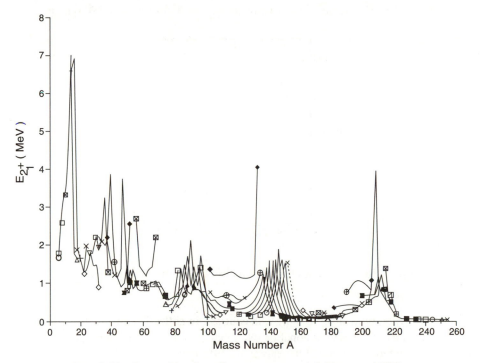

FIG. 2.12. $E_{2_1^+}$ values for all even–even nuclei (Raman, 1987).

sequences of nuclei—that is, to examine nuclear *systematics*. Three of the most telling data are collected in Figs. 2.12–2.16. Figure 2.12 shows the energies of the first excited 2^+ states (2_1^+ levels) in even–even nuclei throughout the periodic table. Figures 2.13 and 2.14 show a more detailed view of the same data in two particular regions: nuclei around mass $A = 100$, and those near $A = 130$. As we have seen near closed shells, $E_{2_1^+}$ is

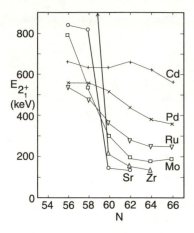

FIG. 2.13. $E_{2_1^+}$ values (in keV) plotted against N for the $A = 100$ region.

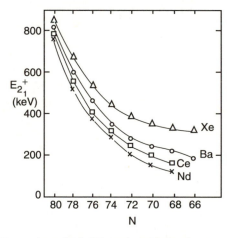

FIG. 2.14. $E_{2_1^+}$ values (in keV) against N for the $A = 130$ region.

rather high lying, typically 1 to 2 MeV. In contrast, in collective nuclei, the 2_1^+ state can be described as either a vibrational or rotational excitation and occurs at much lower energy. These figures highlight the enormous differences in $E_{2_1^+}$ characteristic of these different structures and the transition regions between them as well as the complexity of some individual regions. Figure 2.15 shows a plot of the energy *ratio* of the energy of the 4_1^+ state (first 4^+ state) to the 2_1^+ state in even–even nuclei. As will become evident throughout the subsequent chapters, this ratio is one of the most important structural signatures and, moreover, is one of the few whose *absolute* value is directly meaningful. At first glance, Fig. 2.15 appears to be a semi-random scattering of points. More careful inspection, and some hindsight from subsequent chapters, shows that $E_{4_1^+}/E_{2_1^+}$ tends to fall into three ranges, values below 2.0 near magic nuclei (see Fig. 2.5), between 2.0 and

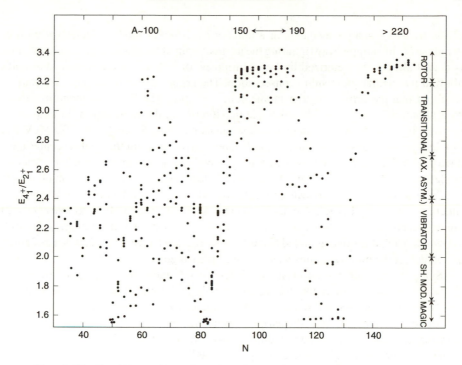

FIG. 2.15. $E_{4_1^+}/E_{2_1^+}$ values plotted against N for the nuclei with $N \geq 30$.

2.4 slightly further away from magic numbers (see Fig. 2.9), and values very close to 3.33 in midshell regions corresponding to rotational motion (see Fig. 2.10). The transitions between the latter two clusters are abrupt indeed.

Most low-lying nuclear states, including essentially all that we will consider, are below the energy threshold for particle emission and hence de-excite primarily by electromagnetic processes. These are usually γ-ray transitions, although electron conversion is an important process. The matrix elements for γ-ray transitions often directly give critical nuclear structure information.

We will have many occasions later on to discuss such radiation from excited nuclear states. This is therefore an appropriate place to deal with the crucial issue of the angular momenta, λ, carried by such radiation and the characteristic probability of each. As is well known, the electromagnetic field can be expanded in multipoles of either electric, Eλ, or magnetic, Mλ, type. We note that the parity carried by such photons is $(-1)^{\lambda}$ for electric and $(-1)^{\lambda+1}$ for magnetic multipoles. Thus, for example, E2 and M1 transitions conserve parity while E1 transitions change it. The probability of emission of radiation of a given multipole is governed by the intrinsic probability of that multipole times a nuclear matrix element. The latter depends on the detailed structure of the initial and final states involved, while the former is a general characteristic of the electromagnetic field and of the "source" of the radiation (the nucleus). We will discuss the properties of the transition matrix elements in a number of different models. Here, we wish to demonstrate in a simple way (ignoring the intrinsic spin of the photon) that such radiation is nearly always characterized by low multipoles, or at least by the lowest multipoles allowed by angular momentum conservation. The argument is very simple. We start by recalling that the orbital angular momentum $l = \mathbf{r} \times \mathbf{p}$. The linear momentum carried by a photon is $p = E_\gamma / c$. Therefore the maximum angular momentum (in units of $\hbar$) is given by $l = E_\gamma R / \hbar c$ where R is the nuclear radius. Since $\hbar c \approx 200$ MeV fm, dipole ($l = 1$) γ-rays emitted at typical distances of 10 fm from the nuclear center must have $E_\gamma \approx 20$ MeV. Since most transitions involved in low-energy nuclear structure are less than 2 MeV, it is obvious, first, that the electromagnetic de-excitation process is relatively slow on a nuclear scale (it must proceed by virtue of the tails of nuclear wave functions extending out to large distances) and, secondly, that high multipoles ($\lambda > 2$) are extremely unlikely. Both these features are very important and are empirically well known. Indeed, it is often assumed that all transitions are E1, M1, or E2, when assigning J^π values, if the multipolarities have not been measured.

Since the ground state of even–even nuclei is 0^+, the first excited state (normally $J^\pi = 2^+$) can only decay by electric quadrupole or *E2 radiation*. Since we shall see that "collective" effects in low-lying states are quadrupole in geometric character (most deformed nuclei are prolate shaped), it should not be surprising that E2 or *electric quadrupole radiation* is of paramount interest. The usual quantitative measure of E2 transition strengths is called a B(E2) value, defined as

$$\text{B}(\text{E2} : J_i \rightarrow J_f) = \frac{1}{2J_i + 1} \langle \psi_f \| \text{E2} \| \psi_i \rangle^2 \tag{2.1}$$

in terms of the reduced E2 matrix element between initial and final states.

It is useful to define a standard for the magnitudes of B(E2) values so that we can assess if a given value is strong (collective) or not. Two (related) standards are used, the Weisskopf unit (W.u.) and the "single particle" rate. They correspond, respectively, to $2^+ \rightarrow 0^+$ and $0^+ \rightarrow 2^+$ transitions. The W.u. is defined by 1 W.u. = $(5.94 \times 10^{-6}) A^{4/3} e^2 b^2$. For $A = 156$, 1 W.u. = $0.005 \, e^2 b^2$. The W.u. represents a rough guide to the B(E2: $2_1^+ \rightarrow 0_1^+$) value corresponding to a $|j^2 J = 2\rangle \rightarrow |j^2 J = 0\rangle$ transition for large j: that is, a transition in which two particles in a shell model orbit j change their angular momentum coupling from 2 to 0. B(E2) values for collective transitions should be large compared to one W.u. As we shall see here (Figs. 2.16, 2.18) and in subsequent chapters, B(E2) values can range up to more than 1000 W.u. for transitions between rotational states in heavy nuclei. Typical collective B(E2) values in spherical vibrational nuclei are ~ 10–50 W.u. Transitions deexciting vibrational excitations in deformed nuclei are likely to range from a few to perhaps 10–30 W.u. Single particle rates (see the $(2J_i + 1)$ factor in Eq. 2.1) are a factor of five larger than W.u. and hence experimental B(E2) values expressed in single particle units (as in Fig. 2.16) are a factor of five smaller than if given in W.u. In reading the literature, the reader is urged to carefully check which value is being quoted in any given case: mistakes (of a factor of five) are often found. W.u. are in far more common use today.

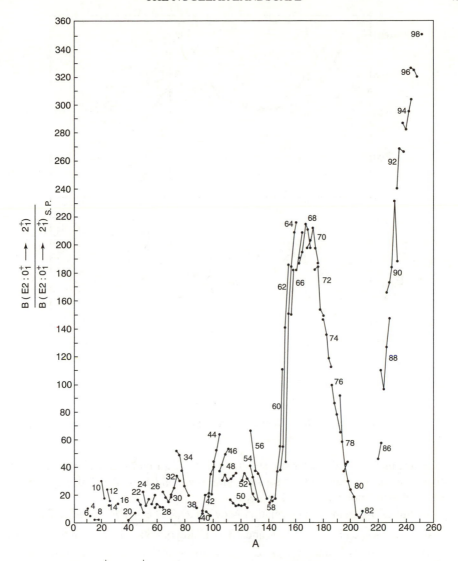

FIG. 2.16. B(E2 : $0_1^+ \rightarrow 2_1^+$) values for even–even nuclei in units of a "single particle" value defined as B(E2: $0_1^+ \rightarrow 2_1^+$)$_{s.p.}$ = 0.00003 $A^{4/3}$ e^2b^2 (Bohr, 1975, 1998). In this book we more frequently express E2 strengths in Weisskopf units (W.u.) which are 1/5 of the single particle unit and are therefore suitable especially for the downward going transitions, B(E2: $2_1^+ \rightarrow 0_1^+$).

Figure 2.16 shows the systematics of B(E2 : $0_1^+ \rightarrow 2_1^+$) values across the nuclear chart. The most obvious feature of the data is the relatively small values near closed shells and the enormous ones that occur in midshell regions, such as those near mass 160 and 240. These peaks offer the most dramatic evidence known for nuclear collectivity. Within

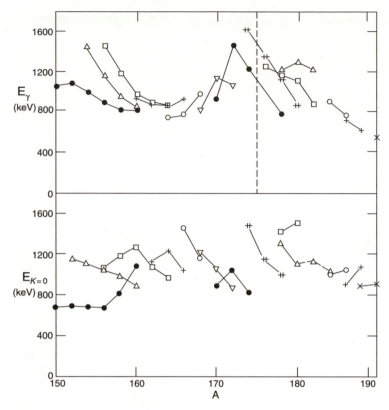

FIG. 2.17. Systematics of $K = 0$ and γ vibrational energies in the rare earth region.

such peaks, there is a characteristic structure: at first, a rapid increase with increasing valence nucleon number and then a saturation near midshell. We shall discuss these points, and their implications, in Chapter 6.

It is interesting to make a correlation between $E_{2_1^+}$, $E_{4_1^+} / E_{2_1^+}$, and B(E2: $2_1^+ \rightarrow 0_1^+$) values (compare Figs. 2.12, 2.15, and 2.16). Using our classification of the $E_{4_1^+}/E_{2_1^+}$ ratios, we see that low values of $E_{4_1^+}/E_{2_1^+}$ near closed shell correlate with low values of B(E2: $2_1^+ \rightarrow 0_1^+$) and high values of $E_{2_1^+}$. As $E_{2_1^+}$ begins to drop as one proceeds through a major shell, $E_{4_1^+}/E_{2_1^+}$ rises slightly to just above 2 and the B(E2) values also begin to increase. Finally, far from magic numbers where $E_{2_1^+}$ drops dramatically and becomes asymptotically constant, $E_{4_1^+}/E_{2_1^+}$ approaches 3.33 and the B(E2) values increase rapidly toward their peak values. As we will study in detail later, the structural transition involved here is one from spherical nuclei near closed shells ($E_{4_1^+}/E_{2_1^+} < 2$) toward spherical, but vibrational, nuclei and culminates in a phase transition to strongly deformed (non-spherical axially symmetric) nuclei whose low-lying states reflect rotational behavior. If we recall the discussion of multistate mixing in Chapter 1 (for example, see Eq. 1.13) we

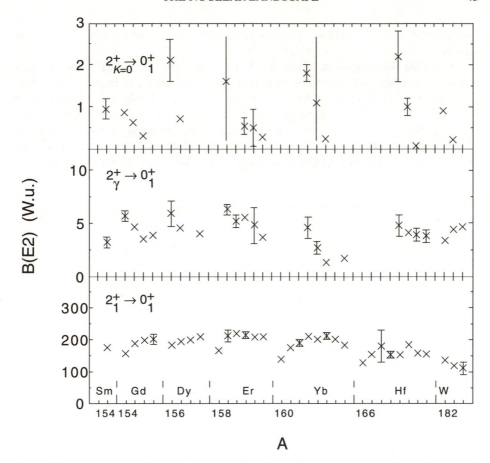

FIG. 2.18. Systematics of some typical B(E2) values relating to rotational and vibrational excitations in the rare earth nuclei.

see that both the drop in $E_{2_1^+}$ and the rise in B(E2) values can be understood in terms of the correlations or collectivity that develop from multistate configuration mixing as one goes from closed shell regions toward midshell. Many of these ideas will be quantified and made more specific later.

Finally, a crucial test of our detailed microscopic understanding of the macroscopic collective shape vibrations is whether we can understand their systematics. Figure 2.17 shows the energies of $K = 0$ and γ vibrations in the rare earth region of deformed nuclei. Figure 2.18 gives some crucial B(E2) values relating to the lowest states in even–even nuclei and to their low-lying $K = 0$ and γ vibrations. The bottom panel shows B(E2: $2_1^+ \rightarrow 0_1^+$) values that describe the matrix elements connecting rotational states. (This is a small subset of those in Fig. 2.16.) The middle and upper panels give the γ-vibrational to ground state and the "$K = 0 \rightarrow g$" B(E2) values. The *intraband values* are a couple of orders of magnitude larger than *interband* B(E2) values and "$\gamma \rightarrow g$" matrix elements dominate "$K = 0 \to g$" values. A successful collective model must account for all these results.

In closing this chapter it is appropriate to summarize the nuclear landscape in a compact form that will later allow us to make instant, a priori, estimates of the likely structure of any given nucleus. We do this by recalling Fig. 2.1, which shows the nuclear chart in an $N - Z$ plane. The magic numbers are indicated by vertical and horizontal lines and the known and expected midshell regions of deformed nuclei are encircled. Much of the rest of this book is devoted to understanding nuclei of each specific type occurring in the chart as well as the evolution of structure from one type to another.

PART II

SHELL MODEL AND RESIDUAL INTERACTIONS

3

THE INDEPENDENT PARTICLE MODEL

In Chapter 1 we discussed some of the basic characteristics of the nuclear force, establishing that, aside from a very short-range repulsive core, it is principally attractive in nature, rather short range, saturates, and is charge-independent (excluding, of course, the Coulomb interaction). We also noted that, while the nuclear force is much stronger than the electromagnetic interaction (indeed, if this were not the case, nuclei would not be bound), it is nevertheless a rather weak interaction when compared to the typical kinetic energies of nucleons inside the nucleus.

In this chapter we discuss the independent particle model, which provides an indispensable theoretical framework for all that follows. It is the basis for the multiparticle shell model, which in turn remains the standard of comparison for other models and provides the justification, rationale, and microscopic basis for macroscopic, collective models.

To be clear from the outset, we define some terminology. By independent particle model we refer to the description of a nucleus in terms of noninteracting particles in the orbits of a spherically symmetric potential $U(r)$, which is itself produced by all the nucleons. Because of this, we immediately anticipate that the resulting orbit energies are mass dependent. The independent particle model is applicable in principle only to nuclei with a single nucleon outside a closed shell and, even then, incorporates certain results from the shell model. By the latter we refer to a model applicable to nuclei with more than one *valence nucleon* that includes residual interactions between these nucleons and allows for the breaking of closed shells.

A central problem of nuclear structure is to describe the motions of the individual nucleons and to deduce observed facets of nuclear excitations from this basis. Ultimately we will present the essential results in terms of the potential of the independent particle model. However, we first need to discuss how the concept of a central potential arises in the first place.

The starting point for nuclear structure calculations cannot be such a *nuclear* potential but rather the interaction between *nucleons*. For simplicity, and as an excellent approximation, we will assume this interaction to be 2-body. The Hamiltonian is given by

$$H = T + V = \sum_{i=1}^{A} \frac{\mathbf{p}_i^2}{2m_i} + \sum_{i>k=1}^{A} V_{ik}(\mathbf{r}_i - \mathbf{r}_k) \qquad (3.1)$$

The interaction V has the form of a nucleon–nucleon potential as sketched in the upper left of Fig. 3.1a. It has a repulsive core and a strongly attractive part extending out to some distance over which the nuclear force acts. [Note that we neglect here the Coulomb potential which is very weak compared to the nucleon–nucleon potential. This is not to say that Coulomb effects are unimportant in nuclei. Since the Coulomb force is long range and scales as $Z(Z - 1)$, it can play a dominant role in heavy nuclei. On the nucleon–nucleon level, however, Coulomb effects are small.] Figure 3.1a (top left) is an extreme simplification of the nucleon–nucleon potential, which depends on spin couplings, has tensor components, and other properties. For a more precise depiction and a detailed example of one form used for this interaction (the Hamada–Johnston potential), see the textbook by Heyde (1990, pp. 83, 84).

The Hamiltonian of Eq. 3.1 has 3A position coordinates (V is a function of the 3 relative position coordinates of each of the particles). It is therefore extremely difficult to deal with, and has been solved only for the lightest few nuclei. To understand nuclear structure for most nuclei, some simplification is necessary. This simplification process leads to the concept of a nuclear (not nucleon) mean field potential in which all the nucleons move.

The transformation from a nucleon–nucleon interaction to a common nuclear potential is our first example of a "paradigm shift". Much of the success of physics in achieving simple and tractable descriptions of nature is due to the exploitation of paradigm shifts. Changes of coordinate systems from Cartesian to cylindrical, the substitution of variables in solving integral or differential equations, or the transformation from the laboratory to center of mass system, are examples of paradigm shifts. Later in this book we will encounter others. Paradigms or paradigm shifts allow us to view a problem from a different perspective so that it becomes simpler or so that new physics becomes clearer or more visible.

One approaches the problem of dealing with Eq. 3.1 by adding and subtracting a 1-body potential $U_i(\mathbf{r})$ to give

$$H = \sum_{i=1}^{A} \left[\frac{\mathbf{p}_i^2}{2m_i} + U_i(\mathbf{r}) \right] + \sum_{i>k=1}^{A} V_{ik}(\mathbf{r}_i - \mathbf{r}_k) - \sum_{i=1}^{A} U_i(\mathbf{r}) \equiv H_0 + H_{\text{residual}} \quad (3.2)$$

The aim is to determine a potential U, which is experienced by all i particles, and which approximates the combined effects of the 2-body interactions in Eq. 3.1, so that the last term on the right side of Eq. 3.2, H_{residual}, is a small perturbation on the Hamiltonian H_0 for a system of nearly independent nucleons orbiting in a common mean field potential. This goal is typically achieved by using Hartree–Fock techniques to give self-consistent solutions for which the potential generated by the A nucleons subject to V_{ik} is the same as the potential U used to obtain these wave functions. A discussion of such techniques is beyond the scope of this book but a schematic introduction to the approach may help the reader who encounters this term in the literature.

Clearly, the potential U generated by all A nucleons will be given by the 2-body interaction convoluted with the nuclear density, and integrated over the nuclear volume

$$U_i(\mathbf{r}) = \int V_{ik}(\mathbf{r}_i - \mathbf{r}_k)\rho(\mathbf{r})d\mathbf{r} = \sum_k \int \psi_k(\mathbf{r})^* V_{ik}(\mathbf{r}_i - \mathbf{r}_k)\psi_k(\mathbf{r})d\mathbf{r}$$

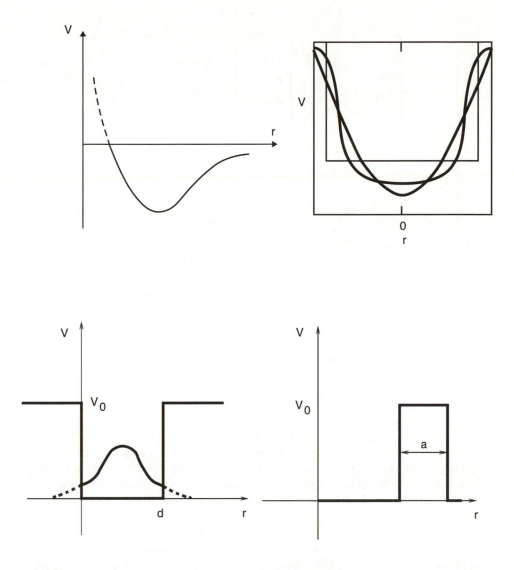

FIG. 3.1a. (top left) Rough illustration of the general behavior of the nucleon–nucleon potential (but see text and the more detailed specification (Heyde, 1990) necessary for actual calculations); (top right) Examples of three typical potentials useful in the discussion of nuclear properties; (bottom left) A 1-dimensional square well potential with walls of height V_0 and width d: (bottom right) A potential barrier used in the discussion of tunneling.

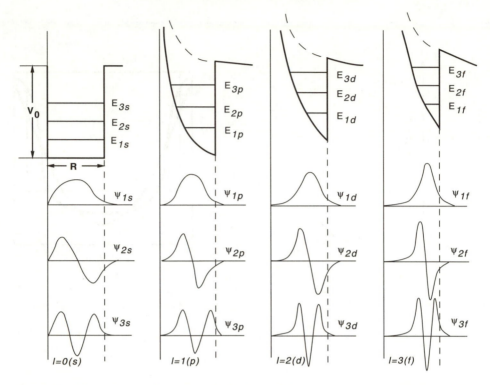

FIG. 3.1b. Illustration of the potential, single particle energies, and wave functions in a 3-dimensional square well with centrifugal force.

where ψ_k is the wave function of a nucleon in an occupied orbit k. [Notes: We are ignoring anti-symmetrization here, corresponding to the simpler Hartree approach without exchange (Fock) terms. Also, if orbits are partially occupied, as happens in the presence of pairing (see Chapter 5), this expression is further modified and the method is called the Hartree–Fock–Bogoliubov approach.] Unfortunately, the procedure faces the paradox that, to know the single particle wave functions, we need to know the potential in which they move but that potential is generated by the particle wave functions themselves. The Hartree–Fock approach for a given $V_{ik}(\mathbf{r}_i - \mathbf{r}_k)$ consists of making an initial guess of the $\psi_k(r)$ and then using these to calculate $U(\mathbf{r})$. Then one uses the resulting $U(\mathbf{r})$ to re-calculate the $\psi_k(\mathbf{r})$. The process continues until self-consistent convergence is obtained.

In this chapter we will discuss the wave functions resulting from the Hamiltonian H_0 which corresponds to the motion of nucleons in a common potential. In Chapters 4 and 5, we will consider residual interactions and their interpretation. The Hamiltonian H_0 for single particle motion is given by

$$H_0 = \sum_{i=1}^{A} \left[\frac{\mathbf{p}_i^2}{2m_i} + U_i(\mathbf{r}) \right] = \sum_{i=1}^{A} H_0^i \qquad (3.3)$$

It has solutions

$$H_0^i \psi_i(\mathbf{r}) = E_i \psi_i(\mathbf{r})$$

where the $\psi_i(\mathbf{r})$ denote wave functions for individual nucleons in the potential $U(\mathbf{r})$ with single particle energies E_i. We will see shortly that, due to the Pauli principle, different nucleons have different individual ψ_i. The total nuclear wave function (again ignoring anti-symmetrization) is a product of the individual wave functions of each particle orbiting in this potential,

$$\psi(\mathbf{r}_1, \mathbf{r}_2, \dots) = \prod_{i=1}^{A} \psi_i(\mathbf{r})$$

and corresponds to a total system energy

$$E_0 = \sum_{i=1}^{A} E_i.$$

It is clear from the foregoing that the basic tenet of the independent particle model is that the nucleons move essentially freely in a central potential that is usually taken as a modified harmonic oscillator or modified square well potential. A little thought, however, raises two apparently serious difficulties before one even attempts this type of approach to the nuclear problem. The first centers on the question of how one can validly speak of independent particle motion in the presence of the strong nuclear interaction and a densely packed nucleus. The answer involves the Pauli principle and the essential weakness rather than strength of the nuclear force referred to earlier. The second question relates to the apparent conflict between a short-range nuclear force and the nature of a harmonic oscillator potential that actually becomes stronger as the distance from the origin (the center of mass) increases. We will return to both issues later.

We saw in Chapters 1 and 2 that the nuclear force is attractive and short range, and that the systematics of certain gross nuclear properties, such as nucleon separation energies, are generally smooth, except at certain specific nucleon numbers, called magic numbers, where they exhibit discrete jumps. The concept of magic numbers and the *shell structure* that they imply is of paramount importance in nuclear physics. Thus we summarize a bit of the voluminous evidence of their existence. Besides the sharp drop in S(n) and S(p) just after magic numbers (see Figs. 1.2 and 1.3), the lowest excited states in nuclei with magic numbers of either protons or neutrons are, on average, extremely high lying. In particular, in nuclei with even numbers of protons and neutrons the first excited state is nearly always a 2^+ state, and its energy is much higher in magic nuclei. This was illustrated by the Ca isotopes in Fig. 2.3: ^{40}Ca and ^{48}Ca correspond to the magic numbers at 20 and 28. Across the even $_{50}$Sn nuclei that have a magic number of protons, the first excited state (2^+) has an energy $E_{2_1^+} \approx 1200$ keV (see Fig. 2.6) as opposed to $E_{2_1^+} \approx 500$ keV for the isotones of Cd or Te (see Fig. 2.8). Even more striking, when Sn becomes doubly magic at ^{132}Sn, $E_{2_1^+}$ suddenly jumps to several MeV. Further support for the idea of magicity stems from the fact that elements with magic proton numbers have higher relative abundance, a larger number of stable isotopes, and relatively low neutron absorption cross sections. The nucleosynthesis of such elements predominantly occurs in stellar supernova explosions where an intense neutron flux leads to rapid, successive neutron capture reactions. This is the so-called r-process ("r" for rapid). The cross sections for such reactions depend mostly on the level density at excitation energies near the neutron separation energy. Such level densities are particularly low in magic nuclei. Therefore, for magic nuclei, the low neutron cross sections imply that, once formed, it is unlikely that a sufficient number of neutron captures take place in the short-lived astrophysical environment to deplete their numbers. In essence, they tend to block the r-process path.

Thus, we see several lines of evidence pointing to the importance of magic numbers. Moreover, we notice a relationship in these lines of evidence; many stem ultimately from the difficulty of exciting a magic or closed shell structure, and the consequent low-level density at low excitation energies.

Combining all the evidence, we can summarize the relevant magic numbers for nuclei as

$$2, 8, 20, 28, 50, 82, 126 \tag{3.4}$$

As we shall see in the discussion of nuclear phase transitions, 40 and 64 are in some cases weakly magic over limited ranges of N and Z.

It is well known in atomic physics that electron binding energies undergo sharp changes just after a closed electron shell. Analogously, it is reasonable to suppose that in the nuclear case, these magic numbers correspond to closed shells of nucleons. Of course, this viewpoint already presupposes a shell model and we will have to see later whether this provides an apt description of nuclear properties for nonclosed shell nuclei. Nevertheless, if one wants to pursue a shell model approach, it is clear that one of the basic features it must reproduce is the particular stability of nuclei with these magic numbers. One would therefore like to construct a nuclear potential that automatically and naturally produces gaps in single particle level energies at the magic numbers.

It is worth noting an often misunderstood point here: One often hears that closed shell nuclei are the most stable nuclei. This is not true, however, as a glance at the chart of separation energies in Figs. 1.2 and 1.3 clearly indicates. As nucleons of a given type (e.g., neutrons) are added, neutron separation energies systematically decrease. Just after a closed shell, the separation energy undergoes a *much larger* drop. Thus, closed shell nuclei are only more stable *relative* to *succeeding* nuclei.

3.1 Fermions in a potential—general properties

In considering an appropriate potential for the nuclear case, a tremendous simplification results if the potential is central, that is, if it depends only on the radial distance from the origin to a given point. This is equivalent to requiring that the potential is spherically symmetric. In discussing the single-particle model, we will discuss a number of different central potentials. These potentials differ in important details but they also share certain properties in common. We will first summarize a few of these and their implications.

In solving the Schrödinger equation for a particle in a central potential we can separate the radial and angular coordinates, writing $\psi = \psi(r)\psi(\theta, \phi)$. We then obtain separate equations in these coordinates. The angular dependence of a particle wave function is independent of the detailed radial behavior of the central potential. The angular variables give quantization conditions (familiar from any elementary quantum mechanics text) on $l^2 = l(l + 1)\hbar$ and the z-component of the angular momentum $l_z = m_z\hbar$ where m_z takes on the values $l, l - 1, \ldots 0, -1, \ldots -l$. We will see in Chapter 4 that much of the low energy spectra of whole classes of nuclei (those with proton and neutron numbers near the magic numbers) can be easily understood in terms of the angular behavior of the wave functions, completely independent of the radial form of the potential. On the other hand, the basic *sequence* of single-particle orbits in the independent particle model depends strongly on the choice of the potential in the radial direction.

Let us start our discussion of nuclear potentials by briefly reviewing the case of a 1-dimensional square well, as sketched in the lower left of Fig. 3.1a where $V = V_0$ for $x \leq 0, \geq d$ and $V = 0$ inside the well. (We use x for r for the 1-dimensional case.) The Schrödinger equation is

$$\frac{-\hbar^2}{2m}\frac{d^2\psi}{dx^2} + V\psi = E\psi$$

or

$$\frac{d^2\psi}{dx^2} = -k^2\psi$$

where

$$k = \frac{1}{\hbar}\sqrt{2m(E - V)} \tag{3.5}$$

The equality (to within a constant) of the wave function and its second derivative implies that ψ has an exponential form.

Inside the well, $(E - V) > 0$ so k is real and k^2 is positive. The second derivative is then equal to a negative number which implies that ψ is of exponential form with argument (ikx). That is

$$\psi(x) = A \sin(kx) + B \cos(kx)$$

If, momentarily, we assume the potential is infinite so that $\psi(0) = 0$, then $B = 0$. From $\psi(d) = 0$, we get $kd = n\pi$ or $k = n\pi/d$ and

$$\psi(x) = A \sin\frac{n\pi x}{d} \tag{3.6}$$

Then, from Eq. 3.5, with $V = 0$,

$$E = \frac{k^2\hbar^2}{2m} = \frac{n^2\pi^2h^2}{2md^2(2\pi)^2} = \frac{n^2h^2}{8md^2} \tag{3.7}$$

where $n = 1, 2, \ldots$ This result is little changed for a deep but finite potential: Since the wave function can "leak out" a little, its wave length is effectively slightly longer and hence its energy a little lower.

Note that our derivation of the energy for an infinite square well really entails nothing more than confinement and the wave-particle relationship. We can see this explicitly as follows. Since the wave function must vanish at $x = 0$ and d, it must have an integer or half integer number of wavelengths in a distance d. Hence

$$n\frac{\lambda}{2} = d \qquad n = 1, 2, \ldots$$

where λ is the particle wavelength given by the de Broglie expression $p = h/\lambda$. But $E = \frac{1}{2}mv^2$ gives $p = (2mE)^{1/2}$ and hence

$$d = \frac{nh}{2p} = \frac{nh}{2\sqrt{2mE}}$$

or

$$E = \frac{n^2h^2}{8md^2}$$

Several features of Eq. 3.7 are worth noting. First, and most important, of course, is the quantization, which results from the boundary conditions (confinement) and is a fundamental property of any bound quantum mechanical wave. Secondly, the energies drop with the square of the dimensions of the well (d) since a larger well size allows, for a given number of nodes in the wave function, a larger wave length. This has direct applications to nuclei where heavier nuclei (i.e., larger nuclei) will have lower energy levels. Finally, we note that E is never zero ($n_{min} = 1$): there is zero point motion.

Zero point motion is a direct reflection of the Uncertainty principle, and of confinement within finite dimensions. This can be seen two ways. If there were no zero point motion, the particle would be stopped but then, from $\Delta x \Delta p \sim \hbar$, $\Delta x \to \infty$ and the particle cannot be localized in the well. Alternately, if there is no zero point motion, the location of the particle is known, that is, $\Delta x = 0$ and $p \to \infty$ so the particle cannot be bound.

In fact, the minimum energy for a particle in a 1-dimensional box can be obtained approximately from the Uncertainty principle alone. From

$$\Delta x \Delta p \geq \frac{\hbar^*}{2}$$

---

*This is the proper expression for the Heisenberg principle as he derived it from the correlated joint probabilities of the fluctuations of conjugate variables (Heisenberg, 1930: I am grateful to J. Ptak for pointing me toward this reference). Expressions with h or $\hbar$ on the right are approximate forms.

we obtain

$$E_{\min} = \frac{p_{\min}^2}{2m} = \frac{(\Delta p)^2}{2m} = \frac{\hbar^2}{8m\,\Delta x^2} = \frac{\hbar^2}{8md^2}$$

We will see another example just below where the ground state or zero point energy can be obtained from the Uncertainty principle when we consider the harmonic oscillator potential.

Outside the well, $(E - V) < 0$, so k is imaginary, and

$$\psi \sim e^{ikx} = e^{-\frac{x}{\hbar}\sqrt{2m(V-E)}} \tag{3.8}$$

which dies out exponentially within the classically forbidden region.

An interesting application of this is to tunneling (important for example, in α-decay processes). Suppose, instead of the potential in the lower left of Fig.3.1a, we have the scenario in the lower right with a finite barrier of width, a. We ask what is the probability, P_t, of tunneling. This is given simply by the probability density of the exponential wave function after traversing the barrier of width a:

$$P_t = \text{penetrability} = \psi^2(a) \sim e^{-\frac{2a}{\hbar}\sqrt{2m(V-E)}}$$

Note the strong dependence of the penetrability on the width, a, of the barrier, the particle mass, m, and the height of the barrier seen by the particle, $(V - E)$. Higher barriers, thicker barriers, and more massive particles have strongly reduced transmission.

This dependence is also easily seen from the Uncertainty principle, $\Delta E\,\Delta t \sim \hbar/2$, and gives a different perspective on the process. The particle can have an energy greater than E by an amount ΔE for a short time Δt. Consider a particle of energy $E = \frac{1}{2}mv^2$ or $v = (2E/m)^{1/2}$. The time required to pass over a barrier of width a is $\Delta t \sim a/v$. In order that E becomes greater than V (for a time Δt), ΔE has to be at least as large as $(V - E)$. From the Uncertainty principle,

$$\Delta E = (V - E) \sim \frac{\hbar}{2\Delta t} = \frac{\hbar v}{2a} = \frac{\hbar}{2a}\sqrt{\frac{2E}{m}}.$$

In this approach tunneling is seen rather as (quickly) "jumping" over the potential barrier. The Uncertainty principle says that a particle of energy E can jump a barrier of height $(V - E)$ given approximately by $(\hbar/2a)(2E/m)^{1/2}$. Hence the thicker the barrier or more massive the particle, the smaller the barrier height that can be jumped, which is consistent with the barrier penetration probability given above.

We can use some empirical properties of the deuteron to estimate the depth of the nuclear potential. The deuteron has one bound state, with a small binding energy of 2.23 MeV. Consider the solutions to the finite square well (as a crude but useful approximation). The energies are given by

$$(E - V) = \frac{n^2h^2}{8mr_d} \tag{3.9}$$

where r_d is the radius of the well of the deuteron. This equation differs from Eq. 3.7 only in that, here, we define the potential to be attractive within the well, that is, $V = -V_0$

inside the well and zero outside, so that we can explicitly look at the role of the quantity $(E - V)$.

Clearly, from Eq. 3.9 the larger the width (r_d) of the well, the smaller is $(E - V)$ and, therefore, for a given energy, E, the smaller is $|V_0|$. From the known deuteron radius, we can therefore estimate V_0. Consider the wave function inside the well, $\psi = e^{ikx}$. If we approximate that the wave function is a maximum at the measured deuteron radius r_d (since that reflects the distance at which the particles tend to orbit relative to each other), that is, $kr_d = \pi/2$, we have

$$kr_d = \frac{r_d}{\hbar}\sqrt{2m(E + V_0)} = \frac{\pi}{2}$$

Sample calculations show that $V_0 >> E$ so we have approximately

$$\frac{r_d}{\hbar}\sqrt{2mV_0} \sim \frac{\pi}{2}$$

or

$$V_0 \sim \frac{\pi^2\hbar^2}{8mr_d^2}$$

Taking $r_d \sim 1.4 + 10^{-13}$ cm and m as the reduced mass, $m_{nucl/2}$, gives

$$V_0 = -50 \text{ MeV}. \tag{3.10}$$

This estimate is quite close to the well depths used, for example, in typical scattering and reaction (optical model) calculations.

Using the simplest possible model for nucleons confined in a potential, we can already deduce some interesting general properties of nucleons in the nucleus including another estimate of the potential well depth. Consider a system of completely non-interacting nucleons in a 3-dimensional box of volume V. This is the so-called Fermi gas model and is equivalent to putting the nucleons in an infinite 3-dimensional square well. Since nucleons are fermions and obey Fermi statistics, the number of states (nucleons) dA with momentum in a 3-dimensional interval dp is four times the phase space volume divided by h^3. [This result can be thought of approximately as a 3-dimensional generalization of the Uncertainty principle, $\Delta p \sim h/\Delta x$, to momentum and coordinate space.] The factor of four stems from the fact that each "state" in momentum space can contain 4 nucleons, two protons (with spins oppositely aligned) and similarly two neutrons. Thus

$$dA = \frac{16\pi p^2 V dp}{h^3} \tag{3.11}$$

Nucleons fill successive quantum levels in the box with each successive level having higher energy and momentum. The last filled state is often called the Fermi level corresponding to the Fermi energy, E_F, and Fermi momentum p_F. Integrating Eq. 3.11 up to the momentum p_F we have

$$A = \frac{16\pi p_F^3 V}{3h^3}$$

or

$$p_F^3 = \frac{3h^3 A}{16\pi V}$$

Substituting $V = (4/3)\pi R^3 = (4/3)\pi r_0^3 A$, where $r_0 = 1.2$ fm, we have

$$p_F^3 = \frac{9h^3 A}{64\pi^2 r_0^3 A} = \frac{9\pi \hbar^3}{8r_0^3} \tag{3.12}$$

or

$$p_F = \frac{\hbar}{r_0}\left(\frac{9\pi}{8}\right)^{1/3} \sim \frac{1.5\hbar}{r_0}$$

The corresponding Fermi energy E_F is

$$E_F = \frac{p_F^2}{2m} = \frac{\hbar^2}{2mr_0^2}\left(\frac{9\pi}{8}\right)^{2/3} = \frac{\hbar^2}{8mr_0^2}(9\pi)^{2/3} \sim \frac{1.2\hbar^2}{mr_0^2} \tag{3.13}$$

Using $\hbar c = 197$ MeV fm, taking the nucleon mass mc^2 as approximately 940 MeV, and using $r_0 = 1.2$ fm gives

$$E_F \sim 34 \text{ MeV}$$

If we recall that the average separation energy of a nucleon is about 6 MeV, and note that the potential well depth, by definition, is the Fermi energy plus the separation energy, we have a reasonable estimate of the nuclear potential $V \sim 40$ MeV which is close to the estimate in Eq. 3.10 from the properties of the deuteron. More sophisticated analysis gives values (~ 50 MeV) not far from this simple estimate.

We can easily take the Fermi gas model one step further and calculate the *average* nucleon energy by integrating the energy up to the Fermi level. This gives

$$E_{\text{ave}} = \frac{\int \frac{p^2}{2m} d^3 p}{\int d^3 p} = \frac{1}{2m}\frac{\int 4\pi p^4 dp}{\int 4\pi p^2 dp} = \frac{1}{2m}\frac{\frac{1}{5}p_F^5}{\frac{1}{3}p_F^3} = \frac{3}{5}\frac{P_F^2}{2m} = \frac{3}{5}E_F \tag{3.14}$$

or $E_{\text{ave}} \sim 20$ MeV. Two useful implications of this result follow immediately. The first is that nucleonic motion in the nucleus is largely non-relativistic since the velocity $v = (2E_{\text{ave}}/m)^{1/2} \sim 0.2c \sim 0.6 \times 10^{10}$ cm/sec. Secondly, the orbit period of a nucleon at the nuclear radius, $r_0 A^{1/3}$, is therefore (taking a typical A value of 125), $t \sim 0.6 \times 10^{-21}$ sec.

Finally, since we have considered the nucleons to be non-interacting (this includes a neglect of the Coulomb repulsion of protons) we could have carried out the analysis leading to Eq. 3.13 separately for neutrons and protons. This would have given additional factors of (N/A) and (Z/A) in Eq. 3.13, respectively. The total nuclear kinetic energy would then be given by

$$E_{\text{tot}} = \frac{Np_{F,N}^2}{2m} + \frac{Zp_{F,Z}^2}{2m} = \frac{3\hbar^2}{10mr_0^2}\left[\frac{9\pi}{4}\right]^{2/3}\left[\frac{N^{5/3} + Z^{5/3}}{A^{2/3}}\right] \tag{3.15}$$

Clearly, E_{tot} will be minimized if $N = Z$, which gives us a result we will see shortly in our discussion of isospin at the end of this chapter. If we write $N - Z = \varepsilon$ where $\varepsilon << A$, which gives $N = (\frac{1}{2})(A + \varepsilon)$ and $Z = (\frac{1}{2})(A - \varepsilon)$, we can expand Eq. 3.15, obtaining after a few trivial steps

$$E_{\text{tot}} \propto A + \frac{5}{9}\frac{(N - Z)^2}{A}$$

This $(N - Z)^2$ dependence is easy to see physically if we imagine protons and neutrons successively filling identical levels in a nuclear potential with two nucleons of a given type per level. For simplicity let us take those levels as equally spaced with an energy spacing (in arbitrary units) of unity. If we convert $\varepsilon = (N - Z)$ protons to neutrons, these new neutrons must go into successively higher orbits and come from successively lower proton orbits. The energy required is given by the number of nucleons moved, ε, times the energy needed for each which is $\varepsilon/2$ in units of the level spacing (if $N - Z = 2$ then 2 nucleons move up one level, if $N - Z = 4$ then 4 nucleons move up 2 levels each, and so on). Thus the cost in energy is $\varepsilon^2/2 \sim (N - Z)^2$. This result plays an important role in understanding nuclear stability and binding. Indeed, if this were the only relevant effect, the $N = Z$ nuclei would be the most stable.

The Fermi gas model is the ultimate independent particle model. In that sense it was the forerunner to the shell model. At the time the shell model was proposed (late 1940s), the idea of independently moving nucleons was not in vogue due to the early success of another approach to nuclear structure, the so-called liquid drop model, which, as its name implies, treats the nucleus basically as an incompressible drop of fluid. Indeed, on the face of it, independent particle motion would seem impossible since the volume of all the individual nucleons in a nucleus is over half of the total volume—hence how could such nucleons execute some 10^{20} undisturbed orbits per second without collision? Of course, we now know the answer: the Pauli principle prohibits most collisions, especially of inner-shell nucleons, since there are no nearby unoccupied states to scatter into. In any case, a droplet picture, which treats the nucleus as a macroscopic object with a more or less well-defined shape, provides a complementary approach to understanding nuclear structure.

In fact, both independent particle and coherent or collective approaches to structure are very successful. Ultimately, they must therefore be mutually consistent. We will explicitly address this issue in Chapter 6, and especially in Chapter 10 where we delve into the microscopic structure of collective vibrations. Later in this chapter, and in Chapters 4 and 5, we will study the independent particle approach to structure. The discussion will include the residual interactions which are a necessary ingredient since the 2-body nucleon–nucleon interactions are not perfectly simulated simply by a 1-body mean field potential. These residual interactions are actually the engine that drives the nucleus to develop correlated motions of nucleons and thereby to the macroscopic behavior that is the taking-off point for droplet or collective models (see Chapter 6). In Chapters 8 and 9, we will combine macroscopic concepts (shapes, rotation) with independent particle motion to discuss a single particle model for deformed nuclei as well as the coupling between these two degrees of freedom (e.g., Coriolis effects).

Here, we first want to briefly apply some very simple droplet concepts to the question of nuclear binding so that we can combine them with the result just obtained from the Fermi gas model to develop a useful expression for binding energies or masses.

The earlier result (see Fig. 1.1) that B.E./A ~ constant tells us that, on average, the nuclear density is constant and the nuclear volume scales with A. Thus one term in the total binding energy is a so-called "volume" term given by a coefficient times A, or $a_v A$. However, the approximately constant average binding of each nucleon resulting from interactions with a surrounding cloud of other nucleons (i.e., the idea behind the result B.E./A ~ constant) is *not* a good approximation at the nuclear surface where there is obviously less binding. Therefore another (negative) contribution to the total binding is represented as a "surface" term, $-a_s A^{2/3}$. In addition, the protons repel each other with the Coulomb interaction. This long range interaction goes as the total number of interacting pairs of protons and depends on their radial separation ($V_c \sim 1/r \sim A^{-1/3}$), giving another negative contribution, the "Coulomb" term, $-a_c Z(Z-1)A^{-1/3}$. These three terms—volume, surface, and Coulomb—comprise the essence of the liquid drop model.

However, recognizing that nucleons do move largely independently in the nucleus, we add to these three binding terms additional contributions reflecting the orbits and interactions of the nucleons themselves. One of the most important of these is the symmetry term just discussed, which can be parametrized as $-a_a(N-Z)^2/A$. We will see this $(N-Z)$ dependence of binding again when we discuss isospin later in this chapter. We note that the Coulomb term opposes the symmetry term leading to the fact that the most stable heavy nuclei no longer have $N = Z$.

Finally, we need a result from Chapter 4, relating to particle interactions not contained in the concept of a mean field potential, namely that pairs of like-nucleons in the same orbit gain energy by coupling their angular momenta oppositely to form a total angular momentum $J = 0$. Relative to an odd–even nucleus, an even–even nucleus will have one extra pair and an odd–odd nucleus one less pair. Hence, the pairing term in the binding energy is often written as $+\delta$ (e–e), 0(o–e or e–o), or $-\delta$ (o–o), where δ is often parameterized as $a_p A^{-1/2}$.

Putting all this together gives an expression for the total nuclear binding energy, sometimes known as the Weizsäcker or semi-empirical mass formula

$$B.E. = a_v A - a_s A^{2/3} - a_c Z(Z-1)A^{-1/3} - a_a(N-Z)^2/A + (\delta, 0, \text{ or } -\delta)$$

(3.16)

Values of the coefficients are obtained by fits to existing data. Typical values are: $a_v \sim$ 15.8 MeV, $a_s \sim$ 18.3 MeV, $a_c \sim$ 0.7 MeV, $a_a \sim$ 23.2 MeV, and $a_p \sim$ 12 MeV (giving $\delta \sim$ 1 MeV in medium and heavy nuclei). The volume term alone would give a total binding of $\sim 16\,\mathrm{MeV}$ per nucleon. The negative terms combine to reduce this to the observed value of $\sim 8\,\mathrm{MeV}$ per nucleon. (See Fig. 7.8 of Heyde, 1994.) The approximate equality of a_v and a_s can be motivated by a simple, highly schematic argument offered to me by S. Moszkowski. If we think of the nucleons arranged on a cubic lattice, each interior nucleon has more interactions than a surface nucleon. Suppose we denote the total number of nucleons A as some number n^3 where n is the number of nucleons on a side. Each of the n^2 nucleons in a horizontal plane interacts with its corresponding nucleon in the adjacent plane. Altogether, there are $(n-1)$ such sets of interactions. The cube is symmetric so this argument applies to each of its three orientations. Hence the total number of interactions is $3n^2(n-1)$. This is $3(n^3 - n^2) = 3(A - A^{2/3})$ nearest neighbor nucleon–nucleon links: that is, the binding energy has a term proportional to $(A - A^{2/3})$ with equal coefficients a_v and a_s.

Over the years mass formulas have been refined to include numerous subtle effects so that they now reproduce empirical nuclear binding energies to better than, on average, an MeV or so. These refinements take into account, among other effects, certain specific aspects of structure and of the distribution in energy of shell model levels. The downside is that they involve a very large number of parameters. Therefore, it is still often useful to use the simple formula above which does give the overall trends and has an obvious intuitive basis.

As we have noted, the Weizsäcker mass formula is an amalgam of liquid drop and independent gas concepts. The former refers to the macroscopic volume, surface, and Coulomb contributions. Terms relating to single particle orbits, such as the symmetry term in $(N - Z)$, terms reflecting non-uniform distributions of single particle energies (so-called shell correction terms, which we have not discussed), as well as the effects of specific residual interactions such as pairing, introduce a "microscopic" aspect. Together, all the terms that contribute to modern mass formulas comprise a sophisticated approach known as macroscopic–microscopic models. Such models have a particular current interest in that they allow predictions of nuclear binding for exotic nuclei far from stability that are now becoming accessible (see Chapter 11).

We now return to our discussion of potentials and consider the properties of solutions to the 3-dimensional potential. In three dimensions the new ingredient is angular momentum and the angular dependence of the wave functions. We denote an arbitrary central potential by $U(\mathbf{r})$ and only require that $U(\mathbf{r})$ is attractive and $U(\mathbf{r})$ and $dU/d\mathbf{r} \to 0$ as $\mathbf{r} \to 0$. The Schrödinger equation for such a potential is

$$H\psi = \left(\frac{\mathbf{p}^2}{2M} + U(\mathbf{r}) \right) \psi_{nlm}(\mathbf{r}) = E_{nlm}\psi_{nlm}(\mathbf{r}) \tag{3.17}$$

This equation is separable into radial and angular coordinates and therefore the solutions ψ_{nlm} can be written

$$\psi_{nlm}(\mathbf{r}) = \psi_{nlm}(r\theta\phi) = \frac{1}{r}R_{nl}(r)\psi_{nl}(\theta\phi) \tag{3.18}$$

Here, n is the radial quantum number, l the orbital angular momentum quantum number and m the eigenvalue of its z-component, l_z. It is conventional in nuclear physics to give names to different l values following the convention:

$$l = 0, 1, 2, 3, 4, 5 \ldots \quad s, p, d, f, g, h \ldots$$

For a given l, m takes the values $l, l - 1, l - 2 \ldots 0, -1, -2 \ldots - (l - 1), -l$. Since $U(\mathbf{r})$ is spherically symmetric the $(2l + 1)$ energies are independent of m and we will usually delete this index.

Turning to the radial solutions, we note that, with angular momentum, there is associated a centrifugal force. Hence, we write

$$U = U(\mathbf{r}) + U_{\text{centrifugal}}$$

where $U(\mathbf{r})$ is the nuclear potential. The centrifugal potential is the integral of the centrifugal force $F_{\text{cent}} = m\omega^2 r$. Hence

$$U_{\text{cent}} = \int m\omega^2 r \, dr = \int \frac{m^2\omega^2 r^4}{mr^3} dr = \int \frac{L^2}{mr^3} dr = \int \frac{l(l+1)\hbar^2}{mr^3} dr = \frac{l(l+1)\hbar^2}{2mr^2}$$

and the radial equation becomes

$$\frac{\hbar^2}{2m}\frac{d^2 R_{nl}(r)}{dr^2} + \left[E_{nl} - U(r) - \frac{\hbar^2}{2m}\frac{l(l+1)}{r^2} \right] R_{nl}(r) = 0 \qquad (3.19)$$

The quantum number n specifies the number of nodes (zeros) of the wave function with the usual, but not universal, convention that one counts one of the two nodes at infinity and at $r = 0$, that is, $n = 1, 2, \ldots$ The detailed specifications of the radial wave functions R_{nl} are of little practical importance in the present discussion. We note for completeness that they are proportional to Laguerre polynomials in r^2.

Let us now consider the solutions to Eqs. 3.17, 19 as a function of n and l. Figure 3.1b illustrates the results for $l = 0$ at the far left, giving the lowest three $l = 0$ wave functions ψ_{1s}, ψ_{2s}, and ψ_{3s} along with their eigenvalues E_{1s}, E_{2s}, and E_{3s}. The lowest energy state has $n = 1$. States with higher n have higher energy. Physically, it is easy to see why this is so. The larger n is, the larger is the number of nodes and the shorter is the wavelength of the oscillations in ψ, and therefore the energy increases. This is a basic consequence of confinement.

Similarly, for two states with the same n value, but two different l values, the wave function with the higher l also has the higher energy. This is also easy to see from Eq. 3.19 since the centrifugal potential is larger for the particle with larger l. Therefore this particle has higher transverse motion and is, on average, further from the nucleus and therefore less bound.

It is useful to look at this a little more explicitly. For $l = 1$ there is a small repulsive centrifugal contribution $(2\hbar^2/2mr^2)$ to the potential. The overall potential is increased and shifted to larger radii. Both effects raise the energy. As the potential is squeezed in radial extent the particle is more localized and hence, from the Uncertainty principle, as we have seen, the momentum and hence the energy increases. This is seen in the radial squeezing of the wave functions in the lower part of the figure which shortens their wave length. This effect is exacerbated for high l as seen in the figure, since the centrifugal contribution goes to $6\hbar^2/2mr^2$ for $l = 2$, $12\hbar^2/2mr^2$ for $l = 3$, and so on.

From these very general and intuitive results one can already deduce an important conclusion: For any well-behaved central potential, the lowest single particle state is always an s state ($l = 0$) and has $n = 1$. In particular, this simple result explains why the ground state of the deuteron is primarily an orbital angular momentum s state. (The small D state admixture is due to non-central potentials, which do not necessarily have the properties just discussed).

We can now use these arguments to deduce a fundamental property of the single particle energies which utterly dominates everything we know about nuclear structure, namely, that, since the energy increases (wavelength decreases) with *both* n and l, it follows that one can *counterbalance* these two effects to create *groupings* of levels at similar energies. That is, two states, one of higher n but lower l (or vice versa) can cluster near the same energy. Note the generality of this result: We have not specified any details about the potential except that it is central and not pathological (e.g., with multiple minima).

This is the basic reason why shell structure exists. That is, it is this inevitable consequence of a central potential and the centrifugal force that leads to level groupings, and hence shell structure and, via the Pauli principle, to the magic numbers which are the empirical benchmarks of structure (at least near the valley of stability; see Chapter 11).

3.2 Nuclear potentials

We now turn to the harmonic oscillator potential. This potential is particularly popular in nuclear physics for two principal reasons: It provides a remarkably good approximate solution to many nuclear problems and it is particularly easy to handle mathematically, thus yielding many results analytically. The harmonic oscillator potential is given simply by

$$V(\mathbf{r}) = \frac{1}{2}k\mathbf{r}^2 = \frac{1}{2}m\omega^2\mathbf{r}^2 \tag{3.20}$$

where k is the spring constant in the classical problem.

The reader should note, however, that the harmonic oscillator has the wrong asymptotic behavior. How does one reconcile the idea that a nucleon at large distances from a real nucleus cannot experience a nuclear force with the fact that a harmonic oscillator potential increases in strength with radius? This point is rather subtle and relates to the problem of separating the internal motion of the nucleons from that of the nuclear center of mass. This problem has been elegantly discussed in de Shalit and Feshbach, and we will not dwell on it here except to comment that in a nuclear potential V, such as the harmonic oscillator, where $V \rightarrow \infty$, the falloff of nuclear wave functions with increasing distance from the origin behaves rather differently (as $\exp[-\alpha r^2]$) than is the case for quantum mechanical tunneling through a finite potential ($\exp[-\alpha r]$). However, this difficulty is primarily a "long distance" problem: the two wave functions are similar within the nuclear volume. Significant errors may accumulate in studying the tails of nuclear wave functions. Such effects can be important, for example, in studying single nucleon transfer reactions that occur primarily in "grazing" collisions at the nucleus surface. Having pointed out this issue, we nevertheless continue with our discussion.

For a 3-dimensional harmonic oscillator, the eigenvalues E_{nl} can be written as $E_{nl} = (2n + l - 1/2)\hbar\omega$ where the lowest n value is $n = 1$. We see that exactly degenerate energies characterize two orbits with

$$\Delta l = -2\Delta n. \tag{3.21}$$

Thus, for example, the energies of a particle in the orbits 2s or 1d are degenerate. Similarly, the larger grouping 1h, 2f, 3p is degenerate. Such degeneracies, as we shall see, are broken for potentials of other shapes but close groupings still remain.

We note that the zero point motion for a harmonic oscillator is $(\frac{1}{2})\hbar\omega$ times the number of dimensions. We can easily obtain this value without actually solving the Schrödinger equation by using the Uncertainty principle, as we did before for the square well. To see this, we write, for one dimension

$$p_{min} = \Delta p = \frac{\hbar}{2\Delta x}$$

Hence

$$E = \frac{p^2}{2m} + \frac{1}{2}m\omega^2 x^2 = \frac{\hbar^2}{8mx^2} + \frac{1}{2}m\omega^2 x^2 \tag{3.22}$$

E_{min} is obtained by setting $dE/dx = 0$, getting

$$\frac{-\hbar^2}{4mx^3} + m\omega^2 x = 0$$

Solving for the radius corresponding to E_{min} gives

$$x^2\Big|_{E_{min}} = \frac{\hbar}{2m\omega}$$

Substituting in eq. 3.22 gives

$$E_{min} = \frac{\hbar^2 2m\omega}{8m\hbar} + \frac{m\omega^2\hbar}{2(2m\omega)} = \frac{1}{2}\hbar\omega \tag{3.23}$$

We again see that confinement naturally leads to zero point motion as a result of the Heisenberg Uncertainty principle. Confinement and the wave-particle relation directly give the quantization of energies, and identical results are obtained with the Schrödinger equation and the Uncertainty principle.

The energy levels of the harmonic oscillator potential are shown on the left in Fig. 3.2. They display the two important properties that are evident from the above discussion and

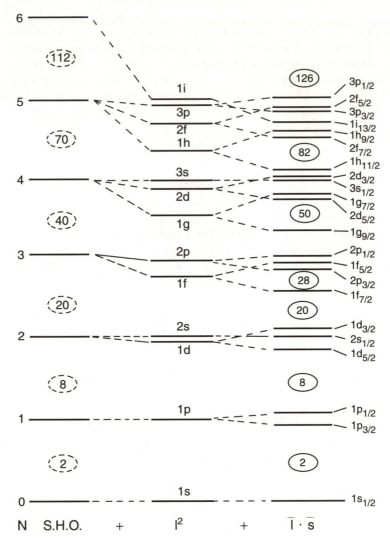

FIG. 3.2. Single-particle energies for a simple harmonic oscillator (SHO), a modified harmonic oscillator with l^2 term, and a realistic shell model potential with l^2 and spin orbit ($l \bullet s$) terms. Here N labels the oscillator shell: $N = 2(n-1) + l$

the expression for E_{nl}. First, the energy levels fall into degenerate multiplets defined by the (integer) values of $2n + l$. Secondly, a given multiplet generally contains more than one value of the principal quantum number n and of the orbital angular momentum l. Thus, for example, as evident in the figure, the levels 3s, 2d, and 1g are all degenerate. It is this grouping of levels that provides the shell structure required of any central potential useful for real nuclei.

We recall that the nucleon has an intrinsic spin s $= (\frac{1}{2})\hbar$. Therefore the total angular momentum quantum number of a nucleon in an orbit l, with total angular momentum $\mathbf{j} = \mathbf{l} + \mathbf{s}$ is $l \pm 1/2$. For example, the d orbit gives rise to states with $j = 3/2$ and $5/2$; h gives $j = 9/2$ and $11/2$. A particle with total angular momentum j has magnetic substates $m_j = j, j - 1, j - 2, \ldots 1, 0, -1, -2, \ldots -j$, or $2j + 1$ in all. The Pauli principle therefore limits the number of nucleons in each orbit (e.g., $2d_{5/2}$, $1h_{11/2}$) to $2j + 1$ (e.g., 6, 12). Thus, a j-level, or a cluster of j-levels, can contain only a certain fixed maximum number of nucleons. If more are added, they must go into a (higher energy) unfilled orbit. Hence, the process of adding nucleons in a potential such as a harmonic oscillator leads to a total system energy that has jumps at certain key nucleon numbers, called magic numbers, where a cluster of levels is filled and where the next particle goes into a substantially higher energy orbit. Since that orbit is less bound, it also follows that the separation energy of that last nucleon will be notably less than that of the nucleons in filled (closed) shells. For the harmonic oscillator, these magic numbers are indicated at the left in Fig. 3.2.

The harmonic oscillator potential, with groupings 1s; 1p; 1d, 2s; 1f, 2p; 1g, 2d, 3s; 1h, 2f, 3p; leads to magic numbers (taking into account that each orbit with $l \neq 0$ has two j-states, each of which can contain $2j + 1$ particles): 2, 8, 20, 40, 70, 112. We note that the first of these indeed reproduce the known magic numbers but that the last two do not. Therefore, while the harmonic oscillator potential is a reasonable first order approximation to the effective nuclear potential, it must be modified to be useful.

It was, in fact, the monumental achievement of Mayer and, independently, of Haxel, Jensen, and Suess, to concoct a simple modification (primarily a spin–orbit term) to the harmonic oscillator potential that enabled it to reproduce the empirical magic numbers. This step revolutionized forever the history of nuclear physics and has led, either directly or indirectly, to essentially all the progress that has been made since. Their achievement, in effect, is the creation of a realistic shell model. Following their discovery in 1948, the extensive and very detailed development of this model has led to an elaborate formalism that provides not only a direct description of many nuclei but also the microscopic basis for many macroscopic models of collective properties of nuclei. Since their work, there have been extensive efforts to derive the nuclear shell model potential from more fundamental data on the nucleon–nucleon interaction. We shall not concern ourselves with such efforts here, but shall consider the potential they proposed, discuss it physically, and here and in the remaining chapters, draw out many of its implications.

It is possible to use some rather general arguments, based on the short-range nature of the nuclear force, to suggest some plausible modifications to the harmonic oscillator potential V. Consider a relatively heavy nucleus with dimensions significantly larger than the range R_N of the nuclear force. Then, as long as a given nucleon lies inside the nuclear surface by a distance greater than R_N, it should be surrounded rather uniformly by nucleons on all sides. It is screened from the asymmetric distribution that appears at the boundary. Therefore, it should experience no net force. In other words, the central part of the nuclear potential should be approximately constant. Thus, from this point of view, a square well potential might be an improvement on the harmonic oscillator. Another possibility is to add an attractive term in l^2 to the harmonic oscillator potential.

It is easy to see why this is equivalent to a flattening of the effective radial shape of the potential. The effects of an l^2 term increase with the orbital angular momentum of the particle. Therefore high angular momentum particles feel a stronger attractive interaction that lowers their energies. However, these are precisely the particles that, because of the centrifugal force, spend a larger fraction of their time at larger radii. Therefore the addition of an l^2 term is equivalent to a more attractive potential at larger radii and comes closer to the desired effect of a more constant interior potential. In fact, it gives a potential intermediate between that of the harmonic oscillator and the square well. A Woods–Saxon potential has a flatter bottom than the harmonic oscillator and also produces effects similar to an l^2 term. In the deformed shell model (Nilsson model), that we will discuss in Chapters 8 and 9, the spherical limit for the single particle energies is explicitly expressed in terms of such an l^2 contribution.

The relation of the single particle levels produced by a harmonic oscillator potential, along with the addition of an l^2 term is illustrated in the middle panel of Fig. 3.2, which shows how the degeneracy of the harmonic oscillator levels is broken as high angular momentum levels are brought down in energy.

Such a nuclear potential is illustrated in Fig. 3.3, which also shows (for protons) the effect of the Coulomb interaction which reduces the potential inside the nucleus but creates a potential barrier outside.

It is clear that neither of these alternatives yet produces the magic numbers observed experimentally. It is easy to do so, however, if one introduces a so-called spin-orbit force. Thus far, we have not discussed the spin quantum number explicitly. Nevertheless, it is well known that the nucleon, either proton or neutron, has an intrinsic spin 1/2, and therefore the total angular momentum of a nucleon in any orbit is given by the vector coupling of the orbital angular momentum l with a spin angular momentum $\mathbf{s} = 1/2$. With a spin-orbit component, the force felt by a given particle differs according to whether its spin and orbital angular momenta are aligned parallel or antiparallel. If the parallel alignment is favored, and if the form of the spin-orbit potential is taken as $V_{l \bullet s} = -V_{ls}(r) l \bullet \mathbf{s}$ so that it affects higher l values more, then its effects will be similar to those illustrated on the far right in Fig. 3.2. Each nl level, such as 1g, will now be split into two orbits, $1g_{9/2}$ and $1g_{7/2}$, with the former lowered and the latter raised in energy. This instantly reproduces all the known magic numbers.

The absolute strength of the spin-orbit force must be substantial (see Fig. 3.2) to produce the correct magic numbers: indeed, the splittings it produces must be comparable to those between adjacent multiplets of the harmonic oscillator potential. Since the constant $\hbar\omega$ of the harmonic oscillator potential is found to be $\hbar\omega = 41/A^{1/3}$ (e.g., $\hbar\omega \approx 8$ MeV for medium and heavy nuclei), it follows that the $V_{ls}(r)$ must attain nearly such magnitudes.

Since the spin-orbit force is an inherently quantum relativistic effect, it is not as easy to give a physical picture for it as for the relation between an l^2 force and the effective change in the behavior of the central potential just discussed. It has been shown, however, to arise naturally, and with the correct sign, from relativistic effects of the nucleonic motion. It is possible, though, to give plausible arguments for the *radial* shape of the spin-orbit potential. These rely again on the notion that, in the interior of the nucleus, a

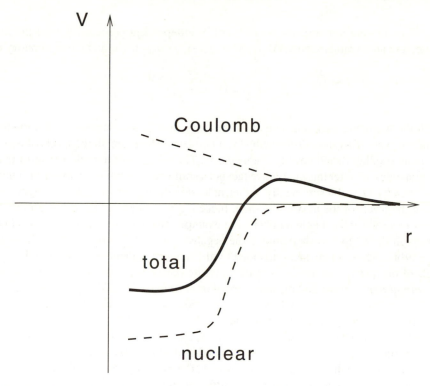

FIG. 3.3. Radial dependence of the nucleon, Coulomb and total 1-body potentials.

nucleon should experience no net force. If the spin-orbit force were large in the nuclear interior there would be a preference for nucleons with spins aligned parallel to their orbital angular momentum rather than vice versa and therefore such a nucleon would not be surrounded by an equal number of nucleons with all spin orientations. This suggests, although it certainly does not prove, that the spin-orbit force is primarily a surface phenomenon. It is therefore customary to write:

$$V_{l \bullet s} = -V_{ls} \frac{\partial V(r)}{\partial r} \mathbf{l} \bullet s \tag{3.24}$$

where $V(r)$ is whatever potential is chosen for the central potential itself and V_{ls} is a strength constant.

It is worth pausing at this point to emphasize the importance of the spin-orbit interaction. It is not merely a device that ensures the appropriate magic numbers. Rather, a significant fraction of nuclear structure research in the last three decades has relied on and exploited the particular consequences of the spin-orbit force. To see this, it is necessary to refer more explicitly to the concept of parity. The parity of a wave function $\psi_{nlm} = R_{nl}(r)Y_{lm}(\theta, \phi)$ is given by the behavior, $(-1)^l$, of the spherical harmonic Y_{lm} under reflection, since the radial wave function does not depend on the sign of r. Thus

$\pi_{nlm} = (-1)^l$. For a multiparticle system of N independent particles, ψ is of the form (neglecting antisymmetrization) $\psi_{nl1}(r_1)\psi_{nl2}(r_2) \ldots \psi_{nlN}(r_N)$, and the total parity is

$$\pi = \prod_{i=1}^{N}(-1)^{l_i} = (-1)^{\sum_{i}^{N} l_i}$$

In the harmonic oscillator potential, the particular $2n + l$ degeneracy led to shells containing sets of l values differing by even numbers. Therefore all the levels of a given harmonic oscillator shell have the same parity. The addition of a modest l^2 term in the potential does not alter this qualitative fact but a spin-orbit potential can lower the energy of the $j = l + 1/2$ orbit sufficiently so that, at least in the higher shells, it is brought down among the levels of the next lower shell. Indeed, this is precisely the effect required in order to reproduce the magic numbers. For example, the positive parity $1i_{13/2}$ orbit now appears in the 82 to 126 shell among the negative parity $2f_{7/2}$, $1h_{9/2}$, $2f_{5/2}$, $3p_{3/2}$ and $3p_{1/2}$ orbits. Thus, a real shell, bounded by the magic numbers, contains a majority of levels of one parity and one level of the opposite parity. It is conventional to call these the *normal parity orbits* and the *non-normal* or *unique parity orbits*, respectively.

The significance of this will only become clear when we consider certain *residual* interactions experienced by the nucleons in the shell model potential. However, the general point can be made easily. In some cases, these residual interactions are diagonal, providing contributions only to the energies of the various levels. In others, however, they have important nondiagonal effects that mix different configurations. Whatever mixing does occur, however, cannot mix levels of different parity. (We neglect here the very weak parity-nonconserving part of the weak interaction.) We recall from the discussion of two- and multistate mixing in Chapter 1 that such mixing effects are strongly dependent on the energy separation of the mixed states. Since the unique parity orbit cannot be admixed with its neighbors in a given major shell and because it is so far separated from the other orbits of the $2n+l$ multiplet from which it originated, it mixes only very weakly with other levels. Therefore, configurations stemming from the unique parity orbit are particularly pure even though they occur amid an enormous complexity of mutually admixed states characteristic of many heavy nuclei. These levels therefore provide an ideal laboratory for testing various nuclear models, since one deals with particularly pure, simple, and well-known wave functions. This fact has been exploited in countless experimental and theoretical studies in recent years. Perhaps the most well known involve the interesting physics of high spin states in deformed nuclei. Indeed, it should not be surprising that the first, and often the majority of tests, of virtually every new approach to the study of high-spin states deals with unique parity levels. Only after whatever new effects are involved are well understood for these levels does one generally dare to look at the normal parity levels.

Unique parity levels are also of particular importance for low-spin states and for reasonably light nuclei. In the next chapter we shall study the effects of residual inter-actions on nuclear level schemes. In general, these are quite complex and can lead to significant *configuration mixing*. However, we shall see that many of the complexities are eliminated, and indeed, rather simple, analytic, physically-reasonable results emerge, if the nucleons are restricted to occupying a "single j" orbit. Clearly the testing ground par excellence for such ideas will again be unique parity orbits. The fact that these are also the highest j orbits in a given shell and thus can contain the most particles, further enriches the phenomena that can be studied.

Now that we have outlined the basic features of the nuclear potential and seen the single particle energy levels to which it leads, we are nearly ready to consider the predictions of this model for various nuclei. The basic idea is simply to fill the levels of the potential sequentially, as one adds nucleons in going from nucleus to nucleus. With the help of some simple arguments we will see that we are immediately able to make many predictions concerning the low-lying levels of a number of nuclei.

Before applying the model there are several rather profound issues, seldom thought about explicitly, that must at least be mentioned. The entire concept of "filling the levels... sequentially" is in fact one of the most important applications of the Pauli principle. Without this dictum, any number of nucleons could go in the lowest orbit. That being the case, there would be no reason (energetically) to favor the addition of a neutron or a proton (neglecting the Coulomb force—see the following). A nucleus with $Z = 2, N = 300$ would be as stable as any other. It is one of the most significant consequences of the Pauli principle that stable nuclei have $N \approx Z$ since a large excess of either would involve the filling of higher-lying levels at an extra cost in energy. Indeed, the shape of the valley of stability (Fig. 2.1) results from the countervailing effects of the Pauli principle that favors $N = Z$ and the Coulomb repulsion that favors a large neutron excess in order to separate the protons further and reduce their average Coulomb repulsion. For low Z, the former effect dominates, while for higher Z the valley of stability curves toward the neutron rich side (see the next paragraph). It is the same effect of the Pauli principle that determines nuclear radii: without it (and the Coulomb force), all nuclei would be comparable in size to an α particle.

This discussion leads naturally into a few comments about the Coulomb force, which is simply an extra, repulsive, long-range potential ($V_{\text{Coul}} \approx 1/r$) felt by the protons. We will not discuss this in detail, but it is obvious that its principal effect is to *raise* the single particle proton energy levels. As we have just seen, in the absence of the Coulomb force, proton and neutron energy levels would fill with equal likelihood, there being no distinction between the two, and the valley of stability would correspond to $N = Z = A/2$. However, because of the Coulomb interaction, there is an extra cost in adding additional protons. This effect is clearly more important for heavier nuclei since the Coulomb force is a long range interaction that scales roughly as $Z(Z - 1)$, the total number of interacting proton pairs. Therefore, as A increases it becomes energetically preferable to add additional neutrons, and thus stable heavy nuclei have a neutron excess.

A second issue of importance concerns the very concept of independent particle motion in a central potential. As noted earlier in this chapter, this seems to be at odds with the idea of a strong nuclear force. Empirically, nuclear radii can be described by the relation $R = 1.2A^{1/3}$. If we take 1 fm as the radius of a nucleon, and imagine the nucleus to be uniformly filled, then the ratio of the nuclear volume to that of a single nucleon is just $(1.2)^3 A$, or hardly more than A times the volume of each nucleon! Therefore, it would seem unlikely that an individual nucleon could execute countless undisturbed orbits in such a densely packed medium. (Note that this is in stark contrast to the emptiness through which atomic electrons follow their orbits. Bohr did not need to address this issue, nor could he, since its solution is yet another consequence of the Pauli principle.)

The resolution of the paradox rests on the conjunction of the Pauli principle with the essentially weak nature of the attractive nuclear force where, by "weak", we mean relative to typical kinetic energies of nucleons within the nuclear volume. This weakness is illustrated by the fact that in the deuteron, the simplest nuclear system, the attractive nuclear interaction is only sufficiently strong to produce one bound state.

To understand the possibility of independent particle motion, consider a central potential with various energy levels, defined by the quantum numbers nl (we ignore spin for this argument), in which the nucleons, either protons or neutrons, sequentially fill each level. Since the central potential is spherically symmetric, the energy levels are independent of the magnetic quantum number m and therefore, each such orbit is $2l + 1$ degenerate (for any l there are $2l + 1$ magnetic substates). By the Pauli principle, which states that no two fermions can occupy the same configuration, such an orbit can contain at most $2l + 1$ particles. Therefore, as more and more nucleons are added to the nucleus, orbits of successively higher n and l values, and therefore higher energies, are filled. Now, imagine a pending collision between two nucleons in relatively inner orbits. By the Pauli principle, unless the impact of the collision is sufficiently strong to raise one of these nucleons to an unoccupied orbit (i.e., above the Fermi surface), the interaction can have no effect on the motion of these two nucleons. Therefore only the outermost nucleons are likely to be affected by such collisions. Thus, just as it is the Pauli principle that prevents the nucleus from collapsing on account of the attractive nuclear potential, it is also the Pauli principle that leads to independent particle motion for most nucleons in the nucleus. This does not mean that there cannot be correlations in two-body nuclear wave functions, but those correlations must be consistent with the Pauli principle and with the strength of the attractive nucleon–nucleon interaction. Later, we shall see that many of those correlations that do occur are an implicit effect of the antisymmetrization of nuclear wave functions associated with the Pauli principle.

As a final preliminary, we must consider the structure and angular momentum of filled shells of nucleons. Consider a system of A nucleons (either protons or neutrons, which, for the present purposes, we treat independently: the interactions between them will be considered later). Each orbit of spin $j = l \pm 1/2$ has $2j + 1$ degenerate magnetic substates. By the Pauli principle, once $2j + 1$ nucleons are in a given orbit, it is filled (closed) and the next one begins to fill. To predict the spins and parities (J^π values) of low-lying levels in nuclei, we need to consider the effects of the (filled) closed shells. We now show that the total angular momentum of a closed single-j shell is identically zero. This is trivial. It is immediately clear that the total magnetic quantum number M of a closed shell is $M = \Sigma m_i = j + (j - 1) + \ldots + (-(j - 1)) + (-j) = 0$. Since this is the only possible M value, it follows that the total angular momentum $J = 0$ since J can never be larger than the largest M. Alternately, for each state j, m there is a state $j, -m$, and therefore the vector sum $\mathbf{J} = \mathbf{j} + \mathbf{j}$ vanishes. Since this is true for each pair of states (with $\pm m$ values) and, of course, for $m = 0$, it is true for the full set of $2j + 1$

states. Furthermore, since it is true for a given j value, it must be true for an entire major shell consisting of several j values.

Note the importance of this result: without it, it would be impossible to apply the independent particle model to any nuclei except hydrogen or a neutron where, of course, the idea of the central potential itself would not be valid. With it, however, one can at least predict the ground state spins and low-lying excited states of any nucleus consisting of one particle, either proton or neutron, outside a closed shell. Since the core—that is, all closed shells (major or single $-j$)—contributes only $J^\pi = 0^+$, the angular momentum of the ground state of a nucleus with one particle in a shell nlj is therefore just $J = j$ with parity $\pi = (-1)^l$.

Any program of predicting ground state J values, or the J values and energies of single-particle excited states, presupposes that we know the order and spacings of the single-particle levels. The reader may perhaps think that we have already dealt with and solved this question by the sequence on the right of Fig. 3.2. However, this ignores the fact that the single-particle energies themselves depend on the number of nucleons in the nucleus since the single-particle potential arises from these same nucleons. In particular, when looking at a sequence of levels in a central potential, as in Fig. 3.2, the overall scale is illustrative only and in fact changes with mass; there is a gradual scale compression for heavier nuclei. Moreover, residual interactions affect the single particle energies themselves. Thus, such sequences are a guide only. In practical use, specific single particle energies for each mass region must be used. We will further discuss this important issue of the N, Z dependence of single particle energies in section 3.4. However, we first want to show how the independent particle model is used to make actual predictions of J^π values and excitation energies for certain classes of nuclei.

3.3 Predictions of the independent particle model

Despite the caveat just made of the variability of single-particle energies, we will, for convenience, use Fig. 3.2 as a single-particle energy reference. (Note: in most of the following discussion we will denote orbits by the lj quantum numbers alone, as in $f_{7/2}$, since omission of n simplifies the notation and seldom causes confusion.)

This simple model works extremely well for a select group of nuclei, namely those with one particle outside a doubly magic core. For example, by inspection of the right-hand panel of Fig. 3.2, we would expect the ground state spin-parity of $^{41}_{20}\text{Ca}_{21}$, which has one neutron beyond the $N = 20$ closed shell, to be $7/2^-$. Similarly, $^{41}_{21}\text{Sc}_{20}$, should have a $7/2^-$ ground state, while the ground states of $^{91}_{41}\text{Nb}_{50}$ and $^{91}_{40}\text{Zr}_{51}$ should be $9/2^+$ and $5/2^+$, respectively. These predictions are verified experimentally, as shown for some of these cases in Fig. 3.4. This situation corresponds to the leftmost panel of Fig. 3.5.

Using the result that $J = 0$ for *any* filled j shell allows us to go a little further. Consider ^{49}Ca, which has one particle outside a filled $f_{7/2}$ shell in the independent particle model. Thus the ground state should have (and does, see Fig. 3.4) $J^\pi = 3/2^-$.

We have just seen how the concept that $J = 0$ for a full j-shell and, by extension, for a full major shell, leads to a simple way of determining the ground state J^π value of a nucleus with one odd nucleon outside a doubly magic core. This is, of course, useful, and an early triumph of the independent particle model but, if this were the extent of its

GROUND STATE J^π VALUES OF
SOME ODD MASS NUCLEI

Z = 20	$\dfrac{3/2^+}{^{37}Ca}$	$\dfrac{3/2^+}{^{39}Ca}$	$\dfrac{7/2^-}{^{41}Ca}$	$\dfrac{7/2^-}{^{43}Ca}$	$\dfrac{7/2^-}{^{45}Ca}$	$\dfrac{7/2^-}{^{47}Ca}$ $\dfrac{3/2^-}{^{49}Ca}$
Z = 40		$\dfrac{9/2^+}{^{87}Zr}$	$\dfrac{9/2^+}{^{89}Zr}$	$\dfrac{5/2^+}{^{91}Zr}$	$\dfrac{5/2^+}{^{93}Zr}$	$\dfrac{5/2^+}{^{95}Zr}$
Z = 39	$\dfrac{1/2^-}{^{85}Y}$	$\dfrac{1/2^-}{^{87}Y}$	$\dfrac{1/2^-}{^{89}Y}$	$\dfrac{1/2^-}{^{91}Y}$	$\dfrac{1/2^-}{^{93}Y}$	$\dfrac{1/2^-}{^{95}Y}$ $\dfrac{1/2^-}{^{97}Y}$
Z = 41		$\dfrac{9/2^+}{^{91}Nb}$	$\dfrac{9/2^+}{^{93}Nb}$	$\dfrac{9/2^+}{^{95}Nb}$	$\dfrac{9/2^+}{^{97}Nb}$	$\dfrac{9/2^+}{^{99}Nb}$
N = 50	$\dfrac{3/2^-}{^{85}_{35}Br}$	$\dfrac{3/2^-}{^{87}_{37}Rb}$	$\dfrac{1/2^-}{^{89}_{39}Y}$	$\dfrac{9/2^+}{^{91}_{41}Nb}$	$\dfrac{9/2^+}{^{93}_{43}Tc}$	$\dfrac{9/2^+}{^{95}_{45}Rh}$

FIG. 3.4. Ground state J^π values for a number of odd-mass nuclei for comparison with
the predictions of the independent particle model (compare Fig. 3.2).

applicability, the model would be of limited utility. We can easily see, however, how it
can be extended in scope.

First, consider a nucleus where the outermost j-orbit has $2j$ nucleons in it (i.e.,
all but one; this is often termed a single "hole" state). From $J_{full\,shell} = 0$, we have
$J_{full\,shell-1} + \mathbf{j} = 0$. Therefore, an orbit with $2j$ nucleons must have $J = j$: a single
particle and a single hole state have the same "spin". Thus, the ground states of both
^{41}Ca and ^{47}Ca, where the last neutrons fill the $1f_{7/2}$ orbit, have $J^\pi = 7/2^-$. Similarly,
^{39}Ca has one hole in the $d_{3/2}$ orbit, giving a $J^\pi = 3/2^+$ ground state, and ^{95}Zr, with a
$d_{5/2}$ hole, has $J^\pi = 5/2^+$. Again, Fig. 3.4 confirms all these predictions.

Secondly, we can use the independent particle model to predict the energies and
J^π values of excited states. Consider the diagrams in Fig. 3.5. That on the left gives
a schematic picture of the filling of nucleons (say, neutrons) into a nucleus which has
neutron number one greater than a magic number. As we have noted, the total angular
momentum (we assume the number of nucleons of the other type—protons in this case—
is magic) is $J = j_1$. The next two panels illustrate the simplest, and generally the lowest
energy, excited states, namely levels with $J = j_2$ and j_3, respectively. The *excitation*
energies of these two states are given by the energy *differences* of the total energy, which,

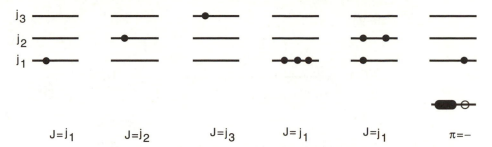

FIG. 3.5. Different independent particle model configurations and J assignments. See text for detailed discussion of each of these possibilities and their respective excitation energies. The thick bar in the rightmost panel indicates a filled major oscillator shell with one hole. J gives the total angular momentum of the ground state.

in the independent particle picture of non-interacting particles orbiting in a potential, are just the energy differences of the last odd neutron. Thus, the excitation energies of these two states are $E_{j_2} - E_{j_1}$ and $E_{j_3} - E_{j_1}$, respectively. This process allows us to predict, for example, the sequence of J values and energies for the low-lying levels of nuclei such as ^{41}Ca, ^{91}Zr, ^{209}Pb, and ^{209}Bi. For example, the first excited state of ^{41}Ca should be $3/2^-$ as is confirmed empirically in Fig. 2.4. Figure 3.6 shows the low-lying levels of ^{209}Bi and confirms the basic content of the proton shell from $Z = 82$ to 126, albeit, as we have hinted, and will discuss at length below, the detailed sequence of single particle levels is not quite in the order given in Fig. 3.2.

Still, applicability only to single particle or single hole nuclei would be a rather severe limitation in scope for the independent particle model. However, if we anticipate one key result from Chapter 4, we can extend its applicability many-fold. Suppose we have three particles in a j-shell. These can couple to several different total angular momenta, $\mathbf{J} = \mathbf{j} + \mathbf{j} + \mathbf{j}$. In the limit of a totally independent particle model, the energies of all these 3-particle configurations would be degenerate. But, in practice they are not. This is our first encounter with the effects of residual interactions, H_{residual}, which must be added to the independent particle model Hamiltonian H_0 of Eq. 3.2 to give the full shell model nuclear Hamiltonian. We will see in Chapters 4 and 5 which angular momentum states are allowed by the Pauli principle. For now we note that one possibility is to anti-align the angular momenta of two of the particles, say particles 1 and 2, so that $J_{12} = 0$ and then independently couple the third particle to this "pair". This gives a 3-particle J-value of $J = j$.

Of all the angular momentum couplings of pairs of nucleons in the same j-orbit, it turns out empirically that $J_{12} = 0$ gives the lowest energy. This result is easily understood in terms of short range, attractive residual interactions and the effects of the Pauli principle, as we shall see in Chapter 4. Simple experimental evidence for it is that the ground states of nuclei with 2 nucleons of one type outside a doubly magic core, such as ^{18}O, ^{42}Ca or ^{210}Pb, are always $J^\pi = 0^+$. (Indeed, the ground states of *all* even–even nuclei are $J^\pi = 0^+$.)

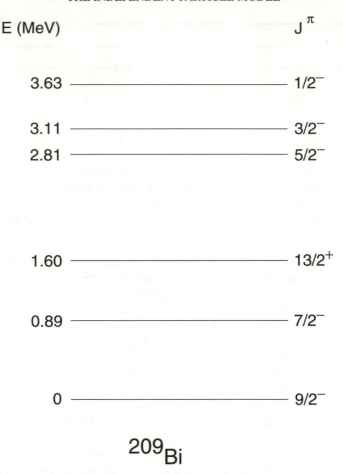

FIG. 3.6. Low-lying single-particle levels of ^{209}Bi.

In assigning the ground state J value of 3-particle nuclei such as ^{43}Ca, ^{93}Zr, ^{103}Sn or ^{211}Pb, one can therefore ignore the J coupling of the first two nucleons (i.e., set $J_{12} = 0$) and, again, conclude that $J_{\text{total}} = j_{\text{last nucleon}}$. For example, in the fourth panel of Fig. 3.5, $J = j_1$. For ^{43}Ca this gives $J^\pi = 7/2^-$ and for ^{93}Zr it gives $J^\pi = 5/2^+$ for their ground states as verified in Fig. 3.4. The argument can be extended to 5 or more particles in any j-shell. All but the last will pair off to $J = 0$ and the total angular momentum $J = j_{\text{last}}$. The same argument also applies to 3-neutron hole nuclei such as ^{87}Zr$(g_{9/2}^{-3})$ which is expected to (and does, see Fig. 3.4) have $J^\pi = 9/2^+$.

Excited states are formed by elevating one or more particles to higher energy orbits. For example, in the fourth panel of Fig. 3.5 if one of the particles in orbit j_1 is elevated to orbit j_2 the two particles remaining in j_1 couple to spin 0 and the total spin J is j_2.

We can now go another step further, as illustrated in the fifth panel of Fig. 3.5. By elevating a particle from a filled orbit (j_1) to the unfilled one (j_2) with an odd number of nucleons, this highest orbit then contains an even number of nucleons, each pair of which couples to $J_{pair} = 0$. Hence the net angular momentum for all the (even number of) nucleons in that orbit is $J_{orbit} = 0$. In the next lower, now partially unfilled, or "single-hole" orbit, all but the last nucleon pair their angular momenta to $J = 0$, leaving the total nuclear angular momentum $J = j_1$.

Of course, j_1 can even be a level in the next lower shell. For example, in ^{41}Ca, a $d_{3/2}$ or $s_{1/2}$ neutron can be elevated to the $f_{7/2}$ orbit. Then we have, in effect, a 3-particle situation. More accurately, we have a 1-hole, 2-particle configuration (e.g., $(d_{3/2}^{-1} f_{7/2}^2)$ in obvious notation). This configuration would have $J^\pi = 3/2^+$, the two $f_{7/2}$ neutrons preferentially coupling to 0^+. Figure 2.4 indeed shows low-lying $3/2^+$ and $1/2^+$ levels that originate in this manner from excitations out of the $d_{3/2}$ and $s_{1/2}$ orbits from below $N = 20$.

The empirical ground and first excited states of ^{69}Ni $(Z = 28, N = 41)$ also nicely illustrate this process. ^{67}Ni $(N = 39)$ has a $J^\pi = 1/2^-$ ground state as expected since it should have a single hole in the $2p_{1/2}$ orbit. This orbit, and the entire $N = 28 - 40$ major shell, are filled in ^{68}Ni. In ^{69}Ni, the 41st neutron should go into the $1g_{9/2}$ orbit giving a $9/2^+$ ground state. A low energy excited state can be formed by raising one of the two neutrons in the $2p_{1/2}$ orbit to the $1g_{9/2}$ orbit. The two neutrons in this orbit would couple to $J(g_{9/2}^2) = 0$ giving a total spin for the state of $1/2^-$ from the single neutron remaining in the $2p_{1/2}$ orbit. This description in fact matches the experimental situation in ^{69}Ni.

Finally, one other interesting feature of the independent particle model is illustrated in the third line in Fig. 3.4 which shows the odd proton $_{39}$Y isotopes. Note that they *all* have a $1/2^-$ ground state. This is the expected orbit for the 39th proton. The interesting point here is that the $1/2^-$ level remains the ground state, regardless of the number of neutrons, over a very wide range of nuclei. The $_{41}$Nb nuclei in the next line exhibit the same feature, except the 41st proton is in the $g_{9/2}$ orbit. Note, however, that this constancy of J when the number of nucleons of the other (even) type is varied will not always be the case, as we shall see momentarily for the Sb isotopes.

Empirically, it is possible to map out the single particle energies of the independent particle model with one-nucleon transfer reactions. These will be discussed extensively for deformed nuclei in Chapter 9. Suffice it to say here that in a reaction like (d, p), illustrated in Fig. 3.7, a neutron is stripped off the incoming deuteron into a specific orbit around the target nucleus, leaving an outgoing proton. Clearly, such a reaction can disclose sequences of states with single-particle structure, each corresponding to different j orbits. Moreover, once the "kinematical" aspects of the reaction collision are removed, the cross sections yield a nuclear matrix element, which relates to the *purity* of the single-particle state. This information is embodied in the so-called spectroscopic factor S(d, p): S(d, p) = 1 corresponds to a pure single-particle neutron wave function coupled to the target nucleus. While a reaction like (d, p) can only populate a state to the extent that the corresponding orbit was empty in the target, "pickup" reactions like (d, t) extract a neutron from an already filled orbit: they produce *hole excitations* and sample single-particle energies below the current valence shell.

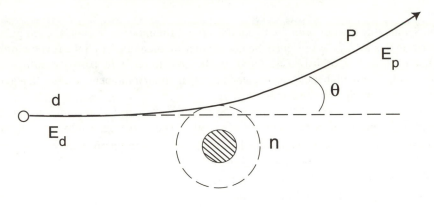

DIRECT (d, p) REACTION

FIG. 3.7. Schematic illustration of a direct (d, p) reaction.

With this introduction to transfer reactions, we return to Fig. 2.4 where we indicated those levels identified in (d, p) or (d, t) and we can see that this "single particle" subset of states in ^{41}Ca is exactly that demanded by the independent particle model. While this confirms some basic predictions of the model, the wealth of other states hints at the greater complexity to come. For ^{209}Pb, more quantitative results are included in Fig. 2.4: the numbers on the levels are the "spectroscopic factors", $S_{lj}(d, p)$. For a 1-particle nucleus like ^{209}Pb the $S_{lj}(d, p)$ values have a maximum theoretical value of unity, which would indicate a pure single particle state (those empirical values slightly exceeding unity may be due to experimental errors or difficulties in removing reaction kinematics from the observed cross sections.) Except for the $15/2^-$ levels, the states have remarkable purity and display an order in basic agreement with the independent particle model.

One last result quickly gives an interesting prediction which is another well known experimental result. Consider a nucleus with a magic number of nucleons. Clearly $J_{gs} = 0^+$ as we have said. But now, suppose we add energy to such a system to lift one of the nucleons in the highest orbit in the filled major shell to orbit j_1 in the next major shell. The total J will be given by the coupling of the j values of the hole (h) and particle (p) configurations, and, at this stage, we cannot predict which J state will lie lowest. However, except for the unique parity orbital in heavier nuclei, the states in each major shell have opposite parity to the ones below and above it. Hence it is clear that *most* of the lowest excited states in doubly magic nuclei will have negative parity. [The total parity is the product of the parities of each nucleon.] So, all paired nucleons in a given orbit give $\pi = +$ and the excited state illustrated in the sixth panel of Fig. 3.5 has

$$\pi = \prod_{i=1}^{A} \pi_i = \pi_h \times \pi_p$$

which in most cases for magic nuclei will be negative. Reference to Fig. 2.2 shows that the lowest excited states of doubly magic nuclei are indeed typically of negative parity.

In the last paragraphs we have now seen how the independent single particle model can be extended to single hole states and, by invoking the favored $J = 0$ coupling of pairs of identical nucleons in the same orbit, we can extend the shell model to many odd-A (singly magic at this stage of our discussion) nuclei. Excited states are formed by shifting the last nucleon to higher shell model orbits or by elevating a nucleon in a filled orbit to a partially occupied orbit and again invoking the $J = 0$ coupling of pairs of particles. It is these simple extensions that give the independent particle model such widespread use and significance.

Although we will not discuss it here, there are many other predictions, for example, magnetic moments or single-particle electromagnetic transition rates, that are reasonably well reproduced by the simple independent particle approach.

Finally, we caution that, while we have used the fact that $J = 0^+$ is the lowest-lying state in a 2-particle configuration of identical nucleons, other J values are possible at higher energy. Empirically, the next energy level nearly always has $J = 2^+$. Therefore it is risky to extend the predictions for odd-mass nuclei above excitation energies comparable to $E(2_1^+)$ in the neighboring even–even nucleus, since configurations of the form $|j \otimes 2^+, J\rangle$ in which a particle in orbit nlj is coupled to a "core" excitation (the first 2^+ state) of the underlying even–even nucleus can then compete in energy with single-particle excitations.

3.4 Mass dependence of single-particle energies

We now return to the question of the interactions of particles in different j shells and differences in the order of single particle levels compared to Fig. 3.2.

The single particle energies (s.p.e.s) of the independent particle model change with N and Z for two reasons. One is inherent in the concept of a potential and its boundary conditions. We have seen, for example, that for a square well the energies go as $E \sim 1/d^2$ where d is the size of the well. The reason is obvious. The lowest energy wave function has nodes at the well boundaries (for an infinite well). If these boundaries expand, the wave function needs to roll over more slowly; that is, with a longer wavelength and, hence, a lower energy.

In the shell model, we will therefore have, approximately,

$$E_{\text{s.p.e.}} \sim \frac{1}{r^2} \sim A^{-2/3} \tag{3.25}$$

This dependence of $E_{\text{s.p.e.}}$ on A is illustrated in Fig. 3.8 which is Fig. 2.30 of Bohr and Mottelson (1975, 1998). This generic scaling will be slightly different for low and high angular momentum orbits since the particles orbit at different radii. But this effect is a small perturbation on a smooth overall decrease. To the extent that all level energies decrease in concert, level sequences in the shell model will not change much.

A much more significant effect is on *relative* shell model energies due to residual interactions. Such effects can be drastic. We start with the effects of closed shells of filled orbits on the energies of single particle levels in the valence space.

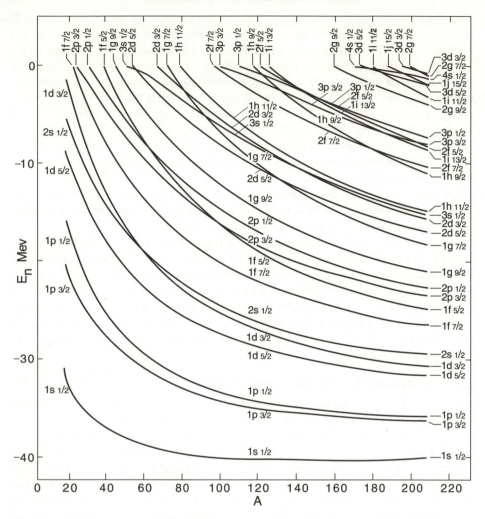

FIG. 3.8. Changes of single particle energies with nucleon number, reflecting the dependence of the energies of confined particles on the size of the containment volume (Bohr and Mottelson 1975, 1998).

The reader may have encountered the idea that the influence of closed shells can be ignored. There is considerable truth in this, but it is not the whole story. Especially in recent years, with the advent of much new data far off stability, this issue takes on real importance. In the next few paragraphs we discuss the effects of closed shells on single-particle energies. An extension to multiparticle configurations will be discussed early in the next chapter.

The basic result is absolutely trivial. Since a closed shell has $J = 0$, its wave function is spherically symmetric. Therefore, imagine a single valence nucleon outside this shell in an orbit j and magnetic substate m. Since the closed shell has no preferred direction in space, its interaction with this nucleon must be independent of m. This does not mean that the interaction can be ignored. It can, and does, exist but it is only equivalent to a *change in the spherically symmetric central potential*. A particle in a particular valence j shell certainly interacts with the closed j shells below it and its *single-particle energy* is altered by that interaction. The preceding argument simply means that this interaction is independent of direction. [Incidentally, to anticipate our later discussion of the multipole expansion of an interaction in $P_k(\cos\theta)$, this shift is due exclusively to the *monopole* ($k = 0$) part of the interaction between the closed and open shell nucleons, since that is the only multipole that is θ independent ($P_0(\cos\theta) =$ constant)].

This interaction of nucleons in an open shell with underlying filled shells has an important consequence. A *major* shell generally consists of several constituent j shells. Each of the j values of a closed major shell can have a different (spherically symmetric) effect on each of the valence j orbits. Thus, the *relative* single-particle energies in a given major shell depend critically on the specific lower-lying, filled closed shells.

Extensive data collected over several decades, much of it from single nucleon transfer reactions, has allowed us to map out the empirical single-particle energies in a number of nuclei that are one particle or hole removed from various major closed shells. The recent extension of such data to nuclei far off stability has greatly expanded the overview of single-particle energies thus provided.

We illustrate some of the results in Fig. 3.9. Each panel gives the observed single-particle or hole energies for two nuclei, along with schematic illustrations of the orbits involved. The point is to compare these energies for different systems, *cores*, and types of nucleon.

The top left panel of Fig. 3.9 shows proton and neutron single-particle energies in the 82 to 126 shell extracted from the particle levels of ^{209}Bi and the hole states of ^{207}Pb. The energies are nearly identical. (The slight expansion of the proton energy scale is probably a Coulomb effect.) This is reasonable. In each of the nuclei, the valence particle or hole "feels" interactions with essentially the same underlying orbits. The only notable difference is that the neutron hole in ^{207}Pb interacts with 43 other neutrons in the 82 to 126 shell, while the proton particle in ^{209}Bi interacts with the full major shell of 44 neutrons. As individual nucleon–nucleon interactions are on the order of a few hundred keV, and j-dependent differences considerably less, this should be a minor effect. On the top right panel in Fig. 3.9, a similar situation is shown for the 50 to 82 shell with identical results.

However, in the bottom panel, the neutron holes in ^{131}Sn interact with the same 50 neutrons and 40 protons as in Zr but, in addition, with the $Z = 40$ to 50 closed proton shell and with other neutrons in their own shell. As a consequence, their energies are very different from ^{91}Zr.

This particular example is actually of great structural consequence. (We will see later that it accounts for the onset of deformation near $A = 100$.) The point is that the interaction of a given shell, j_1, with another (closed or open), j_2, depends on the overlap of the respective wave functions; the only difference if j_2 is closed is that, then, the interaction is angle independent, so it is the *radial* overlap that counts. Orbits with similar

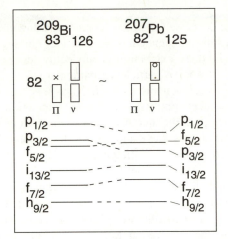
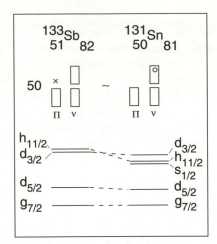
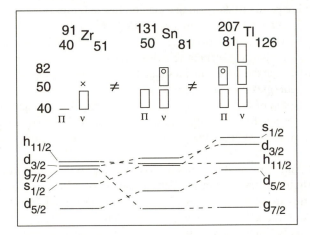

FIG. 3.9. Changes in single-particle energies in different nuclei that illustrate the effect of closed shells on the valence orbits. Since we are interested in relative changes only, one energy is normalized in each box.

quantum numbers nlj have higher overlaps. This is demonstrated in Fig. 3.10, where the dependence of the radial overlaps of various orbits with a $1s$ orbit on $\Delta n = n_1 - n_2$ and $\Delta l = l_1 - l_2$ is illustrated. The falloff with Δn and Δl is clear. (Incidentally, the simple estimate that the interaction goes roughly as $1/(\Delta n + \Delta l)$ is not a bad guide).

In this particular case, as the proton $1g_{9/2}$ orbit fills from $Z = 40$ to 50 it exerts a strong attractive *pull* on the $1g_{7/2}$ neutron orbit, drastically lowering the energy as seen clearly in Fig. 3.9. In contrast, an orbit like $3s_{1/2}$ has poor overlap with $1g_{9/2}$ and, relatively, its energy increases.

The same idea is vividly illustrated in more detail as a function of A in Fig. 3.11 which shows the $2d_{5/2}$ and $1g_{7/2}$ *proton* levels of the Sb ($Z = 51$) nuclei, as a function

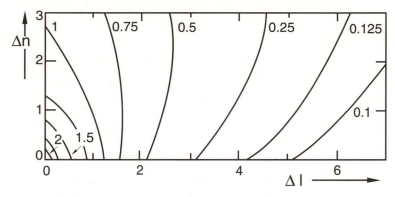

FIG. 3.10. Indication of the dependence of a residual interaction on the difference in principle (n) and orbital angular momentum (l) quantum numbers of the two orbits involved. The contours give constant values of the radial overlap integrals of a $1s_{1/2}$ orbit with orbits of different n, l (Heyde, 1987).

of *neutron* number. Were the independent particle model energies totally robust, the addition of neutrons should not affect the proton levels. However, the strong attractive p–n interaction between the single odd valence proton orbiting the Sn $(Z = 50)$ core and neutrons in the $N = 50 - 82$ shell causes a dramatic change in proton level order. Neutrons filling the $1g_{7/2}$ and $1h_{11/2}$ orbits have larger overlap with a proton in the $1g_{7/2}$ orbit than with a proton in the $2d_{5/2}$ orbit. Hence, the energy of the former is lowered. The effect is strong enough that these two proton orbits switch order at about $N = 70$: thereafter the ground states of odd A Sb nuclei are $7/2^+$ instead of $5/2^+$.

The sensitivity of relative single-particle energies in a given shell to the occupation of different underlying shells is further illustrated in the lower panel Fig. 3.9 by the extension to ^{207}Tl. Here, one observes the proton orbits in the 50 to 82 shell, but now an additional neutron shell (82 to 126) has been filled. Since this extra neutron shell has, on average, higher j values (1/2 to 13/2 instead of 1/2 to 11/2 for 50 to 82 or 1/2 to 9/2 for 28 to 50), the main effect is to further lower the higher j-proton single-particle energies (7/2, 11/2) relative to the lower ones (1/2, 3/2, 5/2).

Another example of this is seen in the bottom row of Fig. 3.4 which gives a sequence of $N = 50$, odd Z nuclei from $Z = 35$ to 45. From Fig. 3.2 we would expect the 35th and 37th protons to fill the $f_{5/2}$ orbit with a low-lying $p_{3/2}$ excitation, the 39th to occupy the $p_{1/2}$ orbit and the 41st to 45th the $g_{9/2}$ orbits. The predictions for $Z = 39$ to 45 are confirmed empirically, but there is an inversion of the $f_{5/2}$ and $p_{3/2}$ levels. The inversion of the ground and first excited states of ^{209}Bi is a third example (compare Figs. 3.2 and 3.6). These are not deficiencies in the model, just a fact of life in its use: one *must* deal with changes in single-particle ordering, especially when the same major shell is inspected in nuclei that are far apart in mass so that the interactions of the intervening nucleons will have altered the single-particle potential. Also, because of the Coulomb potential, proton and neutron single-particle sequences are expected, and found to be, slightly different, especially in heavy nuclei.

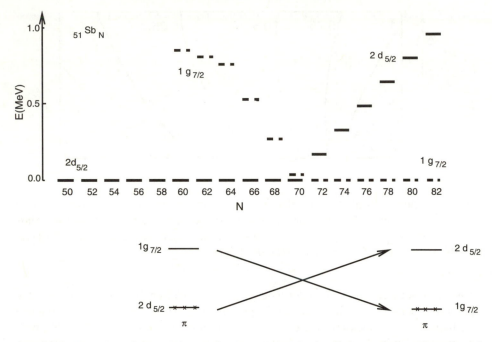

FIG. 3.11. Energies of the $\pi 2d_{5/2}$ and $\pi 1g_{7/2}$ states in the Sb isotopes as a function of neutron number. The lower part sketches the changes in *proton* single-particle energies as a function of neutron number. (Redrawn from a figure provided by W. Walters.)

The rather lengthy discussion in this section is important in order to highlight the three points that (1) a closed shell has the *same* effect on all *magnetic substates* of a particle in an open shell, (i.e., it is equivalent to a change in the spherical potential), (2) the energies of *different* shells may be very differently affected by underlying shells, and (3) different underlying closed shells (j or major) may affect a given open shell *very* differently. It also serves to re-emphasize that an illustration of the "shell model levels," as shown in Fig. 3.2, can only give a semi-quantitative guide: the energies are mass dependent.

Of course, there are also real discrepancies with the predictions of the independent particle model. Figures 2.4 and 2.11 showed the examples of [169]Yb and [161]Dy, where a totally different empirical picture, completely at variance with Fig. 3.2, is observed. We will see the reasons for this later.

3.5 Isospin

We have frequently mentioned and invoked the concepts of the charge symmetry and charge independence of the nuclear force. These ideas can be formalized via the concept of an isospin quantum number which allows one, from the standpoint of the strong interaction (i.e., neglecting Coulomb effects), to treat the proton and neutron as two manifestations (or states) of the same particle, the nucleon. The idea of isospin leads to the

concepts of isobaric multiplets and analogue states or resonances, to an elegant treatment
of 2-body nucleon–nucleon interactions (see Chapter 4) and to a classification of states
that leads to various important selection rules. The isospin approach to nuclear structure
allows physical insight into a number of physical processes and phenomena that would
otherwise be much less intuitively clear. Isospin provides a convenient bookkeeping aid
in carrying out shell model calculations involving both protons and neutrons. Finally, it
has a number of applications to the properties and decay modes of elementary particles.
While it is therefore a very powerful concept, it is likewise true that virtually all results
obtained using it can be equally well obtained by explicitly taking into account protons
and neutrons as separate entities. Often this can be done only with great difficulty or
complexity (hence the usefulness of isospin) but it is always at least possible in principle.

For some reason the concept and applications of isospin in nuclear structure are often
shrouded in mystery to the student. The formalism may seem trivial but yet striking results
emerge. Some of the mystery of isospin stems from the fact that isospin T is generally
not an observable that can be measured. It is like a hidden property of a state, whose
value often seems known only to an inner circle of cognoscenti, who easily proclaim that
certain states have this or that isospin to the consternation and bewilderment of those
outside the pale. This situation is unfortunate since the ideas are really very simple, and
yet powerful. States of different isospin do *not* just differ in that mysterious quantity but
have *real* structural differences in orbit occupancies or spin couplings, and isospin is, in
essence, an elegant shorthand labeling system for these differences. We hope to shed the
veil of fuzziness surrounding isospin in the pages that follow, focusing on the physical
ideas involved. These ideas will be further used in Chapter 4 where they lead to some
beautiful results in the treatment of residual interactions.

In closing these introductory remarks, it is perhaps useful to explicitly note that
the first edition of this book may have inadvertently helped spread some confusion via
two terrible misprints ("misprints" is a clever neutral word meaning either "mistakes" or
"carelessness"), where, on pages 12 and 69 of the first edition, the T_z and T values relevant
for a proton–neutron system were wrong and, more or less, interchanged: Apologies from
the author who may himself have been mystified by the mystery of isospin. The following
pages are therefore not only for the student but represent a catharsis for the author and
a public penance.

We first outline the formalism of isospin, which is identical to that of spin, in order
to prove a few simple results. As stated just above, we consider the proton and neutron
as two states, ψ_p and ψ_n, of the nucleon, distinguished by a projection quantum number
T_z:

$$T_{z_n} = +\frac{1}{2} \qquad\qquad\qquad T_{z_p} = -\frac{1}{2} \qquad\qquad (3.26)$$

Important note: In the literature the definition of ψ_p and ψ_n and hence their T_z values,
are sometimes interchanged, that is, one encounters, especially in older texts, and in texts
on elementary particles, the definitions $T_{z_p} = +1/2$ and $T_{z_n} = -1/2$. Either definition
is acceptable but, of course, they must be used *consistently*.

T_z is the z-component of the total isospin T and, therefore, for multi-particle systems, the individual T_z values add algebraically (like J_z) while the T values add vectorially. Thus, for a nucleus with A nucleons

$$T_z = \sum_{i=1}^{A} T_{zi} \text{ (algebraic sum)}; \quad \mathbf{T} = \sum_{i=1}^{A} \mathbf{T_i} \text{ (vector sum)} \qquad (3.27)$$

More simply

$$T_z = 1/2(N - Z)$$

which is 0 for nuclei with $N = Z$ and ranges upwards and downwards as N and Z differ by larger and larger amounts.

We take, first, as the simplest case, the p–p, n–n, and p–n systems for which one has

$$
\begin{array}{lll}
n\text{–}n & T_z = +1/2 + 1/2 = +1 & T = 1 \\
p\text{–}p & T_z = -1/2 - 1/2 = -1 & T = 1 \\
n\text{–}p & T_z = 1/2 - 1/2 = 0 & T = 0 \text{ or } 1
\end{array} \qquad (3.28)
$$

To see the origin of the T values on the right we start by noting that, as with spin, the total T can never be less than its projection and a given T must have projection quantum numbers $T_z = T, T - 1, 0, \ldots - T$. Since the n–n and p–p systems have $T_z = \pm 1$, T must be ≥ 1. But T cannot be greater than 1 since that would require the existence of $T_z = \pm 2$ values for the 2 nucleon system which is impossible since two projection quantum numbers of $1/2$ cannot add to more than unity. Therefore T must be unity for the p–p and n–n systems. But this requires that there be a $T_z = 0$, $T = 1$ state, which is a p–n system.

We have indicated in Eq. 3.28, however, that the p–n system can have *either* $T = 0$ or 1. How do we know this? The answer is clear by thinking about the physical significance of T. Consider the diagram in Fig. 3.12 which shows allowed spin couplings for systems of two nucleons occupying an s ($l = 0$) orbit, with protons indicated by filled circles and neutrons by open circles (the same convention applies to subsequent isospin figures in this section).

By the Pauli principle an s orbit can only contain at most two identical nucleons, with spins (indicated by arrows in Fig. 3.12) pointing in opposite directions as shown in the *top* row of the diagram. The three spin couplings for the n–n, p–n, and p–p systems in the top row are all identical. If we think of the proton and neutron as the same particle (except for T_z) these three systems are identical (except for Coulomb effects in the p–p case). This triplet, with $T_z = +1, 0, -1$, constitutes the states labeled $T = 1$, as described above. We now see the relation of isospin to charge independence. The triplet of states in the upper row of Fig. 3.12, being identical, have the same energy (always neglecting Coulomb effects).

However, the p–n system can *also* exist with parallel spins (either both up or both down-for definiteness we use the former). At this stage, all we can say is that this coupling has $T_z = 0$ and $T \geq 0$. But, since the p–p and n–n systems *cannot* exist with parallel

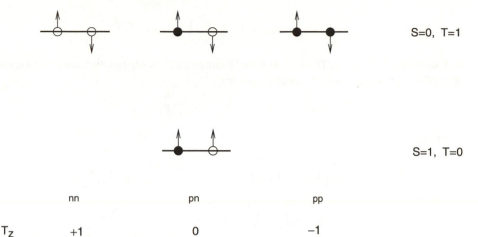

FIG. 3.12. Illustration of level occupation and spin directions for the 2-nucleon system. Protons are indicated by filled circles, neutrons by open circles. The total intrinsic spin and isospin quantum numbers are given on the right.

spins, this state exists *only* as $T_z = 0$. It is an isospin "singlet". If the *only* substate of a spin-like quantum number has a value zero, then the spin itself has that value. Hence this state must have $T = 0$, giving the two p–n couplings in Eq. 3.28.

We already can see that T is not a vague or mysterious property of nuclear systems but actually is just a notation that, here, distinguishes states with nucleon spins coupled in different ways. We will see a generalized meaning of T in higher mass systems shortly.

Let us formalize these ideas a bit, by defining wave function vectors (column matrices) for the proton and neutron, as

$$\psi_n = \begin{pmatrix} 1 \\ 0 \end{pmatrix}, \qquad \psi_p = \begin{pmatrix} 0 \\ 1 \end{pmatrix} \tag{3.29}$$

Then, an appropriate definition of the operator $\mathbf{T_z}$ that results in Eq. 3.26, is

$$\mathbf{T_z} = \frac{1}{2} \begin{pmatrix} 1 & 0 \\ 0 & -1 \end{pmatrix} \tag{3.30}$$

so that $\mathbf{T_z}\psi_n = \frac{1}{2}\psi_n$ and $\mathbf{T_z}\psi_p = -\frac{1}{2}\psi_p$. Analogously to spin we also define

$$\mathbf{T_x} = \frac{1}{2} \begin{pmatrix} 0 & 1 \\ 1 & 0 \end{pmatrix} \quad \text{and} \quad \mathbf{T_y} = \frac{1}{2} \begin{pmatrix} 0 & -i \\ i & 0 \end{pmatrix}$$

Finally,

$$\mathbf{T}^2 = \mathbf{T_x}^2 + \mathbf{T_y}^2 + \mathbf{T_z}^2$$

Consider now a system of two nucleons and define the like-nucleon wave function as

$$\psi_{nn} = \begin{pmatrix} 1 \\ 0 \end{pmatrix}_1 \begin{pmatrix} 1 \\ 0 \end{pmatrix}_2$$

and analogously for ψ_{pp}. The operation on a multiparticle wave function with an isospin operator is given by (using $\mathbf{T_z}$ as an example)

$$\mathbf{T_z}\psi_{nn} = (\mathbf{T_z}\psi_{n_1})\psi_{n_2} + \psi_{n_1}(\mathbf{T_z}\psi_{n_2})$$

Working this out gives

$$\mathbf{T_z}\psi_{nn} = \frac{1}{2}\left[\begin{pmatrix} 1 & 0 \\ 0 & -1 \end{pmatrix}\begin{pmatrix} 1 \\ 0 \end{pmatrix}_1 \begin{pmatrix} 1 \\ 0 \end{pmatrix}_2 + \begin{pmatrix} 1 \\ 0 \end{pmatrix}_1 \begin{pmatrix} 1 & 0 \\ 0 & -1 \end{pmatrix}\begin{pmatrix} 1 \\ 0 \end{pmatrix}_2\right]$$

$$= \frac{1}{2}\left[\begin{pmatrix} 1 \\ 0 \end{pmatrix}_1 \begin{pmatrix} 1 \\ 0 \end{pmatrix}_2 + \begin{pmatrix} 1 \\ 0 \end{pmatrix}_1 \begin{pmatrix} 1 \\ 0 \end{pmatrix}_2\right] = \frac{1}{2}\psi_{nn} + \frac{1}{2}\psi_{nn} = \psi_{nn}$$

Hence, the eigenvalue of $\mathbf{T_z}$ in the state ψ_{nn} is 1. Similarly, $\mathbf{T_z}\psi_{pp} = -\psi_{pp}$ and one has $T_z = -1$. The total isospin is given by

$$\mathbf{T}^2\psi_{nn} = \mathbf{T_x}[\mathbf{T_x}\psi_{nn}] + \mathbf{T_y}[\mathbf{T_y}\psi_{nn}] + \mathbf{T_z}[\mathbf{T_z}\psi_{nn}]$$

$$= \mathbf{T_x}\left[\mathbf{T_x}\begin{pmatrix} 1 \\ 0 \end{pmatrix}_1 \begin{pmatrix} 1 \\ 0 \end{pmatrix}_2 + \begin{pmatrix} 1 \\ 0 \end{pmatrix}_1 \mathbf{T_x}\begin{pmatrix} 1 \\ 0 \end{pmatrix}_2\right] + \text{similarly for } \mathbf{T_y}, \mathbf{T_z}$$

By matrix multiplication, one gets

$$\mathbf{T}^2\psi_{nn} = 2\psi_{nn}$$

Similarly

$$\mathbf{T}^2\psi_{pp} = 2\psi_{pp}$$

These are special cases, for these $T = 1$ systems, of the general expression (analogous to angular momentum)

$$\mathbf{T}^2\psi = T(T+1)\psi.$$

We have just discussed the p–p and n–n systems: What about the p–n wave function. To understand this we need to consider a generalization of the Pauli principle. A multi-particle fermionic wave function must be totally anti-symmetric with respect to formal interchange of all coordinates. If we include now the isospin wave function in defining a state we can write

$$\psi = \psi_{\text{space}} \times \psi_{\text{spin}} \times \psi_{\text{isospin}} \qquad (3.31)$$

Now, consider the p-n system with both particles in a $1s_{1/2}$ orbit. Then $l = 0$ for each particle and hence ψ_{space} is symmetric with respect to the exchange of the spatial coordinates of both nucleons. S for two nucleons is either $S = 0$ or 1. $S = 0$ corresponds to opposite spin couplings and is therefore the anti-symmetric case, while $S = 1$ is symmetric. Therefore, we can obtain a totally antisymmetric p–n wave function in two ways

$$\psi_{\text{spin}} = \text{Anti-Sym} \quad \psi_{\text{isospin}} = \text{Sym} \qquad (\text{Case 1}, S = 0)$$

or

$$\psi_{\text{spin}} = \text{Sym} \qquad \psi_{\text{isospin}} = \text{Anti-Sym} \quad (\text{Case 2}, S = 1)$$

What might seem the most obvious isospin p–n wave function, $\psi_{pn} = \psi_p \psi_n$, does not have manifest symmetric or anti-symmetric character. However, two linear combinations that do have unique isospin symmetry character are:

$$\psi_{\text{isospin}}^{S=0}(pn) = \frac{1}{\sqrt{2}}(\psi_p \psi_n + \psi_n \psi_p) \qquad T = 1, T_z = 0$$

and   (3.32)

$$\psi_{\text{isospin}}^{S=1}(pn) = \frac{1}{\sqrt{2}}(\psi_p \psi_n - \psi_n \psi_p) \qquad T = 0, T_z = 0$$

Clearly the upper isospin wave function is symmetric. Simple matrix multiplication shows that it has $T = 1$ while the latter has $T = 0$. Also, simple matrix calculation gives $T_z = 0$ for both wave functions. These results are indicated at the right in Eq. 3.32.

While this bit of formalism provides more than sufficient background for our needs, it is useful for completeness to continue the analogy to the spin formalism and define isospin raising and lowering operators

$$\mathbf{T}_+ = \mathbf{T_x} + i\mathbf{T_y} = \begin{pmatrix} 0 & 1 \\ 0 & 0 \end{pmatrix}$$

$$\mathbf{T}_- = \mathbf{T_x} - i\mathbf{T_y} = \begin{pmatrix} 0 & 0 \\ 1 & 0 \end{pmatrix}$$

Calculations with ψ_n and ψ_p then give

$$\begin{aligned} \mathbf{T}_+ \psi_n &= 0 & \mathbf{T}_+ \psi_p &= \psi_n \\ \mathbf{T}_- \psi_n &= \psi_p & \mathbf{T}_- \psi_p &= 0 \end{aligned}$$

(3.33)

$T_+(T_-)$ changes a proton (neutron) into a neutron (proton). These operators are clearly very useful for treating processes such as β-decay.

So much for the formal technology of isospin. Let us consider its physical utility and applications. We start with the deuteron and refer again to Fig. 3.12. We also recall that the nuclear force is only strong enough to just barely bind the deuteron which has only one bound state. From experiment, we know that the p–p and n–n systems are unbound. The corresponding (i.e., the $T = 1$) p–n system must then also be unbound (from the isospin concept or charge independence, which are equivalent here). This corresponds to the upper row of configurations in Fig. 3.12. From the spin arrows indicated it is clear that these states have $S = 0$.

The lower configuration is the only other possible configuration for a p–n system and thus must represent the bound state of the deuteron. (Clearly, putting one or both particles in a $1p_{1/2}$ or $1p_{3/2}$ orbit would require more energy). Since the deuteron does have one bound state, it must be this one, with parallel spins and $T = 0$ as indicated at the right of the figure. [Note that we deduced this here without explicitly ever using the experimental fact that $J_d = 1$ and, with both nucleons predominantly in $1s_{1/2}$ orbits, that $L = 0$ and therefore $S = 1$, but we see that these empirical results are fully consistent.]

Hence we conclude, as before, but now specifically in reference to the deuteron, that the isospin label distinguishes different physical wave functions (in this case, with different intrinsic spin couplings), that these different spin couplings (isospins) have different energies, and that the strength of the interaction of two unlike nucleons in a $T = 0$ state is stronger (more attractive) than for a $T = 1$ state.

Finally, Fig. 3.12 gives us our first example of an isospin or isobaric multiplet—namely a set of configurations with the same number of nucleons (hence the name isobaric multiplet), and the same isospin (hence the alternate name isospin multiplet), but different z-components of isospin. We will now extend these ideas to slightly heavier systems and see a generalization of the isospin concept to reflect the orbital wave functions as well.

We start with the mass 3 system, illustrated in Fig. 3.13. Experimentally, ^{3}H (^{3}H, the triton, has a β-decay half-life of 12.3 years) and ^{3}He exist while the tri-proton (^{3}Li) and tri-neutron are unbound. The reason for this difference is clear from the figure. The Pauli principle allows all 3 nucleons to occupy the lowest orbit ($1s_{1/2}$) only in ^{3}H and ^{3}He, where there are at most two nucleons of a given kind, giving it the most spatial symmetry and the highest binding. For both the ^{3}Li and nnn systems, at least one nucleon must go into the next higher energy orbit, which is sufficient to unbind these systems.

With 3 nucleons each, these four systems have T_z values ranging from $+3/2$ to $-3/2$. Clearly, then, $T = 3/2$ is one value for the total isospin, which must characterize the two systems with $|T_z| = 3/2$, namely the ^{3}Li and nnn systems, as well as excited states in ^{3}H and ^{3}He which are likewise unbound. But the ground states of ^{3}H and ^{3}He cannot have the same T value since they cannot be transformed into the ^{3}Li and nnn systems by changing a proton into a neutron (T_+ or T_-) or vice versa. The transmuted nucleon would have to change orbit as well. So, these two states must have a different T. Since only 3-nucleon systems that have *both* protons and neutrons can have all 3 nucleons in the same $1s_{1/2}$ orbit, and since there are only two such systems (^{3}H and ^{3}He), and since these systems obviously have $T_z = 1/2$ and $-1/2$, they must comprise the two z-components of an isospin doublet with $T = 1/2$.

Above, we found (for the $A = 2$ system) that different isospins represented different spin couplings. Now, we see that different isospins can also represent states in which the nucleons occupy different *orbits*.

Turning now to the $A = 12$ system, consider Fig. 3.14. The ideas here are similar, but they highlight two new results of importance—the concepts of extra binding for $N \sim Z$ nuclei (low T_z nuclei) and of isobaric analogue states. The lowest energy configuration for ^{12}B($T_z = +1$) and ^{12}N($T_z = -1$) [with $(Z, N) = (5, 7)$ and $(7, 5)$, respectively],

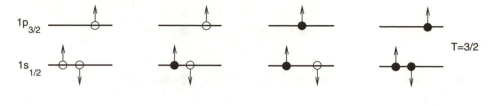

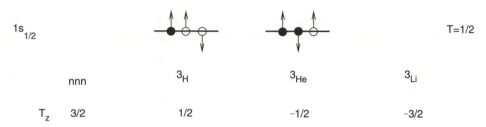

FIG. 3.13. Similar to Fig. 3.12, for the $A = 3$ system.

must look as illustrated on the top segment of the figure. No lower energy configuration is permitted by the Pauli principle. Of course, ^{12}C($T_z = 0$) can *also* exist in such a state (in ^{12}C it is an excited state), as shown, and the three forms of this pattern of orbit occupations are the $T_z = -1, 0, +1$ members of a $T = 1$ isospin triplet. However, ^{12}C (with $N = Z = 6$) can also occupy a lower energy state, illustrated in the lower segment. Since it is the only nucleus with $A = 12$ that can do so, it must be an isospin singlet, with $T_z = 0, T = 0$.

Here now we see an example of a fairly general result (at least in light nuclei where Coulomb effects are small), namely that the highest orbit symmetry and lowest total energy occurs for the smallest T_z nucleus of a multiplet (e.g., $T_z = \pm 1/2$ for $A = 3, T_z = 0$ for ^{12}C). (In fact, the lightest $N = Z$ nucleus, the deuteron, also shows this.) Hence, the most stable bound nuclei have $N \sim Z$ up until a mass $A \sim 30 - 40$ where Coulomb effects begin to shift the valley of stability toward the neutron rich side. More generally, denoting states of lower and higher T by $T_<$ and $T_>$ it is usually the case, for the lowest states of each isospin, that $E(T_<) < E(T_>)$. This is a simple consequence of the Pauli principle which is codified in the isospin approach.

Figure 3.15 shows the level schemes of the $A = 12$ nuclei. A diagram like this highlights the concept of isospin multiplet, that is, a set nuclei with states of the same A but different T_z values, and the level correspondences among them. In comparing the energies of states in a set of isobaric nuclei (either states of the same or different T) it is always necessary to correct for Coulomb effects, that is, for the Coulomb energy displacement because of the differing numbers of protons in the different T_z members of an isospin multiplet. In figures such as Fig. 3.15, such Coulomb corrections have been incorporated. Note that they can be fairly large. For a uniformly charged sphere

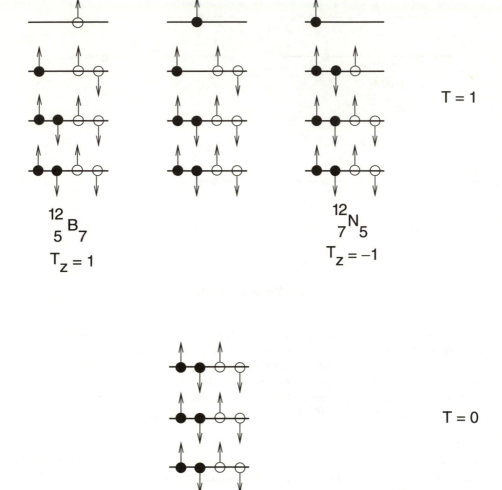

FIG. 3.14. Similar to Fig. 3.12, except for $A = 12$.

of radius $1.44A^{1/3}$ fm, $E_{Coul} \sim 0.6Z(Z-1)/A^{1/3}$ MeV. For $Z = 6$, this is 7.86 MeV and the Coulomb energy difference between ^{12}C and ^{12}N is 3.15 MeV. Deviations from identical energies in Coulomb-corrected states of the multiplet can sometimes give useful and subtle information on the Coulomb corrections (e.g., due to non-spherical shape components) or to other structural effects.

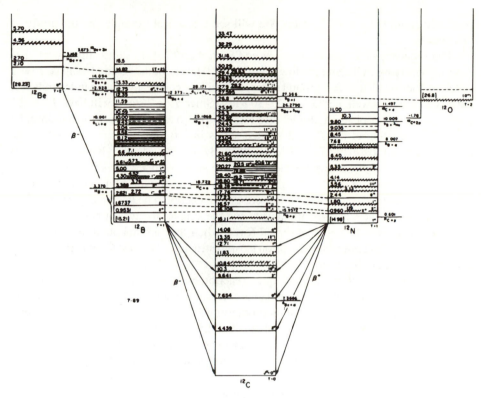

FIG. 3.15. Empirical level schemes of the $A = 12$ nuclei (Ajzenberg-Selove, 1990).

From Fig. 3.15 we see that the ground and low-lying states of ^{12}C have $T = 0$. The ground states of ^{12}B and ^{12}N, with $T = 1$, correspond to a highly excited state of ^{12}C. Indeed, *each* of the low-lying $T = 1$ states in ^{12}B and ^{12}N has a $T = 1$ counterpart, at fairly high excitation energy, in ^{12}C. Such multiplets of states are called isobaric analogue states (IAS) since their wave functions are exactly analogous (except for Coulomb effects): they differ only in the interchange of a proton with a neutron or vice versa in the same orbit. Isospin multiplets are useful because they focus attention on these isobaric analogue states. If an IAS state is unbound, it is sometimes called an isobaric analogue resonance (IAR).

Notice again that such $T = 1$ IAS or IAR states in the $T_z = 0$ nucleus are at high energy where they will lie, in general, amidst a wealth of other states (with $T = 0$). Normally, when high-lying excited states lie very close in energy they will mix and their wave functions will be complex linear combinations of many configurations. The IAS states, though, have the same simple structure as the ground states of the higher $|T_z|$ members of the $T = 1$ multiplet and they differ considerably from states that are nearby in excitation energy. They therefore tend not to mix with the surrounding (mainly $T = 0$) sea of states and to remain isolated islands of purity floating (so to speak) amidst (but protected from) the sea of $T_<$ states. We will see a practical example and the utility of this shortly.

The sequence of $T_>$ states does not end with the $(T_< + 1)$ states. Just as ^{12}C has $T = 1$ states lying well above the low energy $T = 0$ levels, both ^{12}B and ^{12}N have $T = 2$ states well above their $T = 1$ ground states. ^{12}C will, of course, by charge independence, have such states as well, perhaps at 25–30 MeV. Since $T = 2$ states will have $T_z = \pm 2$ members, a larger isobaric multiplet (see Fig. 3.15) also exists, comprising a quintuplet of nuclei consisting of the $T = 2$ excited states in ^{12}B, ^{12}C, and ^{12}N as well as the ground state and low-lying $T = 2$ levels of the $|T_z| = 2$ systems ^{12}Be $(Z, N = 4, 8)$ and ^{12}O $(Z, N = 8, 4)$. A given nucleus can have states with T values from $|N - Z|/2$ up to $A/2$, although, of course, not all such states will be bound, and usually only states with low values of $T - |T_z|$ will be low enough in energy to be known experimentally.

Nevertheless, the relatively high purity and the simple structure or parentage of IAS states allow them to be easily observed (even when they are unbound resonances) in certain reactions that selectively focus on isospin structure. The best known and simplest example is the (p, n) reaction.

As we discussed earlier in this chapter, reaction cross sections depend on both kinematic and structural factors. The kinematics enters, for example, through the reaction Q value or the difference in incoming and outgoing kinetic energies. For example, for a $A(d, p)A + 1$ direct reaction, the Q value is the difference between the binding energy needed to break up the deuteron (-2.23 MeV) and the binding energy gained (the neutron separation energy) when the transferred neutron falls into the target nucleus. Since, typically, $S_n \sim 6$ MeV, (d, p) Q values to ground or low-lying states of the $(A + 1)$ system are $\sim +3$ or 4 MeV. The outgoing proton receives an energy boost.

Direct reactions are sensitive to structure since reaction cross sections are proportional to an overlap of initial and final states. In the above example, this overlap would be schematically written $\langle \psi_A \psi_d | \psi_{A+1} \psi_p \rangle$: As we discussed in the context of spectroscopic factors for single-particle levels, σ (d, p) depends on how much the final nucleus looks like the initial one with a neutron simply dropped into an unfilled orbit. If, instead, a given state in the $(A+1)$ nucleus has a configuration in which, say, there is one additional neutron but where three other neutrons and two protons also change orbits, the overlap will likely be extremely small and so will be the cross section.

Now let us apply this to the (p, n) reaction which connects initial and final nuclei that are isobars by transferring a proton and removing a neutron. (The ideas to follow are illustrated in Fig. 3.16.) If we form a state in the final nucleus in which the proton is in the *same* orbit previously occupied by the neutron, the overlap integral will be large. Also, barring large Coulomb energy differences, the Q value should be favorable. Hence the cross section should be large.

This process of exchanging a proton for a neutron in the same orbit is exactly the process of going from one member of an isospin multiplet to another: That is, such a process conserves isospin T, changes T_z by -1 (as any (p, n) reaction must do), and forms a state in the final nucleus which is the IAS of the ground state of the target. But this IAS will *not* in general be the ground state of the final nucleus since the ground state of the nucleus $A(Z, N)$ does not have the same isospin as the ground state of the nucleus

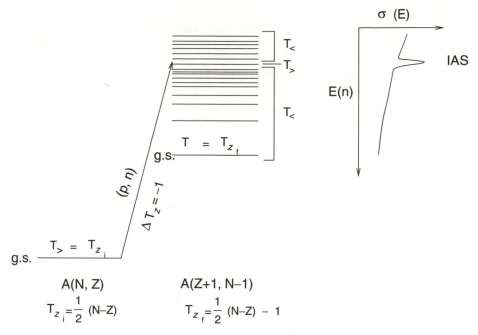

FIG. 3.16. Illustration of the enhanced (resonance) population of an IAS state with the (p, n) reaction (see text).

$A(Z + 1, N - 1)$, as we just saw for $A = 12$. The $\Delta T_z = -1$ (p, n) reaction would preferentially go, for example, from the ground state of ^{12}B to its $T = 1$ analogue state in ^{12}C, not the $T = 0$ states of ^{12}C.

The other ($T_<$) levels in the same energy region will be complicated (relative to the target ground state) states with weak cross sections.

One final point is useful in understanding these reactions. Often, the $T_>$ IAS, and $T_<$ neighbors, will be in a particle unbound excitation energy region of the final nucleus. The background of $T_<$ states is featureless, therefore, not only because the (p, n) cross sections are low but also because these states have short lifetimes, and hence large natural widths. The intermingled $T_>$ states, though, are often nearly pure, mixing little with their neighbors. Their decay by neutron emission is slow because it involves a large change in isospin and a substantial rearrangement of filled and unfilled orbits. This is an example of an isospin selection rule on reactions (see below). With only proton and γ-decay possible, the widths therefore are much smaller. Hence, not only are the (p, n) cross sections large but they are peaked in a narrow excitation energy range.

Thus, these (p, n) reactions should be characterized by neutron spectra (whose energies are directly related—via the Q value—to the excitation energy produced in the final nucleus) with a nearly featureless background of low cross sections, punctuated by spikes (or resonances) corresponding to (and therefore useful for identifying) the IAS of the target ground state (see Fig. 3.16). We said early on in the discussion of isospin that T is generally not an "observable". The case of these isobaric analogue resonance reactions is almost a counter-example: We do not directly "measure" T but we do locate states with the *same* T as the target ground state.

As we have just noted, the concept of isospin is sometimes useful in obtaining selection rules for certain decay processes and reactions. In general, assuming charge independence, isospin must be conserved. For example, a single nucleon transfer reaction can only change T by $1/2$; on an even–even $T = 0$ target nucleus it can only make states with $T = 1/2$. Higher-lying $T = 3/2$ states would be forbidden.

As another example, consider a 2-particle pickup reaction where a deuteron is incident on a ^{16}O nucleus. In the simple concept of a direct process, a proton and a neutron can be pulled out of ^{16}O, leading to a final state consisting of ^{14}N and an α-particle: that is, in standard notation, ^{16}O $(d, \alpha)^{14}$N. Since ^{16}O, the deuteron, and the α particle all have $N = Z$, they have $T_z = 0$ and $T = 0$ in their ground states. Therefore, by conservation of isospin T, only $T = 0$ states in ^{14}N can be populated. Studies of ^{14}N and the $A = 14$ isospin multiplet have shown that the low-lying states (below ~ 5 MeV) of ^{14}N are all $T = 0$, except the 0^+ state at 2.3 MeV which has $T = 1$ and is the analogue of the ^{14}O and ^{14}C ground states. The (d, α) reaction to the 2.3 MeV level is therefore forbidden. Experimentally, it is populated with cross sections more than an order of magnitude weaker than the ground state. However, selection rules such as this need to be used with care as there could be other (structural) reasons for a low cross section. The isospin formalism also leads to electromagnetic selection rules (e.g., one forbidding $\Delta T = 0$ E1 transitions in $N = Z$ nuclei) that are sometimes useful.

Lastly, and perhaps most importantly, though we have not discussed it yet, the isospin formalism often enormously simplifies complex shell model calculations, such as those discussed in Chapters 4 and 5, in nuclei with both valence protons and neutrons. Such calculations can always be done by separately keeping track of the protons and neutrons but, sometimes, at considerable cost in complexity.

To summarize our discussion of isospin, we focus on just a couple of key points. Isospin is a labeling scheme, founded in the concept of charge independence, that allows an elegant and often simpler treatment of proton–neutron degrees of freedom. The isospin formalism is identical to that of ordinary spin, and isospin operators such as T_+ and T_- are useful in treating processes such as β-decay.

States with different T values but the same T_z (i.e., states within the same nucleus) differ in the orbits occupied by the protons and neutrons or in the spin couplings. States in different nuclei of an isospin multiplet with the same T but different T_z differ in the interchange of protons into neutrons in the same orbit or vice versa. Generally, states of the lowest isospin lie lowest and are the most proton–neutron symmetric states in the space and spin coordinates: that is, the Pauli principle favors low T. Coulomb effects, especially in medium and heavy nuclei can offset these generalities and, in fact, isospin is generally less useful a concept in such nuclei (except for IAS in nuclei with simple configurations near closed shells). Mass formulas often contain a term in $(N - Z)$, representing an isospin dependence that reflects Pauli and Coulomb effects.

Hopefully it is now clear that there is no great mystery to isospin as long as it is recalled that states of different isospin have real structural differences and that isospin is a way of linking states of similar structure that might not otherwise be recognized as similar, and of interpreting cross sections and decay or deexcitation processes involving such states.

4

THE SHELL MODEL: TWO-PARTICLE CONFIGURATIONS

In Chapter 3 we discussed independent particle motion corresponding to the potential $U_i(\mathbf{R})$ in the Hamiltonian H_0 in Eq. 3.2. Now we discuss the effects of residual inter-actions. These lead to all the correlations, configurations mixing, and collectivity that characterizes most nuclei.

In order to proceed, we must deal more deeply and systematically with the problem of multinucleon configurations. By this we mean "valence" configurations of two or more particles outside a core, which is usually assumed to consist of inert closed shells. There are really two issues here: First, which J values are allowed by the Pauli principle, and, second, what are the effects of the interactions. The first question can be answered without reference to any discussion of a central potential or residual interaction, while the second depends on the details of those interactions, although many properties can be deduced from the simple fact that the radial and angular parts of a wave function in a central potential can be separated and that predictions for the former are independent of J value, while for the angular part, many results can be obtained by rather general arguments based on its short- or long-range character or by considering a multipole expansion of it. This chapter addresses these issues. We will start by considering the simplest case of two identical nucleons in the same or different orbits. Then we will turn briefly to the case of two nonidentical nucleons (proton and neutron) and the role of isospin and exchange terms in the residual interaction. In the next chapter we will consider the case of larger multinucleon configurations of the form j^n.

A complete treatment of many of these issues involves extensive and sophisticated familiarity with the formalism of angular momentum and tensor algebra. Indeed, many of the results can only be proved by rather formal manipulation of the various angular momentum coefficients ($3-j$ symbols, Racah coefficients, coefficients of fractional parentage, and so on) and a deep understanding of their symmetry properties. There are numerous existing texts that deal with this at great length in superb fashion. Three of the best are those of de Shalit and Feshbach, de Shalit and Talmi, and Heyde. While using their results, our purpose is to provide a physical understanding and motivation for the results and to present arguments for their plausibility. The excellent book by Heyde is complementary to the present treatment and is highly recommended.

Before discussing the particular characteristics of multinucleon configurations under the influence of residual interactions, it is crucial to discuss the role of closed shells on the valence nucleons. As with the energy levels of single nucleons in shell model orbits, it is only if the effects of closed shells can be neglected that the study of multinucleon configurations will be applicable to other than the lightest nuclei.

Since a closed shell is spherically symmetric, the interaction of a valence nucleon in the state jm cannot depend on m: hence, the closed shell particles have the same effect on all valence nucleons in a given j orbit. In multiparticle configurations, the effects of the closed shell nucleons are independent of the way (the relative orientations) that the individual jm values are coupled to the total J. Therefore, the closed shell can have no effect on energy *differences* of these different J states. As discussed in Chapter 3, this does not rule out an additive energy for the entire group of states and different j values can lead to different shifts. Thus, the rules we will develop for spacings refer to those within a configuration, not to the relative energies of different j shells in a major shell.

With these preliminaries in hand, we can now discuss two-particle configurations. First, we need to determine which J values result from coupling of a nucleon in orbit j_1 with one in orbit j_2. If $j_1 \neq j_2$, these are simply the integer values from $|j_1 - j_2|$ to $|j_1 + j_2|$. If the orbits are equivalent, $n_1 l_1 j_1 = n_2 l_2 j_2$ (which we shall often abbreviate to $j_1 = j_2 = j$ when no confusion should arise) we distinguish two cases—identical and nonidentical nucleons. For the latter (proton–neutron) case, J takes on all integer values from 0 to $2j$, as there is no Pauli principle restriction on the occupation of identical m states. However, for identical nucleons, one must explicitly consider the effects of the Pauli principle, which requires the *total* wave function to be antisymmetric.

For identical nucleons, the isospin projection $T_z = T_z(1) + T_z(2) = \pm 1$ and hence the total isospin $T = 1$. (Clearly, T cannot be less than its projection.) This is a symmetric wave function. Hence, the space-spin part must be antisymmetric. When we impose this requirement, only certain J values are allowed. To see this, we have to look at the particular m states occupied by the particles, 1 and 2. The Pauli principle requires that

$$\psi_1(jm)\psi_2(jm') = -\psi_2(jm)\psi_1(jm') \tag{4.1}$$

A properly antisymmetrized wave function that has this property is given by

$$\psi(j^2 JM) = N \sum_{m,m'} (jmjm'|jjJM) \left[\phi_1(m)\phi_2(m') - \phi_1(m')\phi_2(m) \right]$$

$$= N \sum_{m,m'} \left[(jmjm'|jjJM) - (jm'jm|jjJM) \right] \phi_1(m)\phi_2(m')$$

where N is a normalization factor. The relation between the two Clebsch–Gordan coefficients on the right-hand side is well known and given by the phase factor $(-1)^{2j-J}$. Hence

$$\psi(j^2 JM) = N \left[1 - (-1)^{2j-J} \right] \sum_{m,m'} (jmjm'|jjJM)\phi_1(m)\phi_2(m') \tag{4.2}$$

Since $2j$ is odd, this vanishes unless J is even: the only *allowed* J states for *two identical* fermions in *equivalent* orbits are those with even total angular momentum $J = 0, 2, 4, \ldots (2j - 1)$. In Chapter 5, we will see that the same result is even simpler to obtain in the m-scheme, which provides a general, though sometimes tedious, way of finding the possible J states for *any* multiparticle configuration.

With this result for identical nucleons in hand, we have an alternate way to look at the complete set of J values $0 - 2j$, available in the p–n system. We consider the isospin structure of this system and use the result just obtained. The p–n system has $T_z = 0$ with $T = 0$ or 1. The $T = 1$ case is identical in all respects to the preceding p–p and n–n $T = 1$ cases and therefore consists of the even J values $0, 2, 4, \ldots (2j - 1)$. Together, the p–n system has all j values from 0 to $2j$ so that the $T = 0$ p–n system contains the odd J values. (Note that, for $j_p \neq j_n$, each two-particle J state is now a mixture of $T = 0$ and $T = 1$ parts: a total nuclear wave function of good isospin is constructed by coupling this two-particle state to the core.)

4.1 Residual interactions: the δ-function

Residual interactions affect the energies of multi-particle configurations and the scattering of particles from one orbit to another (that is, 'configuration mixing'). Here we mostly focus on the former, only briefly discussing off-diagonal mixing (see also Section 5.4 and Chapters 6, 8).

So, we consider the energies $E(j_1 j_2 J)$ of a two-particle configuration of identical nucleons denoted $|j_1 j_2 JM\rangle$, in the presence of a residual interaction. We have noted earlier that for any central or scalar interaction, the wave function can be separated into radial and angular parts. Since the radial behavior of the two interacting particles does not differ for different J, many results are independent of the detailed specification of the radial nature of the interaction. This is certainly true for the ordering and *relative* spacings of different J states: absolute values, of course, depend on detailed integrations over the radial coordinates and on the strength of the interaction.

Of course, with no residual interaction, all J states of the two-particle configuration are degenerate, as shown on the left in Fig. 4.1. In the presence of a residual interaction, the energy *shifts* (Fig. 4.1 right) relative to the degenerate case are

$$\Delta E(j_1 j_2 J) = \langle j_1 j_2 JM | V_{12} | j_1 j_2 JM \rangle = \frac{1}{\sqrt{2J + 1}} \langle j_1 j_2 J || V_{12} || j_1 j_2 J \rangle \qquad (4.3)$$

where the last step utilizes the Wigner–Eckart theorem. We have already stated that it is possible to separate the radial and angular coordinates for many residual interactions. To illustrate this, it is useful to consider the simple δ-function interaction: by definition, this interaction vanishes unless the particles occupy the same spatial position. The reason for choosing a δ-function residual interaction is not simply mathematical convenience. More importantly, it is preeminently a short-range interaction, and we know that the nuclear force, including residual interactions, has just this character. At least qualitatively, a δ-function residual interaction reproduces many observed properties of nuclei.

Moreover, it can be shown that a δ-interaction in a j^n configuration is equivalent to an odd tensor interaction. Such interactions are diagonal (see Chapter 5), in the so-called seniority scheme and as such are particularly useful for treating multiparticle configurations, since many important results reduce to the two-particle case. Thus, a discussion of the δ-interaction in $|j_1 j_2 J\rangle$ configurations has profound implications throughout the study of nuclear structure.

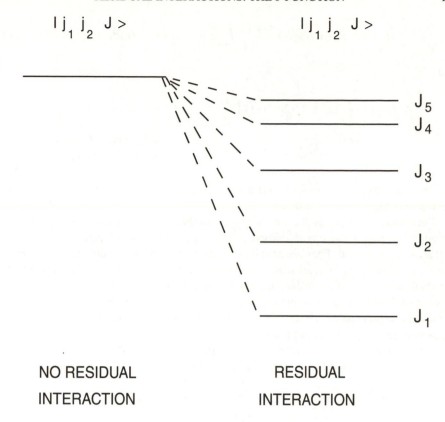

FIG. 4.1. Schematic illustration of the energy shifts in a configuration with particle 1 in orbit j_1 and particle 2 in orbit j_2 coupled to various spins J.

We can write the δ-interaction as

$$V_{12}(\delta) = -V_0 \delta(\mathbf{r}_1 - \mathbf{r}_2)$$
$$= \frac{-V_0}{r_1 r_2} \delta(r_1 - r_2) \delta(\cos\theta_1 - \cos\theta_2) \delta(\phi_1 - \phi_2) \tag{4.4}$$

where the second form expresses the interaction in polar coordinates. This is particularly useful when separating the angular and radial parts. The $1/r_1 r_2$ factor is necessary because the integration over the angular coordinates introduces a factor $4\pi r^2$.

Using the polar coordinate form and performing some straightforward but tedious angular momentum algebra (see de Shalit and Feshbach, Chapter 5) we obtain for the energy shifts in the identical particle configuration $|j_1 j_2 J\rangle$

$$\Delta E(j_1 j_2 J) = -V_0 F_R(n_1 l_1 n_2 l_2) A(j_1 j_2 J) \tag{4.5}$$

where

$$F_R(n_1 l_1 n_2 l_2) = \frac{1}{4\pi} \int \frac{1}{r^2} R_{n_1 l_1}^2(r) R_{n_2 l_2}^2(r) dr \qquad (4.6)$$

and

$$A(j_1 j_2 J) = (2j_1 + 1)(2j_2 + 1) \begin{pmatrix} j_1 & j_2 & J \\ \frac{1}{2} & -\frac{1}{2} & 0 \end{pmatrix}^2 \quad \text{(if } l_1 + l_2 - J \text{ is even)}$$

$$= 0 \qquad\qquad\qquad \text{(if } l_1 + l_2 - J \text{ is odd)} \qquad (4.7)$$
$$\text{(Non-equivalent orbits)}$$

F_R depends only on the *radial* coordinates, while the quantity A results from an integration over the *angular* coordinates.

Therefore, one obtains the extremely important result that the *relative* splittings depend *only* on *universal angular functions A*, which are totally independent of the nature of the central potential. They are also independent of the principle quantum numbers n_1 and n_2 (except for an overall scale incorporated into the factors F_R); for this reason, we shall not specify the n values unless it is necessary in a particular case. Although Eqs. 4.6–4.7 apply in detail only to a δ-function interaction, the general separability into radial and angular parts is valid for *any* residual interaction that depends only on the separation $(\mathbf{r}_1 - \mathbf{r}_2)$ of the two nucleons.

We will soon discuss the meaning and implications of Eq. 4.5 at considerable length. First, however, it is useful to give the expression for the specific case of *equivalent* orbits $l_1 j_1 = l_2 j_2$. Hence $l_1 + l_2$, must be even. Moreover, proper normalization of the wave functions introduces a factor of 1/2 in the energies that are now given by

$$\Delta E(j^2 J) = -V_0 F_R(nl) A(jJ) \qquad (J \text{ even}) \qquad (4.8)$$

where

$$F_R(nl) = \frac{1}{4\pi} \int \frac{1}{r^2} R_{nl}^4(r) dr \qquad (4.9)$$

and

$$A(j^2 j) = \frac{(2j+1)^2}{2} \begin{pmatrix} j & j & J \\ \frac{1}{2} & -\frac{1}{2} & 0 \end{pmatrix}^2 \quad (J \text{ even}) \qquad (4.10)$$
$$\text{(Equivalent orbits)}$$

Note that for $J = 0$, $A \propto (2j + 1)/2$, so that $\Delta E(j^2 J = 0) = \text{const} (2j + 1)/2$, that is, the energy lowering of 0^+ states is larger for large j and is in fact proportional to the number of magnetic substates in the orbit j. This property is identical to that defined for a pairing interaction. (See Section 4.3.) Here, its physical basis is that, for high j, there are more magnetic substates spanning the same angular range of orbit orientations. Hence, the wave function for a given substate is more localized in angle. Two particles with the same $|m|$ value thus have greater overlap and hence a larger interaction. Similar overlap arguments will help us understand many of the effects of residual interactions. They will be formalized later and will pervade this chapter.

Returning to the general case of any j_1, j_2, we note that, while the relative energies of individual J values depend only on the angular structure of the wave functions, the overall *scale* of the interaction and the *average* interaction strength for particles in orbits j_1 and j_2 depend on the radial integral (Eq. 4.6). As we stated in Chapter 3, this will be largest for similar n_1l_1 and n_2l_2 values.

Shortly, we will calculate explicit numerical values of $\Delta E(j_1 j_2 J)$, or rather, of $A(j_1 j_2 J)$. First, however, it will help us understand two-particle configurations and, later, multiparticle situations, if we momentarily ignore the analytic formula and try to understand the basic results for the energies $\Delta E(j_1 j_2 J)$ from simple physical arguments. It is remarkable how far this will take us.

We start with the obvious statement that the attractive δ-function interaction can only be large when there is large spatial overlap between the orbits of the particles. As we have seen, for a given j_1 and j_2, the overlap for different J values depends on the *orientation* of the orbits in space. We shall see in Section 4.2 the explicit relation between the $3-j$ symbols in Eqs. 4.7 and 4.10 and the relative angular orientation of the semiclassical orbit planes. For now, we proceed more qualitatively. From overlap considerations alone, one might think that the interaction would be largest either for $J = 0$ or $J = J_{\max}$, which are simply the two cases for which the angular momentum vectors j_1 and j_2 are most nearly antiparallel or parallel, respectively, and therefore those in which the nucleons orbit the nucleus most nearly in the same plane.

While this simple view has an element of truth to it, the requirements of antisymmetrization refine it considerably. Antisymmetrization, or the Pauli principle, has enormous and profound implications throughout the study of nuclear structure. We have seen how it determines the valley of stability, validates the fundamental concept of independent particle motion in a dense sea of nucleons, and gives the magic numbers and attendant shell structure. These (or at least the first and last) are straightforward and obvious effects. Others are subtle, even unexpected, and are certainly seldom appreciated. Our case of two nucleons interacting via a short-range residual interaction is just such a case. We start by considering the configuration of two identical particles (two protons or two neutrons) in orbits j_1, j_2.

In this case, the two-particle wave function must be totally antisymmetric, that is, antisymmetric in space, spin, and isospin. A careful understanding of this requirement leads to some beautiful, remarkable, and profound results.

Since we are dealing with two identical particles, the states involved must have total $T_z = \pm 1$ and hence $T = 1$: they are symmetric with respect to the isospin coordinates. Therefore they must be antisymmetric in space and spin coordinates.

Thus far we have carried out the discussion in what is known as a jj coupling scheme, in which the total angular momentum j_1 of particle 1 is coupled to the total angular momentum j_2 of particle 2 to produce a final total angular momentum J. One can also think of the problem in terms of the so-called LS coupling scheme, in which the orbital angular momenta of the two particles l_1, l_2 are first coupled to total L and the intrinsic spins $(1/2\hbar)$ s_1 and s_2 are coupled to $S = 1$ or 0. Generally the jj coupling

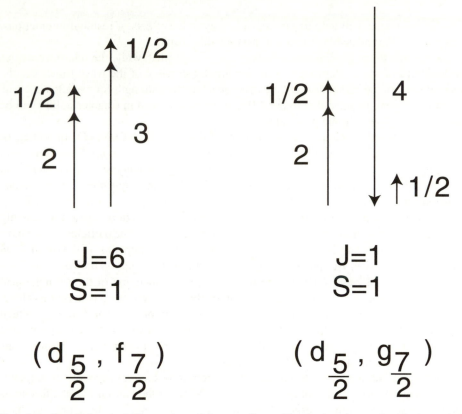

FIG. 4.2. Semi-classical illustration of the coupling of intrinsic spins and orbital angular momenta for two configurations $(d_{5/2}f_{7/2})$ and $(d_{5/2}g_{7/2})$.

scheme is more useful in most nuclear structure applications, but in the present case the *LS* coupling scheme separates out the angular and spin coordinates nicely and allows a simpler understanding of the effects of antisymmetrization.

To see how this works, let us take a particular example that gives a simple result. Consider the case of two particles in a $|d_{5/2} \, f_{7/2} J\rangle$ configuration as shown in Fig. 4.2. The *l* values are $l_1 = 2$ and $l_2 = 3$, respectively, and in both orbits, $j = l + 1/2$. The allowed *J* values range from 1 to 6. If we picture the vector coupling of j_1 and j_2 to form various *J* values, it is clear that the orbital planes of the two particles will overlap the most when $J = J_{max}$ and J_{min}. We therefore expect one or both of these to be the most affected (lowered) by an attractive short-range interaction. To proceed further, we note that J_{max} is greater than $l_1 + l_2 = 5$. Therefore, the J_{max} state can *only* be formed by aligning the orbital angular momenta to $L = 5$ *and* the intrinsic spin angular momenta to $S = 1$, and then aligning *L* and *S* to $J = 6$.

Since $S = 1$, the two intrinsic spins point in the same direction and thus the spin part of the wave function is symmetric. Therefore, antisymmetrization of the total wave function requires that the spatial part be antisymmetric. Denoting the angular part of the ith particle wave function by $\phi_i(\mathbf{r}_i)$, the requirement of antisymmetrization in the angular coordinates is equivalent to $\phi_{j1}(\mathbf{r}_1)\phi_{j2}(\mathbf{r}_2) = -\phi_{j1}(\mathbf{r}_2)\phi_{j2}(\mathbf{r}_1)$. However, the δ-function interaction is only effective when the particles are in contact, when $\mathbf{r}_1 = \mathbf{r}_2$. At this point, this expression for the antisymmetrization condition, however, requires that the wave function equal its negative $[\phi_{j1}(\mathbf{r})\phi_{j2}(\mathbf{r}) = -\phi_{j1}(\mathbf{r})\phi_{j2}(\mathbf{r})]$, which of course can only happen if each side vanishes. Therefore the wave function *vanishes* at the *only point in space* where the δ-function interaction acts, so the residual interaction has no effect whatsoever in this particular $J = 6$ state. Turning the argument around, a δ-function interaction between identical nucleons can *only affect states through amplitudes* in which $S = 0$, in which case the spin part of the wave function is antisymmetric and the spatial part is symmetric and need not vanish at $|\mathbf{r}_1 - \mathbf{r}_2| = 0$. Interestingly, although the δ-function interaction has no explicit spin dependence, its effects depend critically on the relative orientations of the spins of the two nucleons. In other words, antisymmetrization introduces an implicit spin dependence.

Clearly, to determine which states are affected by a δ-interaction we would like to know which have $S = 0$. Unfortunately, except for the configuration $|s_{1/2}^2 J = 0\rangle$, where S obviously must be zero, all two-particle configurations that can have $S = 0$ amplitudes will also have $S = 1$ amplitudes. For example, it might be thought that the $J = 1$ state of the $|d_{5/2}\,f_{7/2}J\rangle$ configuration would be pure $S = 0$ since this state is made by antialigning $j_1 = 5/2$ and $j_2 = 7/2$, each of which is $j = l + 1/2$. Therefore the two intrinsic spins are also antialigned, giving $S = 0$. Though this argument does identify the main component of the configuration, it is a bit naive: one must be careful of mentally mixing the LS and jj schemes. In fact, the $J = 1$ state can be made in three ways—by coupling l_1 and l_2 to $L = 1$, with $S = 0$ as just stated, but also with the same L value and vector coupling $S = 1$ to again give $J = 1$ and, finally, by coupling l_1 and l_2 to give $L = 2$ and then antialigning a $S = 1$ vector to give $J = 1$.

In general, the relative amplitudes of $S = 0$ and 1 in a given J value for the configuration, $|(l_1s_1)j_1\,(l_2s_2)j_2;\,JM\rangle$ are given by the $9 - j$ symbol

$$\begin{Bmatrix} l_1 & \tfrac{1}{2} & j_1 \\ l_2 & \tfrac{1}{2} & j_2 \\ L & S & J \end{Bmatrix}$$

Returning to the issue of determining which state is most affected by a δ-interaction without actually carrying out the calculation of Eqs. 4.7 or 4.10, there are three practical methods. All depend on the fact that the J_{min} and J_{max} states are most nearly coplanar, and therefore one of these will be the most lowered, while the other is unaffected. Determining either, then, answers the question.

The most straightforward approach is the one we have just used in the $|d_{5/2}f_{7/2}J\rangle$ case, namely to look for a state that is pure $S = 1$: this state will be unaffected, as we have seen.

There will be such a state (either J_{min} or J_{max}) with pure $S = 1$ only when there is an allowed J value that is greater than $(l_1 + l_2)$ or less than $(l_1 - l_2)$. This occurs, for example, for $J = 6$ in $|d_{5/2}f_{7/2}J\rangle$, or $J = 1$ in $|d_{5/2}g_{7/2}J\rangle$ where $L_{min} = 2$ (see Fig. 4.2), for $J = 8$ in $|d_{5/2}h_{11/2}J\rangle$, but not for any state in $|g_{7/2}h_{11/2}J\rangle$ or for the allowed (even) J values in any configuration of the type $|j^2J\rangle$. A useful rule is that a state with pure $S = 1$ always exists if $j_1 \neq j_2$ and j_{lower} is $l + 1/2$.

When no such state exists, a second method uses the fact that a δ-interaction only affects half of the states in a multiplet; either those with even J or those with odd J. We have seen an example of this in the $|d_{5/2}f_{7/2}J\rangle$ configuration where the $J^\pi = 6^-$ level is unaffected, and the 1^- level is lowered the most. In contrast, in the $|d_{5/2}g_{7/2}J\rangle$ configuration, J_{min} again equals 1. But here $L_{min} = |l_1 - l_2| = 2$ and therefore the 1^+ state can only be found by antialigning $S = 1$ to $L = 2$. Thus, here it is the 1^+ state that is unaffected, while the 6^+ state is lowered the most. These two cases are just examples of the general rule that for positive parity configurations only even J levels are lowered, while for negative parity only odd J levels are lowered. This rule is clear from the restrictions on the right in Eq. 4.7: for positive parity $l_1 + l_2$ is even, so $l_1 + l_2 - J$ is even only for J even and vice versa for negative parity. We use this rule, which arises simply from the parity of the interaction matrix elements that involve the spherical harmonics $Y_l^m(\theta)$ where θ is the angle between the orbital planes (see Section 4.2), as follows. For any $j_1 j_2$, one of J_{min} or J_{max} will be odd and the other even. That state that falls into the class of states affected by the interaction will be lowered the most, and the others in the same class successively less so as their J values deviate more from J_{max} or J_{min}. Let us consider a couple of examples. Starting with the familiar $|d_{5/2}f_{7/2}J\rangle$ case, J_{min} and J_{max} are 1 and 6, the parity is negative and thus only the odd J states will be affected, that is $J^\pi = 1^-, 3^-, 5^-$. Therefore the $J = 1^-$ state is lowered the most, the $J = 3^-, 5^-$ states successively less so, and the $J = 2^-, 4^-$ and 6^- states not at all. For the $|d_{5/2}g_{7/2}J\rangle$ configuration, $J_{min} = 1$, $J_{max} = 6$, as before but here $\pi = +$ so that $J = 6$ is lowered the most and the order is (from lowest to highest) $6^+, 4^+, 2^+$ ($1^+, 3^+, 5^+$ degenerate at the unperturbed position). Finally, for $|g_{7/2}h_{11/2}J\rangle$, $J_{min} = 2$, $J_{max} = 9$, $\pi = -$, and so the order (again, lowest to highest) is 9, 7, 5, 3 (with 2, 4, 6, 8 unperturbed). We note an interesting point, namely that, just as with S values, although the δ-function has no explicit parity dependence, the resulting energies of states $|j_1 j_2 J\rangle$ are, in fact, different for the two different parities.

Although no states (other than $|s_{1/2}^2 J = 0\rangle$) have pure $S = 0$, a third method, which is formally incorrect, emphasizes the jj coupling picture and does always give the right answer. To illustrate this procedure, in $|g_{7/2}h_{11/2}J\rangle$, all states $J = 2$ to 9 can be formed with $S = 0$ or 1. The naive argument would say that, since $j_1 = l_1 - 1/2$ and $j_2 = l_2 + 1/2$, the $J_{min} = 2$ state formed by antialigning j_1 and j_2 would correspond to $S = 1$ and therefore be unaffected by a δ-interaction. The same argumentation would imply that the $J_{max} = 9$ state is $S = 0$, and therefore would be lowered the most.

This approach can be used for equivalent orbit cases to obtain a very important result. Consider a configuration $|j^2J\rangle$ such as $|(f_{7/2})^2J\rangle$. The j value corresponds to $l + 1/2$, and therefore the antialigned J_{min} state, $J = 0$, should have $S = 0$ and be most affected. Again, although the argument is not really correct, the result is. Since the overlaps of the particle orbits are reduced with increasing J, the excitation energies increase monotonically with J. Note that the same result applies for any case of identical particles in equivalent orbits: the $J^{\pi} = 0^+$ state will always lie lowest. Note also that, in this respect, the δ-function interaction resembles the well-known effects of the pairing force that is designed (defined, actually) to lower states in which pairs of identical particles are coupled to spin 0. As pointed out just after Eq. 4.10, and as seen in the upper panels in Fig. 4.3, this effect of the δ-interaction in $|j^2 J\rangle$ configuration is greater for higher j (proportional to $(2j + 1)$).

This lowering of the 0^+ state is an extremely important result. Ultimately, it is the underlying reason why all even–even nuclei have 0^+ ground states and, often, large spacings to the 2_1^+ level. It directly explains this result only for nuclei two nucleons away from closed shells. However, we shall see in Chapter 5 that our two-particle result can be generalized to j^n configurations, and by extension, to wave functions that are linear combinations of several j_i^n configurations. Further generalizations to vibrational excitations and deformed nuclei will be seen in later chapters.

There is an easy geometrical way of viewing this case of identical nucleons in equivalent orbits that provides a physical rationale for the lowering of the 0^+ states. Intuitively, it might seem that the overlap of the two particles in the J_{min} and J_{max} states would be comparable: the two orbits are essentially coplanar in both cases. Once again, however, the Pauli principle plays a key role in distinguishing these situations. We show a schematic illustration in Fig. 4.4. In the $J = J_{max}$ state, the near alignment of the two j values implies nearly coplanar orbits in which the two particles orbit in the same direction. The Pauli principle, however, forbids contact. In effect, this means that the two particles must repel each other at short distances: therefore they track each other around the nucleus on opposite sides of an orbit so that they always remain apart. Thus, there can be little δ-function interaction between them. For $J^{\pi} = 0^+$, the orbits once again are coplanar but now the two particles orbit in opposite senses. As they do so, their separation will vary but the average separation will clearly be much less than in the J_{max} state, "contact" situations occur, and a large δ-function interaction results. The actual values of $A(j_1 j_2 J)$ for a number of different spin combinations are summarized in Table 4.1, and Fig. 4.3 shows several examples (including those used most often in the preceding discussion) of $\Delta E(j_1 j_2 J)$ values under the influence of a δ-function interaction. In studying Table 4.1, recall that the interactions are attractive so that larger values of $A(j_1 j_2 J)$ correspond to lower-lying levels. Perusal of the table and figure shows that the preceding rules are always satisfied.

It is possible to summarize these results succinctly, as is done in Table 4.2. (The trigonometric functions in column 6 of Table 4.2 will be explained in Section 4.2.) To illustrate the construction of the table, let us consider the top row. A little thought, or working out a few examples of the preceding rules, will convince the reader that, for positive parity, the state most affected will be J_{min} if $(j_1 - j_2)$ is even (as in $|d_{5/2} g_{9/2} J\rangle$). Note that even values of $(j_1 - j_2)$ are equivalent to odd values of $(j_1 + j_2)$. Moreover, since only even J states are affected for positive parity states, odd values of $(j_1 + j_2)$ are the same as odd values of $(j_1 + j_2 + J)$. The consistent element in the table is that, if $j_1 + j_2 + J$ is odd (even) for the affected states, then the lowest-lying state will be $J_{min} = |j_1 - j_2|$ $(J_{max} = j_1 + j_2)$. We will encounter the important role of the quantity

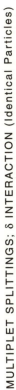

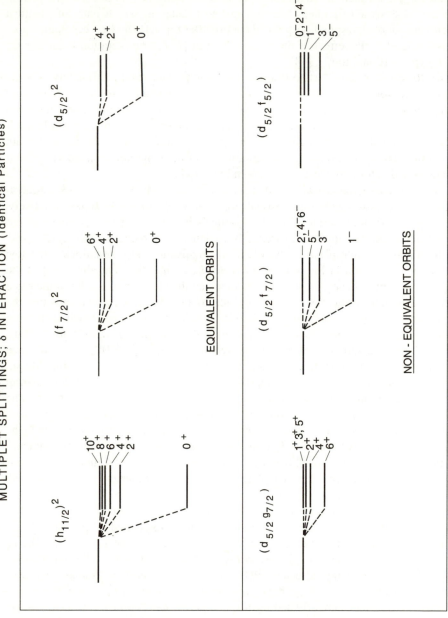

FIG. 4.3. Energy shifts for a δ-function residual interaction for identical nucleons in several different orbit combinations.

Table 4.1 *Relative J state energy values for various identical two-particle configurations $|j_1 j_2 J\rangle$ with an attractive δ-interaction.** *

j_1	j_2	$(lj)_1$	$(lj)_2$	π	J											
					0	1	2	3	4	5	6	7	8	9	10	11
5/2	5/2	$d_{5/2}$	$d_{5/2}$	+	3.00	–	0.685	–	0.286	–	–	–	–	–	–	–
		$\{f_{5/2},\ d_{5/2}\}$	$\{f_{5/2},\ f_{5/2}\}$	–	0	0.173	0	0.457	0	1.29	–	–	–	–	–	–
7/2	7/2	$g_{7/2}$	$g_{7/2}$	+	4.00	–	0.95	–	0.467	–	0.233	–	–	–	–	–
		$\{f_{7/2},\ g_{7/2}\}$	$\{f_{7/2},\ f_{7/2}\}$	–	0	0.127	0	0.312	0	0.599	0	1.52	–	–	–	–
11/2	11/2	$h_{11/2}$	$h_{11/2}$	+	6.00	–	1.47	–	0.785	–	0.493	–	0.318	–	0.180	–
		$\{i_{11/2},\ h_{11/2}\}$	$\{i_{11/2},\ i_{11/2}\}$	–	0	0.084	0	0.196	0	0.329	0	0.509	0	0.818	0	1.89
3/2	5/2	$\{d_{3/2},\ p_{3/2}\}$	$\{d_{5/2},\ f_{5/2}\}$	+	–	0	0.343	0	1.14	–	–	–	–	–	–	–
		$\{d_{3/2},\ p_{3/2}\}$	$\{f_{5/2},\ d_{5/2}\}$	–	–	2.4	0	0.686	0	–	–	–	–	–	–	–
5/2	7/2	$\{d_{5/2},\ f_{5/2}\}$	$\{g_{7/2},\ f_{7/2}\}$	+	–	0	0.230	0	0.518	0	1.40	–	–	–	–	–
		$\{d_{5/2},\ f_{5/2}\}$	$\{f_{7/2},\ g_{7/2}\}$	–	–	3.41	0	1.14	0	0.518	0	–	–	–	–	–
5/2	9/2	$\{d_{5/2},\ f_{5/2}\}$	$\{g_{9/2},\ h_{9/2}\}$	+	–	–	2.85	0	1.04	0	0.490	0	–	–	–	–
		$\{d_{5/2},\ f_{5/2}\}$	$\{h_{9/2},\ g_{9/2}\}$	–	–	–	0	0.258	0	0.558	0	1.47	–	–	–	–
5/2	11/2	$\{d_{5/2},\ f_{5/2}\}$	$\{i_{11/2},\ h_{11/2}\}$	+	–	–	–	0	0.279	0	0.587	0	1.52	–	–	–
		$\{d_{5/2},\ f_{5/2}\}$	$\{h_{11/2},\ i_{11/2}\}$	–	–	–	–	2.65	0	0.979	0	0.470	0	–	–	–

* The table gives values of $A(j_1 j_2 J)$ from Eqs. 4.7 and 4.10. Dashes indicate J values not allowed for the given configuration. States that are unaffected by the interaction (i.e., J even for $\pi = -$, J odd for $\pi = +$ configurations) are given a value of 0. Other numbers are proportional to the *decrease* in energy of the state J.

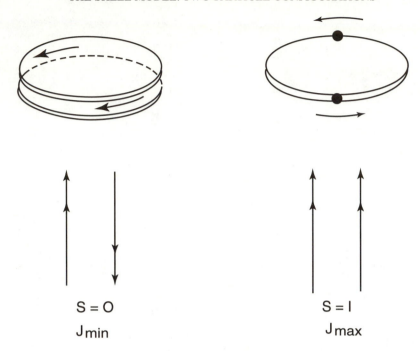

IDENTICAL NUCLEONS
EQUIVALENT ORBITS

FIG. 4.4. Pictorial illustration of the motions of two identical nucleons in equivalent orbits for the cases of maximum and minimum total spin J, showing the effects of the Pauli principle.

$j_1 + j_2 + J$ later in dealing with interactions in p–n multiplets.

To illustrate how these ideas relate to empirical data on nuclei with two valence nucleons, we recall Fig. 2.5, which showed the yrast states for the typical shell model nuclei ^{210}Po, ^{210}Pb, and ^{134}Te. We repeat those data here in Fig. 4.5, along with the predictions for a δ-interaction normalized to the $E_{2_1^+}$ energy for the relevant configuration [$(h_{9/2})^2$ for ^{210}Po, $(g_{9/2})^2$ for ^{210}Pb (these two cases are actually identical as seen in Eq. 4.10), and $(g_{7/2})^2$ for ^{134}Te.] The agreement is quite good. Note the strong lowering of the 0^+ ground state. From data such as this, known for many nuclei, it is possible to estimate the absolute strength of the interaction. One obtains, as a rough guideline, that $\Delta E\,(j_1 j_2 J) \sim (30\ \text{MeV/A})\ A(j_1 j_2 J)$, where the A in the denominator is the mass number. Recalling that $A(j^2 J = 0) = (2j + 1)/2$, this gives a typical lowering of the ground state of several MeV in light nuclei and of 1 to 1.5 MeV in heavy nuclei where the interaction strength is only ~ 200 keV. Both of these are well-known features of the data (e.g., Fig. 2.6), and again show that a δ-function naturally produces the famous "energy

Table 4.2 *Rules for the effects of a δ-function interaction on two-particle identical nucleon configuration*

Parity	Configuration category		States affected	Lowest state	Semi-classical dependence on θ^*
	$j_1 + j_2$	$j_1 + j_2 + J$			
Positive	odd	odd	even J	J_{min}	$\tan \theta/2$
Negative	even	odd	odd J	(antialigned)	
Positive	even	even	even J	J_{min}	$\cot \theta/2$
Negative	odd	even	odd J	(aligned)	

$^*\theta$ is defined by j_1, j_2, J as $\theta = \cos^{-1}\left[\frac{J^2 - j_1^2 - j_2^2}{2|j_1||j_2|}\right] = \cos^{-1}\left[\frac{J(J+1) - j_1(j_1+1) - j_2(j_2+1)}{2\sqrt{j_1(j_1+1)}\sqrt{j_2(j_2+1)}}\right]$

gap" in even–even nuclei usually associated with the pairing force (see the following).

In all these examples, there is another important feature we have not yet commented on: the lowest level for a given multiplet is substantially lowered, but the *differences* in interaction strength for the others monotonically decrease. That is, there is a relative compression of levels near the *unperturbed* position. This is not just an accident of the $3 - j$ symbols, but has a simple physical origin that we shall discuss shortly. As we noted in Fig. 2.5 and, here in Fig. 4.5 for the case of $j_1 = j_2$, it is also a well-known empirical effect characteristic of the low levels of many "shell model" nuclei.

We have been discussing the effects of a δ-interaction between identical nucleons. Such states have $T_z(1) = T_z(2)$, hence $T_z = \pm 1$, and $T = 1$. The proton–neutron system also exists in a $T = 1$ state. By charge independence of the non-Coulomb part of the nuclear force, the p–n $T = 1$ system must then also satisfy Eqs. 4.5–4.7. Indeed, the familiar statement of charge independence that p–p, n–n, and n–p forces are equal applies specifically (and only) to the $T = 1$ mode for the p–n system.

As we have seen, however, the p–n system can also exist in a $T = 0$ state for which there is no need for equality to the p–p or n–n forces. Empirically, in fact, the $T = 0$ interaction seems to be significantly stronger than the $T = 1$. This $T = 0$ coupling is extremely important in nuclear structure, as it is now thought to be most responsible for single-particle configuration mixing and the onset of collectivity, phase transitions, and deformations. We shall return to these points in later chapters. For now, we are interested in simple two-particle p–n configurations in shell model (noncollective, nondeformed) nuclei under the action of a δ-interaction.

In order to address this issue, we must deal with a specific complication that arises in the p–n system. Suppose we imagine such a system occupying levels a and b as shown on the left in Fig. 4.6a. Then, if we treat the proton and neutron as two states of the same particle (the nucleon), the orbits or the charges can be exchanged indistinguishably. Thus, the wavefunction for the two-particle system will have components of all the types illustrated. An interaction matrix element will then contain, for example,

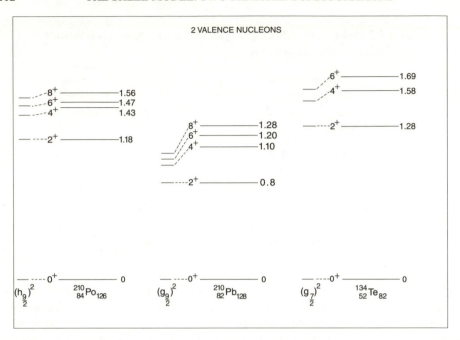

FIG. 4.5. Comparison of experimental and calculated low-lying even spin (yrast) states in three nuclei with two valence nucleons. The orbits used for the two identical nucleons are indicated in each case.

$$\langle \phi_1^a(p)\phi_2^b(n) \,|V_{12}|\, \alpha\phi_1^a(p)\phi_2^b(n) + \beta\phi_1^b(n)\phi_2^a(p)$$
$$+ \gamma\phi_1^b(p)\phi_2^a(n) + \delta\phi_1^a(n)\phi_2^b(p)\rangle \tag{4.11}$$

The four terms on the right correspond to the four cases of particles 1 and 2, each occupying levels a or b, and each identified as either a proton or a neutron. The term in α in Eq. 4.11 is a *direct* term that should be relatively large since the overlap of the wave functions on the two sides is unity. The last term keeps the particles in the same orbits (1 in a, 2 in b) as on the left side but exchanges their type (p $\rightarrow$ n, n $\rightarrow$ p). This can also be large if the interaction contains terms that can change a proton into a neutron and vice versa. The second and third terms involve overlaps in which the particles change orbits: therefore they are normally small. In principle, however, all must be taken into account. Clearly, this is both complicated and tedious. The isospin formalism for doing so is well known and is discussed in standard texts, so we will not consider this situation. Here we are largely concerned with medium and heavy nuclei, which greatly simplifies the problem since protons and neutrons are usually filling different major shells. This situation is illustrated in Fig. 4.6b, where the neutron shell corresponding to the proton shell is already filled: hence, exchange matrix elements such as those in the γ and δ terms in Eq. 4.11 (the rightmost two terms in Fig. 4.6a) are impossible ("blocked"), since the neutron orbits in that shell are filled. The only remaining exchange term is one of those expected to be small. We can consider only the direct term to good approximation in such

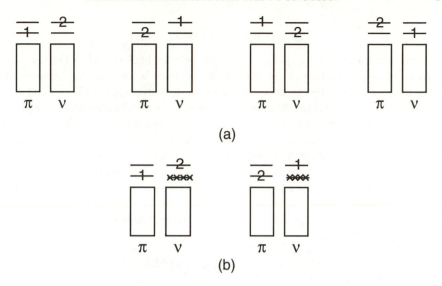

(a)

(b)

FIG. 4.6. "Direct" and "exchange" configurations for protons and neutrons treated as indistinguishable particles (a) filling the same shell, and (b) where the neutron shell corresponding to the valence protons is already filled.

a case. This is equivalent to treating the protons and neutrons as *distinguishable particles*. Note that, while we can choose to do so in this case, because of the blocking of exchange terms, it is not mandatory: we could (and will momentarily) use the isospin formalism as well. The proton–neutron and isospin formalisms are equivalent ways of obtaining the same result. Which one is used in a given case is a choice based on practical simplicity. When exchange effects are known, a priori, to be inconsequential, the proton–neutron formalism may be simpler.

In the present case, treating the protons and neutrons as distinguishable means that the Pauli principle places no restrictions on their coupling. Their wave functions need not be antisymmetrized. We can write the two-particle p–n wave function as $\psi_{pn} = \phi_p \phi_n$. To see the relation to the isospin approach, this can be rewritten as $\psi_{pn} = 1/2[(\phi_p \phi_n + \phi_n \phi_p) + (\phi_p \phi_n - \phi_n \phi_p)]$. The first term is symmetric with respect to interchange of protons and neutrons, the second antisymmetric. These two terms therefore correspond to $T = 1$ and $T = 0$, respectively, and the energy shifts in the presence of a residual interaction can be written as the average of their values in the two isospin channels, that is, as $1/2[\Delta E(T = 1) + \Delta E(T = 0)]$. In the like-nucleon case, antisymmetrization requires that the $S = 1$ (symmetric) spin coupling be accompanied by an antisymmetric spatial part of the wave function (which vanishes when the two particles are in contact): such states do not feel the δ-interaction. Here, there is no such limitation. Both $S = 0$ and $S = 1$ states can be accompanied by symmetric spatial wave functions and will be shifted by the interaction. Generally, the interaction need not be of the same strength in the two cases, and so we specify two interaction strengths $V_{S=0}$ and $V_{S=1}$

$$V_{12} = (V_{S=0} + V_{S=1})\delta(|\mathbf{r}_1 - \mathbf{r}_2|) \tag{4.12}$$

where it is implicit that the $V_{S=0}$ $(V_{S=1})$ term acts only on the $S = 0$ $(S = 1)$ components.

We can get the same result in terms of isospin. We saw before that the unsymmetrized two-particle wave function can be written in terms of an average over $T = 1$ and $T = 0$ parts. In this case, the wave function for each isospin term must be separately antisymmetric. Specifying to a δ-function interaction, the only states affected must have symmetric spatial wave functions. Therefore, the isospin–spin part must be antisymmetric: hence, the $T = 0$ part goes with $S = 1$ and the $T = 1$ with $S = 0$. Again, we get $S = 0$ and 1 terms as in Eq. 4.12. (Recall that the total nuclear wave function must have good isospin, which is obtained by coupling the isospin of the two-particle system to that of the $(T \neq 0)$ core.)

Thus, the analogues of Eqs. 4.5–4.7, for a p–n system under the action of a δ-interaction given by Eq. 4.12 become:

$$\Delta E(j_p j_n J) = F_R(n_p l_p n_n l_n) A(j_p j_n J) \tag{4.13}$$

where

$$F_R(n_p l_p n_n l_n) = \frac{1}{8\pi} \int \frac{1}{r^2} R^2_{n_p l_p}(r) R^2_{n_n l_n}(r) dr \tag{4.14}$$

and

$$A(j_p j_n J) = (2j_p + 1)(2j_n + 1) \begin{pmatrix} j_p & j_n & J \\ \frac{1}{2} & -\frac{1}{2} & 0 \end{pmatrix}^2 \left\{ V_{S=0} \frac{(1 + (-1)^{l_p + l_n + J})}{2} \right.$$

$$\left. + V_{S=1} \left[\frac{(1 - (-1)^{l_p + l_n + J})}{2} + \frac{[(2j_p + 1) + (-1)^{j_p + j_n + J}(2j_n + 1)]^2}{4J(J + 1)} \right] \right\} \tag{4.15}$$

The $3 - j$ symbol is identical to that appearing in the like-nucleon case, but now there are two terms with different J, j_p, j_n dependencies. Note that, in the first, or $V_{S=0}$, term in Eq. 4.15, only half of the levels are affected, namely those with even J for positive parity or with odd J for negative parity. This is the same condition we saw for like nucleons, as it must be since this is the $S = 0$ $(T = 1)$ term. The other J values are then affected by the first of the $V_{S=1}$ terms. For $j_p \neq j_n$, the second $V_{S=1}$ term itself affects all J values, and small J values the most.

As in the like-nucleon case, this equation simplifies for equivalent orbits ($j_p^\pi = j_n^\pi = j$) to

$$\Delta E(jjJ) = F_R(nl)(2j + 1)^2 \begin{pmatrix} j & j & J \\ \frac{1}{2} & -\frac{1}{2} & 0 \end{pmatrix}^2$$

$$\times \left[V_{S=0} \, \delta_{J,\text{even}} + V_{S=1} \, \delta_{J,\text{odd}} \left(1 + \frac{(2j + 1)^2}{J(J + 1)} \right) \right] \tag{4.16}$$

Here we can easily see the explicit relation to the isospin formalism. The $T = 1$ p–n interaction must be identical to the p–p and n–n interactions. We saw earlier that, for equivalent orbits, only even J states are allowed for the $T = 1$ p–p and n–n systems, and hence, for a p–n system, which has all J values from 0 to $2j$, the even J values must have $T = 1$ and the remaining levels, namely those with odd J, must be $T = 0$. Thus, in Eq. 4.16, the first term corresponds to the interaction in the $T = 1$ channel and the second term to the $T = 0$ channel.

We now need to consider the relative strengths $V_{S=0}$ and $V_{S=1}$. As we noted in Chapter 1, from the fact that the deuteron has an $S = 1$ ground state, it is clear that $V_{S=1}$ is stronger than $V_{S=0}$. However, there is additional evidence for this from such simple data as neutron separation energies that is directly applicable to nuclei with all A values. As we have just seen (Eq. 4.16) the $T = 1$ and 0 interactions can be associated with the $S = 0$ and 1 terms, respectively. We saw in Chapter 1 from the separation energy data that the nonpairing, like-nucleon ($T = 1$) residual interaction is, on average, repulsive, where by the phrase "on average" we mean averaged over all final J states and by "nonpairing" we mean excluding the 0^+ state (if any). So, by charge independence, the p–n $T = 1$ interaction must on average also be repulsive. Yet, we also noted in Chapter 1 that both S(p) and S(n) *increase* with increasing numbers of particles of the *opposite* type. The interaction between protons and neutrons has both $T = 0$ and $T = 1$ components. So, on balance, the total ($T = 0 + T = 1$) p–n interaction must be attractive. This can only occur if the $T = 0$ component is both *attractive* and *stronger* than the $T = 1$; that is, if the $|V_{S=1}|$ is greater than $|V_{S=0}|$.

Of course, the strength of the two isospin components of the interaction can also be obtained by fitting actual p–n multiplets (groups of states with pure proton and neutron configurations j_p and j_n and J values ranging from $|j_p - j_n|$ to $j_p + j_n$). Schiffer and True and Molinari and co-workers have carried out extensive surveys of this type near all closed shells from ^{16}O to ^{208}Pb. We will discuss their results in Section 4.2 in terms of a simple geometrical analysis. Here it is useful to convey a feeling as to how the data on individual isospins can be deduced. The nuclei near ^{208}Pb offer a nice example. Consider, for example, the states of $^{210}_{84}$Po$_{126}$ in a $|h_{9/2}i_{13/2}J\rangle$ two-proton $T = 1$ multiplet. (These can be found from the ^{209}Bi (^{3}He, d) ^{210}Po reaction since ^{209}Bi has a single proton in the $h_{9/2}$ orbit.) The energy shifts in this multiplet can be used to extract the ($h_{9/2}\,i_{13/2}$) $T = 1$ interaction. The same multiplet exists in $^{208}_{83}$Bi$_{125}$ as a particle–hole p–n multiplet. The energy shifts $\Delta E(j_1 j_2^{-1} J)$ can be converted (see end of chapter) to an equivalent set of particle–particle shifts and the total p–n interaction obtained for each J state. The difference of the $T = 1$ and total interactions then yields the net $T = 0$ strengths. Extraction of $T = 1$ and $T = 0$ strengths is even simpler in the case of equivalent orbits ($j_p = j_n$), of course, where the even and odd J states directly give the $T = 1$ and 0 interactions, respectively. This approach is useful in light nuclei where the protons and neutrons are filling identical orbits (e.g., the $f_{7/2}$ orbit in ^{42}Sc).

To illustrate the application of these ideas, we consider the classic example of $^{38}_{17}$Cl$_{21}$. Since the $N = 8$ to 20 neutron shell is filled, this is an appropriate case to ignore exchange terms. In the lowest-lying states, the configuration is ($d_{3/2p}f_{7/2n}$) giving states $J = 2^-, 3^-, 4^-$, and 5^-. Since $V_{S=1} > V_{S=0}$, the second group of terms in Eq. 4.15 will generally dominate and the overall ordering of levels in the p–n system will tend

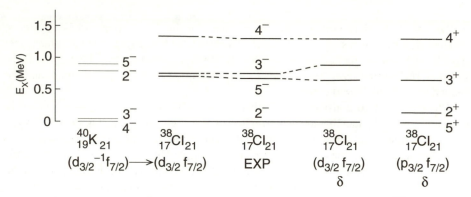

FIG. 4.7. Comparison of low-lying empirical and calculated energies for ^{38}Cl. The two panels on the right correspond to calculations with a two-body δ-function residual interaction, assuming two different orbits for the proton. Clearly, the $(d_{3/2}f_{7/2})$ configuration is favored. The calculation on the left uses the empirical levels of the $(d_{3/2}^{-1} f_{7/2})$ particle–hole configuration in ^{40}K in conjunction with Eq. 4.34 to predict the particle–particle levels of ^{38}Cl. (See deShalit, 1974.)

to be contrary to that in the like nucleon case. Moreover, whereas only half the states are affected for like nucleons (J odd for $\pi = -$; J even for $\pi = +$), all states will be shifted in the p–n case. We therefore may expect the lowest level to be the even J state with highest overlap, the $J = 2^-$ level. The ^{38}Cl experimental spectrum and that calculated with $V_{S=1} = 2V_{S=0}$ are shown on the middle right in Fig. 4.7. (The part on the left describes an alternate approach to calculating ^{38}Cl, to be discussed near the end of this chapter.) The 2^- level does in fact occur lowest, and the agreement is reasonable. The figure also shows that the calculated levels for an alternate configuration with the same j values, $(p_{3/2p}, f_{7/2n})$, have a rather different pattern since, here, the orbital phase factors in Eq. 4.15 are different ($l_p + l_n$ is now even) and the $J = 5, 3$ set is lowered relative to the $J = 2, 4$ pair in disagreement with the data. This indicates how one can even sometimes suggest j configurations and J^π values by examining energy sequences and spacings in p–n multiplets.

4.2 Geometrical interpretation

Having dealt extensively now with both like and unlike two-particle configurations under the influence of a δ-function interaction, we have gained a feeling for the physics behind the analytic results that can be obtained. The physics revolves around the overlaps of the two-particle wave functions. It is possible to approach this entire subject from an alternate viewpoint and actually derive the typical behavior of the $3 - j$ symbol in Eqs. 4.7, 4.15, and 4.16 from a simple geometrical analysis, which will give us additional insight into the interactions in two-particle configurations.

We commented earlier that the characteristic and typical behavior of that $3 - j$ symbol is a gradual reduction in the spacings as the excitation energy increases (as the

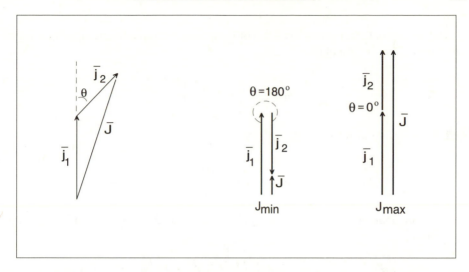

FIG. 4.8. Definition and schematic illustration of some of the ideas used in the geometrical analysis of short-range residual interactions.

interaction weakens). This is not one of those annoying "accidental" effects of Clebsch–Gordan coefficients that plague many students, but rather it has a very simple physical origin. In pursuing this we will better understand why this $3 - j$ symbol behaves as it does. Moreover, we will see that the energies $\Delta E(j_1 j_2 J)$ exhibit the same basic pattern for any $j_1 j_2$ and that this pattern simply reflects the spatial overlaps of the particles and Pauli principle effects.

We start with the semiclassical concept of the angle, θ, between the angular momentum vectors $\mathbf{j}_1$ and $\mathbf{j}_2$ (hence between the orbital planes) of the two particles as illustrated schematically in Fig. 4.8. Then

$$\mathbf{J}^2 = \mathbf{j}_1^2 + \mathbf{j}_2^2 + 2|\mathbf{j}_1||\mathbf{j}_2| \cos \theta$$

or

$$\cos \theta = \frac{\mathbf{J}^2 - \mathbf{j}_1^2 - \mathbf{j}_2^2}{2|\mathbf{j}_1||\mathbf{j}_2|} = \frac{J(J+1) - j_1(j_1+1) - j_2(j_2+1)}{2\sqrt{j_1(j_1+1)j_2(j_2+1)}} \qquad (4.17)$$

From here on, for simplicity, we take the case of identical particles in equivalent orbits ($j_1 = j_2 = j$) and assume that $j, J \gg 1$ so that terms like $J(J+1)$ can be approximated by J^2. Then,

$$\cos \theta = \frac{J^2 - 2j^2}{2j^2} = \frac{J^2}{2j^2} - 1$$

Note that $\theta = 0°$ corresponds to high J and $\theta = 180°$ corresponds to low J. Thus, for $j_1 = j_2, \theta = 180°$ corresponds to $J = 0$ and $\theta_{\min} \to 0°$ to $J = J_{\max} = 2j - 1$.

Before proceeding, we first make use of some simple trigonometric equations. From $\sin^2 \theta = 1 - \cos^2 \theta$, we obtain

$$\sin \theta = \frac{J}{j} \left[1 - \frac{J^2}{4j^2} \right]^{\frac{1}{2}}$$

And, from $\sin \theta/2 = [(1 - \cos \theta)/2]^{1/2}$, we get

$$\sin \frac{\theta}{2} = \left(1 - \frac{J^2}{4j^2} \right)^{\frac{1}{2}}$$

We also note that $\tan \theta/2 = (1 - \cos \theta)/ \sin \theta$. Now, the $3 - j$ symbol in Eq. 4.10 can be written

$$\begin{pmatrix} j & j & J \\ \frac{1}{2} & -\frac{1}{2} & 0 \end{pmatrix}^2 = \frac{(2j + 1 + J)(2j - J)}{(2j + 1)^2} \begin{pmatrix} j - \frac{1}{2} & j - \frac{1}{2} & J \\ 0 & 0 & 0 \end{pmatrix}^2$$

A good approximation to this for large j, J is

$$\begin{pmatrix} j & j & J \\ \frac{1}{2} & -\frac{1}{2} & 0 \end{pmatrix}^2 = \frac{(2j + 1 + J)(2j - J)}{(2j + 1)^2} \frac{1}{2\pi} \frac{1}{(\frac{1}{2}J)} \frac{1}{\sqrt{(j - \frac{1}{2})^2 - \frac{J^2}{4}}}$$

Neglecting quantities of the order of unity compared to j, J we get

$$\begin{pmatrix} j & j & J \\ \frac{1}{2} & -\frac{1}{2} & 0 \end{pmatrix}^2 \approx \frac{(2j + J)(2j - J)}{4j^2} \frac{1}{\pi} \frac{1}{Jj \left(1 - \frac{J^2}{4j^2} \right)^{\frac{1}{2}}}$$

$$\approx \frac{4j^2 \left(1 - \frac{J^2}{4j^2} \right)}{4j^2} \frac{1}{\pi} \frac{1}{Jj \left(1 - \frac{J^2}{4j^2} \right)^{\frac{1}{2}}}$$

$$\approx \frac{\sin^2 \frac{\theta}{2}}{\pi j^2 \sin \theta}$$

Hence,

$$\Delta E(j^2 J) \approx - \left(\frac{V_0 F_R}{2} \right) \frac{(2j + 1)^2 \sin^2 \frac{\theta}{2}}{\pi j^2 \sin \theta} \approx \frac{-2V_0 F_R \sin^2 \frac{\theta}{2}}{\pi \sin \theta}$$

Using the relations for $\sin \theta/2$ and $\tan \theta/2$, we have

EQUIVALENT ORBITS

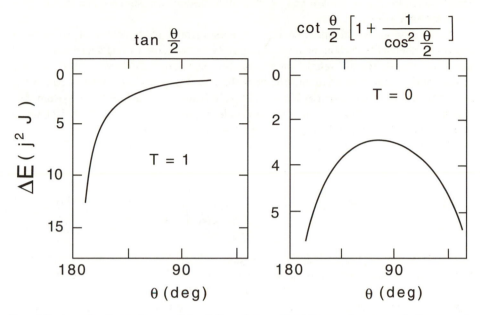

FIG. 4.9. Angular dependence of the δ-function residual interaction strength (lower values correspond to more attractive residual interactions) for two particles in equivalent orbits. (Left) The $T = 1$ (J even) states. (Right) The $T = 0$ (J odd) states. The analytic expressions are indicated above their respective plots.

$$\Delta E(j^2 J) \approx \frac{-2V_0 F_R}{\pi} \frac{\dfrac{1 - \cos\theta}{2}}{\sin\theta}$$

or, finally

$$\Delta E(j^2 J) \approx \frac{-V_0 F_R}{\pi} \tan\frac{\theta}{2} \quad (T = 1, J \text{ even}) \tag{4.18}$$

This extremely simple result expresses the shifts in different J states for a δ-interaction between two identical particles in equivalent orbits. It was derived for large j, J, but is remarkably accurate even for low spins (e.g., as low as $j = 3/2$ and $J = 1$ but specifically not for $J = 0$). The function $\tan\theta/2$ is plotted against θ in Fig. 4.9. Since $\tan\theta/2 \approx \theta/2$ for small θ and goes to infinity for $\theta \to 180°$, Eq. 4.18 simply states that the energy shifts become large (and negative since the force is attractive) for $\theta \approx 180°$ (for small J where the two angular momenta are antialigned), while the smallest effect occurs when $J = J_{\max} \approx j + j$, since $\theta \approx 0°$. Moreover, the curve $\tan\theta/2$ becomes asymptotically flat for large J, giving a geometrical interpretation to the compression in spacings discussed above and illustrated for ^{210}Po, ^{210}Pb, and ^{134}Te in Fig. 2.5.

Note that this formula automatically reflects the Pauli principle arguments discussed earlier, in which the δ-interaction affects only $S = 0$ states. The Pauli principle appears here through the $3 - j$ symbol and, in particular, through the spin angular momenta of $\pm 1/2$ appearing in it.

As we have noted, identical nucleons in equivalent orbits have $T = 1$, so Eq. 4.18 applies to them. For proton–neutron configurations we have to consider both $T = 1$ and $T = 0$ parts. We still restrict ourselves, though, to equivalent orbits. A $T = 1$ p–n system, which cannot be distinguished from the $T = 1$ p–p and n–n systems, consists of even J states and is described by Eq. 4.18 as well. However, in a $T = 0$ p–n system (odd J states) $\Delta E_{pn}(j^2 J)$ has a different J dependence, reflecting the different J behavior of the second ($V_{S=1}$) term in Eqs. 4.15 and 4.16 compared to the first term.

For the odd J, $T = 0$ case, a similar analysis gives (again for $j_p = j_n$)

$$\Delta E \left(j_p j_n J \right)_{j_p=j_n} = \frac{-V_0 F_R}{\pi} \left(\cot \frac{\theta}{2} \right) \left[1 + \frac{1}{\cos^2 \left(\dfrac{\theta}{2} \right)} \right] \qquad (T = 0, J \text{ odd}) \quad (4.19)$$

The behavior of $\Delta E(j^2 J)$ for $T = 0$ is also indicated in Fig. 4.9. We recall that this only applies to equivalent orbits $j_p = j_n$. While the $T = 1$ interaction is largest (lowest-lying on the plot) for $J_{\min} (\theta \approx 180°)$ and is smallest for $J = J_{\max} (\theta \to 0°)$, the $T = 0$ expression is large for *both* $J_{\min}$ and $J_{\max}$. For $T = 0$ and $\theta \approx 180°$, $\Delta E(j^2 J_{\min}) \approx 1/\cos \theta/2 \to \infty$; for $\theta = 0°$, $\Delta E(j^2 J_{\max}) \approx \cot \theta/2 \to \infty$. Both the $T = 0$ and $T = 1$ expressions are small for $\theta \approx 90°$.

All these features can be easily understood physically. The interaction should be small for $\theta = 90°$ for *both* $T = 0$ and $T = 1$, since the particles are orbiting in nearly perpendicular planes and are seldom close enough to interact. For $T = 1$ (which, by charge independence, means we can re-use the identical-particle arguments), the interaction is strong when the two nucleons orbit in opposite directions ($J = 0, \theta = 180°$). However, it vanishes when they orbit in the same direction ($J_{\max}, \theta = 0°$) since, then, the two particles have identical quantum numbers and the spatial wave function is required to be antisymmetric: it must vanish if the nucleons "touch." The Pauli principle effectively introduces a short-range repulsion. The only way the particles can orbit in the same direction and yet not touch is if they circulate out of phase at opposite ends of an orbit diameter. This gives an interaction that is small for small θ, but large for large angles in agreement with Fig. 4.9. The basic idea is the same as for the identical particle $T = 1$ case (Fig. 4.4). For the $T = 0$ case we treat the particles as distinct and, for both the small and large J extremes, the orbits are nearly coplanar. Since we need not worry about antisymmetry, there is no restriction on phasing, and "contact" is abundant, leading to a strong interaction for both $\theta \approx 0°$ and $\theta \approx 180°$.

Empirically, these effects are well documented as shown by the examples in Fig. 4.10 taken from the aforementioned empirical analyses of p–n multiplets throughout the periodic table by Schiffer and True. Note the interesting point that for even J, $T = 1$, the empirical interaction is actually slightly positive (repulsive) for small θ (high J). A δ-function interaction cannot give this: at best, it vanishes near $J = J_{\max}$. Such an analysis clearly shows the need for a separate repulsive component in the residual interaction.

EQUIVALENT ORBIT SPECTRA

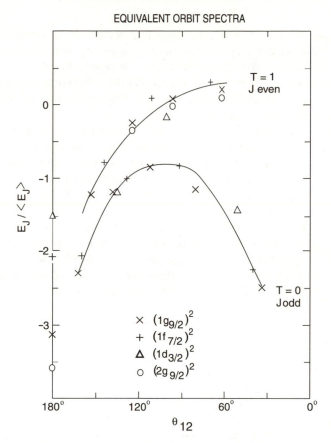

FIG. 4.10. Empirical proton–neutron multiplets for two particle equivalent orbit configurations for comparison with the behavior shown in Fig. 4.9. The curves are drawn
through the data (Schiffer, 1971).

Several studies have successfully carried out multipole analyses of these effective residual
interaction, incorporating dipole, quadrupole, etc., components. Evidence for a sizable
quadrupole component has been found. This multipole varies as $P_2(\cos\theta)$ where, again,
θ is the angle between the two orbits. As is well known, this function crosses zero at
$\theta \approx 55°$ so that even for an overall attractive quadrupole term, the interaction is actually
repulsive for angles between $55°$ and $125°$. This is just the region where Fig. 4.10 shows
positive (repulsive) empirical $T = 1$ interactions. This repulsive aspect should not be
surprising. We have already encountered it. We noted in our discussion of separation
energies in Chapter 1 that the like nucleon ($T = 1$) nonpairing residual interaction was,
on balance, repulsive ($S(n)$ decreases with increasing N). From this empirical fact, we
also deduced that the $T = 0$ interaction is on balance stronger (more negative) than the
$T = 1$. This is also evident in Fig. 4.10.

Finally, note that for $J^\pi = 0^+$, the interaction deviates from the geometric expression. The 0^+ behavior, however, is physically reasonable. As with the like-nucleon case, the interactions are ordered by j: they are largest for large j. The larger the j value, the more magnetic substates there are, and the smaller the permissible angular range of an orbit for a given m. Thus the orbit planes are more tightly defined and the overlaps of particles in $\pm m$ substates are greater.

Thus far, for simplicity, we have carried out the geometrical analysis for the simple case of $j_1 = j_2$. For $j_1 \neq j_2$ and identical particles, we saw in Section 4.1 that either the $J_{\min}(\theta \to 180°)$ or the $J_{\max}(\theta \to 0°)$ state will be lowest depending on the particular j values and their $j = l \pm 1/2$ character. Table 4.2 summarized the different cases leading to these two situations. These two categories of two-particle configurations should be and are reflected in the geometrical analysis. One obtains two curves now, of which one is identical to Eq. 4.18 ($\Delta E \approx \tan \theta/2$) giving the lowest energy for the antialignment of the two values ($J_{\min}$, θ close to 180°), and the other curve goes as $\cot \theta/2$ so that the lowest energy occurs for parallel alignment ($J_{\max}$) and θ close to 0°. The correspondence of these two trigonometrical functions and different sets of j_1, j_2 values is made explicit in the sixth column of Table 4.2. Finally, note that the equivalent-orbit situation is actually a special case of this. Here, $\pi = +$, $j_1 + j_2 = 2j$ is always odd and so the $\tan \theta/2$ dependence applies and the $J_{\min}$ (in this case 0^+) state is lowest.

For the p–n case, a similar analysis again leads to two distinct curves as in the $T = 1$ and $T = 0$ cases for equivalent orbits. However, the classification is slightly different. Recall that, for equivalent orbits, the $T = 1$ states are even J only. Thus $j_1 + j_2 + J = 2j + J$ is odd. The $T = 0$ case, with J odd, has $j_1 + j_2 + J = 2j + J$ even. It is this distinction that persists when $j_1 \neq j_2$ for a p–n multiplet. Again, we obtain two curves, but distinguished according to the odd or even character of $j_1 + j_2 + J$ and describable by geometrical functions of θ very similar to Eqs. 4.18 and 4.19. This is beautifully illustrated by the data for several multiplets collected in Fig. 4.11, and exemplified in depth for ^{210}Bi in Fig. 4.12. In all these cases the empirical energy distributions within the p–n multiplets follow the expected energy patterns quite well.

One last point worth mentioning is that extensive surveys of empirical p–n interaction multiplets show that the strength of the interaction, especially in $T = 0$ states, smoothly decreases with increasing mass. This is quite plausible since the average radius of shell model orbits increases with higher oscillator numbers, while the interaction range is constant so that the average interaction strength decreases. In heavy nuclei, typical V_{pn} interaction matrix elements are ≈ 200 to 300 keV but of course this depends on the orbits involved.

What is perhaps most important to emphasize in concluding this part of the discussion is that, without ever having dealt with the radial parts of the wave functions, or indeed, calculating anything, it has been possible to predict the qualitative energy ordering of the different J states in two-particle configurations. Moreover, exact quantitative results for the relative spacings involve only the evaluation of a single $3 - j$ symbol. (Of course, the absolute spacings depend on the radial integrations and the strength of the interaction.) This is but one example of how far one can go in a shell model treatment of multiparticle configurations by invoking only very general arguments.

NONEQUIVALENT p – n SPECTRA

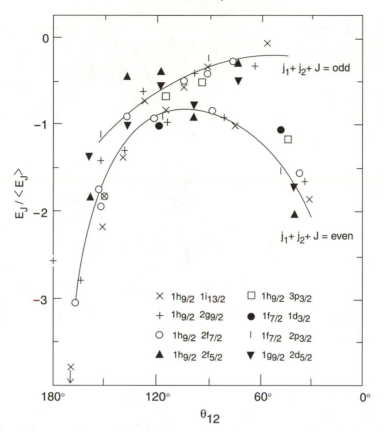

FIG. 4.11. Comparison of empirical and calculated multiplet splittings for two-particle
   configuration nonequivalent orbits (Schiffer, 1971).

4.3 Pairing interaction

We re-emphasize that these results for the δ-function interaction are of more than pass-
ing interest since, representing the short-range interaction par excellence, this interaction
simulates in many respects results from more realistic short-range interactions. In particular, it is closely allied to the pairing interaction specifically designed to mock up
a strong, attractive interaction in the $J = 0$ configuration of two identical nucleons.
The motivation is similar to that for the δ-function interaction—the pairing interaction
is only effective when the particles have extremely high spatial overlaps. Formally, one
can define the pairing interaction by

$$\langle j_1 j_2 J | V_{\text{pair}} | j_3 j_4 J' \rangle = -G \sqrt{\left(j_1 + \frac{1}{2} \right) \left(j_3 + \frac{1}{2} \right)} \delta_{j_1 j_2} \delta_{j_3 j_4} \delta_{J 0 J' 0} \qquad (4.20)$$

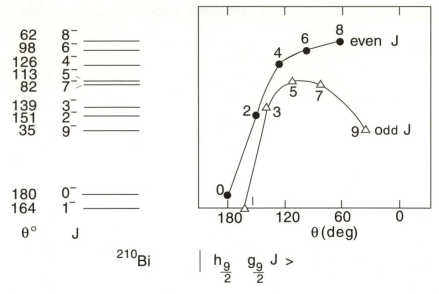

FIG. 4.12. A geometrical analysis of the $|h_{9/2}g_{9/2}J\rangle$ p–n multiplet in ^{210}Bi. The empirical levels are shown on the left along with the semiclassical angle between the orbits of the two nucleons. The right side shows that the levels split into two families, according to J even or J odd. The solid lines are drawn to connect the points.

where G gives the overall strength in the interaction. Note that this interaction is attractive and, by definition, only effective for 0^+ states of identical nucleons in equivalent orbits. It is not, however, limited to diagonal matrix elements $\langle j_1^2 0^+|V_{\mathrm{pair}}|j_1^2 0^+\rangle$, but rather allows nondiagonal scatterings, $\langle j_1^2 0^+|V_{\mathrm{pair}}|j_3^2 0^+\rangle$, in which the pair of particles switches to another orbit as a pair. This feature is critical to the build-up of pairing correlations and the so-called pairing gap in even–even nuclei, and will be treated in more detail in a later chapter. For diagonal matrix elements $|j_1 = j_2 = j_3 = j_4\rangle$ the pairing interaction strongly lowers the 0^+ state without affecting the others.

Both the δ-function force and the pairing force are intended to represent the short-range component of the nuclear interaction. However, the residual interaction also contains a long-range component, that, as we shall see, is crucial in producing collective properties and nonspherical nuclei. It is common to mock this up by the so-called quadrupole interaction. (In recent years higher order multipoles, such as hexadecapole interactions, have also been considered, but we shall ignore their effects here.) The combination of these two forces, the so-called pairing plus quadrupole interaction, has been perhaps the most widely used simulation of nuclear interactions in heavy nuclei in the last couple of decades. It can be written as:

$$V_{ppq} = V_{\mathrm{pair}} + \kappa \, r_1^2 r_2^2 P_2 (\cos\theta) \tag{4.21}$$

where θ is the angle between the radius vectors to each particle (see Fig 4.8) and κ is the strength of the quadrupole part. Clearly, while V_{pair} is short-range, the quadrupole

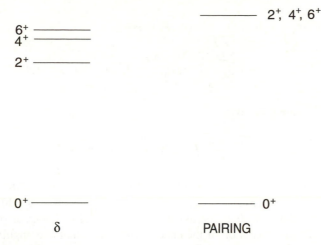

FIG. 4.13. Comparison of levels of a $(j = 7/2)^2$ configuration for a δ-function and a pairing interaction.

component simulates at least part of the long-range aspect of the residual interaction (see the following section).

The popularity of the pairing interaction, or of any other that reproduces the low energy of $J = 0$ coupled pairs of nucleons, clearly lies in the fact that all even–even nuclei have 0^+ ground states. The quadrupole interaction is motivated, empirically, by the fact that nearly all nuclei more than a few mass numbers away from closed shells display properties that can be described in terms of quadrupole distortions of the spherical shape. Figure 4.13 shows the differences between the pairing and δ-function interactions. For the case of identical nucleons in equivalent orbits, both produce a low-lying 0^+ state and a grouped cluster of states with higher angular momentum. With the pairing interaction, this latter group is unaffected, remaining degenerate. The δ-function interaction seems to be a better approximation, as is clear in Fig. 4.5, as well as Fig. 2.6 for the Sn isotopes. For many-particle configurations, the quadrupole force dominates both the pairing or δ-function interactions and the differences in their properties are washed out.

4.4 Multipole decomposition of residual interactions

It is useful at this point to consider a more general approach to calculating nuclear interactions and their effects on energy levels in two-particle configurations. This will help us understand, in a simple manner, the different effects of different residual interactions. It will lead to a deeper appreciation of the δ-function results, in particular the origin of the two curves in Figs. 4.9–4.12, and it will shed more light on the relation between the pairing and δ-interactions. We can expand any interaction that depends on the separation of the two particles as

$$V(|\mathbf{r}_1 - \mathbf{r}_2|) = \sum_k v_k(r_1 r_2) P_k(\cos\theta)$$

$$= \sum_{km} v_k(r_1 r_2) \frac{4\pi}{2k+1} Y_{km}^*(\theta_1\phi_1) Y_{km}(\theta_2\phi_2) \qquad (4.22)$$

where the $v_k(r_1 r_2)$ are given by

$$v_k(r_1 r_2) = \frac{2k+1}{2} \int V(|\mathbf{r}_1 - \mathbf{r}_2|) P_k(\cos\theta) d\cos\theta \qquad (4.23)$$

In the following we do not consider the effects of antisymmetrization explicitly, but the results are still valid for identical particles and for p–n systems involving different major shells. If we write the wave function as a product of radial and angular parts, we can then achieve a similar separation of the energy shift into radial and angular parts and, moreover, separate the angular parts of particles 1 and 2. We obtain, as in Eq. 4.5.

$$\Delta E(j_1 j_2 J) = \sum_k F_R^k A_k \qquad k \le \min(2j_1, 2j_2, 2l_1, 2l_2) \qquad (4.24)$$

where

$$F_R^k = \int |R_{n_1 l_1}(r_1)|^2 |R_{n_2 l_2}(r_2)|^2 v_k(r_1 r_2) dr_1 dr_2 \qquad (4.25)$$

and

$$A_k(J) = \frac{(-)^{j_1+j_2+J}(4\pi)}{(2k+1)} \langle l_1 j_1 \| \mathbf{Y}_k \| l_1 j_1 \rangle \langle l_2 j_2 \| \mathbf{Y}_k \| l_2 j_2 \rangle \begin{Bmatrix} j_1 & j_2 & J \\ j_2 & j_1 & k \end{Bmatrix} \qquad (4.26)$$

As before, F_R is a purely radial integral and depends on the details of the specific interaction chosen, while A_k depends only on the angular coordinates and is therefore *completely independent* of the interactions involved.

We now note a significant point. The summation in Eq. 4.24 is limited to even k values from 0 to $\min(2j_1, 2j_2, 2l_1, 2l_2)$. These limitations arise simply by angular momentum conservation in the reduced matrix elements of Y_k and as specified by the triangle conditions involving the $6-j$ symbol. This severely constrains the number of multipoles that go into the calculation of any given interaction. For example, for *any* interaction $V(\mathbf{r}_1 - \mathbf{r}_2)$ in a $(d_{3/2}f_{7/2})$ configuration, $k_{\max} = 2$ and hence k takes on *only* the values 0 and 2. Moreover, since the monopole part is constant over all (angular) space $[P_0(\cos\theta) = \text{constant}]$, it is the same for all relative orientations of j_1 and j_2, that is, all J values: it does *not* contribute to splittings of a multiplet but only gives an overall shift to the multiplet. Hence, in the $(d_{3/2}, f_{7/2}, J)$ configuration, the energy splittings of the $J^\pi = 2^- - 5^-$ levels are *only* given by the $k = 2$ quadrupole force. In fact, for the larger set of configurations $(d_{3/2} + \text{anything})$ or $(p_{3/2} + \text{anything})$, the splittings for *any* force (even a δ force! – see below) are identical to those for a quadrupole force.

Equation 4.24 applies to any interaction that can be written in terms of the separation of the two particles; that is, as $V(|\mathbf{r}_1 - \mathbf{r}_2|)$. However, one sometimes encounters spin-dependent interactions. Then we obtain a result similar to Eq. 4.24, with the same limitation on k_{max}, but including odd k. This point is very important, since the δ-interaction can be shown to be equivalent to an interaction $V(|\mathbf{r}_1 - \mathbf{r}_2|)$ times a spin-dependent operator. Thus, a δ-function interaction is equivalent to an odd-tensor interaction, even though its multipole expansion contains only even k. The reason is beyond the scope of our treatment but involves the fact that the interaction takes place only at "contact," and therefore a δ-function interaction can be multiplied by various "exchange" interactions that have odd-tensor character. This class of interactions is crucial in nuclear structure since they have the special property of conserving seniority (see Chapter 5).

For a δ-interaction, $v_k(r_1r_2)$ is given by

$$v_k(r_1r_2) = \frac{2k + 1}{4\pi} \frac{\delta(\mathbf{r}_1 - \mathbf{r}_2)}{r_1r_2} \tag{4.27}$$

Note that, despite its short-range, the δ-interaction has a monopole component ($k = 0$) corresponding to a part of the interaction that is *independent* of angle. This is a long-range piece par excellence and it says that a "contact" interaction has a component that pays no heed to the angular separation of the particles! In practical calculations, the situation in regard to allowed multipoles can be even more bizarre. Consider a δ-interaction between two $d_{3/2}$ particles. As we have said, not only does the $k = 0$ multipole contribute, but the triangle conditions on k limit the multipoles to only the values $k = 0, 2$. No really short-range multipoles appear. Even a δ force is equivalent to a long-range quadrupole force.

This seemingly paradoxical situation of an infinitely short-range interaction being simulated only by relatively long-range multipoles is actually easy to understand. This understanding reveals much about the relationship between forces and the orbits and wave functions of the interacting particles. More accurately, it clarifies the way particles from different orbits can "probe" different interactions. It relates in a general philosophical way to how one determines structure in any physical system. The general rule is that in order to sample the structure of a given scale, the probe must be comparable to or smaller than that scale. (It is difficult to distinguish between a potato and a carrot by bombarding either with a truck.)

Imagine a proton and a neutron in an $s_{1/2}$ orbit. They have only one m state and are spherically symmetric wave functions: the orbit is uniformly spread out, at a given radius, over time, in a spherical shell. The two particles are therefore always in "contact" and are simply unable to sense *any* details of the residual attraction. To them, a δ-force is identical to a constant force over all space. The higher the j value of the interacting particles, the more sensitive they are to the details of the force simply because higher j values have more magnetic substates. This set of magnetic substates spans the same angular range around the body of the nucleus and hence, each magnetic substate is restricted to a narrower angular range. Thus, two particles in $j = 13/2$ orbits coupled to $J = 12$ can sense the fine details of an interaction: each substate M samples a different angular range of the force. In contrast, even though the residual interaction may be a

δ-force, two low j (e.g., $s_{1/2}, p_{1/2}, p_{3/2}, \ldots$) orbits cannot "know" this. They are the wrong probe.

There is another interesting way these ideas have been used. Consider again a $(d_{3/2}, f_{7/2})$ configuration. Regardless of the interaction, the energies of the four states $J = 2, 3, 4, 5$ are given by only two equations. Indeed, as discussed by de Shalit and Feshbach, one can show that

$$E(5^-)-E(2^-): E(3^-)-E(2^-): E(4^-)-E(2^-): = 1:2.5:3.5$$

independent of the interaction. The experimental spacings in ^{38}Cl ($\pi d_{3/2}, \nu f_{7/2}$) are in the ratios 1:1.1:2.0. They have the same sequencing of levels as in the calculation $(2^-, 5^-, 3^-, 4^-)$, but the magnitudes of the spacings disagree. Without ever having specified an interaction, or evaluated any radial wave functions, we can unambiguously conclude either that an accurate description of these ^{38}Cl states requires more complex configurations, or that the interaction depends on something other than $(\mathbf{r}_1 - \mathbf{r}_2)$ (e.g., spin). Note that this is the same nucleus that we treated earlier with a δ-function interaction (Fig. 4.7). The earlier results are different than these precisely because we used different strengths (i.e., a spin dependence) for the $S = 0$ and $S = 1$ terms.

Coupled with the practical historic fact that most shell model calculations have dealt with light nuclei and, therefore, low j states, we now see why different interactions are often used to account for the same data, and why it often is difficult to determine the details of the interaction.

The other extreme of k values (multipole order) can also be discussed. Neglecting the limitation given by the $j_1 l_1$, $j_2 l_2$ quantum numbers (or, equivalently, assuming a large j shell), we can relate the largest relevant k values to the range of the force. Consider the classical picture shown in Fig. 4.8. The distance between the two particles is given by $r^2 = r_1^2 + r_2^2 + 2r_1 r_2 \cos\theta$. If we approximate the range of the force by a single number r_r, the integration over θ will be limited to $\theta \leq r_r/r$. The Legendre polynomial $P_k(\cos\theta)$ oscillates more rapidly as k increases (see Fig. 4.14a). If P_k oscillates many times within the allowed integration range, the integral will be small because of cancellation effects. In practice, therefore, for a given force range, k is limited to values satisfying $\pi/k \ll r_r/r$. Clearly, then, as the range decreases, more and more k values are required. The limiting case is of course, the δ-interaction, although other factors (l, j values) relating to the ability of a given wave function to "sample" an interaction will come into play to limit the allowed range of k values as we have discussed.

The idea of the multipole expansion of a residual interaction can be used to obtain some earlier results in a physically transparent way that offers new insights. This idea can also be used to derive a famous result known as the *parabolic rule* for energies of states in p–n multiplets. We have already seen how one can understand the J-dependence of the residual interaction in terms of the angle between the two orbits involved (see Eqs. 4.18, 4.19 and Figs. 4.9–4.12). We can apply similar arguments to specific multipoles.

Each multipole, k, has an angular dependence $P_k(\cos\theta)$. The lowest few are $P_0 = $ constant, $P_2 = 1/2(3\cos^2\theta - 1)$, $P_4 = 1/8(35\cos^4\theta - 30\cos^2\theta + 3)$. The P_k were shown in Fig. 4.14a. As Eq. 4.17 indicates, for a given j_1, j_2 there is a specific relation between the total angular momentum J and the angle θ: thus, one can plot $P_k(\cos\theta)$

equally well against J or $J(J+1)$ (as is sometimes done). This alternate scale is included in Fig. 4.14a for the example of $j_1 = 9/2$, $j_2 = 11/2$ (e.g., a $(g_{9/2}h_{11/2})$ configuration).

Obviously, since each multipole is proportional to $P_k(\cos\theta)$, the interaction in the kth multipole is strongest when P_k is largest. Thus, we can tell directly which spin states will be most affected for each multipole.

As an aid in this, Table 4.3 gives the semi-classical angles θ for the pairs of orbits indicated and several total J values. Comparison of these angles with the maxima and minima of each $P_k(\cos\theta)$ gives a qualitative guide to which levels (which J values) will be most shifted by residual interactions of any given multipolarity and, for non-diagonal matrix elements, which configurations will tend to mix.

The monopole ($k = 0$) component is constant, affects all states equally, and simply gives an overall shift to the entire multiplet. This is an important effect, altering the relative excitation energies of different multiplets, but it contributes nothing to the splitting.

It is the principal origin of the mass dependence of relative single-particle energies that we discussed at the end of Chapter 3. In particular, the monopole p–n interaction leads to the changes noted there (Fig. 3.4) in neutron single-particle energies as a function of proton number and vice versa. In this chapter, however, we are focusing on energy *splittings* due to residual interactions and so we turn now to other multipoles.

The quadrupole ($k = 2$) component is strongest when $\theta = 0°$ and $180°$ and actually changes sign around $90°$. (Recall that there is an overall minus sign to be applied to Fig. 4.14a for an attractive interaction so that negative values on the figure refer to repulsive effects.) Thus, a quadrupole interaction lowers the energies of the extreme J values the most, but can in fact raise the energies of intermediate spins. The general behavior of the quadrupole term is similar to the $j_1 + j_2 + J$ even curve in Fig. 4.11 but this is somewhat accidental. We shall return to the relationship between these results.

It is interesting to discuss a simple expression for the energy shifts $\Delta E(j_1 j_2 J)$ for a quadrupole interaction. The relevant matrix elements involve couplings of spherical harmonics (see Eqs. 4.22, 4.23). This leads to the $6 - j$ symbol in Eq. 4.26. An evaluation similar to the geometric derivation in Section 4.2 leads to an expression for the energy shift in terms of trigonometric relations such as Eq. 4.17. The result is

$$\Delta E(j_1 j_2 J)$$
$$\propto \frac{[J(J+1) - j_1(j_1+1) - j_2(j_2+1)]^2 - [J(J+1) - j_1(j_1+1) - j_2(j_2+1)]}{4[j_1(j_1+1)j_2(j_2+1)]^{1/2}}$$
$$\propto A[J(J+1)]^2 + BJ(J+1) + C$$

$$(4.28)$$

where A, B, and C are functions of j_1 and j_2 but are independent of J. Using just the approximation $\Delta E \sim \cos^2\theta$ gives almost the same expression, namely the first term in Eq. 4.28 alone, which dominates the effects discussed below. This result for $\Delta E(j_1 j_2 J)$ is a parabola in $J(J+1)$ and is the *parabolic rule* discussed frequently by Paar. By differentiating we can find the vertex, J_v:

$$J_v(J_v + 1) = j_1(j_1+1) + j_2(j_2+1) - \frac{1}{2}$$

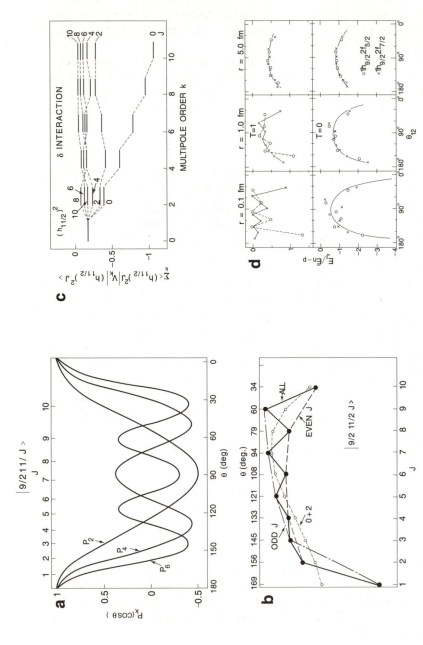

FIG. 4.14. Multipole decomposition of residual interactions: (a) Legendre polynomials $P_k(\cos\theta)$ plotted against θ and J value for a (9/2, 11/2) configuration. (b) Contribution of different multipoles to $\Delta E(g_{9/2}h_{11/2}J)$ with a δ function residual interaction. The short-dashed and solid curves correspond to the multipoles $k = 0, 2$ and $k = 0, 2, 4, 6,$ and 8 respectively. For the complete ($k = 0 - 8$) case, the even and odd J states are separately connected by broken lines. (c) Effect of the addition of each successive multipole on the energies of an $(h_{11/2})^2$ configuration. (Parts (b) and (c) are based on Heyde, 1989.) (d) Effect of forces of different ranges on the relative energy shifts as a function of θ, that is, J, for two different two-particle configurations. (Schiffer and True, 1976.)

Table 4.3 *Semi-classical angles for the configurations* $| j_i j_k J >$

	1/2	3/2	5/2	7/2	9/2	13/2
$J^\pi = 2^+$						
1/2		63.4	133.1			
3/2	63.4	101.5	124.6	151.4		
5/2	133.1	124.6	131.1	142.0	159.1	
7/2		151.4	142.0	144.0	150.9	
9/2			159.1	150.9	151.5	166.5
13/2					166.3	159.8
$J^\pi = 4^+$						
1/2				59.4	129.7	
3/2			49.1	88.1	116.2	
5/2		49.1	81.8	101.0	117.3	155.2
7/2	59.4	88.1	101.0	111.4	121.3	143.4
9/2	129.7	116.2	117.3	121.3	126.6	140.4
13/2			155.2	143.4	140.4	142.6
$J^\pi = 6^+$						
1/2						128.3
3/2					45.5	112.8
5/2				41.8	73.2	112.0
7/2			41.8	70.5	87.8	114.0
9/2		45.5	73.2	87.8	98.7	117.0
13/2	128.3	112.8	112.0	114.0	117.0	124.7
$J^\pi = 8^+$						
1/2						
3/2						43.9
5/2						69.4
7/2					37.1	82.2
9/2				37.1	63.0	91.2
13/2		43.9	69.4	82.2	91.2	105.2

* I am grateful to Walid Younes who kindly provided this table from his thesis
(W. Younes, Ph. D. thesis, Rutgers University, 1996, unpublished) and to C. Barton
for additional values.

and, solving the quadratic,

$$J_\nu = \left[j_1(j_1 + 1) + j_2(j_2 + 1) - \frac{1}{4} \right]^{1/2} - \frac{1}{2} \tag{4.29}$$

Note that J_ν is normally the *highest*-lying member of a downward pointing (inverted bowl) parabola since the overlap of the particles is largest for $J_{\max}$ and $J_{\min}$. In the

$g_{9/2}h_{11/2}$ case in Fig. 4.14a, this gives $J_\nu \approx 7$, which is indeed where the $P_2(\cos\theta)$ minimum occurs. For this multiplet, the $J = 1$, 10 states are lowest, $J \approx 7$ highest.

The relationship of these ideas to the J-dependence in Figs. 4.9–4.12 is informative, particularly the split into two curves*. In general, points falling alternately (for odd and even J) on these two curves cannot be described by a single parabola in $J(J + 1)$ (we will see one exception later). To generate the δ-function results, the higher multipoles must be added in. From Fig. 4.14a it is clear that these oscillate more rapidly in θ or J, and this introduces an irregularity or zig-zag pattern to the spin dependence. For example, a hexadecapole component would raise the $J = 7$ state and lower the $J = 4$ and 9 states. It also broadens the minimum and tends to shift it toward the spin 8 level. A $P_6(\cos\theta)$ term will add further perturbations. When all the multipoles are summed, with amplitudes appropriate to the δ-interaction, two curves similar to those of Fig. 4.11 are reproduced.

This is evident in Fig. 4.14b, which shows the strength of the interaction as a function of J for different combinations of multipoles (with amplitudes appropriate to a δ-function). For the $|g_{9/2}h_{11/2}J\rangle$ configuration, the $k = 0 + 2$ curve is smooth and minimizes at $J = 7$, as we have just seen. If this curve is replotted against $J(J + 1)$, an exact parabola is obtained. Adding in all the multipoles, $k = 0, 2, 4, 6, 8$, gives the solid curve. If we connect just the odd J states along this curve, we get a dependence just like the $j_1 + j_2 + J$ odd curve of Fig. 4.11. If we connect the even J values we get the $j_1 + j_2 + J$ even curve of Fig. 4.11. Thus we see that the separation of the two curves comes from the higher multipoles and reflects the Pauli effects since these multipoles have shorter ranges.

Figure 4.14c shows this in a particularly illuminating way for the identical particle $|(h_{11/2})^2 J\rangle$ configuration. Here, the energies obtained with the addition of each successive multipole in a δ-interaction are displayed in level scheme form. For $k = 0, 2$ only, a parabolic behavior is observed: the $0^+, 2^+, 4^+$ levels are lowest, the intermediate spins $6^+, 8^+$ highest and the largest spin, 10^+, lower again. As shorter and shorter range multipoles are added, the levels shift toward the characteristic δ-function sequence shown on the right. We have discussed how such a sequence, particularly the weak effects of a δ-interaction for the $J_{\max}$ states (since they have $S = 1$), is a specific Pauli principle effect. Now we see exactly how this comes in through the *shortest*-range multipoles, which are in fact repulsive for these spin states, reflecting the Pauli prohibition against "contact."

Figures 4.14a and 4.14c also dramatically show why the 0^+ state (or, generally, the lowest state of a given multiplet) is lowered so much. This state corresponds to orbit planes closest to $\theta = 180°$ (or $0°$, depending on the j values if nonequivalent orbits are involved). Here, *all* the multipoles contribute coherently because $P_k(\cos 180°$ or $0°)$ is always unity. Only for other spins do cancellation effects enter.

Finally, we can use Fig. 4.14c to better understand the relation between δ- and pairing forces. Imagine continuing this figure further to the right. The $0^+ - J$ separation would grow and the separation among the $J^\pi \neq 0^+$ states would diminish relative to

*I am indebted to K. Heyde for much helpful advice in this section and for Figs. 4.14b and 4.14c.

their separation from the 0^+ level. The limit is the pairing picture shown in Fig. 4.13. Although we do not write the pairing interaction in terms of multipoles, we see that, in effect, it corresponds to the dominance of very high ones. This, incidentally, is the origin and basis for the phrase occasionally encountered that the pairing force is of even "shorter range" than the δ-interaction, a statement that sounds paradoxical.

We have now discussed, in several ways, the relation between multipoles and the (angular) ranges of the forces they describe and their effects on different J states of two-particle multiplets: briefly, attractive short-range (higher multipole order) forces tend to lower especially the J_{max} and J_{min} states (subject always to Pauli principle constraints). We can see this relation even more explicitly by using forces that are finite (radial) range. To this end, Fig. 4.14d shows the energy shifts calculated for two configurations for a Yukawa type force with range parameter $r = 0.1$, 1.0 and 5.0 fm. In general, the shorter the range, the stronger the effects on the extreme J (or θ) states where the orbits are most nearly coplanar and the particles, on average, closer. The different patterns for $r = 0.1$ fm for the $T = 1$ interaction just reflect the $\cot \theta/2$ or $\tan \theta/2$ dependences discussed for the two different configurations (see Table 4.2). This difference, a Pauli effect, is washed out for longer-range forces, as is most of the J (or θ) dependence, since the force becomes rather insensitive to the separation of the particles within the nuclear volume.

With this discussion of multipole contributions and force ranges, the reader should now be in a position to estimate, at least qualitatively, the effect of any particular interaction in a given configuration with little or no explicit calculation.

An interesting special case of the multipole decomposition is that of a $(p_{3/2}d_{3/2})$ p–n configuration. For this configuration, Eq. 4.24 only allows $k = 0, 2$ values, and the $k = 0$ multipole is irrelevant for splittings, so the δ-interaction in this case should give exactly the parabolic dependence on $J(J+1)$ characteristic of $P_2(\cos \theta)$. Yet, from our previous discussion, we know that the $J = 0, 2$ ($j_1 + j_2 + J$ odd) states would occur on the upper curve in Fig. 4.11 and the $J = 1, 3$ states on the lower curve. Although this hardly seems to yield a parabola, careful inspection of the curves in Fig. 4.11 for the angles appropriate to the $|p_{3/2}d_{3/2} \, J\rangle$ configuration shows that these four points have an exact parabolic form against $J(J+1)$.

Actually, this result suggests why a quadrupole interaction is more important than it might at first seem: because of the angular momentum constraints on k_{max}, many higher multipoles that might normally contribute are eliminated. Moreover, the δ-function is not necessarily the best choice of interaction. Other, "finite range" interactions (e.g., Gaussian $e^{-\alpha(|r_1 - r_2|)}$) are often used and have relatively larger low-k amplitudes. Finally, there are often other residual interactions besides the p–n interaction, such as particle–core coupling contributions that are dominated by quadrupole components. Indeed, we saw in Section 4.2 that analyses of empirical two-particle multiplets do suggest evidence for enhanced quadrupole interactions. (We shall see in Chapter 6 that quadrupole core "vibrations" are the dominant low lying "collective" modes in nuclei.)

Thus, the parabolic rule is often an excellent approximation. Figure 4.15 shows a few examples taken from Paar and introduces one final but important point. The $(g_{7/2}d_{3/2})$ multiplet of $J = 2^+ - 5^+$ states in ^{122}Sb is well reproduced by a simple parabola in

$J(J+1)$, as is the $(g_{7/2}h_{11/2})$ multiplet. ^{48}Sc and ^{116}I also show multiplets with beautiful empirical parabolic behavior, except that they are inverted! The reason is well understood. In ^{48}Sc$_{27}$, the $(f_{7/2p}f_{7/2n})$ multiplet is really a particle–*hole* p–n configuration $(f_{7/2p}f_{7/2n}^7) \equiv (f_{7/2p}f_{7/2n}^{-1})$, as is the $(g_{9/2p}^{-1} h_{11/2n})$ configuration in $^{116}_{49}$In$_{67}$. In this case, the residual interaction has the *opposite* sign of a particle–particle or hole–hole multiplet. In other words, it is repulsive, and the J states with high p–n overlap (J_{max}, J_{min}) are *raised* in energy, while J_n is lowered the most. We note that this change in sign is not always the case; in Chapter 5 we will see it as a characteristic of even multipole interactions such as the quadrupole interaction we are considering here. Interestingly, it does not apply to the δ-interaction (even though we have discussed its expansion in even multipoles). For odd multipole interactions ΔE(p–p) or ΔE(h–h) is identical to ΔE(p–h).

In Chapter 5, we will discuss at length the pairing interaction and the concept of the quasi-particle, which is a state only partially occupied, neither fully particle nor fully hole. Anticipating that discussion, the *occupancy* of an orbit is given by a probability, denoted V_j^2 (the number of particles in the orbit is $(2j + 1)V_j^2$). The orbit *emptiness* is U_j^2 and $U_j^2 + V_j^2 = 1$. Thus, we can rewrite Eq. 4.28 for the general situation of quasi-particles as

$$\Delta E_{qp}(j_1 j_2 J) \sim \Delta E(j_1 j_2 J)(U_1^2 - V_1^2)(U_2^2 - V_2^2) \qquad (4.30)$$

Thus, for a given proton number (particle "1"), $(U_1^2 - V_1^2)$ is constant but $(U_2^2 - V_2^2)$ changes from $+1$ to -1 as the neutron orbit ("2") is filled over a sequence of isotopes. Therefore, the parabolic rule actually refers to a whole family of possible parabolas for given j_1, j_2, ranging from bowl-shaped (p–h case) to nearly flat to a convex upward parabola (p–p or h–h cases). The theoretical shapes, for several values of $V_{h11/2n}^2$ in a $(g_{9/2p}^{-1}h_{11/2n})$ multiplet are illustrated in Fig. 4.16.

The fact that different interactions have different multipole composition can sometimes be used to gain information on the nature of the effective interaction. This has been extensively pursued by Schiffer and Molinari and co-workers, who, for example, deduced effective multipole coefficients applicable to broad ranges of nuclei, and by Heyde and co-workers who studied Gaussian and other interactions. In general, stronger quadrupole components are required than given by a δ-force, signaling the need for "finite" range interactions.

4.5   Implications (low-lying spectra of even–even nuclei, hole, and particle–hole configurations)

In closing the discussion of two-particle systems, we note a few other important results. Consider the case of two identical particles in equivalent orbits and an odd-tensor interaction. Only odd k terms appear in Eq. 4.24 in this case. The total summed interaction energy ΔE_{tot} is given in general by

$$\Delta E_{tot} = \sum_J (2J + 1)\Delta E(j^2 J) \qquad (J \text{ even}) \qquad (4.31)$$

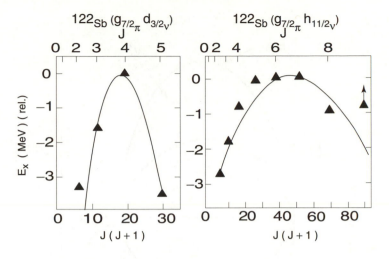

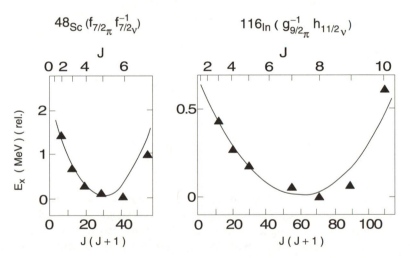

FIG. 4.15. Illustrations of the parabolic rule for a quadrupole residual interaction for several p–n multiplets (Paar, 1979).

From the properties of sums over $6 - j$ symbols, without ever considering the radial functions F^k, we find that by restricting ourselves to *odd-tensor interactions*,

$$\Delta E_{\text{tot}} = \sum_{J \text{ even}} (2J + 1)\Delta E(j^2 J) = \frac{2j + 1}{2}\Delta E(j^2 J = 0) \qquad (4.32)$$

That is, the total shift of all the levels in a configuration of two identical particles in the same orbit is simply related just to the shift of the $J = 0$ level, for any odd-tensor interaction. Since the δ-function interaction is equivalent to an odd-tensor interaction,

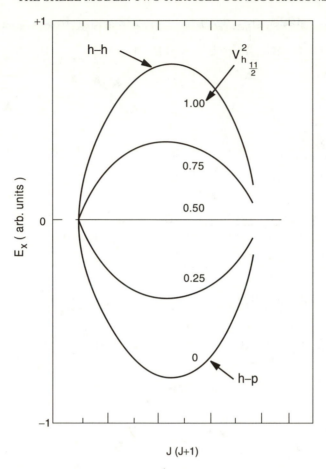

FIG. 4.16. Dependence of the shape of $(g_{9/2p}^{-1}h_{11/2n})$ multiplet parabolic splittings as a function of the occupancy. The uppermost curve corresponds to a nearly full $h_{11/2n}$ orbit and therefore a hole–hole configuration. The lowest (h–p) curve corresponds to an $h_{11/2n}$ orbit with only a few particles and therefore a hole–particle configuration (based on Van Maldeghem, 1985).

it can be expected that Eq. 4.31 will be approximately satisfied in real nuclei. Finally, we re-emphasize that it is the separation into angular and radial coordinates that allows results that are totally independent of the details of the interaction (except, in this case, for the assumption of odd-tensors).

Thus far, we have restricted the discussion to two valence particles. Multiparticle configurations will be dealt with shortly. It is useful, however, to consider one simple result that allows a nice generalization of the present discussion. Using the formalism of the seniority scheme (see Chapter 5) it is possible to relate the matrix elements of any two-body operator in the j^n configuration to matrix elements in configurations with fewer particles. For the large class of interactions that conserve seniority (including a δ-function that is a prototype of short-range interactions) it can be shown that in a multiparticle, j^n configuration, the more nucleons that are paired off to $J = 0$, the lower-lying the state will be. Thus, for n even, the lowest state of the j^n configuration is a $J^\pi = 0^+$ state with all nucleons paired to $J = 0$. (The $\Delta E(j_1 j_2 J)$ for the levels of the configuration $|j^2 J\rangle$ is just a special case of this). This result is sometimes called the pairing property, and is characteristic of any interaction that conserves seniority. Another key result is that the order and spacing of the $J \neq 0$ states in the j^n configuration with two unpaired particles will be exactly the same as in the j^2 configuration. The more realistic situation in which the valence nucleons occupy several j orbits, $j_1, j_2, \ldots$ in a multiparticle configuration $|j_1^{n_1}, j_1^{n_2}, \ldots, J\rangle$ with n_i even, is just a generalization. By the first result, the $J = 0$ combination will lie lower than one in which a pair of nucleons is coupled to $J \neq 0$. Thus we can generalize these conclusions and state that the ground state of even–even nuclei will always be 0^+. Moreover, since the odd J levels of $|j^n J\rangle$ are unaffected by a δ-interaction (n even), the low-lying *positive* parity states should all have *even* spin and should increase in energy with spin. Remarkably, in view of the simplicity of the argument, this is an almost universally observed situation.

Furthermore, we noted that in $\pi = -$ two-particle configurations of identical particles $|j_1 j_2 J\rangle$, it is the odd-spin levels that are lowered. A simple generalization using the seniority scheme shows that the lowest-lying *negative* parity levels in even–even nuclei should have *odd* spin. This is also almost always observed, and is known as the Talmi–Glaubman rule.

One can go even further. Consider an arbitrary shell in heavy nuclei consisting of several states (j orbits) from one shell and one unique parity orbit from the next shell. Recall from Fig. 3.2 that the highest normal parity l value, l_{max}, in each succeeding major shell increases by one. The corresponding highest j value is $j_{max} = l_{max} - 1/2$. (The $j = l_{max} + 1/2$ is the orbit brought down by the spin-orbit interaction into the next lower shell.) The next highest j is $l_{max} - 3/2$. The unique parity orbit is the $l + 1/2$ coupling from the next highest shell, and so has $j_{unique} = l_{max} + 1 + 1/2 = l_{max} + 3/2$. For example, in the 50–82 shell, $l_{max} = 4$, so the highest j normal parity orbits are the $g_{7/2}, d_{5/2}$ and the unique parity orbit is $h_{11/2}$. In the $g_{7/2} - h_{11/2}$ configuration, our rules (see Table 4.2) show that the $J = j_1 + j_2 = 7/2 + 11/2 = 9$ state is brought lowest and the 2^- is high-lying at the unperturbed energy. In the $d_{5/2} - h_{11/2}$ case, the $J^\pi = 3^-$ is lowest. A little thought shows that this relationship of j values is generally true for any shell in heavy nuclei. Therefore, we can obtain, without calculation, two general predictions. First, the lowest-*spin*, negative parity, two-particle state in an even–even nucleus will be a 2^- level: there is no simple spin combination that gives a 1^- level. However, since this 2^- level always results from a $j = l - 1/2$ normal parity orbit coupled antiparallel to a $j = l' + 1/2$ unique parity orbit, it will always occur rather high in energy. Therefore, the lowest-*lying*, low-spin, negative parity state will be the 3^- level. This prediction is borne out in nearly all heavy even–even nondoubly magic nuclei (in closed shell nuclei two *different* shells are involved in $\pi = -$ excitations). Examples were shown in Figs. 2.5 (^{210}Pb) and 2.6 (Sn). The main exceptions to all these predictions are themselves illuminating: they concern nonspherical, or deformed, nuclei in which, to borrow terminology from a later chapter, one sometimes encounters two-quasi-particle, negative parity excitations involving a unique parity Nilsson orbit (e.g., the low-lying 4^- state in ^{168}Er). But, as is well known, the nonspherical character of these nuclei is itself induced by a strong residual quadrupole interaction: being an even rank tensor, it does not conserve seniority and therefore, a priori, one would not expect the preceding arguments to apply.

A final topic to deal with before considering more complex multiparticle systems is the spectra of hole and particle–hole configurations, where, by the former, we mean those of the type $(j^{-n}) = (j^{2j+1-n})$. We have briefly mentioned these in our discussion of multipole forces. Here, we obtain a few more explicit results.

For hole states, a diagonal matrix element of any *single particle* operator acting on an n-hole state $|j^{-n} JM\rangle$ will be equal in magnitude to its value in the n-particle system $|j^n JM\rangle$. The sign relation will depend on the odd- or even-tensor character of the operator. This is easy to see by considering all the possible m states of the n-hole configuration. There will be n unoccupied m states. The expectation value of a single-particle operator $\mathbf{O}_m$ (which we denote $\mathbf{O}$ for simplicity) will be the sum of its expectation values over all the particles:

$$\left\langle m_i^{-n} \left| \sum_{i=1}^{2j+1-n} \mathbf{O} \right| m_i^{-n} \right\rangle$$

where m^{-n} simply indicates the n-hole state, but where the calculation is carried out over the $2j + 1 - n$ particles. This is clearly equal to

$$\left\langle m_i \left| \sum_{i=1}^{2j+1} \mathbf{O} \right| m_i \right\rangle - \left\langle m_i^n \left| \sum_{i=2j+1-n}^{2j+1} \mathbf{O} \right| m_i^n \right\rangle$$

that is, equal to the sum over the entire shell minus that for the n missing particles. But, we have shown above that the first term must vanish since the closed j shell can have no preferred direction in space and, therefore

$$\left\langle m_i^{-n} \left| \sum_{i=1}^{2j+1-n} \mathbf{O} \right| m_i^{-n} \right\rangle = -\left\langle m_i^n \left| \sum_{i=1}^{n} \mathbf{O} \right| m_i^n \right\rangle$$

Note that the n-particle and n-hole states are not exactly equivalent. They have different total M values. For n particles in a given j state occupying m states $m_1, m_2, \ldots m_n$, $M_p = {}^n\sum_{i=1} m_i$. For n holes in the same orbit, one clearly must have $M_h = -M_p$ (e.g., $1/2 + 3/2 = -[5/2 + (-1/2) + (-3/2) + (-5/2)]$). So the above matrix element can be written (in simplified notation)

$$\langle J, -M|\mathbf{O}|J, -M\rangle_h = -\langle JM|\mathbf{O}|JM\rangle_p$$

where the subscripts indicate the n-hole and n-particle configurations. Using the Wigner Eckart theorem for a tensor operator $\mathbf{O}^k$ of rank k gives

$$\langle J, -M | \mathbf{O}^k | J, -M \rangle = (-1)^k \langle J, M | \mathbf{O}^k | J, M \rangle$$

and hence we can now relate the matrix elements of any single-particle operator in states of the *same JM* for n-hole and n-particle configurations:

$$\langle j^{-n} J | \mathbf{O}^k | j^{-n} J \rangle = (-1)^{k+1} \langle j^n J | \mathbf{O}^k | j^n J \rangle \quad (k \neq 0) \tag{4.33}$$

This gives the critical result that the expectation value of any *odd-tensor* single-particle operator (e.g., magnetic moment) in an *n-hole* state is the same as the corresponding *n-particle* expectation value, while for even rank tensors (e.g., quadrupole moment), the two expectation values are the negatives of each other.

In Chapter 5 we shall encounter more general results of this special case. This has immediate consequences of great importance. It implies, for example, that magnetic moments or dipole transition rates are the same for corresponding particle and hole configurations, while quadrupole moments or E2 matrix elements change sign. Therefore, such matrix elements must vanish at mid-shell. Of course, these features are well known empirically, although they are partly obscured by configuration mixing effects.

A nice extension of this is to the interaction energies in particle–hole configurations relative to those in particle–particle ones. Compare a proton–neutron particle–particle configuration with a proton hole–neutron particle configuration. The interaction is a product of proton and neutron tensor operators. The effect of the neutron operator on the neutron wave functions (a particle in both cases) is the same for both configurations. Only the proton operator acts on different configurations (particle in one case, hole in the other) and the above results for one-body operators then apply. Thus, for an odd-tensor interaction (product of odd-tensor one-body operators), the energy shifts $\Delta E(j_p^{-1} j_n J) = \Delta E(j_p j_n J)$ while, for an even-tensor interaction, $\Delta E(j_p^{-1} j_n J) = -\Delta E(j_p j_n J)$. The U, V dependence of a quadrupole force leading to the up-and-down pointing parabolas for p–p (or h–h) and p–h configurations in Eq. 4.30 is an example of this sign change. For the more general case of a mixed-interaction, the results are slightly more complicated and are given by

$$\Delta E(j_p^{-1} j_n J M) = \sum_{J'} (2J' + 1) \begin{Bmatrix} j_p & j_n & J \\ j_p & j_n & J' \end{Bmatrix} \Delta E(j_p j_n J' M) \tag{4.34}$$

Note that this gives the energy *differences* for different J values. The energies themselves can have a constant additive term that is J-independent (but that can depend on j). Thus, while Eq. 4.34, or the simpler relations for odd- and even-tensor interactions, relates p–h and p–p spectra for a *given j* shell, it does not relate such spectra for different j orbits in a major shell.

Since, physically, any very short-range interaction cannot give results qualitatively different from that of a δ-function (equivalent to an odd-tensor interaction), one expects that the odd-tensor result will be closer to that observed empirically. In other words, the energies (order and spacing) will be identical in the p–h and p–p configurations. Nevertheless, Eq. 4.34 is very important because it is so general. It is valid for any two-body interaction in a jj-coupling scheme. Moreover, it has a practical use from a different, empirical, point of view. Without knowing anything about the actual interaction, the knowledge of the order and spacings in a p–p configuration allows one to predict those in the corresponding p–h configuration. A classic example of this is the comparison of $^{38}_{17}\text{Cl}_{21}$ and $^{40}_{19}\text{K}_{21}$: the former is expected to be a $(d_{3/2p}f_{7/2n})$ configuration, while the latter is $(d^3_{3/2p}f_{7/2n}) = (d^{-1}_{3/2p}f_{7/2n})$. The comparison is included in Fig. 4.7, where the empirical ^{40}K energies were used to calculate the level energies in ^{38}Cl. The agreement is excellent.

Finally, we consider the interaction energies of the $(j^{-n}J)$ configuration relative to those in the $(j^n J)$ configuration. By arguments similar to the preceding in the m-scheme, it is easy to show that the *interaction energy* for any two-body operator in an n-hole state will be equal to that in the n-particle state plus an additive constant. Therefore all *spacings* in n-hole configurations $|j^{-n}J\rangle$ should be identical to those in the corresponding n-particle states $|j^n J\rangle$. Note an important point analogous to a comment we stressed in regard to the effect of closed shells on open shells in Chapter 3. As used here, the term shell refers to a single j shell, not a *major* shell consisting, generally, of several j values. The additive constant relating n-hole to n-particle configuration need not be independent of j. Therefore, the full set of spacings in a configuration such as $|j_1^{-n_1} j_2^{-n_2} \ldots J\rangle$ will not generally equal those in $|j_1^{n_1} j_2^{n_2} \ldots J\rangle$. The equality of spacings applies to each j, individually. If the realistic wave functions contain admixtures of several j values, there is no simple general relation of particle and hole energies.

A final point: thus far we have only discussed diagonal matrix elements (energy shifts) due to a δ-interaction, but off diagonal effects are equally important. These, of course, give rise to the mixed wave functions characteristic of realistic shell model calculations. Expressions for the off-diagonal matrix elements $\langle j_1 j_2 |\delta| j_3 j_4 \rangle$ are similar to Eq. 4.7, except there is now a second $3 - j$ symbol involving j_3, j_4. The strength of a given matrix element then depends on the angular correlation of the two particles in both initial and final states, as we shall as on the radial overlaps.

It is worthwhile giving a perspective on what we have done in this chapter. We first considered a δ-function residual interaction in a rather quantitative way, for both like and unlike particles. Then, we noted the geometrical relation of the orbits in two-body interactions. With a corresponding geometrical understanding of the force between these two particles, it is then trivial to deduce (intuitively or quantitatively) the effects of that force on the various total angular momenta J. We did this in detail for the δ-interaction, recovering the earlier results. We then developed the idea of a multipole expansion of any interaction and showed how it can be used to estimate the effects of an arbitrary two-body residual interaction between nucleons in any pair of single particle orbits. All this was carried out for two-particle configurations. Now we turn to see how these results can be used for the multiparticle case.

5

MULTIPARTICLE CONFIGURATIONS

We now turn to a more systematic treatment of multiparticle ($n > 2$) configurations of valence nucleons. We have already anticipated a few results from this discussion in Chapter 4. Without having done so, the two-particle results would have been completely useless. We needed to understand, for example, the result that closed shells could be ignored in considering residual interactions, along with some of the results relating p–h to p–p spectra. This allowed us to apply the two-particle discussion to the case of two valence particles or holes relative to any closed shell. We also quoted the enormously important conclusion that (if seniority is conserved) many predictions for $|j^n J\rangle$ configurations are identical to those for $|j^2 J\rangle$ configurations. (The requirement of seniority conservation basically means that the other $(n-2)$-particles couple pairwise to $J_{n-2} = 0$.) Thus the two-particle results could be used for nearly all spherical nuclei.

5.1 *J* values in multiparticle configurations: the *m* scheme

Clearly, our first task in the systematic study of multiparticle systems is to consider which values of the total angular momenta J are permissible for an n-particle system. The principal consideration here is, of course, the restrictions imposed by the Pauli principle. There are several ways of approaching this issue. The most physically transparent and easy to use is the so-called m-scheme, to which we now turn. Let us start by working out an explicit case. Consider three identical nucleons in a $d_{5/2}$ orbit. They cannot couple to a $J = 15/2$, since such a state must have an $M = 15/2$ substate. The only way such a substate can be made is by placing all three particles in $m = 5/2$ substates, violating the Pauli principle. Similarly, $J = 13/2$ or $11/2$ states are impossible since the former would require two of the particles to be in $m = 5/2$ substates and $J = 11/2$ would require two particles in $m = 3/2$ states. However, a $J = 9/2$ state is indeed possible since it can be formed by placing three particles in the states $j_1 m_1 = 5/2, 5/2, j_2 m_2 = 5/2, 3/2, j_3 m_3 = 5/2, 1/2$.

It is easy to see that a slight generalization of this example gives the following result: in the configuration j^n the maximum angular momentum is given by

$$J_{\max} = nj - \sum_{k=2}^{n} (k-1) = nj - \frac{n(n-1)}{2}$$

For the $(d_{5/2})^3$ case just discussed, this gives $J_{\max} = 3j - 3 = 9/2$.

Thus, we see an example of how a consideration of the possible m substate occupations can give information on permissible total angular momenta J. Basically, the m-scheme is a systematic set of procedures for doing this in a general case.

The m-scheme is best described with a detailed example. Earlier we showed that only even J values were allowed in the j^2 configuration of identical nucleons. This was argued in terms of the symmetry properties of spherical harmonics. However, it can also be seen by inspecting the possible magnetic substates. Table 5.1 summarizes the allowed magnetic substates for a two-particle configuration $(7/2)^2$. (Actually, the $M < 0$ cases are omitted since they are completely symmetric to the $M > 0$ cases.) We construct such a table starting with the highest magnetic substate for particle 1 and list all of the possible substates for particle 2 allowed by the Pauli principle, then carrying out the same procedure for the next lower magnetic substate (in this case 5/2) for the first particle. It is necessary to recall that the two nucleons are indistinguishable so a combination 5/2, 7/2, for example, is not allowed, since the 7/2 and 5/2 combination has already been listed. Continuing in this way, we obtain all of the possible m values for the two–particle system. Since a given total angular momentum J must have magnetic substates $M = J, J-1, \ldots -(J-1), -J$, it is now trivial to deduce the possible final J values using the table. In the present example, there must be a $J = 6$ state, since there is a $M = 6$ configuration. This $J = 6$ state has magnetic substates $M = 6, 5, 4, 3, 2, 1, 0$, and the corresponding negative values. Thus the top seven magnetic substates listed in the table must be used up for this single $J = 6$ state. There is no $M = 5$ magnetic substate left over and thus there cannot be a $J = 5$ state. There must, however, be a $J = 4$ configuration to consume the $M = 4, 3, 2, 1, 0$ magnetic substates listed next in the table. By continuing this argument, one sees that there is no $J = 3$ state, but there are $J = 2$ and $J = 0$ states, thus proving in a different way the result obtained earlier that two identical particles in the same orbit can couple only to even total angular momenta J.

The case for a $(5/2)^3$ configuration is shown in Table 5.2, which will not be discussed in detail although the reader may go through the example and verify the results just as we did for the $(7/2)^2$ case. Clearly, for multinucleon configurations where n is large, this procedure can be lengthy. Other techniques are available. However, the m-scheme is important because it shows in a transparent way how the physical effects of the Pauli principle arise.

We noted above that the m-scheme gives a rule for the maximum permissible J value in a j^n configuration of identical particles very simply. It also gives, in an equally simple way, the result that a j^n configuration can never have a state with $J = J_{max} - 1$. We will show this only for the case of two particles, but the generalization is straightforward. (Although the following considerations are general, reference to the specific example in Table 5.1 will clarify the arguments.) The maximum J in a j^2 configuration is $J_{max} = 2j - 1$. A state $J = J_{max} - 1$ would have (if it existed) $J = 2(j - 1)$. One value of $M = J_{max} - 1$ must be used for the $J = J_{max}$ state. Therefore, in order to have a state with $J = J_{max} - 1 = 2(j - 1)$, there must be a second permissible $M = J_{max} - 1 = 2(j - 1)$ state. This cannot involve a particle in an $m = j$ state, since that state is already consumed for the $J = J_{max}$ level. Therefore, the only way to make another magnetic substate $M = 2(j - 1)$ is to have two particles with $m = j - 1$. But this violates the Pauli principle, and therefore is impossible, proving that a $J = J_{max} - 1$ state never exists in a j^2 configuration. As noted, this can be generalized to the j^n configuration.

Table 5.1 *m scheme for the configuration* $|(7/2)^2 J\rangle^*$

$j_1 = 7/2$ m_1	$j_2 = 7/2$ m_2	M	J
7/2	5/2	6	
7/2	3/2	5	
7/2	1/2	4	
7/2	−1/2	3	6
7/2	−3/2	2	
7/2	−5/2	1	
7/2	−7/2	0	
5/2	3/2	4	
5/2	1/2	3	
5/2	−1/2	2	4
5/2	−3/2	1	
5/2	−5/2	0	
3/2	1/2	2	
3/2	−1/2	1	2
3/2	−3/2	0	
1/2	−1/2	0	0

* Only positive total M values are shown. The table is symmetric for $M < 0$.

Table 5.2 *m scheme for the configuration* $|(5/2)^3 J\rangle^*$

$j_1 = 5/2$ m_1	$j_2 = 5/2$ m_2	$j_3 = 5/2$ m_3	M	J
5/2	3/2	1/2	9/2	
5/2	3/2	−1/2	7/2	
5/2	3/2	−3/2	5/2	
5/2	3/2	−5/2	3/2	
5/2	1/2	−1/2	5/2	
5/2	1/2	−3/2	3/2	5/2 9/2
5/2	1/2	−5/2	1/2	
5/2	−1/2	−3/2	1/2	
3/2	1/2	−1/2	3/2	
3/2	1/2	−3/2	1/2	3/2

$$J = 9/2,\ 5/2,\ 3/2$$

* The full set of allowable m_i combinations that give $M > 0$ are obtained by the conditions $m_1 > 0$, $m_3 < m_2 < m_1$ and no two m_i values identical.

5.2 Coefficients of fractional parentage (CFP)

Now that we have a feeling for those J values that can be obtained for any multiparticle configuration, we can discuss the effects of various interactions on the energies of such J states and dynamic matrix elements involving such configurations.

Clearly, when considerations such as those discussed in Chapter 4 are attempted for multiparticle configurations, the situation rapidly becomes much more complex. To see this in a simple example, consider a $(d_{5/2})^3$ configuration of identical nucleons. A $J = 5/2$ state can be made in three distinct ways by first coupling two particles to an intermediate $J' = 0, 2, 4$:

$$\left[(d_{5/2})^2 J' = 0, j = 5/2 \right]^{\frac{5}{2}}$$

$$\left[(d_{5/2})^2 J' = 2, j = 5/2 \right]^{\frac{5}{2}}$$

and

$$\left[(d_{5/2})^4 J' = 4, j = 5/2 \right]^{\frac{5}{2}}$$

However, the $(5/2)^3$ configuration has only one $J = 5/2$ state (see Table 5.2). Its wave function must therefore be a totally antisymmetric linear combination of these three basis states. The normalized coefficients in this linear combination are called *coefficients of fractional parentage* (CFP): their squares give the probability that a given final state is constructed from a specific "parent" configuration—in this case, a two-particle state. The relative magnitudes of the three CFP's for the $(d_{5/2})^3$ configuration are not arbitrary, but are given by certain angular momentum coupling coefficients. CFP coefficients can be constructed not only for three-particle configurations, but for any n.

Unfortunately, the complexity of the possible couplings makes the notation for the coefficients rather complex and this has deterred many nuclear physicists from delving into the subject; the formalism can be terrifying. Formal textbooks can be filled with page after page of long, daunting expressions involving sequences of CFP coefficients, angular momentum coefficients, summations over them, and the like. Here, we attempt to cut through much of this by summarizing some of the essential results and their motivations with a few simple examples. This is not entirely satisfactory, since it deprives the reader of an appreciation of the beauty and power of the formalism. Moreover, as the author can personally attest, while a simple presentation of the final results saves the reader from the tedium of struggling through their derivation, it also confers on them an element of mystery—the reader is left with a sense of wonder at how one can start with such general interactions and general configurations and end up with very simple final results. He or she may glance back over the imposing derivations in the hopes of seeing where some simplifying assumption or some restrictive case has been invoked. The real power and beauty of the method, however, is that such assumptions are usually not required: very general results that enormously simplify the treatment of many-particle shell model configurations can often be obtained.

In any case, we will introduce the formal notation for CFP coefficients, but will avoid their manipulation as much as possible. We will derive one simple result that illustrates their power and economy.

Consider a configuration of identical nucleons in equivalent orbits, the $j^n \alpha J M$ state (α specifies any additional quantum numbers needed to describe the states). This n-particle state can be written in terms of $(n-2)$-particle wave functions by using the two-particle CFP coefficients (there are also analogously defined one-particle CFP coefficients). The CFP is denoted

$$\langle j^n J \{ | j^2 (J_2) j^{n-2} (J_{n-2}) J \rangle$$

and determines, as in the three-particle case, the probability that the wave function $|j^n JM\rangle$ can be written in terms of the $(n-2)$-particle configuration $|j^{n-2}(J_{n-2}M_{n-2})\rangle$ coupled to a two-particle configuration $|j^2 J_2 M_2\rangle$. The defining equation is therefore

$$|j^n JM\rangle = \sum_{J_2 J_{n-2}} \langle j^n J \{ | j^2 (J_2) j^{n-2} (J_{n-2}) J \rangle | j^2 (J_2) j^{n-2} (J_{n-2}) JM \rangle \qquad (5.1)$$

Clearly, then,

$$\sum_{J_2, J_{n-2}} (\text{CFP})^2 = 1$$

We can see how the concept of CFP coefficients and the parentage of n-particle configurations is useful. Consider a configuration $|j^n J\rangle$ and ask what the energy shifts, $\Delta E(j^n J)$, are for each final J value for an arbitrary interaction. (Note that, as usual, we drop the magnetic quantum numbers to simplify the notation.) First, we note that, since the particles are indistinguishable, the *total* interaction energy in any final state J is given simply by the interaction energy for any pair of nucleons (say, particles 1 and 2) times the total number of possible pairs $n(n-1)/2$. However, the two-particle matrix element V_{12} can only depend on J_2, and not on the way in which J_2 is coupled with J_{n-2} for the other $n-2$ nucleons to give the final J. The total interaction energy for particles 1 and 2 is just the sum of the interaction energies for each two-particle angular momentum J_2 multiplied by the probability of each J_2 in the state $|j^n J\rangle$. We denote this probability $W(j^n J J_2)$. Thus, we can immediately write the interaction energy from particles 1 and 2 in the state $|j^n J\rangle$ as:

$$\left\langle j^n J \left| \sum_{i>j}^{n} V_{ij} \right| j^n J \right\rangle = \sum_{J_2} W (j^n J J_2) \langle j^2 J_2 |V_{12}| j^2 J_2 \rangle \qquad (5.2)$$

where the W coefficient is the sum of the squares of the CFP coefficients for a given J_2 over all possible values J_{n-2}. That is,

$$W (j^n J J_2) = \sum_{J_{n-2}} \left| \langle j^n J \{ | j^2 (J_2) j^{n-2} (J_{n-2}) J \rangle \right|^2 \qquad (5.3)$$

The interesting point here is that there are in general *fewer* values of J_2 than there are of J. For example, in the $(7/2)^3$ configuration, J can be 15/2, 11/2, 9/2, 7/2, 5/2, 3/2 while $(7/2)^2$ can only couple to $J_2 = 0, 2, 4$, and 6. Thus, by Eq. 5.2, the six energies of the configuration $j^n J$ are given in terms of the four matrix elements

$$\langle j^2 J \,| V_{12}| \, j^2 J \rangle, \quad J = 0, 2, 4, 6$$

The beauty of this is that these matrix elements are usually easy to calculate for a known interaction and, even when the interaction is not known, empirical values for them can be obtained from the neighboring even–even nucleus (with $n = 2$). This can then be used to calculate the energy levels of the adjacent odd mass nucleus.

We have discussed the $(7/2)^3$ example here because it is treated in detail in de Shalit and Feshbach, where the low-lying $(f_{7/2})^3$ energy levels of ^{51}V are calculated in terms of the empirically known $(f_{7/2})^2$ levels of ^{50}Ti ($0^+ : 0, 2^+ : 1.55, 4^+ : 2.68, 6^+ : 3.2$ MeV). The results are shown in Fig. 5.1; the agreement is remarkably good for such a simple approach. Note once again that nowhere in this discussion has *any* aspect of the *interaction* been specified, except to assume that it is two-body only. We could also have calculated ^{51}V with the same formulas using a δ-function interaction to simulate ^{50}Ti, that is, to define the $(f_{7/2})^2$ matrix elements. Normalizing the δ-function strength to the 0–6 spacing in ^{50}Ti gives calculated ^{50}Ti energies of $0^+ : 0, 2^+ : 2.68, 4^+ : 3.0$, and $6^+ : 3.2$ MeV. These have a different distribution than the empirical levels and, when applied to ^{51}V, give the fit on the right of Fig. 5.1. Clearly, this approach is not nearly as successful. The point is that the *empirical* ^{50}Ti spectrum automatically includes *all* relevant interactions in the $(f_{7/2})^2$ system. The CFP techniques relate this directly to ^{51}V, independent of a knowledge or guess of the interaction. Thus, an understanding of the makeup of an n-particle configuration in terms of its $(n-2)$-particle structure can greatly simplify the treatment of nuclear spectra in complex systems. The present results can be generalized to $n > 3$, and provide comparable, and even greater, simplifications.

5.3 Multiparticle configurations j^n: the seniority scheme

When there are numerous particles outside closed shells, they can enter different shell model orbits. For example, in $_{40}^{99}$Zr$_{59}$ the nine valence neutrons might be in a configuration $(d_{5/2})^6(g_{7/2})^3$. Here, the $d_{5/2}$ shell is filled and the earlier arguments on the effect of closed shells on the values of $\Delta E(j^{2j_1+1} j^n{}_2 J)$ tell us that the $d_{5/2}$ orbit can be neglected, so this configuration is equivalent to $(g_{7/2})^3$. Now, consider ^{95}Zr. In this case, the lowest expected configuration would be $(d_{5/2})^5$. By the particle–hole equivalency discussed earlier, this is exactly equivalent to a single neutron in the $d_{5/2}$ orbit, leading to a one-state configuration with $J = j = 5/2$ and, indeed, the ground state of ^{95}Zr is $5/2^+$. However, one could also imagine excited states in ^{95}Zr of the form $(d_{5/2})^3(g_{7/2})^2$. Normally, at least near closed shells, such configurations are rather high-lying excited states: our primary interest is usually in the lowest-lying levels in which as many particles as possible are packed into the lowest accessible j value. Thus, at least in simple shell model treatments, one is frequently interested in j^n configurations. Moreover, even though realistic shell model calculations will often involve important components coupling two j values, an understanding of the single j case greatly helps to interpret and even anticipate such calculations. So far, we have ignored the possibility of both valence

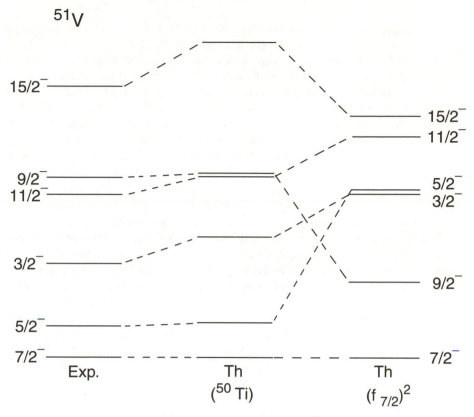

FIG. 5.1. Comparison of the low-lying empirical levels of ^{51}V with calculations obtained by coupling an $f_{7/2}$ proton to an $(f_{7/2})^2$ two-particle configuration (right) and by coupling an $f_{7/2}$ proton to the empirical levels of ^{50}Ti. (See deShalit, 1974.)

protons and neutrons. This clearly complicates the situation, as seen in the discussion earlier of the δ-function interaction for p–n systems. Moreover, as we shall see later, once one has nucleons of both types outside of closed shells, collective effects rapidly accumulate and other models provide alternate, and often better, approaches. Therefore, it is appropriate to again stress the j^n configuration of identical nucleons. Despite this restriction, the following considerations have extremely wide applicability.

The tendency of particles to pair to $J = 0^+$ leads to a scheme in which this property is explicitly recognized and exploited. Consider the j^n configuration. We ask what is the *smallest* value of n that can produce a given J value. Denoting this value by ν, it is clear that there can be no particles coupled in pairs to $J = 0$ in the configuration $j^\nu J$. (Otherwise, a $j^{\nu-2}$ configuration would have a spin J.) Such a state is then said to have *seniority* ν. From a configuration $j^{\nu+2}$ we can make a state of the same spin J by coupling one pair of particles to $J = 0$. This state is also said to have seniority ν. Physically, ν is simply the number of *unpaired* particles in a state of angular momentum

J in the configuration j^n. The number of paired particles is $(n - v)$ and the number of such pairs is $(n - v)/2$. For $v = 0$, all particles are paired and $J = 0$.

Let us further illustrate this concept with a simple example. Consider the $(f_{7/2})^4$ configuration. From the m-scheme and the simple formula derived earlier, $J_{max} = 4j - 4(3)/2 = 8$. This state can only be made by maximizing the alignment of all $j=7/2$ angular momenta as allowed by the Pauli principle. The $J = 8$ state therefore has seniority 4; there are no particles coupled in pairs to $J = 0$. On the other hand, $J = 2, 4$, and 6 states can be made by first coupling one pair of particles to $J = 0$ and then using the remaining $|(7/2)^2 J\rangle$ configuration to produce angular momenta of 2, 4, or 6. Such states have seniority $v = 2$. Finally, the $J = 0$ state of the $(f_{7/2})^4$ configuration obviously has seniority 0, that is, all particles are coupled in pairs to $J = 0$. (Note that there may be other $J = 0, 2, 4, 6$ states of the $(f_{7/2})^4$ configuration, all with $v = 4$.) What we have shown is that $J = 0, 2, 4, 6$ states of $v = 0$ or $v = 2$ can be constructed.

The seniority concept is important for several reasons. First, it leads to many simple, powerful results under very general conditions. For example, various interactions and matrix elements can be classified in terms of whether or not they conserve seniority. As will be seen, these matrix elements have very different properties as the number of particles in a shell increases. Secondly, and perhaps most importantly, it seems that many realistic residual interactions conserve seniority, so this scheme gives reasonable predictions for actual nuclei. It is impossible within the scope or philosophy of this book to derive all the results of the seniority scheme without adding an undesirable complexity. Such derivations are available in many detailed textbooks on the shell model. The complexity of these derivations often tends to obscure some of the simple ideas lying behind them. It is these ideas that we wish to emphasize here. We will derive or motivate a few crucial results; the others can be obtained by analogous, though more tedious, manipulations.

Perhaps the most important ingredient in understanding the results of the seniority scheme is the following: consider the j^2 configuration and the matrix element of any *odd tensor* interaction. (The introduction of the concept of tensors and their rank here should not be intimidating. The spherical harmonics of order k, Y_{km}, simply form the $2k + 1$ components of a tensor of rank k. An example of an odd rank tensor is the magnetic dipole operator. The quadrupole operator is an even rank tensor. As commented earlier, the δ-function interaction is equivalent to an odd-tensor interaction.)

For the case of a one-body odd-tensor operator acting in the j^2 configuration

$$\langle j^2 J \, |\mathbf{O}_k| \, j^2 J = 0\rangle = 0 \tag{5.4}$$

The proof of this is trivial. We recall that in the two-particle configuration only even J values are allowed. Therefore, J on the left side must be even and, by conservation of angular momentum, there is no way that $J = 0$ can be coupled to an even J by an operator carrying odd multipolarity.

Equation 5.4 simply states that *all* matrix elements of one-body odd-tensor operators vanish in the j^2 configuration. This includes the $J = 0$ case. Odd tensor operators cannot "break" a $J = 0$ coupled pair, nor can they contribute a diagonal "moment". The significance of this simple equation cannot be overemphasized.

In many-particle systems, it has three enormously important consequences. For such configurations, one-body operators are normally expressed in terms of sums over operators acting on each particle. A one-body odd-tensor operator acting in a j^n configuration is given by $\mathbf{U}^k = {}^n \sum_{i=1} \mathbf{U}_i^k$. Since an odd tensor operator cannot change *any* $\left| j^2 J = 0 \right\rangle$ pair to one with $J \neq 0$ (J even), odd-tensor operators *must conserve seniority*. Equation 5.4 shows that there is no contribution to $\mathbf{U}^k$ from such pairs of particles coupled to $J = 0$. Thus matrix elements of one-body odd-tensor operators in j^n configurations with seniority v, can be reduced to those in the j^v configuration. Moreover, they are independent of n (for $n \geq v$).

These results follow so trivially from Eq. 5.4 that the preceding comments essentially constitute a derivation. However, they are so important and basic that it is worthwhile to go through the arguments more explicitly. Consider a matrix element such as $\langle j^n v J' \left| {}^n \sum_{i=1} \mathbf{U}_i^k \right| j^n v J \rangle$. For $v < n$, the left side can be rewritten in terms of wave functions of the configuration $\left| j^{n-2} v(J') j^2 (J = 0) J' \right\rangle$ and similarly for the right side. For simplicity, we take the particles thus separated off as the $(n-1)$th and nth particles. Application of Eq. 5.4 to these two particles contributes nothing to the overall matrix element, and we can replace the operator ${}^n \sum_{i=1} \mathbf{U}_i^k$ with ${}^{n-2} \sum_{i=1} \mathbf{U}_i^k$ extending over n–2 particles. Since the matrix element is now independent of the last two particles, we can integrate over them. Since they are in the same state $\left| j^2 J = 0 \right\rangle$, this integral is unity by orthonormality. If $(n - v) \geq 4$, we can repeat this procedure for another pair of particles. We continue this procedure for any even v until we are dealing with an operator ${}^v \sum_{i=1} \mathbf{U}_i^k$ acting on the states $\left| j^v J \right\rangle$ and $\left| j^v J' \right\rangle$. Thus we obtain

$$\left\langle j^n v J' \left\| \sum_{i=1}^{n} \mathbf{U}_i^k \right\| j^n v J \right\rangle = \left\langle j^v v J' \left\| \sum_{i=1}^{v} \mathbf{U}_i^k \right\| j^v v J \right\rangle \tag{5.5}$$

which shows both the reduction of a matrix element in the j^n configuration to one in j^v and the independence of n.

The other result, conservation of seniority, is equally obtainable. Suppose the two wave functions in the above matrix element have different seniorities v, $v' < n$. There is some point in the successive reduction (the successive peeling off of pairs of particles) where an overlap integral over the wave functions $\left| j^2 J = 0 \right\rangle$ and $\left| j^2 J \neq 0 \right\rangle$ appears. Clearly, by orthogonality, this vanishes. To reiterate, we have the absolutely critical results:

- Odd-tensor single-particle operators conserve seniority in j^n configurations.
- The matrix elements of odd-tensor single-particle operators in j^n configurations in the seniority scheme can be reduced to ones in the j^v configuration.
- These matrix elements are independent of n.

These rather abstract results have many practical applications. They imply, for example, that the magnetic moment of the 7/2 state of an $(f_{7/2})^3$ configuration is identical to that in the single particle $f_{7/2}$ configuration: in general, magnetic moments in odd mass nuclei where the valence particles occupy a given j orbit should be independent of the (odd) number of valence nucleons. Similar arguments cannot be applied to even-tensor operators like the quadrupole operator. It turns out that these operators are not diagonal in the seniority scheme, but rather can connect states with seniorities ν and $\nu \pm 2$. Using arguments such as these, it is therefore clear why M1 transitions in even mass nuclei are rare—they can only connect states of the same seniority—while E2 transitions dominate even in near-closed shell nuclei. Therefore this dominance is not necessarily a demonstration of collectivity, but a reflection of the seniority structure of low-lying states in $|j^n J\rangle$ configurations.

Thus far in our discussion of seniority, we have considered single-particle operators representing moments or transitions. Equally important are two-body interactions, which can be either diagonal or nondiagonal. Both are important, although we will emphasize the former since they determine the contribution of residual interactions to level energies. A key example is the δ-function interaction. Clearly, interactions can be written as *products* of single-particle operators. We saw an example of this earlier in discussing multipole expansions of arbitrary interactions. We now turn to consider the properties of various interactions in the seniority scheme.

Consider an arbitrary *odd-tensor two*-body interaction V_{12}. This can be taken as a product of one-body operators, $\sum_{k\text{ odd}} f_1^k f_2^k$. As with one-body operators, it is extremely useful to be able to relate the two-body interaction matrix elements of seniority ν states in the j^n configuration (n even) to the matrix elements in a j^ν configuration. Deriving this desired result is trivial. Consider the matrix element (a subscripted k labels particles, not rank)

$$\left\langle j^n \nu \alpha J \left| \sum_{i<k}^n V_{ik} \right| j^n \nu \alpha' J \right\rangle$$

where the sum is over the n-particles, and where α and α' denote any additional quantum numbers needed. Since the states have seniority ν (even), there are $(n-\nu)$ particles paired off to $J = 0$. By the same reasoning that led to Eq. 5.4, the terms in $^n \sum_{i<k} V_{ik}$ that act on these particles cannot change their coupling. All that this part of the sum can do is contribute a diagonal matrix element of the form $\langle j^2 J = 0 | V_{ik} | j^2 J = 0\rangle$. But this is just the lowering of the 0^+ energy in a j^2 two-particle configuration. We define this energy lowering by V_0. The sum contributes this for each such $(J = 0)$-coupled pair, of which there are $(n - \nu)/2$. Having thus separated off these particles, we are left with a sum over ν particles of the same interaction. Thus, we obtain,

$$\left\langle j^n \nu \alpha J \left| \sum_{i<k}^n V_{ik} \right| j^n \nu \alpha' J \right\rangle = \left\langle j^\nu \nu \alpha J \left| \sum_{i<k}^\nu V_{ik} \right| j^\nu \nu \alpha' J \right\rangle + \frac{n-\nu}{2} V_0 \delta_{\alpha\alpha'} \qquad (5.6)$$

This interaction matrix element may be either diagonal or nondiagonal (in α), but it cannot change ν since it is of odd-tensor character. In either case, it is of *absolutely* central importance in nuclear spectroscopy. As with the case of one-body odd-tensor operators, we have an equation relating matrix elements of a two-body interaction in the j^n configuration to those in the j^ν configuration. Here, however these matrix elements are not constant across a shell, but *linear* in $(n - v)/2$, the number of nucleons paired off to $J = 0$. Such matrix elements *peak at midshell*. This feature is sometimes known as the *pairing property*.

To understand other important implications of this, let us first consider diagonal matrix elements where $\alpha = \alpha'$. The second term on the right in Eq. 5.6 is simply the number of pairs of particles coupled to $J = 0$ multiplied by the interaction energy, V_0, for each pair. Recalling that we are dealing with attractive residual interactions (larger matrix elements imply lower-lying states), then states with lower seniority v will lie lower in energy. The $v = 0$ states, which must have $J^\pi = 0^+$, will lie lowest. Immediately, this accounts for the well-known empirical property that the ground states of (spherical) even–even nuclei all have $J^\pi = 0^+$.

Similarly for odd mass spherical nuclei, the ground state will usually be a $v = 1$, $J = j$ state in which all but one nucleon is paired off in $|j^2 J = 0\rangle$ combinations.

It is worthwhile to explicitly write Eq. 5.6 for a j^n configuration in the $v = 0$, $J = 0$, and $v = 1$, $J = j$ states. For both situations the first term vanishes since there cannot be a two-body interaction in a $j^{v=0}$ (no particle) or $j^{v=1}$ (one particle) system. Therefore, the energies are given by the second term:

$$\left\langle j^n J = 0 \left| \sum_{i<k}^n V_{ik} \right| j^n J = 0 \right\rangle = \frac{n}{2} V_0 \qquad (n \text{ even}, v = 0) \qquad (5.7)$$

$$\left\langle j^n J = j \left| \sum_{i<k}^n V_{ik} \right| j^n J = j \right\rangle = \frac{n-1}{2} V_0 \qquad (n \text{ odd}, v = 1) \qquad (5.8)$$

These equations simply state that the ground state energies in the respective systems depend solely on the numbers of pairs of particles coupled to $J = 0$. In the odd particle case, the unpaired nucleon is, from this point of view, just a spectator. Indeed, as de Shalit and Feshbach emphasize, the nuclear force effectively *measures* the number of pairs of particles coupled to $J = 0$, at least insofar as it can be approximated by odd tensor interactions.

One of the most crucial uses of Eqs. 5.6 and 5.7 concerns the energies of seniority $v = 2$ states (the following argument applies to higher seniority states as well, but these are less often identified experimentally). Let us consider the energy difference $E(j^n v = 2, J) - E(j^n v = 0, J = 0)$. Simplifying the notation by denoting the interaction by V, Eq. 5.6 and 5.7 give

$$E\left(j^n, v = 2, J\right) - E\left(j^n, v = 0, J = 0\right) = \left\langle j^2 J |V| j^2 J \right\rangle + \frac{n-2}{2} V_0 - \frac{n}{2} V_0$$

$$= \left\langle j^2 J |V| j^2 J \right\rangle - V_0 \qquad (5.9)$$

Therefore, the energies of the $v = 2$ states are independent of n. Let us also calculate the *spacings* within the $v = 2$ configuration. These are given by

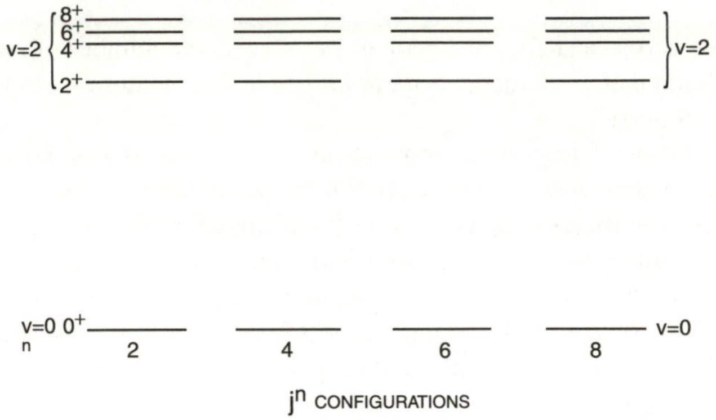

FIG. 5.2. Schematic illustration of the constancy of $\nu = 2$ levels in j^n configurations.

$$
\begin{aligned}
E\left(j^n, \nu = 2, J\right) - E\left(j^n, \nu = 2, J'\right) &= \left[\left\langle j^2 J | V | j^2 J\right\rangle + \frac{n-2}{2} V_0\right] \\
&\quad - \left[\left\langle j^2 J' | V | j^2 J'\right\rangle + \frac{n-2}{2} V_0\right] \\
&= \left\langle j^2 J | V | j^2 J\right\rangle - \left\langle j^2 J' | V | j^2 J'\right\rangle \\
&= E\left(j^2, \nu = 2, J\right) - E\left(j^2, \nu = 2, J'\right)
\end{aligned}
\tag{5.10}
$$

Thus, *all* energy differences of seniority $\nu = 0$ and $\nu = 2$ states in the n-particle configuration are *identical* to those in the two-particle system and are independent of n. This is illustrated in Fig. 5.2. This result is crucial because its absence would make it virtually impossible to apply the shell model in a simple way to nuclei other than those within one or two nucleons of closed shells. Indeed, this result was anticipated in the previous chapter in arguing that the shell model has broad applicability. The low-lying levels of good seniority in a j^n configuration are independent of n. In practice, more than one orbit j will be occupied by the valence nucleons. Nevertheless, the present result can be approximately generalized if one writes the wave function in the schematic form

$$
\psi = \sum_i \alpha_i \left| j_i^{n_i}\right\rangle
$$

In fact, the incorporation of such two-body configuration mixing is essentially equivalent to a modification of the interaction itself, and thus Eq. 5.6 is widely applicable. The Sn nuclei (see Fig. 2.6) provide a classic example of Eq. 5.10 and its generalization to the multi-j case: the entire known set of $\nu = 2$ levels, $J = 0^+, 2^+, 4^+, 6^+$, is virtually constant across an entire major shell. The Ca isotopes (Fig. 2.3) provide another example that will be discussed later.

To recapitulate, for the matrix elements of *odd*-tensor operators and interactions between states of good seniority in j^n configurations:

- One-body *matrix elements*(e.g., dipole moments) are independent of n and therefore constant across a j shell.   (5.11)
- Two-body *interactions* are linear in the number of paired particles, $(n - v)/2$, peaking at midshell.   (5.12)

It is possible to derive analogous results for even-tensor operators and interactions. The derivations involve the manipulation of, and recursion relations for, CFP coefficients. These are tedious but straightforward. We will cite two important results. Once again, we look first at one-body operators. As noted, even-tensor operators do not necessarily conserve seniority and can link states with $\Delta v = \pm 2$. The expression for such matrix elements is given by:

$$\left\langle j^n v \alpha J \left\| \sum_{i=1}^{n} \mathbf{U}_i^k \right\| j^n v - 2, \alpha' J' \right\rangle = \left[\frac{(n - v + 2)(2j + 3 - n - v)}{2(2j + 3 - 2v)} \right]^{\frac{1}{2}}$$

$$\times \left\langle j^v v \alpha J \left\| \sum_{i=1}^{v} \mathbf{U}_i^k \right\| j^v v - 2, \alpha' J' \right\rangle \qquad (5.13)$$

Once again, the power of the seniority scheme allows us to link matrix elements in the configuration j^n to those in the configuration j^v. The square of Eq. 5.13 gives the behavior of the transition rates induced by the operator $\mathbf{U}$ throughout a shell. For large j and $n(j, n >> v)$, this transition probability goes as $(f(1-f))$ where $f = n/(2j+1)$ is the fractional filling of the shell. This expression at first increases as f, then flattens out, peaking at midshell. Moreover, it is clearly symmetric about the midshell point. Probably the most common and important application of this concerns E2 transition rates induced by the operator $\mathbf{Q} = r^2 Y_2$. The important quantity $\langle 2_1^+ ||\mathbf{Q}||0_1^+ \rangle^2$ is proportional to the E2 transition rate from the first 2^+ state to the ground state in an even–even nucleus, and can be written for the j^n configuration as [assuming the $2_1^+(0_1^+)$ state has $v = 2(v = 0)$]:

$$\langle j^n J = 2_1^+ \|\mathbf{Q}\| j^n J = 0_1^+ \rangle^2 = \left(\frac{n(2j + 1 - n)}{2(2j - 1)} \right) \langle j^2 J = 2_1^+ \|\mathbf{Q}\| j^2 J = 0_1^+ \rangle^2$$

$$(5.14)$$

For shells that are not too filled, so that $(2j \pm 1) >> n$, this becomes

$$\langle j^n J = 2_1^+ \|\mathbf{Q}\| j^n J = 0_1^+ \rangle^2 \approx \frac{n}{2} \langle j^2 J = 2_1^+ \|\mathbf{Q}\| j^2 J = 0_1^+ \rangle^2 \qquad (5.15)$$

That is, in the j^n configuration, the B(E2) value, defined as

$$B\left(E2 : J_i \to J_f\right) = \frac{1}{2J_i + 1} \langle J_f \|\mathbf{Q}\| J_i \rangle^2$$

is just proportional to the number of particles n in the shell, for small n. For large $n, n \rightarrow 2j + 1$, it falls off, vanishing, as it must, at the closed shell. For $j, n >> 2$, we see that, as given in the general case above,

$$B\left(E2 : 2_1^+ \rightarrow 0_1^+\right) \propto n\left(1 - \frac{n}{2j}\right) \approx f(1 - f) \tag{5.16}$$

This behavior is commonly observed in real nuclei, with B(E2: $2_1^+ \rightarrow 0_1^+$) values rising to midshell and falling thereafter. Data beautifully illustrating this are shown for the $Z = 50$ to 82, $N = 82$ to 126 region in Fig. 5.3. (The peak regions of the B(E2) values in Fig. 2.16 are additional examples of this in condensed form.) In part, this behavior is due to coherent effects involving single-particle configuration mixing of different j values in the wave functions for each particle, but the overall behavior still reflects a generalization of this simple result for the seniority scheme.

For transitions induced by even-tensor operators of rank $k > 0$ that do *not* change seniority, the expression corresponding to Eq. 5.13 is

$$\left\langle j^n v\alpha J \left\| \sum_{i=1}^n \mathbf{U}_i^k \right\| j^n v\alpha' J' \right\rangle = \left(\frac{2j + 1 - 2n}{2j + 1 - 2v}\right) \left\langle j^v v\alpha J \left\| \sum_{i=1}^v \mathbf{U}_i^k \right\| j^v v\alpha' J' \right\rangle \tag{5.17}$$

This equation again expresses an n-particle matrix element for states of seniority v in terms of the v-particle matrix element. It has an interesting behavior as a function of n, as given by the factor outside the matrix element. In terms of f (the fractional filling of the shell), for low v the numerator goes simply as $(1 - 2f)$. It therefore has opposite signs in the first and second halves of the shell and hence must vanish identically at midshell. This is, of course, an extremely important result, indicating that, for example, quadrupole moments of j^n configurations in even–even nuclei change sign in midshell. The generalization to many-j shells suggests that such moments should have opposite signs at the beginning and end of major shells. Although the trends in realistic cases are complicated by the different j shell degeneracies, this qualitative feature is a well-known empirical characteristic of heavy nuclei, and it contrasts markedly with that for odd-tensor operators that are independent of n (see Eq. 5.5).

We shall see another important application of Eq. 5.17 in our discussion of the p–n interaction in Chapter 6. Moreover, the decrease toward midshell and symmetry about that point will see important reflections even in deformed nuclei (systematics of β: Fig. 6.11) where seniority is strongly broken.

Finally, we turn to two-body interactions for even-tensor operators. Some of these interactions can change seniority, connecting states with v and $v - 2$. For this case, the result is trivial to derive. An even-tensor two-body interaction connecting states with seniorities v and $v - 2$ must be a product of two one-body operators—one that conserves seniority, another that connects v and $v - 2$. In the reduction to a matrix element in the j^v configuration, the first gives a factor identical to that in Eq. 5.17, the second gives the factor in Eq. 5.13. Thus, their product yields the result

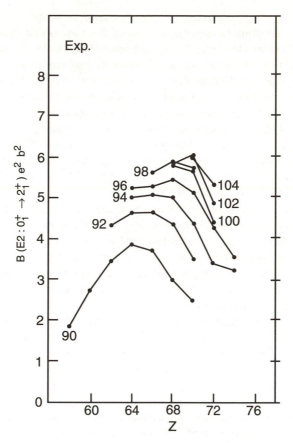

FIG. 5.3. Saturation of empirical B(E2) values in the rare earth region that illustrates Eqs. 5.13 and 5.16. The numbers on each line give the neutron number.

$$\left\langle j^n v\alpha J \left\| \sum_{i<k}^{n} V_{ik} \right\| j^n, v-2, \alpha' J \right\rangle = \left(\frac{2j+1-2n}{2j+1-2v} \right) \left[\frac{(n-v+2)(2j+3-n-v)}{2(2j+3-2v)} \right]^{\frac{1}{2}}$$

$$\left\langle j^v v\alpha J \left\| \sum_{i<k}^{v} V_{ik} \right\| j^v, v-2, \alpha' J \right\rangle \tag{5.18}$$

Once again, we note the factor $(2j+1-2n)$, which vanishes at midshell and has opposite signs in the first and second halves. For $j, n \gg v$, this interaction energy varies across a shell as $(1-2f)\sqrt{f(1-f)}$: at first this increases with f but it peaks well before the shell is one-quarter filled, tapers off, and crosses zero at midshell; in the second half of the shell, it is symmetric to the first half except for a change in sign.

We will not give the general expression for seniority-conserving matrix elements in the j^n configuration since they are more complex, involving not only matrix elements of the interaction in the j^ν but in the $j^{\nu+2}$ configuration as well.

It is useful at this point to summarize some of these important results. This is done in Fig. 5.4, which shows the behavior of both seniority conserving and nonconserving matrix elements for one-body operators and two-body interactions across a j shell under the assumption (where applicable) that j and n are large and much greater than ν. For the $\nu \to \nu - 2$ even-tensor case the square of the matrix element is given since it is directly proportional to the most common example of such behavior, B(E2: $2_1^+ \to 0_1^+$) values. Each panel also gives the (sometimes approximate) analytic formula.

To recapitulate: one-body odd-tensor operators (e.g., magnetic moments) conserve seniority and are constant: one-body even-tensor operators may change seniority, with $\nu \to \nu - 2$ transition matrix elements (e.g., B(E2 : $2_1^+ \to 0_1^+$) values) peaking at midshell, while seniority conserving matrix elements (e.g., quadrupole moments) vanish at midshell and are negatives of each other for particles and holes; odd-tensor two-body interactions (e.g., δ-interactions) behave as $(n - \nu)/2$ (the number of pairs of nucleons, or holes, in $J = 0$ couplings), and therefore peak at midshell.

Finally, to relate this discussion to an earlier one regarding the behavior of various matrix elements for particle configurations and the corresponding hole configurations, we note that the results in Eqs. 5.5 and 5.17 are consistent with Eq. 4.33, derived for diagonal matrix elements of one-body operators.

Thus far, the discussion has focused on the δ-interaction. A very popular alternative is the surface δ-interaction (SDI), which, as its name implies, acts only at the nuclear surface. It is equivalent to the *angular* part of the δ-interaction and to the assumption that all radial integrals are equal. Though this is a simplifying assumption, it permits an important generalization of one of the key preceding results: for *degenerate* orbits, the SDI conserves seniority in the multi-j configuration $|j_1^{n_1} j_2^{n_2} \ldots, \nu, J\rangle$. (In a single j shell the δ and SDI interactions are identical.) The SDI also has off diagonal matrix elements in multi-j situations that give rise to mixed wave functions. These matrix elements are generally larger than for the volume δ-interaction: the reason is simply that, when the interaction can occur throughout the nuclear volume, the effect of non-complete overlap of the particle wave functions is larger.

5.4 Some examples

With this theoretical background, it is interesting to consider an example that reflects some of these properties of the seniority scheme. Figure 2.3 showed the energies of the first 2^+ state, $E_{2_1^+}$ for the Ca isotopes as well as the B(E2 : $2_1^+ \to 0_1^+$) values. $^{40}_{20}$Ca$_{20}$ is doubly magic. The lowest orbit beyond the closed shells is $1f_{7/2}$ and the ground states of the nuclei from ^{41}Ca to ^{48}Ca are formed by adding nucleons in this orbit successively. The low-lying states in the even Ca isotopes can then be viewed as (primarily) an $(f_{7/2})^n$ configuration.

There is an interesting theorem, which we shall not prove, that states that *any* two-body interaction in the j^n configuration is diagonal in the seniority scheme, provided it is diagonal in the j^3 configuration; that is, if there are no finite matrix elements connecting $\nu = 3$ with $\nu = 1$ states. Clearly, since two-body interactions only connect states of the

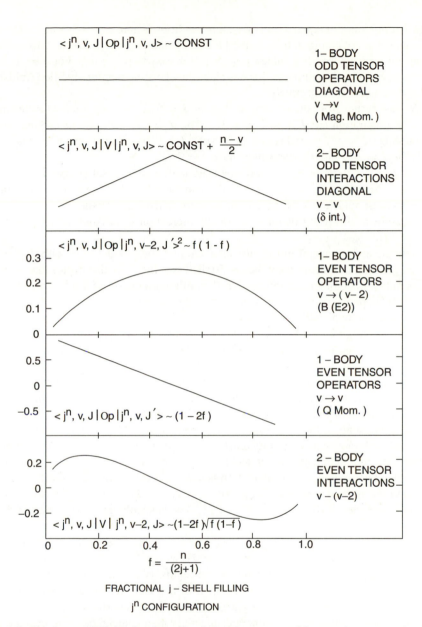

FIG. 5.4. Summary of the behavior of various operators and interactions across a shell in the seniority scheme. Note that the middle panel gives the square of the matrix element since this corresponds to the physically interesting case of B(E2) values. For binding energies, the line in the second panel should continue linearly upward throughout the shell.

same J (only states of equal J can mix), this condition is automatically satisfied for any J value that is not common to both the $|j^3 v = 3\rangle$ and $|j^3 v = 1\rangle$ states. This is useful because it happens for *all* j values $j \leq 7/2$. It is easy to prove this. For $j = 1/2$ and $3/2$ it is trivial: they have no $v = 3$ states since they become maximally filled (midshell) at $n = 1$ and $n = 2$, respectively.

We saw earlier from the m-scheme that for $j = 5/2$, the only allowed states in the $(5/2)^3$ configuration are $J = 9/2, 5/2, 3/2$. The $J = 5/2$ state can clearly be formed by coupling a $j = 5/2$ particle to a $|(j = 5/2)^2 J = 0\rangle$ configuration and so has $v = 1$, while the $J = 3/2$ and $9/2$ states must have $v = 3$.

For $j = 7/2$ there is, again, only one state with $J = 7/2$. Of course, it has $v = 1$ and, since there are no $v = 3$ states with the same J value, the above mentioned matrix elements must vanish. So the preceding theorem is trivially satisfied for all $j \leq 7/2$, giving us the useful result that *any* two-body interaction is *diagonal* in the j^n scheme for $j \leq 7/2$.

Note the importance of this result: since the first shell model orbit with $j \geq 9/2$ is the $1g_{9/2}$ orbit at the upper end of the 28–50 shell (see Fig. 3.2), this means that seniority is generally a good quantum number for j^n configurations for *all nuclei* with $A < 80$! These just happen to be the nuclei where shell model calculations are most feasible (the spaces to be diagonalized are not yet too large). Of course, this theorem does not apply to mixed-j configurations $|j_1^{n_1} j_2^{n_2} J\rangle$, but it certainly shows why the seniority scheme is so important and why seniority is often a reasonably good quantum number even in rather complicated configurations.

In any case, the theorem implies that seniority *must* be a good quantum number for the $(f_{7/2})^n$ states in the Ca isotopes, *independent* of the interaction. This does not mean that the interaction energies are independent of the interaction since, as we saw, the matrix elements of these interactions depend on their odd- or even-tensor character. Nevertheless, knowing that seniority must be a good quantum number, we can confidently choose some reasonable interaction and inspect the predictions of the seniority scheme. We assume that an odd-tensor interaction is a good choice since the δ-function can be written in this form. Then, from Eq. 5.6 or, more explicitly, from Eqs. 5.9 and 5.10, the excitation energies of the $J = 2^+, 4^+$, and 6^+ levels should be constant across the Ca isotopes.

Let us inspect the data in Fig. 2.3. The yrast energies are high at ^{40}Ca and ^{48}Ca since both of these nuclei are doubly magic. However, $^{42, 44, 46}$Ca have, as expected, much lower 2_1^+ energies and these are indeed relatively constant. The B(E2) values in the Ca isotopes are also roughly consistent with the seniority picture since they peak near midshell, although the symmetry about midshell is not particularly evident.

The constancy of $v = 2$ level energies in singly magic nuclei has long been emphasized by Talmi. His classic example is the Sn isotopes, whose energies were shown in Fig. 2.6.

One can also use the seniority scheme to look at nuclear binding energies, B.E.(j^n), (n even) for a series of nuclei in which an orbit j is filling. These binding energies are the absolute energies of the $|j^n, v = 0, J = 0^+\rangle$ ground states (we assume an odd-tensor interaction). Of course, this is only the residual interaction energy, to which must be added the n single-particle energies E_{nlj}. To be slightly more general, there is one other interaction that is diagonal in the seniority scheme—the trivial case of a scalar interaction. Since this is a constant, the interaction matrix element will be equal to that constant multiplied by the total number of pairs of nucleons that can interact, $n(n-1)/2$. Combining these three terms, one then has for the binding energy in a j^n configuration, relative to the closed shell,

$$\text{B.E.} \left(j^n \right) = n E_{nlj} + \frac{n(n-1)}{2} B + \frac{n}{2} V_0 \qquad (5.19)$$

This formula, which is valid for any interaction that is diagonal in the seniority scheme (and therefore, *for any* interaction for $j \leq 7/2$) displays the well-known parabolic behavior of nuclear masses. We note that the parameter B always turns out to be negative: the quadratic term is always repulsive. Evidence of this can be seen empirically from the data presented in Chapter 1, which showed that binding energies are approximately proportional to A. If $B > 0$ (attractive), binding energies would increase quadratically, instead of linearly, with A.

This is the same point we have made before, that the nonpairing, residual interaction between *like* nucleons is *repulsive*. We can now carry this one step further. Since the quadratic term stems from the scalar part of the interaction, which is obviously long range, it is the *long-range* component of the interaction between like nucleons that must be repulsive.

This conclusion, based on the simple form of the binding energies in the seniority scheme and the empirical behavior of separation energies, forces us to conclude that deformation does not arise simply from an abundance of valence nucleons outside closed shells, but must specifically involve *both* valence protons and neutrons. We will see later how the properties of the proton–neutron interaction can indeed lead to deformation through the effects of one-body configuration mixing.

Most of the examples of the seniority scheme so far in this chapter have concerned even nuclei. However, the scheme is equally powerful in treating odd mass nuclei near closed shells. We will illustrate this with a simple calculation that is useful in considering low-lying energy levels in sequences of odd A nuclei.

As we pointed out, the lowest state of the j^n configuration in an odd-mass nucleus normally has $J = j$ and $v = 1$. The n-dependence of its interaction energy is $[(n-1)/2]V_0$ where V_0 is the interaction energy in the $J = 0$ state of the j^2 configuration. For an odd-tensor interaction, the excitation energies of the $v = 3$ states can be obtained from Eqs. 5.6 and 5.8:

$$E(j^n, v = 3, J') - E(j^n, v = 1, J = j)$$

$$= \left[\left\langle j^3, v = 3, J' | V | j^3, v = 3, J' \right\rangle + \frac{(n-3)}{2} V_0 \right] - \left[\frac{(n-1)}{2} V_0 \right]$$

$$= \left\langle j^3, v = 3, J' | V | j^3, v = 3, J' \right\rangle - V_0$$

$$= E(j^3, v = 3, J') - E(j^3, v = 1, J = j) \qquad (5.20)$$

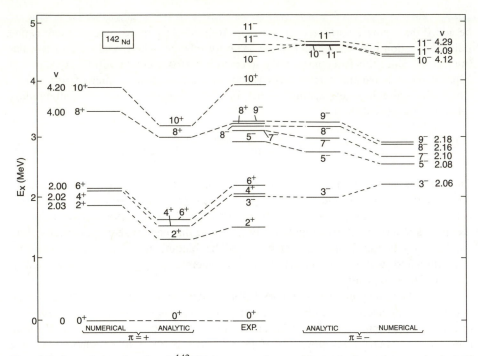

FIG. 5.5. Low-lying levels of ^{142}Nd in comparison with analytic and numerical shell calculations. (The empirical level scheme is based mainly on Wirowski, 1988.)

Thus, these excitation energies are identical to those in the j^3 configuration and are independent of n—that is, they are constant across a shell. Clearly this also means that the spacings *between* $v = 3$ states are n-independent. This can of course be seen by explicit calculation:

$$E\left(j^n, v = 3, J'\right) - E\left(j^n, v = 3, J\right)$$
$$= \left[\left(\left\langle j^3, v = 3, J'|V|j^3, v = 3, J'\right\rangle - \frac{(n-3)}{2}V_0\right)\right.$$
$$\left. - \left(\left\langle j^3, v = 3, J|V|j^3, v = 3, J\right\rangle - \frac{(n-3)}{2}V_0\right)\right]$$
$$= E\left(j^3, v = 3, J'\right) - E\left(j^3, v = 3, J\right) \tag{5.21}$$

In closing this discussion of multiparticle configurations, it is interesting to take a more complicated level scheme as an example and see how far we can go in interpreting it by exploiting the simple results in this chapter. Consider then the nucleus $^{142}_{60}$Nd$_{82}$ with ten protons outside doubly magic ^{132}Sn (see Fig. 5.5). Although the level scheme seems rather complex, nearly every feature can be easily understood and, indeed, derived analytically, without any complex calculations.

To start, we note that the lowest shell model orbits in the $Z = 50$–82 shell are $2d_{5/2}$ and $1g_{7/2}$. There are many ways that the protons can be distributed over the two positive parity orbits. If we assume that the splitting of states of a given seniority v is small compared to the spacings between states of different seniorities (we are neglecting the first term relative to the second in Eq. 5.6), then the $v = 0$ $J = 0^+$ state will, as usual, lie lowest, by an amount V_0 below the $v = 2$ states. The $v = 2$ states, in turn, occur, on average, this same distance below the $v = 4$ states. Therefore, we can assume that all of the $J \neq 0$ *low*-lying positive-parity states ($J^\pi \leq 6^+$) are $v = 2$. Since we are dealing with (positive parity) j values $\leq 7/2$ we know that seniority is a good quantum number *regardless* of the interaction. Since $|(d_{5/2})^n v = 2\rangle$ does not give a 6^+ state, let us assume for simplicity that the $d_{5/2}$ shell is filled, leaving four protons in the $g_{7/2}$ orbit. By Eq. 5.6 and the preceding discussion, the $J = 2^+, 4^+, 6^+ v = 2$ excitation energies should be identical to those in a $(g_{7/2})^2$ configuration, and as such can be estimated theoretically using the results for a δ-function interaction. We postpone doing so for a moment until we determine how best to estimate the absolute strength of the interaction. The next positive parity state is 8^+ and must have $v = 4$. By our earlier arguments it should lie higher than the $v = 2$ states by roughly $|V_0|$. The lowest-energy way to construct it is to couple the two $v = 2$ states with $J^\pi = 6^+$ and 2^+ together. Similarly, the 10^+ level must also have $v = 4$. The easiest way to form it is by coupling the $J^\pi = 6^+ (v = 2)$ and $J^\pi = 4^+ (v = 2)$ states.

The negative parity states must involve the $h_{11/2}$ orbit. The two lowest configurations should involve eight protons coupled to $J = 0$ in the $d_{5/2}$ and $g_{7/2}$ orbits and an $(h_{11/2}g_{7/2})$ or $(h_{11/2}d_{5/2})$ pair. The former gives spins $J = 2^- - 9^-$ while the latter yields $J = 3^- - 8^-$. From the rules developed in Chapter 4 for the ordering of different J states in two-particle configurations under the influence of a δ-interaction, we find that the 9^- state should be the lowest in the $|h_{11/2}g_{7/2}J\rangle$ configuration while the 3^- should lie lowest in the $|h_{11/2}d_{5/2}J\rangle$ configuration. In both cases the even-spin negative-parity states are unaffected by the interaction. Empirically, the lowest negative parity state is indeed 3^-, suggesting the $(h_{11/2}d_{5/2})$ assignment. The lowest 7^- and 8^- levels must also belong to this multiplet (since, if they were part of the $(h_{11/2}g_{7/2})(J = 2^- - 9^-)$ multiplet they would lie *above* the 9^-). The 8^- level gives the unperturbed position of the $(h_{11/2}d_{5/2})$ multiplet, and the energy difference $8^- - 3^-$ gives the absolute scale of the interaction strength, thus allowing us to predict (using Table 4.1) the 5^- and 7^- energies, as well as the spacings among the positive parity levels. The $J = 9^-$ state is then the lowest member of the $(h_{11/2}g_{7/2})$ multiplet. (The lower-spin members would not have been detected in experiments carried out for ^{142}Nd.) Finally, the $J^\pi = 10^-$ and 11^- states cannot arise from either $(h_{11/2}g_{7/2})$ or $(h_{11/2}d_{5/2})$ couplings. Since, empirically, the latter lies lower, a reasonable configuration for the 10^- level is $|(h_{11/2}d_{5/2}))J = 8^- \otimes 2_1^+\rangle_{10^-}$, meaning an $(h_{11/2}d_{5/2})$ pair coupled to $J^\pi = 8^-$ built on the seniority $v = 2$ 2^+ level of the remaining $(n - 2)$ particle system. The energy difference $E(10^-) - E(8^-) \approx E_{2_1^+}$ is approximately satisfied experimentally. The lowest 11^- levels can be made either by coupling the $|h_{11/2}g_{7/2}J = 9^-\rangle$ state to the $v = 2$ 2_1^+ level or the $|h_{11/2}d_{5/2}J = 7^-\rangle$ level to the $v = 2$ 4_1^+ level.

All these results are incorporated now into Fig. 5.5, where it is evident that the agreement of this extremely simple calculation with experiment is actually remarkably good. The only puzzling aspect of our analysis is that the positive parity states required the $g_{7/2}$ orbit to lie above the $d_{5/2}$ while the negative parity orbits required the opposite. This suggests that the $d_{5/2}$ and $g_{7/2}$ energies are quite close. Figure 5.5 also shows a detailed 10-particle shell model calculation. Such calculations proceed by choosing an interaction and a set of basis configurations, which, after diagonalization (see Sections 1.4, 1.5), give mixed configuration wave functions. We use a surface δ-interaction (which gives relative spacings, within a configuration $|j^n J\rangle$, which are the same as for a volume δ-interaction but which has convenient property that its strength is independent of j) with strength 0.4 MeV and with the single-particle energies (in MeV): $\varepsilon_{g7/2} = 0$, $\varepsilon_{d5/2} = 0.7$, $\varepsilon_{h11/2} = 2.5$. The calculation also shows reasonable agreement with the empirical scheme, but more importantly, it shows that our simple analytic interpretation is a remarkably accurate approximation of a complex shell model diagonalization. The principal difference in the numerical diagonalization is that additional components, such as $(d_{5/2})^2$ and $(d_{5/2}g_{7/2})$ or even $(d_{5/2}g_{7/2}^3)$, come into play. Similar analyses can be applied to other nuclei (e.g., Sn) and greatly help to understand the results of complex realistic calculations.

Improved calculations, compared to Fig. 5.5, have also been carried out. They allow $(h_{11/2})^2$ as well as $(d_{5/2})^{n_1}$ and $(g_{7/2})^{n_2}$ configurations in the positive parity levels. The inclusion of these amplitudes mainly affects the required $h_{11/2}$ single particle energy and, of course, changes the strength of the interaction needed to fit the data. The reason is easy to see from Eq. 4.10. Since the lowering of the 0^+ state in a configuration $|j^2 J\rangle$ is proportional to $(2j + 1)/2$ multiplied by the interaction strength, V_0, the inclusion of $(h_{11/2})^2$ means that a smaller $|V_0|$ is required to maintain the same $J_{\text{yrast}} - J_{0_1^+}$ spacing. The extra lowering of the 0_1^+ level effectively raises the excitation energy of all the others. To regain a fit to the data for the negative parity states, a lower $\varepsilon_{h11/2}$ is required.

This illustrates two important points. First, the choice of effective residual interaction, single-particle energies and the shell model space are intimately linked. One should be wary of conclusions regarding any of these if there is not supporting evidence for the choices concerning the others. Secondly, despite the beauty of analyticity, a realistic treatment of complex nuclei still requires detailed explicit calculations if really quantitative results are desired.

Finally, note that we have completely ignored core excitations of the protons or neutrons. These can be significant and their effects can vary significantly for different states. In particular, even in singly magic nuclei (e.g., ^{142}Nd or Sn), the lowest 2^+ and 3^- states are often rather "collective" with a number of major components, including core excited particle–hole components (e.g., $(g_{9/2}^{-1} h_{11/2})J = 3^-$ for protons in Sn). To some average extent, their ignored effects are mocked up by the choices of the single-particle energies of the valence orbits and of the interaction strength. It is no wonder that residual interactions are often called "effective interactions" and that an extensive theory of such interactions has been built up. Indeed, as discussed at the end of Chapter 4, a number of alternates to the δ-function interaction are often used. Often, these are more structured, finite-range, interactions such as Gaussian forms. Various so-called Skyrme interactions are also popular, as are interactions defined explicitly in terms of sets of two-body empirical matrix elements.

5.5 Pairing correlations

In closing this chapter, we want to turn to another interaction, similar in many ways to the δ-interaction, which is very important in understanding the structure of heavy nuclei. We saw in Chapter 1 that proton and neutron separation energies exhibited an odd–even effect, indicating an extra binding in the 0^+ ground states of even–even nuclei. We have also seen that the δ-function force acting on two identical nucleons in equivalent orbits produces a strong lowering of the $J = 0^+$ state. This behavior is typical of any very short-range interaction and is a direct consequence of the Pauli principle, which allows such interactions to affect only *spatially* symmetric wave functions. Although the δ-function force is particularly easy to deal with (and has specific tensor properties that imply conservation of seniority), it has become traditional in heavy nuclei to describe such behavior more directly in terms of the so-called pairing interaction. This is *defined* (see Eq. 4.20) to be an attractive interaction acting *only* on two identical particles in total angular momentum 0^+ states. States with $J^\pi \neq 0^+$ are unaffected. A comparison of the pairing interaction with a δ-function interaction for an $(f_{7/2})^2$ configuration was shown in Fig. 4.13. The sequence of energy levels for the δ-function interaction is the familiar result we have been studying and simply reflects the properties of the $3 - j$ symbol in Eq. 4.10. The pairing force presents a similar overall pattern but with a degenerate multiplet of $J^\pi \neq 0^+$ states at the unperturbed energy. As we have seen, inspection of the empirical level schemes of two-particle states in singly magic nuclei such as ^{210}Po, ^{210}Pb, ^{134}Te (Fig. 4.5), and many others, shows that the δ-interaction reproduces empirical spectra much more accurately. Even for multiparticle (but still singly magic) cases, such as the Sn isotopes, the δ-function provides a more realistic interpretation (see Fig. 2.6). Nonetheless, it has become nearly universal practice to invoke a pairing interaction and to refer to the lowering of the 0^+ state as a pairing effect. (Indeed, we already referred to the second term in Eq. 5.6 as the "pairing property" in analogy with the effect of the pairing force.) Part of the appeal of the pairing force is the ease with which it can be extended to multiparticle systems, where it leads to the desired result of 0^+ ground states in even–even nuclei without the need for the seniority apparatus and angular momentum algebra we have discussed.

There are a number of experimental facts that motivate the introduction of a pairing force and the concept of pairing correlations. We have mentioned some, but it is useful to summarize them here. The best known is the simple fact that the ground states of all even–even nuclei has $J^\pi = 0^+$. A related point is that this 0^+ state is normally far below other noncollective intrinsic states. This is the so-called pairing gap.

Perhaps the most direct evidence for a pairing interaction is the so-called odd–even mass difference. This simply refers to the fact that when nucleons are added to a nucleus, the gain in binding energy is greater when an even–even nucleus is formed than when the neighboring odd-mass nucleus is formed. This empirical fact can be inferred from the data of Figs. 1.2 and 1.3 by comparing the absolute values of S(n) or S(p) for adjacent odd and even nuclei. An extra attractive interaction that couples pairs of like-nucleons together can accommodate this fact, and indeed, the separation energy data suggest a strength for the pairing interaction of ~ 1–2 MeV.

There are three other features that can be seen, at least in retrospect, as clearly pointing to the need for a pairing interaction. The pairing interaction clearly favors sphericity since it favors the formation of pairs of particles coupled to a total magnetic substate $M = 0$. Therefore, near closed shells, the presence of a strong pairing interaction will inhibit the tendency to deform. Instead of the smooth transition toward deformation that would normally occur as valence nucleons are added, one typically sees, empirically, a sequence of more or less spherical nuclei followed by a rather rapid transition region to deformation. Secondly, for a deformed nucleus of a given shape, it is easy to calculate the moment of inertia. This in turn determines the spacing of rotational levels (see Chapter 6). Empirical moments of inertia extracted from those energies are systematically lower than calculated in the shell model. The inclusion of pairing correlations essentially removes the discrepancy, as we shall see in Chapter 6. Finally, the energy distribution of Nilsson orbits near the ground state in deformed odd mass nuclei is, on average, denser than expected from the Nilsson diagram. This will be seen to be a "compression" effect that occurs in the transformation from particle–hole to quasi-particle energies.

The pairing interaction was defined in Chapter 4 in terms of its matrix elements as:

$$\langle j_1 j_2 J \, |V_{\text{pair}}| \, j_3 j_4 J' \rangle = -G \left(j_1 + \frac{1}{2} \right)^{\frac{1}{2}} \left(j_3 + \frac{1}{2} \right)^{\frac{1}{2}} \delta_{j_1 j_2} \delta_{j_3 j_4} \delta_{J0} \delta_{J'0}$$

where G is the so-called strength of the pairing force and the rest of the expression has an obvious meaning. Note that the pairing force is independent of orbit but, since it is identical for each magnetic substate, scales for an orbit j as $(2j + 1)$. It is therefore stronger in high j orbits. Although G is orbit independent, it decreases with A in heavier nuclei where the outer nucleons are generally further apart and so spatial overlaps are likely to be less. G may be different for protons and neutrons, being lower for the former because of Coulomb repulsion. Commonly used prescriptions are

$$G_p = \frac{17}{A}, \qquad G_n = \frac{23}{A} \qquad \text{MeV}$$

It is frequently imagined that the pairing force is an interaction only between two particles in the same j state, coupled with their angular momenta antiparallel, to form a $J^\pi = 0^+$ state. This, of course, is an important facet of the pairing interaction. However, there is another vital ingredient that is evident from its definition. The pairing force is equally strong for matrix elements connecting a $|j^2 J = 0^+\rangle$ state with a $|j'^2 J = 0^+\rangle$ state. That is, the pairing force has *nondiagonal* as well as diagonal components that can "scatter" pairs of particles from one orbit to another. Note the importance of this point. If the force were purely diagonal, then, while $J = 0^+$ pairs would still be tightly bound, excited states could be formed simply by raising both particles in such a pair to the next unoccupied orbit. On average, this would require twice the energy needed to raise a single particle from orbit j to orbit j' in the absence of pairing. The "energy gap" in even–even nuclei would then be only twice the average spacing of low-lying levels in the adjacent odd nucleus and not 5 to 10 times as large, as observed experimentally. Rather, by scattering particles from one j orbit to another, the pairing force mixes 0^+ states and creates partial occupancies near the Fermi surface. Hence it builds up a coherence in the pair wave functions, which further lowers the lowest 0^+ state and thereby enlarges the gap: if we picture the pairing force in a perturbation theory context, the amplitudes for scattering a pair of particles from orbit j to orbit j' will be proportional to the matrix element $\langle j^2 J = 0 | V_{\text{pair}} | j'^2 J = 0 \rangle / (\varepsilon_j - \varepsilon_{j'}) = G / (\varepsilon_j - \varepsilon_{j'})$. In the absence of pairing, all levels would be occupied up to some point (the Fermi energy) while those above this energy would be completely empty. With pairing, however, many orbits can be partially occupied. This in turn, radically alters the concept of hole excitations, and thereby, the levels schemes of both odd and even nuclei. We shall see all this a bit more formally when we outline the basic results of pairing theory. We will not derive these standard results but will try to highlight their key effects.

To proceed, we refer to Fig. 5.6 where several quantities important in discussing the pairing interaction are defined: the Fermi energy, denoted λ; the single-particle energies, ε_i, with ε_0 being reserved for that level closest to the Fermi surface; and Δ, the so-called gap parameter, defined in terms of a sum over orbits i, j as

$$\Delta = G \sum_{i,j} U_i V_j \tag{5.22}$$

where the usual U and V factors are the so-called emptiness and fullness factors that pervade the study of heavy nuclei. They are given by:

$$U_i = \frac{1}{\sqrt{2}} \left[1 + \frac{(\varepsilon_i - \lambda)}{\sqrt{(\varepsilon_i - \lambda)^2 + \Delta^2}} \right]^{\frac{1}{2}}, \, V_i = \frac{1}{\sqrt{2}} \left[1 - \frac{(\varepsilon_i - \lambda)}{\sqrt{(\varepsilon_i - \lambda)^2 + \Delta^2}} \right]^{\frac{1}{2}} \tag{5.23}$$

These equations can be solved for Δ, U_i, and V_i for a given set of single-particle energies ε_i. Or, Δ can be estimated from empirical mass differences between adjacent nuclei with odd and even numbers of nucleons. The behavior of V_i^2 against the ratio $(\varepsilon_i - \lambda)/\Delta$ is shown in Fig. 5.6. We see that $V_i \rightarrow 1$ for $(\varepsilon_i - \lambda) << 0$ and vanishes for levels far above the Fermi surface. The opposite applies to U_i^2. Both fall off rapidly for single-particle energies $\varepsilon_i \sim \lambda$. Also, from Eqs. 5.23, $V^2 + U^2 = 1$. U_i^2 is the probability that the orbit i is empty, whereas V_i^2 is the probability that it is filled. Pairing smooths out the level occupancies near the Fermi surface over a range $\sim \Delta$, which, in turn, is proportional to G, the "strength" of the interaction. If there were no pairing, the Fermi surface λ would coincide with the last orbit being filled and $(\varepsilon_i - \lambda)$ would be the excitation energy required to excite one of the nucleons in this last orbit to one of the higher orbits ε_i. In the presence of pairing, however, this single-particle excitation energy $(\varepsilon_i - \lambda)$ is replaced by a *quasi-particle* energy E_i given by

$$E_i = \sqrt{(\varepsilon_i - \lambda)^2 + \Delta^2} \tag{5.24}$$

$$V_i^2 = \frac{1}{2}\left[1 - \frac{\varepsilon_i - \lambda}{\sqrt{(\varepsilon_i - \lambda)^2 + \Delta^2}}\right]$$

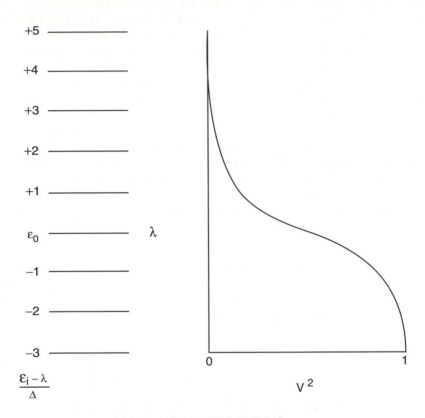

PARTIAL OCCUPANCY DUE TO PAIRING

FIG. 5.6. Definition of several quantities used in the calculation of pairing correlations. (Left) Idealized set of equally spaced single-particle shell model levels. (Right) Resultant orbit occupancies V^2. The calculation is general since the single-particle energies are given in units of Δ.

Thus, particles and holes are replaced by quasi-particles representing partially filled levels, and a particle–hole excitation is replaced by the creation of one quasi-particle and the destruction of another.

The transformation from particles to quasi-particles is of interest not only because the pairing interaction happens to lead to partial occupancy, but because it allows an enormous simplification in shell model calculations with many valence nucleons. It is basically a transformation from a viewpoint based on the closed shell to one based on the Fermi surface. Thus, instead of having to deal with *all* possible ways of constructing each J state within a major shell, it naturally produces a physically intuitive truncation scheme directly keyed to the scale of excitation energies one is interested in. For low-lying states, one needs to consider only low energy quasi-particles. In the shell model, in contrast, a near-midshell nucleus would involve single-particle energies lying rather high in the shell regardless of the excitation energy. Moreover, pair scattering would give rise to extremely complex wave functions in the shell model. In the quasi-particle picture, all the complexity due to partial pair occupancies induced by the pairing force is effectively absorbed into the ground state (quasi-particle vacuum) and we need only consider quasi-particle excitations relative to the Fermi surface.

The behavior of E_i is interesting and has important consequences. If $\varepsilon_i \sim \lambda$, that is, if the ith single-particle level is near the energy where the occupation probabilities fall off rapidly, $E_i \sim \Delta$. This value is also clearly the minimum value of E_i.

In an *odd* mass nucleus, however, this is not the minimum energy for an excited state: Rather, excited levels are obtained by *replacing* the quasi-particle defining the ground state by one corresponding to a different single-particle level. Thus the excitation energy E_{xi}^0 for the ith quasi-particle in an odd-mass nucleus is given by

$$E_{x_i}^o = E_i - E_0 = \sqrt{(\varepsilon_i - \lambda)^2 + \Delta^2} - \sqrt{(\varepsilon_0 - \lambda)^2 + \Delta^2} \qquad (5.25)$$

where $E_0(\varepsilon_0)$ is the quasi-particle (single-particle) energy of that orbit nearest the Fermi energy λ. Thus we see that $E_{x_i}^0$ can take on arbitrarily small values. Indeed, since all $E_i \sim \Delta$ for $(\varepsilon_i - \lambda) << \Delta$, the effect of pairing is actually to *decrease* the excitation energies of the low-lying states, compressing the excitation energy spectrum. Figure 5.7 shows this compression of $E_{x_i}^0$ as a function of $(\varepsilon_i - \lambda)$ in units of Δ. At higher energies $[(\varepsilon_i - \lambda) >> \Delta]$, the effect is to lower all states by an amount $\sim \Delta$ since $E_i \rightarrow (\varepsilon_i - \lambda)$ and $E_{xi}^0 = E_i - E_o \sim (\varepsilon_i - \lambda) - \Delta$.

In *even* nuclei, the effect of pairing is nearly the opposite: instead of a compression of excited quasi-particle levels, there are no simple excitations below $E_x^e = 2\Delta$. This is easy to show. The simplest excitation consists of breaking one pair and raising a particle to the next higher orbit. Without pairing, this is a particle–hole excitation. In the presence of pairing, it appears as a two quasi-particle excitation, one quasi-particle being the hole left behind and the other being the particle excitation newly created. Thus, the excitation energy is given by

$$E_{x_{ij}}^e = \sqrt{(\varepsilon_i - \lambda)^2 + \Delta^2} + \sqrt{(\varepsilon_j - \lambda)^2 + \Delta^2} \qquad (5.26)$$

where i and j label the two quasi-particles involved. This is plotted in Fig. 5.7 as well. It is immediately clear that the minimum energy is $E_x^e = 2\Delta$, giving the famous "pairing gap" that is a nearly universal feature of even–even nuclei in which few two quasi-particle excitations appear below $\sim$1.5–2 MeV. In fact, this empirical energy gap is one way of extracting Δ experimentally. As noted, typical values of Δ range from about 700 keV–1 MeV.

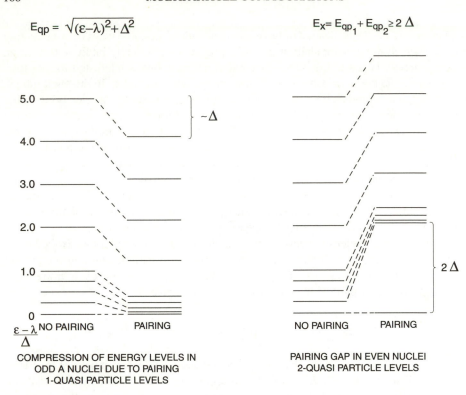

$$E_{qp} = \sqrt{(\varepsilon - \lambda)^2 + \Delta^2}$$

$$E_x = E_{qp_1} + E_{qp_2} \geq 2\,\Delta$$

COMPRESSION OF ENERGY LEVELS IN
ODD A NUCLEI DUE TO PAIRING
1-QUASI PARTICLE LEVELS

PAIRING GAP IN EVEN NUCLEI
2-QUASI PARTICLE LEVELS

FIG. 5.7. Effects of pairing in odd and even nuclei. (Left) Level compression near the ground state for odd mass nuclei. (Right) The energy gap in even–even nuclei.

Although there is a large gap between the ground state of an even–even nucleus and the first excited two-quasi-particle state (2Δ), above this energy the quasi-particle level density will be at least as large as in a neighboring odd-mass nucleus.

The energy gap and the high-level density just above it are evident in virtually any even–even nucleus. Figures 2.5, 2.6, and 2.10 show several examples. However, there is another very simple and elegant way of demonstrating the pairing gap empirically without the need for in depth study of each individual excited state. The (n, γ) reaction proceeds by the capture of a neutron of energy E_n into a target nucleus, forming a residual nucleus at an energy $E_c = S(n) + E_n$. This "capture" state then decays by the emission of γ-rays to low-lying levels. Each γ-ray thereby defines an excited state by the equation $E_{xi} = E_c - E_\gamma$. Under appropriate experimental conditions the (n, γ) reaction provides an a priori guarantee of observing all final states in certain spin and excitation energy regions. Thus, a spectrum of γ-ray transitions will reveal a sequence of peaks, each corresponding to a specific excited level, and together, displaying all the excited states with certain J^π values. Because of the "completeness," the pairing gap should be immediately identifiable by a sudden increase in peak density. Figure 5.8 shows such a spectrum for the case of ^{196}Pt, and includes an insert portraying the reaction process

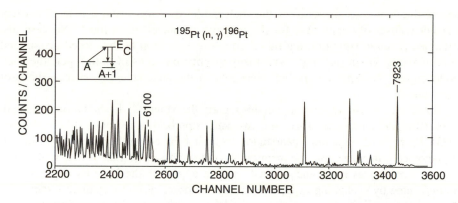

FIG. 5.8. Spectrum of primary γ-rays following average resonance capture (ARC) into ^{196}Pt. The inset schematically illustrates the reaction process. Two peaks in the spectrum are labeled by their γ-ray energies in keV: the rightmost peak corresponds to a transition to the ground state, while that at 6100 keV populates a state at 1823 keV. The increased level density at the pairing gap is immediately obvious at approximately this excitation energy (Cizewski, 1979).

schematically. One sees the ground state, a few low-lying excited states, and then, at $E_x \sim 1.8$ MeV, a dramatic increase in level density. By inspection, one can immediately point to an approximate value for 2Δ.

We also note that, seemingly contrary to the preceding discussion, there are, in fact, a few peaks below the pairing gap. We shall discuss these in considerable detail in later chapters. They correspond to macroscopic "collective" vibrations and/or rotations of the nucleus as a whole and can be viewed as resulting from the interactions among a number of two-quasi-particle states in which the lowest such state is pushed down by mixing, and therefore occurs below the energy gap.

Finally, the defining equation for Δ is interesting since it indicates something about the origin of the pairing gap. Clearly Δ should vanish in the absence of pairing, which the proportionality to G in the Eq. 5.22 reflects. Secondly, the range and number of energy levels over which U and V are different from 1 and 0 increases with G, as does the sum in Eq. 5.22.

To summarize the discussion of pairing, it is an extremely important contribution to the structure of medium and heavy nuclei, affecting energy levels, γ-ray transition rates, particle transfer cross sections, moments of inertia, and the microscopic structure of virtually all collective excitations. It modifies most nuclear matrix elements. It is also, as we commented in Chapter 4, one of the two ingredients in the so-called pairing-plus quadrupole (or PPQ) interaction, which is a standard basis for the study of collective excitations (see Chapter 10). We shall see the effects of pairing throughout the rest of this book.

To conclude this chapter and lead into the next part of the book we emphasize that, despite the numerous successes of the shell model over the years, it is apparent that it is not realistically applicable for the majority of nuclei. Even the ^{142}Nd case just discussed required simplification by ignoring the excitation of the closed proton and neutron shells. When one deals with many valence nucleons of both types, the shell model rapidly becomes intractable, especially in the larger multi-j valence shells of heavy nuclei.

However, a recent development in this regard, the Monte Carlo shell model method, holds substantial promise. There are two such approaches (Koonin, 1997; Honma, 1995). Both use an operator linearization technique based on the so-called Hubbard-Stratanovich transformation (we mention this detail only so that the reader will recognize such references when encountered). In one approach, the large basis problem is circumvented by evaluating expectation values of observables by statistical sampling. This method gives ground state properties and thermally averaged descriptions of excited states. In the other, the Monte Carlo sampling is used to select an optimum relatively small number of basis states for subsequent diagonalization, giving approximate wave functions, eigenvalues, and transition rates. Both methods are in their infancy and are attracting much current interest.

Regardless of such advances, however, as we shall see, many nuclei display characteristics that clearly point to nonspherical or deformed shapes. Thus, they are better discussed in the context of geometric (and other) models where such shapes arise naturally. We now turn to a discussion of such models. Later we will show how the collective features that characterize these models can, in fact, be obtained (and therefore justified) microscopically from the shell model.

PART III

COLLECTIVITY, PHASE TRANSITIONS, DEFORMATION

6

COLLECTIVE EXCITATIONS IN EVEN–EVEN NUCLEI: VIBRATIONAL AND ROTATIONAL MOTION

6.1 An introduction to collectivity, configuration mixing and deformation

The shell model is generally considered the fundamental nuclear model. Historically, it was the first model to have considerable, detailed success. Of course, this is not accidental. The shell model works best for light nuclei. Much of the crucial information on nuclear structure comes from nuclear reactions that require the incident projectile to have energies near or above the Coulomb barrier in order to penetrate the nucleus. This requires higher and higher energies for heavier and heavier elements and, consequently, most early studies with low-energy accelerators also focused on light nuclei. The development of nuclear models is intimately connected to the historical progress of experimental techniques.

More fundamental to the shell model's central position in nuclear physics is that it provides a well-defined procedure for the calculation of basic nuclear observables. The principle ingredients needed for any given calculation are a choice of the single-particle energy levels, of the number of these that should be included in the space to be diagonalized, and the residual interaction to be used. Moreover, since the shell model is the only broadly applicable *microscopic* model available, it is the standard against which others are compared, and it is the source and rationalization of macroscopic models. A collective model that can be shown to be inconsistent with the shell model is discarded with little delay.

Unfortunately, the use of the shell model is, in practice, rather severely limited. In previous chapters we considered some simple applications, mostly those in which the particles were confined to a single j shell. Except for very light nuclei, however, and those very near closed shells, the valence nucleons occupy more than one j shell. (The pairing interaction assures this if the single-particle spacing is less than or comparable to its "spreading" parameter Δ.)

While shell model calculations for multi-j-configurations are certainly possible, the size of the matrices in which the residual interaction must be diagonalized rapidly becomes enormous. For even a few valence nucleons in several j orbits, it is easily possible to construct hundreds of states of a given J^π value. Even if such calculations are possible using high speed computers, the results are difficult to interpret physically and the consequences of agreement or disagreement with the data are much less intuitively informative. The situation becomes totally intractable when more valence nucleons are added. To quote the famous example of Talmi, in $^{154}_{62}\text{Sm}_{92}$, approximately 3×10^{14} 2^+ states can be constructed with protons and neutrons in the $Z = 50$–82 and $N = 82$–126

major shells. In such a case, no shell model calculation is even remotely feasible, nor could one even begin to understand the resultant wave functions. Since the vast majority of nuclei fall into the many-valence nucleon category, any advance in our understanding of nuclear structure beyond a very select group of nuclei must depend either on some truncation or simplification (e.g., Monte Carlo methods) of the shell model or on the development of alternatives to it. This problem is, of course, a common one in physics. One develops a simple model. Later, one introduces first-order perturbations to it to refine its predictions. Further improvements involve incorporating additional degrees of freedom, which enlarge the basis. When the required basis becomes too large or the wave functions too complex in that basis, one looks for a new *physical* picture that allows one to approach the problem from an alternate basis.

The development of nuclear structure in medium and heavy nuclei has followed two related paths to avoid the large basis problem, both of which we shall discuss extensively in the next three chapters. Both involve the concept of a nonspherical shape, but one emphasizes the single-particle motion of nucleons in a field of that shape, while the other stresses the macroscopic motions and excitations of a nucleus having that shape. Ultimately, the former approach (a "deformed shell model") can be shown or must be shown to be the microscopic justification for the latter, but in practical terms, the two models are often used complementarily in a microscopic–macroscopic combination that has proved to be very powerful. The nonspherical, or deformed, shell model approach is usually called the Nilsson model, while the macroscopic one is a fundamentally *collective* model, generally known as the collective model of Bohr and Mottelson although many others have contributed in essential ways to it. Each of these approaches, of course, has numerous offshoots, extensions, and refinements.

The assumption of a nonspherical macroscopic nuclear shape is a phenomenological or ad hoc one: while the enormous successes of the collective model, its demonstrated predictive power, and the numerous progeny it has spawned over the last three decades leave little doubt that it aptly describes the nuclear structure of perhaps the majority of nuclei, one is left with an uneasiness about the apparent incompatibility of an independent particle picture such as we have been discussing and the clearly collective and coherent motion involved in macroscopic rotations and vibrations. It was a theoretical achievement of basic importance in the early 1960s when it was demonstrated that macroscopic collectivity could indeed result from the shell model with appropriate and reasonable realistic residual interactions. The essence of the method involved is the so-called random phase approximation or RPA method, or its somewhat simpler cousin the Tamm–Dancoff approximation (TDA). We shall discuss these two methods in some formal detail in Chapter 10.

In the last two decades or so, a rather different approach to collective behavior in nuclei has been developed. It exploits the dynamical symmetry structure of nuclei, and utilizes powerful group theoretical techniques to obtain many nuclear properties by simple algebraic techniques, often in analytic form. There are many such algebraic models in use today. However, beyond a doubt, the most popular, successful, and widely tested to date has been the interacting boson approximation (IBA) model of Iachello and Arima. This model will also be discussed later in this chapter. Other algebraic models, which emphasize the fermions (nucleons) directly, offer interesting alternatives.

Since the shell model picture of nucleons orbiting in a *spherically* symmetric central field must ultimately be relied upon for the microscopic justification of collective behavior in *nonspherical* nuclear shapes, it is worthwhile to see how this model is capable of producing nonspherical configurations. To many students this is a mysterious point: the explanation is actually quite simple while the confusion stems from a semantic misunderstanding. This discussion will form at least a qualitative justification for the basic concepts of the Nilsson model—the *assumption* of a deformed shell model potential—as well as highlight the central importance of the proton–neutron (p–n) interaction in the development of collectivity and deformation in nuclei. And finally, it will help to foster an appreciation of the key role played by the *distribution* of valence nucleons between protons and neutrons, as opposed to a simple consideration of the *total* number of valence nucleons.

There is really nothing mysterious in the idea of generating deformed shapes from the *spherical* shell model. The model is *spherical* in the sense of the *shape* of the central potential, not the resulting shape of the nucleus. A spherical potential allows no distinction between magnetic substates of a given orbit, all of which are degenerate. A single-particle in a shell model orbit occupies all such substates equally, and overall, its wave function will be spherically symmetric. However, *any* mechanism that yields an *unequal* occupation of m states gives a nonspherical shape. (An orbit may be confined (in a semi-classical sense, of course) to a specific *plane* and therefore gives the nucleus a "bulge" in that plane. Of course, shell model wave functions are not circular either: they are described by spherical harmonics Y_{lm} and linear combinations thereof. But it remains true that the wave function for a given m-state is largely confined to a planar region.) Such mechanisms abound in the shell model without introducing anything fancy (even without residual interactions). Consider, for example, a configuration of two identical nucleons in equivalent orbits, say $|(f_{7/2})^2 J\rangle$, outside a magic core. The 0^+ ground state is spherical, of course, but let us look at the construction of the 2^+ state in terms of m substates. From Table 5.1 we see that many possible m_1, m_2 values, such as $m_1 = 7/2, m_2 = 1/2$ or $m_1 = 5/2, m_2 = 3/2$ give $M = m_1 + m_2 > 2$ and cannot contribute to the 2^+ state. This state is then formed from a nonuniform distribution of m_1, m_2 components and must be nonspherical. Indeed, this is why it has a quadrupole moment (see Eq. 5.17).

A nonuniform magnetic substate distribution is in fact so characteristic of deformation that one of the best known features of the Nilsson (deformed shell) model is a filling of orbits based on their m values instead of their j values. But that jumps ahead of the discussion. We must first discuss how and why j itself may not always be a good quantum number. This is an essential point since "configuration mixing" of single-particle j values *ensures* an unequal m substate distribution and is therefore tantamount to ensuring deformation.

To do this we will show that there is a fundamental difference between the occupation of valence orbits by *like* nucleons (e.g., two protons or two neutrons) and *unlike* nucleons (one proton and one neutron). Consider then, and by way of example, a nucleus with two valence nucleons filling the lower part of the 8–20 shell in the $1d_{5/2}$ and $1d_{3/2}$ orbits, which are separated by ≈ 5 MeV.

We want to consider matrix elements that can admix $d_{5/2}$ and $d_{3/2}$ components. In Chapter 4 we primarily discussed *diagonal* matrix elements of short-range two-body residual interactions. Now, we are dealing with nondiagonal ones (although the diagonal elements still play an important role in modifying the unperturbed energies of the states that mix). Nevertheless, the basic idea is the same: if the particles are not close to one another in the two two-particle configurations, the matrix element will be small. The Pauli principle must also be considered. In addition, as opposed to the diagonal case, here we also need to consider the unperturbed (initial) energy spacing of the two configurations. The whole issue then is just one of two-state mixing. We consider the possible matrix elements:

$$\left\langle d_{3/2}^2 J \,|\, V \,|\, d_{3/2}^2 J \right\rangle, \left\langle d_{5/2}^2 J \,|\, V \,|\, d_{5/2} d_{3/2} J \right\rangle, \left\langle d_{3/2}^2 J \,|\, V \,|\, d_{5/2} d_{3/2} J \right\rangle$$

and concentrate on qualitative effects, ignoring complexities due to angular momentum coupling coefficients.

For *like nucleons* ($T = 1$) the unperturbed energies of the $(d_{5/2})^2$ and $(d_{3/2})^2$ configurations are ≈ 10 MeV apart. Though the individual J states of each are lowered by the diagonal residual interaction (see Fig.4.3 (top)), this lowering is roughly similar in the two configurations. The spacing thus remains ≈ 10 MeV, and it is unlikely that strong mixing will occur. (Recall the pairing discussion: $\Delta \approx 1$ MeV and states are admixed only over an energy range of that magnitude.)

The mixing is also small for matrix elements like $\langle d_{5/2}^2 J | V | d_{5/2} d_{3/2} J \rangle$, since although the unperturbed separation is now only ≈ 5 MeV, the Pauli principle enters in an important way. The configuration $(d_{5/2})^2$ for like nucleons only exists in $J = 0, 2, 4$ states. The $(d_{5/2} d_{3/2})$ configuration does not exist as $J = 0$. Therefore, the strong $J = 0$ interaction is forfeited and we are left with only the $J = 2$ and 4 cases. But, we can apply our geometric analysis of Chapter 4 to these cases. For $J = 4$, for example, the angle θ between the two orbits in a $(d_{5/2})^2$ configuration is $\approx 82°$, while it is $\approx 49°$ for $(d_{5/2} d_{3/2})$. We can see the effect of this if we imagine that one orbit in each configuration is fully aligned with one in the other (the optimum case). Then, to couple these two states (that is, for a finite matrix element), the short-range residual interaction must act over an angular "distance" $\Delta\theta \approx 33°$. We have seen in Chapter 4 that such matrix elements are small. For all three matrix elements, then, the like nucleon configuration mixing amplitudes will be small.

For *unlike nucleons*, the large energy difference (≈ 10 MeV) between $(d_{5/2})^2$ and $(d_{3/2})^2$ configurations again leads to small mixing. However, the *single nucleon mixing* induced by the $\langle d_{5/2}{}^2 J | V | d_{5/2} d_{3/2} J \rangle$ matrix element is not necessarily small. First, the spacing is only ≈ 5 MeV. Second, both configurations exist in $J = 1, 2, 3$, and 4 states and, third, the angles involved favor large matrix elements. For example, for $J = 1$, the angle between the two nucleons in $(d_{5/2})^2$ is $\approx 152°$ while for $(d_{5/2} d_{3/2})$ it is $\approx 156°$. Another way of saying this is that in the $J = 1$ states, the main difference between $(d_{5/2})^2$ and $(d_{5/2} d_{3/2})$ is a flip of one intrinsic spin—the matrix element corresponds to the strong ${}^3S$ interaction.

Therefore, we conclude that $T = 1$ configurations of *identical* nucleons are *not* very mixed by short-range attractive residual interactions, because of the large energy differences between such configurations, because of the consequences of antisymmetrization in determining which spin states are allowed, and because of the magnitude of the matrix elements that do exist. In contrast, configurations of *nonidentical* nucleons can be *strongly* admixed. Moreover, the mixing is a *single nucleon* effect. Therefore, such excitations cannot be absorbed into an effective two-body interaction. The strong admixtures of different *single* nucleon wave functions, in this case $d_{5/2}$ and $d_{3/2}$, implies that the spherical symmetry of the wave functions is lost since the resultant wave functions must have nonuniform m state distributions (e.g., $(d_{5/2})^2$(p–n) has a component $M = 5/2 + 5/2 = 5$ while $(d_{5/2}d_{3/2})$(p–n) has $M_{\max} = 5/2 + 3/2 = 4$. Thus, one can write the single-particle nuclear functions as $\psi = C_{5/2m}d_{5/2} + C_{3/2m}d_{3/2}$. As we shall see, this is exactly the form of Nilsson wave functions for a deformed shell model potential.

Although we have illustrated the idea for a particular case, the argument is general. It is also interesting to note that the C_{jm} coefficients *must* depend on the substate m. In this example, $C_{3/2\ m=3/2}$ can be nonzero, but $C_{3/2\ m=5/2}$ must be zero. Looking ahead for a moment, this, in essence, explains why Nilsson wave functions are m-dependent (m is often called K in the Nilsson model) and also why they are purest for the highest K values since, in that case, no other orbits can contribute admixed amplitudes.

In closing these introductory pages it is worthwhile to re-emphasize that our arguments for the existence of deformation and configuration mixing arose as a consequence of the Pauli principle, which led to a different behavior of $T = 0$ and $T = 1$ configurations, and of the short-range attractive nature of the nucleon residual interaction. Nowhere was it necessary to specify the interaction in detail. Of course, the choice of a specific residual interaction will affect the detailed wave functions that result, but the possibility of nonspherical wave functions is a rather general feature resulting from the particular configurations and interactions allowed by the Pauli principle when nonidentical nucleons are involved.

6.2   Collective excitations in spherical even–even nuclei

One of the most characteristic empirical facts of nuclear systematics is that the shell model picture of nearly independent particle motion under the influence of weak residual interactions in simple configurations breaks down as one adds more and more valence nucleons past magic numbers. Simply put, the residual interactions among a growing number of valence nucleons build up to such an extent that they obliterate much of the underlying shell structure. Products of single particle wave functions become a poor first-order approximation to the real nuclear wave functions. In short, they no longer serve as the most appropriate basis states. In general, in a physical system, one always searches for some suitable set of basis states such that the realistic wave functions are dominated by one or a few components and any admixtures of basis states can be treated as relatively small perturbations. This is not to say that the shell model cannot provide a valid microscopic description of such collective excitations. Indeed, we shall see in a later chapter that the widely used and extremely important RPA and TDA techniques are just such descriptions. Nevertheless, an alternate viewpoint, that approaches the nuclear structure more macroscopically, emphasizes the nuclear shape and excitations of that shape, providing a much simpler, physically transparent approach.

In this chapter we shall discuss a sampling of the most important models for collective excitations in even–even nuclei. As always, the emphasis will be on the physical ideas.

To begin, we recall some of the systematics shown in Chapter 2. Figure 2.8 showed the energy levels of the Sn, Xe, Te, and Cd nuclei. Sn, with $Z = 50$, is singly magic and displays a typical shell model behavior regardless of the number of valence neutrons. The 2_1^+ energy remains high and the 4_1^+, 6_1^+ levels cluster. As soon as valence nucleons are added, for example in Te and Cd (where the two valence protons are counted as holes), $E_{2_1^+}$ drops sharply. The decrease grows as the number of valence neutrons increases. The drop is even faster for Xe, which has four valence protons. Figure 2.15 showed the systematics of the energy ratio $E_{4_1^+}/E_{2_1^+}$. It ranges from values < 2 for shell model nuclei through ~ 2 for nuclei reasonably close to closed shells, then increases sharply towards the limiting value of 3.33 near midshell. As we shall discuss, values near 2.0, 2.5, and 3.33 are all typical of different types of macroscopic collective shapes: spherical harmonic vibrator, axially asymmetric rotor, and axially symmetric rotor, respectively.

Generally, there is a smooth progression from one to another of these idealized collective limits. However, inspection of Figs. 2.13 and 2.14 shows that the systematics is anything but simple. In the next chapter we shall see some easy, physically transparent, ways of understanding this complexity and of parameterizing the behavior of heavy nuclei. Appropriately enough, this approach will be based on a recognition of the importance of the residual p–n interaction among the valence nucleons.

Here, though, we discuss models for each type of behavior, turning later to their evolution from one into another. We start the discussion with the least collective nuclei, which occur soon after closed shells: spherical-vibrational nuclei. The generic concept of vibrational motion in nuclei is widespread and encompasses a great richness of phenomena.

Suppose we expand the residual interaction among the valence nucleons in multipoles; the first few terms will correspond to monopole, dipole, quadrupole, octupole, and hexadecapole components. Each of these carries a parity $\pi = (-1)^\lambda$ where λ is the multipolarity involved. The electric dipole mode corresponds, geometrically, to a shift in the center of mass, and therefore plays little role in the low-lying spectrum of even–even nuclei. At higher energies, however, it induces the well-known giant dipole resonance, which can be pictured as an oscillation of the proton distribution against the neutron distribution. As this mode involves a rather large scale displacement of major components of the nucleus, it requires considerable energy, typically between 8 and 20 MeV. Since it is also a negative parity excitation, it involves excitations from one major shell to the next, again requiring rather high energy. Giant quadrupole (E2) and monopole (E0) modes also exist. The isoscalar (protons, neutrons in phase) monopole resonance energy is a measure of nuclear incompressibility. We shall not discuss the giant resonances in more detail now, only because the emphasis here is on low energy modes.

There is, however, a low-lying magnetic dipole excitation discovered in electron scattering and γ ray inelastic scattering experiments. It occurs, for example, in heavy

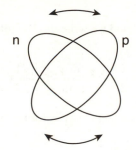

M1 SCISSORS MODE

FIG. 6.1. The M1 scissors mode.

deformed nuclei at roughly 3 MeV and corresponds to an isovector dipole vibration in which the proton and neutron distributions oscillate with respect to each other with a scissors type of motion, as opposed to the linear vibrational motion of the giant electric dipole resonance. The idea is illustrated in Fig. 6.1.

The characteristic excitation is a 1^+ state found in many nuclei at about 3 MeV. It is often fragmented into several components by mixing with the high density of nearby 1^+ states at such excitation energies. This 1^+ mode, characterized by strong M1 electromagnetic transitions to the ground state and first 2_1^+ state, is now known in a number of nuclei and an interesting systematics has been established.

It has been found, for example, that there is an excellent correlation between the B(M1) values from the ground state to the 1^+ state and the B(E2) values from the ground state to the first 2^+ state. Since the latter (see below) scale with the deformation β^2, so do the B(M1) values. This result and other features of these levels have been studied from both geometric and algebraic (IBM-2) viewpoints. We briefly return to isovector modes at the end of Section 6.5.

Another proton–neutron mode is 'magnetic rotation' which gives regular spectra but is not collective in the traditional sense. Here, initially perpendicular single particle proton and neutron angular momentum vectors close like a shears, leading to successively higher J states.

6.2.1 Quadrupole vibrations

The next vibrational mode, that we shall consider in detail, is the electric quadrupole or E2 vibrational mode. It appears in different guises in different categories of nuclei. Near closed shells, where the nuclei are spherical in their ground state, the action of a quadrupole residual interaction causes the nucleus to oscillate in shape, taking on a range of quadrupole distortions as a function of time.

We will first discuss the vibrator using a coordinate representation and then turn to the powerful formalism of second quantization. For a nucleus that oscillates through a series of shapes of multipole order λ we can write the radius as

$$R = R_0[1 + \sum_{\mu} \alpha_{\lambda\mu} Y_{\lambda\mu}(\theta, \phi)] \qquad (6.1)$$

For quadrupole distortions $\lambda = 2$, and we write the Hamiltonian in terms of the $\alpha_{2\mu}$ and their time derivatives as

$$H = T + V = \frac{1}{2} B \sum_\mu \left| \frac{d\alpha_{2\mu}}{dt} \right|^2 + \frac{1}{2} C \sum_\mu |\alpha_{2\mu}|^2 \tag{6.2}$$

B plays the role of a mass parameter and C is a restoring force. Differentiating H, which is a constant of the motion, gives

$$B \frac{d^2 \alpha_{2\mu}}{dt^2} + C\alpha_{2\mu} = 0$$

This is identical in form to the differential equation of motion of a harmonic oscillator and hence we obtain the solution that (each of the) $\alpha_{2\mu}$ undergoes oscillations with a frequency

$$\omega = \sqrt{C/B} \tag{6.3}$$

and the vibrational energy is $\hbar\omega$.

Many elegant results for a vibrator, especially when one considers multi-phonon states (see below), are obtained using the formalism of second quantization. In this approach one defines operators that create and destroy vibrational quanta and one then writes a general Hamiltonian for such a system in terms of those operators. As we shall see, an appropriate Hamiltonian is

$$H = E_0 + \hbar\omega \sum_\mu \left(\mathbf{b}_{2\mu}^\dagger \mathbf{b}_{2\mu} + \frac{1}{2} \right) + \sum_L C_L \left\{ \left[\mathbf{b}_2^\dagger \times \mathbf{b}_2^\dagger \right]^{(L)} \cdot \left[\mathbf{b}_2 \times \mathbf{b}_2 \right]^{(L)} \right\}^{(0)} \tag{6.4}$$

where E_0 is the zero-point energy and the operators $\mathbf{b}_2^\dagger$ and $\mathbf{b}_2$ create and destroy this quadrupole vibration: $\psi_{\text{ph}} = \mathbf{b}_2^\dagger |0\rangle$.

For simplicity of notation, and to keep the essential physics to the fore, we shall henceforth usually drop the subscripts on the operators $\mathbf{b}_{2\mu}$. In the same spirit, summations over the components μ will usually be implied rather than explicit. Since we shall make frequent use of phonon or boson creation and destruction operators, we pause for a moment to recall some key properties of such operators in the formalism of second quantization. The basic defining rules for arbitrary creation and destruction operators $\mathbf{b}^\dagger$ and $\mathbf{b}$ are:

$$\mathbf{b}|n_b\rangle = \sqrt{n_b}|n_b - 1\rangle \tag{6.5}$$

and

$$\mathbf{b}^\dagger|n_b\rangle = \sqrt{n_b + 1}|n_b + 1\rangle \tag{6.6}$$

where $|n_b\rangle$ is a state with n_b bosons. Here $\mathbf{b}$ refers to quadrupole phonons; later, in the discussion of the IBA, $\mathbf{b}$ will refer to either s or d bosons. From these definitions

Table 6.1 *m scheme for two-quadrupole phonon states**

$J_1 = 2$ m_1	$J_2 = 2$ m_2	$M = \sum m_i$		J
2	2	4	⎤	
2	1	3		
2	0	2		4
2	−1	1		
2	−2	0	⎦	
1	1	2	⎤	
1	0	1		2
1	−1	0	⎦	
0	0	0	⎤⎦	0

*Only positive total M values are shown: the table is symmetric for $M < 0$. The full set of allowable m_i values giving $M \geq 0$ is obtained by the conditions $m_1 \geq 0$, $m_2 \leq m_1$.

$$\mathbf{b^\dagger b}|n_b\rangle = \mathbf{b^\dagger}\sqrt{n_b}|n_b - 1\rangle = \sqrt{n_b}\sqrt{(n_b - 1) + 1}|n_b\rangle$$

or

$$\mathbf{b^\dagger b}|n_b\rangle = n_b|n_b\rangle \tag{6.7}$$

So, $\mathbf{b^\dagger b}$ simply counts the number of b-type bosons. Thus from the second term in H we see that $\hbar\omega$ is just the energy, relative to the ground state energy E_0, needed to create the quadrupole phonon excitation, which naturally carries a spin and parity 2^+.

There is no reason, except the limitations provided by the Pauli principle when the microscopic structure of these vibrations is considered (see below and Chapter 10), that prevents more than one phonon excitation from simultaneously existing. These multiphonon states $\psi_{N\text{ph}} = (\mathbf{b^\dagger})^N|0\rangle$ will correspond to higher and higher nuclear levels. From Eq. 6.7, the second term in H is the product of the number of quadrupole phonons and the energy of each. Clearly, at this stage in the Hamiltonian one has a purely harmonic vibrational spectrum, where the excitation energy is linear in the number of phonons: for an N_{ph}-phonon state, $E_x = \hbar\omega(N_{\text{ph}} + 5/2)$, since the quadrupole mode is a 5-dimensional oscillator.

To continue, we must first turn to the question of which spin states are allowed in multiphonon excitations. For the two-quadrupole phonon case, it is clear that the maximum possible spin is 4^+. But it turns out that only a triplet of levels with spins $J^\pi = 0^+, 2^+, 4^+$ is allowed. There are many ways to derive this result. Perhaps the most elegant is the use of Young tableaux, but here we shall use the simpler and more straightforward, though more tedious, method of the m-scheme. The essential difference between the use of the m-scheme for phonon excitations and for single-particle excitations is the recognition that phonons, having integer spins, behave as bosons. Therefore, the Pauli

Table 6.2 *m scheme for three-quadrupole phonon states*[*]

$J_1 = 2$ m_1	$J_2 = 2$ m_2	$J_3 = 2$ m_3	M	J
2	2	2	6	
2	2	1	5	
2	2	0	4	
2	2	−1	3	
2	2	−2	2	
2	1	1	4	
2	1	0	3	
2	1	−1	2	
2	1	−2	1	
2	0	0	2	
2	0	−1	1	
2	0	−2	0	
2	−1	−1	0	
1	1	1	3	
1	1	0	2	
1	1	−1	1	
1	1	−2	0	
1	0	0	1	
1	0	−1	0	
0	0	0	0	

$$J = 6, 4, 3, 2, 0$$

[*]Only positive total M values are shown; the table is symmetric for $M < 0$. The full set of allowable m_i values giving $M \geq 0$ is obtained by the conditions $m_1 \geq 0$, $m_3 \leq m_2 \leq m_1$.

principle is not applicable and the wave functions must be totally *symmetric*. This means that all combinations of m states are allowed. Table 6.1 shows the m-scheme counting of substates for the case of two quadrupole phonons and shows that the allowed spins are as stated previously. It was the recognition of this result and the experimental fact that, in many nuclei not far from closed shells, $R_{4/2} \equiv E(4_1^+)/E(2_1^+) \sim 2.2$ that led to the proposal of a vibrational interpretation for these nuclei in the 1950s.

The m-scheme analysis for the three-phonon case is given in Table 6.2, which shows that this excitation comprises a quintuplet of levels (at three times the single phonon energy), with spins $J^\pi = 0^+, 2^+, 3^+, 4^+, 6^+$. This harmonic picture of single- and multi-phonon excitations is illustrated in Fig. 6.2.

To pursue the study of multiphonon states, it is necessary to delve more deeply into their structure. Consider the three-phonon levels. As Fig. 6.3 illustrates, the 6^+ state can only be made in one way: by aligning the angular momentum of a single phonon

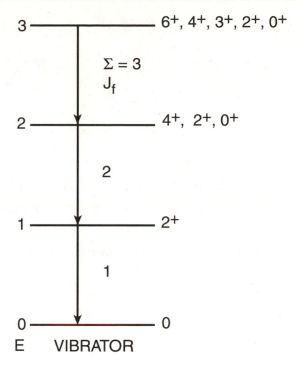

FIG. 6.2. Low-lying levels of the harmonic vibrator phonon model.

state with that of the 4^+ two-phonon level. Similarly, the 0_3^+ three-phonon level can only be constructed by antialigning a single quadrupole phonon with a 2^+ two-phonon state. However, the other three-phonon levels can be constructed in more than one way. For example, the 2^+ level can be made by coupling a quadrupole phonon with the 2^+ two-phonon state or by antialigning a quadrupole phonon with the 4^+ two-phonon state. In similar fashion, the 3^+ and 4^+ three-phonon states can be made by coupling the third quadrupole phonon to more than one of the two-phonon states. The wave functions for the 2^+, 3^+, and 4^+ three-phonon states are therefore linear combinations of two terms, and the relative amplitudes are simply *phonon* coefficients of fractional parentage whose squares give the relative likelihoods that the three-phonon state is made in a certain way. It is often useful to know these coefficients, so we give them for the $N_{ph} = 3$ states in Table 6.3. We will encounter two applications of these coefficients momentarily.

An important aspect of the vibrational model centers on electromagnetic transition rates since they are particularly sensitive to coherence properties in nuclear wave functions. We saw, for example, in Fig. 2.16, the systematics of B(E2 : $0_1^+ \rightarrow 2_1^+$) values throughout the periodic chart. Although small and comparable to single-particle estimates in light nuclei, they attain values orders of magnitude larger in heavy deformed regions. Intermediate values characterize the realm of spherical-vibrational nuclei we are presently considering.

In general, radiation can be given off when any nucleon changes its orbit. For example,

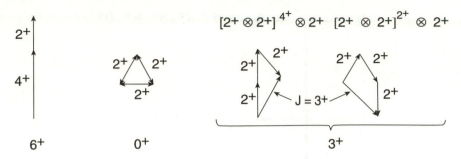

CONSTRUCTION OF SOME 3–PHONON STATES

FIG. 6.3. Two-phonon composition of three-phonon states.

Table 6.3 *Relative coefficients of fractional parentage for three-phonon quadrupole vibrator states*[†]

J_{3ph}	0^+	J_{2ph} 2^+	4^+
6^+			$\sqrt{3}$
4^+		$\sqrt{11/7}$	$\sqrt{10/7}$
3^+		$\sqrt{15/7}$	$-\sqrt{6/7}$
2^+	$\sqrt{7/5}$	$\sqrt{4/7}$	$\sqrt{36/35}$
0^+		$\sqrt{3}$	

[†] The normalization is to $\sqrt{N}$. For the three-phonon states, J_{3ph}, the squares of the coefficients for each J_{2ph} also give the relative values of B(E2 : $J_{3ph} \rightarrow Ji_{2ph}$). For example, the B(E2 : $4^+_{3ph} \rightarrow 2^+_{2ph}$) and B(E2 : $4^+_{3ph} \rightarrow 4^+_{2ph}$) values are in the ratio $11/10 = 1.1$. See text.

changes in single-particle orbits in shell model nuclei are often accompanied by the emission of γ-radiation. While collective excitations are clearly not of single-particle nature and the destruction of one does not correspond to a single change of orbit by an individual nucleon, we will see in Chapter 10 that their wave functions can be represented as coherent linear combinations of single-particle–hole (or, equivalently, two quasi-particle) excitations. Therefore, not only are γ-ray transitions between phonon levels permitted, but the coherence can make them particularly strong. Since a two-phonon excitation involves a superposition of two linear combinations of one-body excitations, the destruction of two-phonons would require a simultaneous destruction of two particle–hole excitations or four quasi-particles. Therefore, such transitions are forbidden and one has the characteristic phonon model selection rule,

$$\Delta N_{ph} = \pm 1 \qquad (6.8)$$

where N_{ph} is the number of phonons. The argument for this selection rule (obtained

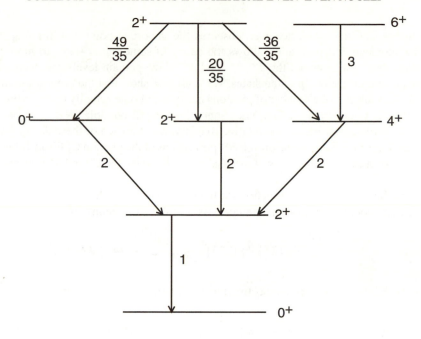

B (E2) VALUES FOR DECAY OF
MULTI –PHONON STATES

FIG. 6.4. B(E2) values in the harmonic vibrator model.

here for quadrupole vibrations of spherical nuclei) is rather general and applies to any phonon structure described as a linear combination of one-particle excitations. We will encounter it repeatedly in various applications.

Let us now consider the magnitude of these B(E2) values between phonon states as illustrated in Figs. 6.2 and 6.4, where we assign a value of unity for the decay of the one-phonon 2^+ state to the ground state. Since, in first order, multiphonon excitations simply consist of the piling on or superposition of more than one identical phonon, it might seem that the B(E2) value for the decay of the two-phonon state would also be unity. However, this neglects the fact that there are two phonons in the initial state and that either one of them may be destroyed. This gives twice as many decay possibilities and therefore $B(E2 : (N_{ph} = 2) \rightarrow 1) = 2$, as indicated in Fig. 6.2. Continuing this, one can state a general expression for the decay of the N_{ph}-phonon state to the $(N_{ph} - 1)$-phonon state. A transition $N_{ph} \rightarrow N_{ph} - 1$ must be accomplished by an E2 operator of the form **b**, that is, a one-phonon destruction operator. By Eq. 6.5.

$$\mathbf{b}|N_{ph}\rangle = \sqrt{N_{ph}}|N_{ph} - 1\rangle$$

and so the B(E2) value is proportional to N_{ph}.

This general statement, however, obscures the important point that, for $N_{ph} \geq 3$, angular momentum conservation allows the decay of some initial states to more than one final state. For example, the $2^+, 3^+$, and 4^+, three-phonon levels can each decay to two or more of the two-phonon states. We shall now show that the proportionality of these B(E2) values to the number of phonons in the initial state actually refers to the *sum* of the B(E2) values from a given N_{ph}-phonon level to all possible ($N_{ph} - 1$)-phonon levels. It is trivial to work out the relative B(E2) values for each of these decay routes: this exercise is in fact one of the promised applications of the phonon CFP's in Table 6.3. Consider as an example the decay of the $N_{ph} = 3, 4^+$ level to the 2^+ and 4^+ two-phonon states.

From Table 6.3, the wave function for the three-phonon 4^+ level can be written, in obvious notation, where subscripts indicate the number of phonons, as

$$\psi_3(4^+) = \frac{1}{\sqrt{3}} \left[\sqrt{\frac{11}{7}} \left[\psi_2\left(2^+\right) \psi_1\left(2^+\right) \right]^{4^+} + \sqrt{\frac{10}{7}} \left[\psi_2\left(4^+\right) \psi_1\left(2^+\right) \right]^{4^+} \right]$$

Then, the E2 matrix element connecting this level to the 2^+ two-phonon state is

$$\langle \psi_2\left(2^+\right) \| E2 \| \psi_3\left(4^+\right) \rangle = \frac{1}{\sqrt{3}} \left[\sqrt{\frac{11}{7}} \langle \psi_2\left(2^+\right) \| \mathbf{b} \| \left[\psi_2\left(2^+\right) \psi_1\left(2^+\right) \right]^{4^+} \rangle \right.$$
$$\left. + \sqrt{\frac{10}{7}} \langle \psi_2\left(2^+\right) \| \mathbf{b} \| \left[\psi_2\left(4^+\right) \psi_1\left(2^+\right) \right]^{4^+} \rangle \right]$$

But the second term vanishes because $\psi_2\left(2^+\right)$ and $\psi_2\left(4^+\right)$ are orthogonal. Hence, using Eq. 6.5 and setting $B(E2 : 2_1^+ \rightarrow 0_1^+) = 1$, we get

$$B\left(E2 : 4_{3ph}^+ \rightarrow 2_{2ph}^+\right) = 3/3\,(11/7) = 11/7$$

Similarly,

$$B\left(E2 : 4_{3ph}^+ \rightarrow 4_{2ph}^+\right) = \frac{10}{7}$$

and we see that the three-phonon → two-phonon B(E2) values are proportional to the squares of the three-phonon CFP coefficients in Table 6.3. The table can be used to obtain the B(E2) values we have not worked out. The results for the decay of the 6^+ and 2^+ three-phonon states are illustrated in Fig. 6.4. We also see an example of our general result, namely,

$$\sum_{J_{2ph}} B\left(E2 : 4_{3ph}^+ \rightarrow J_{2ph}^+\right) = 3B\left(E2 : 2_{1ph}^+ \rightarrow 0_{gs}^+\right)$$

and similarly for the other three-phonon levels.

The reader is cautioned to bear these results in mind, since one occasionally encounters statements such as that the B(E2) for the decay of a three-phonon state to a two-phonon state is three times that for the decay of the one-phonon state to the ground state. The proper relation involves the sum of the decays to the possible final states.

Residual interactions between phonons can break the degeneracy of the multiplets in multi-phonon states. In general, the calculation of such anharmonicities is complex and depends on the specific j shells and interactions involved. However, there is one situation in which a very simple and elegant result, easily tested experimentally, can be obtained essentially by inspection of the structure of the multi-phonon wave functions.

This brings up the second application of the CFPs in Table 6.3. Suppose we assume that degeneracy breaking is caused by residual two-body interactions only. This means that the level energies in a two-phonon state are not simply twice the one-phonon energy, but differ because of a residual interaction between the two phonons. This, in fact, is the effect of the third term of H in Eq. 6.4: it represents interactions between two phonons.

In the three-phonon states, the same residual interactions apply and our assumption simply states that there are no mutual interactions among the three phonons at a given time. In this case, without *ever* specifying the residual interaction, the nature, or the microscopic structure of the phonon, one can immediately deduce the anharmonic energies of the three-phonon states from those of the two-phonon levels. This situation is illustrated in Fig. 6.5, where the energies of the $0^+, 2^+$, and 4^+ two-phonon states are written in terms of the harmonic value $2E_{2_1^+}$, plus a perturbation ε_j ($\varepsilon_0, \varepsilon_2, \varepsilon_4$ are the anharmonicities). Consider now the three-phonon levels. As we have seen, there is only one way to make the 6^+ level: by aligning a single 2^+ quadrupole phonon with a pair of phonons coupled so as to produce the 4^+ two-phonon state. In the three-phonon 6^+ state, there are three possible pairs of (indistinguishable) phonons that can couple and interact in forming the intermediate 4^+ plus state. Therefore, the anharmonicity (the deviation of E_{6^+}, from $3E_{2_1^+}$) is three times the anharmonicity in the 4^+ two-phonon state, or $3\varepsilon_4$. In the same fashion, the 0^+ 3-phonon state can only be made by antialigning one phonon with the 2^+ two-phonon state. Again, there are three ways to do this, and the anharmonicity in the three-phonon 0^+ energy will be triple the anharmonicity in the two-phonon 2^+ level, or $3\varepsilon_2$. The other three-phonon states, which can be made in two or more ways from the 2-phonon levels, will have total energy anharmonicities given by the relative proportions of their wave functions arising from the various two-phonon states. These relative proportions are given by the CFP coefficients of Table 6.3, and the resulting energy anharmonicities are shown in Table 6.4 and illustrated in Fig. 6.5.

It is worthwhile to reiterate what has been derived here. We have never specified the structure of the phonon itself. We have also never specified the nature of the residual interaction except to state that it is two-body. Nevertheless, from the observed anharmonicities in the two-phonon states, we have been able to derive predicted anharmonicities for the three-phonon levels.

This model can be tested by observation of the three-phonon levels. We will discuss the experimental situation in multi-phonon states just below. For now, we note that, if discrepancies with the relations in Table 6.4 are found, then we immediately know that they must arise either from three-body interactions, from Pauli principle effects in the multi-phonon states (see below), or from interactions with other types of (nearby)

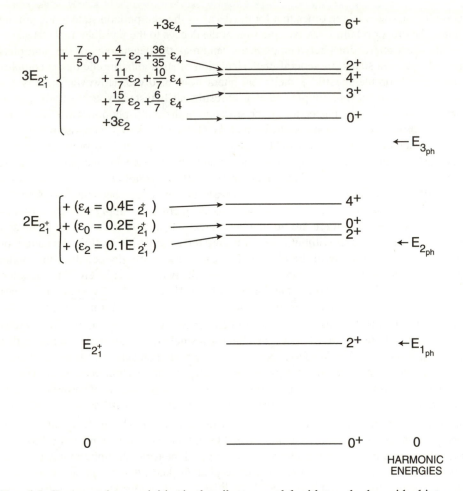

FIG. 6.5. Energy anharmonicities in the vibrator model with two-body residual interactions.

excitations. Our analysis helps to isolate the specific mechanisms leading to the observed anharmonicities.

The vibrational model was originally proposed in the early 1950s, and it was thought that numerous examples of such a structure were observed empirically. Triplets of levels with spins $0^+, 2^+, 4^+$, were known in nuclei near closed shells at slightly more than twice the energy (typically, $E_{2ph}/E_{1ph} \sim 2.2$) of the first 2^+ state. However, it was commonly found that the energy splitting among the two-phonon levels is comparable to the one-phonon energy; basically, the structural effects leading to anharmonicities are comparable to those involving the phonon itself. Moreover, the predicted phonon model B(E2) values were significantly violated. Perhaps most importantly, over the last

Table 6.4 *Energies of the three-phonon quintuplet states in terms of the two-phonon anharmonicities.*

N_{ph}	J	Energy* (relative units)
0	0	0
1	2	1
2	0	$2 + \varepsilon_0$
2	2	$2 + \varepsilon_2$
2	4	$2 + \varepsilon_4$
3	0	$3 + 3\varepsilon_2$
3	2	$3 + 7/5\varepsilon_0 + 4/7\varepsilon_2 + 36/35\varepsilon_4$
3	3	$3 + 15/7\varepsilon_2 + 6/7\varepsilon_4$
3	4	$3 + 11/7\varepsilon_2 + 10/7\varepsilon_4$
3	6	$3 + 3\varepsilon_4$

*ε_0, ε_2, and ε_4 are defined as the deviations of the 0^+, 2^+, and 4^+ level energies of the two-phonon triplet from $2E_{2_1^+}$.

decades, additional low-lying levels near the two-phonon states have been detected in many nuclei.

While in a number of these nuclei there may be an *underlying* vibrational structure, it seems that it is often significantly perturbed and perhaps admixed with other degrees of freedom. To understand these issues, we need to delve more deeply into the nature of, and the experimental search for, multi-phonon vibrator levels. This discussion will bring up several fundamental aspects of these states and of their experimental elucidation.

The vibrations of spherical nuclei, and other vibrational excitations such as the intrinsic excitations of deformed nuclei known as γ and β vibrations or octupole excitations (see Section 6.4), and the various giant resonances in both spherical and deformed nuclei, are all fundamental collective modes of the nuclear quantal system. As we have said, they carry integer spins and can be described as boson excitations. However, nuclei are composed of fermions, not bosons, and fermions must obey the Pauli principle. Since nuclear phonons can be described microscopically (see Chapter 10) as linear combinations of particle–hole excitations (or 2 quasi-particle excitations) relative to the Fermi surface, the interplay of the Pauli principle and the single particle structure of the phonons comes into conflict in multi-phonon states. To see this, suppose that a given phonon consists (as is often the case in practical calculations) of only a few individual particle–hole components, each with a substantial amplitude. To create a 2 (or higher) phonon state, the same particle–hole excitation must be replicated. Since the Pauli principle limits the occupation of shell model orbits of spin j to $2j + 1$ identical nucleons, one may easily encounter situations where either a given "source" (hole) orbit is emptied or a "destination" (particle) orbit is filled by a phonon excitation (or with the creation of n-phonons). It would then be impossible to superpose yet another phonon excitation (e.g., to create an $(n + 1)$-phonon mode). This particular amplitude in the multi-phonon wave function would be "blocked".

Another way of saying this is that, at some point in the creation of successive n-phonon states, the Pauli principle forces an alteration in the structure of the creation operator $\mathbf{b}^+$ to account for this blocking. If this effect is small, the multi-phonon mode is anharmonic–that is the energy will not be exactly n-times the one phonon energy, the degeneracies in multi-phonon multiplets may be broken, and deexcitation transition rates (e.g., B(E2) values) will not follow the exact predictions or selection rules of the phonon model. If the blocking is severe, the multi-phonon state may not exist in any meaningful sense—that is, the collective strength may be so fragmented as to be virtually destroyed. The question of the existence and collectivity of multi-phonon modes is thus of fundamental importance in understanding the many-body, strongly interacting fermionic quantal system that is the atomic nucleus.

Experimental searches for multi-phonon states are beset by two problems: the high level density in the excitation energy regions where they are expected and the phase space dependence of γ ray transition rates, in conjunction with the selection rules for the decay of multi-phonon states.

As we have seen, an E2 transition can destroy only a single phonon. Therefore, an n-phonon state should have large E2 decay *matrix elements* to the $(n - 1)$-phonon states, and not to the ground state. However, E2 *transition rates* are proportional to E_γ^5 times the square of the appropriate matrix elements. Therefore, if a 2-phonon level is at roughly twice the energy of the 1-phonon state, then, separately from the relative values of the intrinsic matrix elements, the E2 transition rate to the ground state is enhanced relative to that to the 1-phonon state by a factor of approximately $2^5 = 32$. This enhancement may easily more than compensate the larger $\Delta N_{ph} = 1$ B(E2) value and hence the allowed γ-ray transition is likely to be weaker in intensity than higher energy, forbidden decays.

This problem is exacerbated since γ ray detectors exhibit a response function to γ rays of a given energy that consists of a full energy photopeak plus lower energy processes related to phenomena such as Compton scattering. Higher energy γ ray transitions contribute a spectral background at lower energies. Therefore the lower the energy of an allowed deexcitation transition the higher the background on which it is superimposed. The Compton suppression techniques commonly used in modern γ ray spectroscopy only partially alleviate this problem.

Nevertheless, as of the mid-1980s, a number of 2-quadrupole phonon multiplets of 0^+, 2^+, and 4^+ states in spherical nuclei had been identified. As noted, these 0^+, 2^+, 4^+ triplets tend to occur at about 2.2 times the single phonon energy. Examples are known in the Se region, in the Ru–Cd nuclei, and in Te: a few of these were shown in Fig. 2.9.

A particularly interesting case study is the Cd isotopes shown in Fig. 6.6: These nuclei, especially ^{114}Cd, historically have been considered the best prototypes of vibrational behavior. However, it is clear from Fig. 6.6 that there are four and sometimes five levels clustered together in the two-phonon energy region.

To understand these states and the structure of these intriguing nuclei in fact turns out to be a challenging task. There is clearly a vibrator aspect. However, there have also been suggestions that the "extra" levels in Cd are so-called "intruder" states. It is not

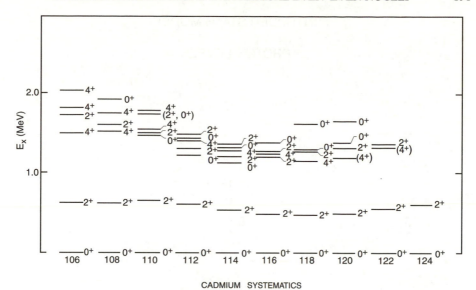

CADMIUM SYSTEMATICS

FIG. 6.6. Energy systematics in the Cd isotopes. Note the intermingling of two-phonon triplet states with extra levels (Aprahamian, 1984).

yet clear which picture gives a better description. However, intruder states are known in other regions and a discussion of the roles of multi-phonon and/or intruder levels in the Cd isotopes provides an ideal introduction to both subjects, and is therefore quite useful. Moreover, the discussion will introduce some basic ideas relating to the p–n interaction, which we shall return to later. We have already alluded several times to the idea that this interaction is essential to the development of collectivity and deformation in nuclei. Here we will encounter our first specific example of this.

Cd has $Z = 48$, and therefore two proton holes relative to the $Z = 50$ magic number. It is of course possible to excite the Cd nuclei by elevating two protons from the $Z = 28$–50 valence shell into the next higher, $Z = 50$–82 shell. The idea is sketched in Fig. 6.7. Normally this requires considerable energy since it involves raising two nucleons across a major shell gap. However, the residual p–n interaction is strong and attractive, and as such introduces a major modification to this first-order energy. In a simple picture, one can view the "normal" states of Cd as consisting of two proton holes interacting with some number of valence neutrons. In the *intruder* state, there are four proton holes in the $Z = 28$–50 shell *plus* two proton particles in the $Z = 50$–82 shell. In a sense, there are six valence protons that can now interact with the same number of valence neutrons. In this picture the intruder states in Cd are analogous to the normal states in Ba, as suggested in Fig. 6.7.

We have already seen that the more valence nucleons there are of both kinds, the "softer" the structure will be. Sufficient numbers of valence protons and neutrons lead to deformed shapes. Therefore, in this schematic view, intruder levels should be more deformed than the "normal" levels. Moreover, since the attractive interaction is three

INTRUDER STATE MODEL

PROTON LEVELS

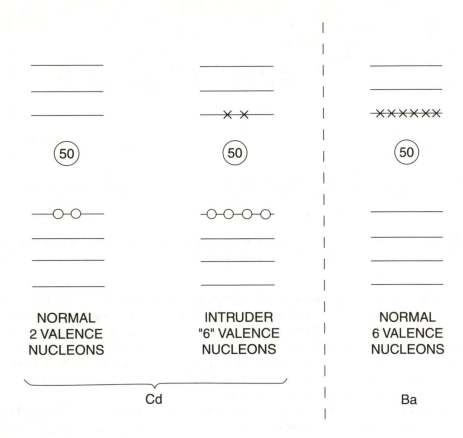

FIG. 6.7. Schematic illustration of intruder excitations and normal states in Cd. The normal states of Ba are shown for comparison with the Cd intruders.

times greater than in the normal levels, the intruder state excitation energies are lowered relative to their unperturbed (no p–n interaction) value. This lowering increases approximately linearly with the number of neutrons, and therefore, one expects the energies of these intruder states to drop from approximately twice the shell gap for $N \sim 50$ or 82 toward midshell, where the p–n interaction strength is maximum.

This intruder state model is quite general and such excitations, once thought to be rare, are now known to abound throughout the periodic table. Perhaps the best known example is in the Pb region, whose systematics are shown in Fig. 6.8.

Most intruder levels observed to date are proton excitations. This is related to the role of a strong p–n interaction in lowering these levels. Since there is a neutron excess

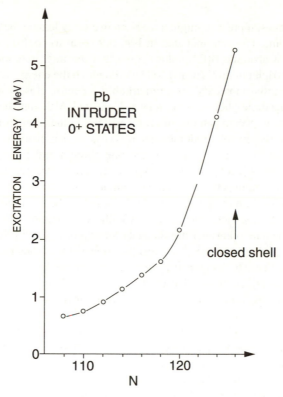

FIG. 6.8. Systematics of 0^+ intruder levels in the Pb isotopes (Van Duppen, 1985).

in heavy nuclei, the excited valence protons in the intruder state occupy the same shell as the neutrons, thus enhancing the p–n interaction (see Fig. 3.10).

The concept of intruder states is far more important than the explanation of a few bothersome levels. It is closely connected with the origin of deformation itself, as we shall discuss in Chapter 7.

The discovery of good candidates for a full 3-phonon quintuplet of states in ^{118}Cd in 1984 added new information relevant to the vibrator-intruder issue and radically altered the experimental status of searches for multi-phonon excitations. At the time of that work, the study of, and search for, multi-phonon states had actually withered as an active topic in nuclear structure, due both to the above mentioned experimental difficulties as well as to theoretical pre-dispositions to doubt their existence. The ^{118}Cd work therefore came as a surprise and has led to a revival and rebirth of interest in multi-phonon modes so that their study (whether it be low-lying quadrupole excitations, quadrupole-octupole modes, or giant resonances) is now more active than ever. This spurt in activity has also been facilitated due to a remarkable growth in experimental technology, especially in the area of γ-ray detectors.

The ^{118}Cd results are shown in Fig. 6.9. Since this nucleus has $N = 70$ one can expect that the intruder states, if present, might have risen in energy leaving behind a reasonable vibrational spectrum. The data included in Fig. 6.9 seem to confirm this expectation, although studies of absolute B(E2) values show that the interpretation is not quite so simple. There is a triplet of levels near 1200 keV in which the energy separation is much less than $E(2_1^+)$. Furthermore, and most remarkably, an entire closely spaced quintuplet of candidates for the three-phonon multiplet was identified. Although there are significant deviations from the expected patterns of relative B(E2) values compared to the phonon model predictions, the overall predominance of $\Delta N_{ph} = 1$ transitions is well satisfied. On the right of Fig. 6.9, the average relative one-phonon and two-phonon changing transition B(E2) values $[(\Delta N_{ph} = 1)/(\Delta N_{ph} = 2)]$ are indicated: there is at least an order of magnitude preference for the phonon model selection rule. On the left are the predicted energies for the three-phonon quintuplet based on the anharmonicities observed in the two-phonon triplet. Here, the agreement with experiment is only qualitative. The deviations could be due to intruder-normal state mixing or to Pauli principle effects.

In the interim period since the ^{118}Cd results extensive new data have appeared and, now, by far, the best case to compare the vibrator with a mixed vibrator-intruder scheme is ^{112,114}Cd. The energies of possible multi-phonon levels in these nuclei are highly anharmonic, which suggest a strongly perturbed vibrator scenario, yet the B(E2) values actually agree extraordinarily well with the harmonic vibrator model predictions, including candidates even for 5-phonon states. These B(E2) values for ^{114}Cd are summarized in Table 6.5 in comparison with the predictions for a harmonic vibrator. Note that, after normalizing to the $2_1^+ \rightarrow 0_1^+$ B(E2) value, there are no free parameters. The agreement is astonishingly good. Nearly all the (few) discrepancies concern rather weak transitions. Nevertheless, as we have indicated, alternate interpretations are possible, in particular, that some levels are proton intruder configurations from above the $Z = 50$ shell gap, lowered in energy by the proton–neutron interaction. It is generally estimated (Heyde, 1988) that the intruder-normal mixing matrix elements are ~ 100 keV and, in fact, calculations by Jolie and co-workers (Délèze, 1993a,b) that incorporate such mixing have been quite successful. Such a description gives good predictions for the energies and almost as good results as the vibrator for the B(E2) values. Of course, it also involves more parameters.

The interpretation of these data in terms of multi-phonon states and/or intruder configurations is therefore, in the end, still a matter of debate: the alternative scenarios account nearly equally well for the data. What does seem clear, however, is that at least some states with 4 and possibly 5 phonons survive intact. To understand how this can occur, and at the same time, why the energies are so anharmonic, is a real challenge for microscopic theories.

Multi-phonon states of different multipolarities have been studied. Among the best established cases is a double phonon $3^- \otimes 3^-$ multiplet in doubly magic ^{208}Pb (see Yeh, 1998). The 3^- level occurs at 2.6 MeV and decays to the ground state with a B(E3) value of 34 W.u. [An E3 W.u. is defined as W.u. (E3) $= (1/16\pi)(1.2A^{1/3})^6 e^2$ fm^6.] The 2-phonon multiplet should consist of $J^\pi = 0^+, 2^+, 4^+$, and 6^+ levels. At almost exactly the harmonic value, three possible 2-phonon states have been identified which decay by strong E3 or E1 transitions to the 3^- level, namely a 0^+ state at 5241 keV, a 4^+ level

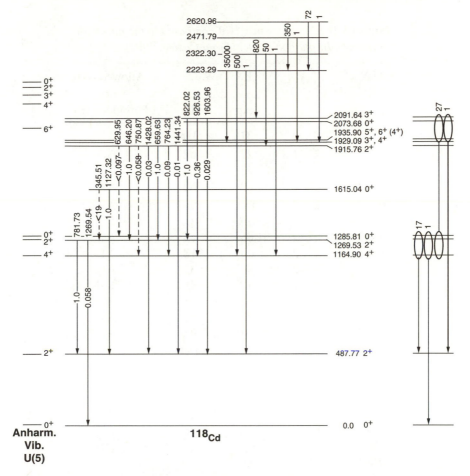

FIG. 6.9. Level scheme of ^{118}Cd showing the one-, two-, and three-phonon states as well as an intruder 0^+ level at 1615 keV and possible candidates for four-phonon excitations above 2.2 MeV. On the right are shown the average $(\Delta N_{\mathrm{ph}} = 1)/(\Delta N_{\mathrm{ph}} = 2)$ branching ratios. On the left are the predictions for the three-phonon states assuming the empirically observed anharmonicities in the two-phonon states. These are the same predictions one would obtain in the U(5) limit of the IBA (Aprahamian, 1987).

at 5216 keV and a tentative 2^+ state at 5286 keV. (Note that E1 decays are, besides E3 transitions, also a likely signature of the decay of $3^- \otimes 3^-$ double octupole states when $|J_{2\mathrm{ph}} - 3^-| = 1$.)

In ^{147}Gd at 997 keV there is a $13/2^+$ level with assigned structure $7/2^- \otimes 3^-$ (outside a ^{146}Gd doubly magic core–treating $Z = 64$ as magic). A $19/2^-$ level at 2572 keV has been assigned as $7/2^- \otimes 3^- \otimes 3^-$ and decays to the $13/2^+$ level with a very collective B(E3) value of 57 W.u. (see Kleinheinz, 1982).

Table 6.5 B(E2) *values in* ^{114}Cd *compared to the harmonic vibrator multi-phonon model (From Casten, 1992)*

$J_i^\pi \, N_{ph}$	$J_f^\pi \, N_{ph}$	B(E2) W.u.[a)]		Allowed (A) or Forbidden (F)
		Exp	Th	
2^+ 1	0^+ 0	31	31[a]	A
0^+ 2	2^+ 1	27	62	A
2^+ 2	2^+ 1	22	62	A
4^+ 2	2^+ 1	62	62	A
0^+ 3	2^+ 2	136	93	A
	2^+ 1	0.3	0	F
2^+ 3	0^+ 2	65	43	A
	2^+ 2	34	18	A
	4^+ 2	46	32	A
	0^+ 1	0.3	0	F
3^+ 3	2^+ 2	67	66	A
	4^+ 2	30	27	A
6^+ 3	4^+ 2	119	93	A
0^+ 4	2^+ 3	79	124	A
	2^+ 1	11	0	F
4^+ 4	2^+ 3	117	80	A
	3^+ 3	< 22	4	A
	2^+ 2	32	0	F
	4^+ 2	16	0	F
	2^+ 1	0.5	0	F
8^+ 4	6^+ 3	86	124	A
6^+ 5	4^+ 4	129	114	A

a) Normalized to the data for the $2_1^+ \to 0_1^+$ transition: 1 W.u. $= 0.00000594$ $A^{4/3}e^2b^2$.

It is also possible to have "mixed" multi-phonon excitations. The most likely low-lying case of this would be a $2^+ \otimes 3^-$ mode. This gives a quintuplet of levels, of spins $1^-, 2^-, 3^-, 4^-$, and 5^-. The Pauli effects should be less severe in such cases than for identical phonons precisely since identical particle–hole amplitudes are not at issue. These states should decay by collective E3 transitions to the one-phonon 2^+ state and by collective E2 transitions to the one-phonon 3^- state, comparable to the $3_1^- \to 0_1^+$ and $2_1^+ \to 0_1^+$ transitions, respectively.

Possible cases have been suggested (by Herzberg, 1995 and Wilhelm, 1996) in ^{142,144}Nd and ^{144}Sm, and other nuclei near $N = 82$. In ^{144}Nd, $1_1^-, 3_2^-$, and 5_1^- levels are thought to comprise three of the five expected members of the $2^+ \otimes 3^-$ multiplet.

The strengths of the $1_1^- \rightarrow 3_1^-$, and $5_1^- \rightarrow 3_1^-$ E2 transitions are about 20 W.u. compared to 24 for the $2_1^+ \rightarrow 0_1^+$ transition, and the $5^- \rightarrow 2^+$B(E3) value is roughly 34 W.u. which equals the $3_1^- \rightarrow 0_1^+$ B(E3) value. In ^{142}Nd a 1^- level at 3424 keV has a B(E2) value to the $1-$ phonon 3^- level of 16 W.u. compared to 12 for the $B(E2 : 2_1^+ \rightarrow 0_1^+)$ value itself. Similar results are obtained in the $N = 82$ isotone ^{144}Sm.

Finally, at higher energies, there are candidates for multi-phonon giant resonances. The status of these states has recently been summarized by Bertulani (1998), by Bortignon, Bracco, and Broglio (1998), and by Aumann, Bortignon, and Emling (1998) and will not be further discussed here except to say that there is now rather abundant evidence for 2-phonon giant resonances of a variety of types in both spherical and deformed nuclei.

We will discuss phonon excitations of deformed nuclei in the next sections and then approach this issue from a rather different perspective in the following chapter where we study nuclear structure from a "horizontal" perspective rather than the usual "vertical" one of excitation energy. Then, the *evolution* of structure with A, Z, and N is the critical ingredient and we will develop a very simple method of correlating the data over large regions of the nuclear chart in a way that points to a pervasive phonon structure.

Despite the above discussion of spherical vibrators, near-harmonic vibrational motion is the exception rather than the rule. The reason seems to be that it takes only a few valence protons and neutrons to soften the nucleus to deformation to such an extent that the simple scheme of quadrupole surface vibrations of a spherical shape loses applicability.

6.3 Deformed nuclei: shapes

Further from closed shells, therefore, the accumulating p–n interaction strength leads to additional configuration mixing and deviations from spherical symmetry even in the ground state. We now turn to nuclei with stable and permanent deformations. The lowest applicable shape component is a quadrupole distortion. There can also be octupole and hexadecapole shapes. These are schematically illustrated in Fig. 6.10a. Nuclei with quadrupole shapes abound throughout the periodic table in midshell regions.

As with the dynamic shape oscillations we considered above, we can write the radius for a nucleus with static quadrupole deformation as

$$R = R_0 \left[1 + \sum_\mu \alpha_\mu Y_{2\mu}(\theta, \phi) \right] \tag{6.9}$$

where R_0 is the radius of the spherical nucleus of the same volume.

The $Y_{2\mu}$ are spherical harmonics of order 2 and the α_μ are expansion coefficients. It is convenient to change notations here and write the five α_μ in terms of three Euler angles and two intrinsic variables. By writing $\alpha_0 = \beta \cos \gamma$ and $\alpha_2 = \alpha_{-2} = \beta \sin \gamma$ we can specify the nuclear shape in terms of β and γ. β represents the extent of quadrupole deformation, while γ gives the degree of axial asymmetry. Most nuclei are axially symmetric, or close to it, at least in their ground states. For an axially symmetric nucleus, the potential has a minimum at $\gamma = 0°$. [It is unfortunate that no single notation for deformation parameters exists. β is quite common, but we shall also encounter ε and δ,

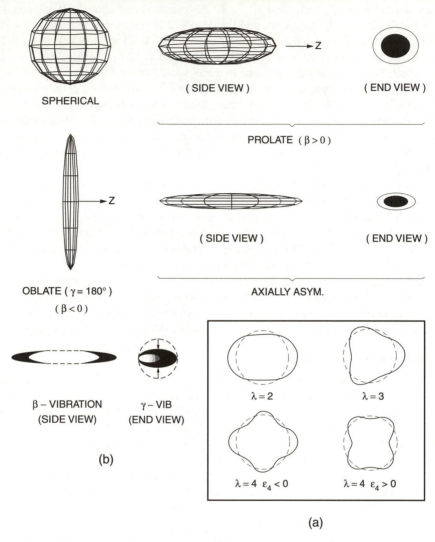

FIG. 6.10. (a) Equal potential surfaces for different multipole distortions. (b) Schematic illustrations of various quadrupole shapes (prolate, oblate, axially asymmetric) as well as of γ and β vibrational motions.

especially in Chapters 8 and 9. Often, a subscript "2" is appended to explicitly denote quadrupole deformation.]

A common convention (the Lund convention) for the ranges of the β and γ variables is that $\beta > 0$, $\gamma = 0°$ for an axially symmetric prolate nucleus (American football) and that $\beta > 0$, $\gamma = -60°$ gives an axially symmetric oblate nucleus (disc or frisbee—note:

Table 6.6 *Changes in the radius of a quadrupole ellipsoid in the x, y, z directions for several γ values and fixed β.**

	γ			
	0°	30°	60°	180°
δR_z	+1	+0.866	+1/2	−1
δR_x	−1/2	0	+1/2	+1/2
δR_y	−1/2	−0.866	−1	+1/2

*All numbers are in units of $\sqrt{5/4\pi}\, R_0\,\beta$.

the frisbee is said to have been invented by Yale students around 1920 who discovered that the plates used by Mrs Frisbie for her pies were aerodynamic and could be sailed across the New Haven Green). An alternate convention is that γ ranges from 0° (axially symmetric) to 30° (maximum axial asymmetry) and that prolate nuclei have $\beta > 0$, oblate nuclei have $\beta < 0$.

The relation between β, γ, and the nuclear radii can be seen by evaluating the *change* in radius ($R_{x,y,z} - R_0$) in Cartesian coordinates as a function of β and γ.

$$\delta R_z = \sqrt{\frac{5}{4\pi}}\, R_0 \beta \cos \gamma$$

$$\delta R_x = \sqrt{\frac{5}{4\pi}}\, R_0 \beta \cos\left[\gamma - \frac{2}{3}\pi\right]$$

$$\delta R_y = \sqrt{\frac{5}{4\pi}}\, R_0 \beta \cos\left[\gamma - \frac{4}{3}\pi\right]$$

To see the shapes implied by these expressions, Table 6.6 gives the values of these correction terms to a spherical shape for four γ values in units of $\sqrt{5/4\pi}\, R_0\beta$. Values greater than zero in the table indicate an elongation in the direction concerned; those less than zero indicate a compression. Note that, for γ values that are a multiple of 60°, two δR_i values are always identical since the nucleus is axially symmetric for these γ values. For $\gamma = 0°$, the nucleus is extended in the z-direction and compressed in x and y. This is a prolate (American football) shape. Oblate (disc-like) nuclei correspond to $\gamma = 60°$ and 180° and are extended in two directions.

The essential difference between prolate and oblate shapes is thus that the former is extended in one direction and squeezed in two, while oblate shapes are extended in two and compressed in one. Intermediate values of $\gamma\,(\gamma \neq n\pi/3)$, such as the $\gamma = 30°$ example in the table, correspond to axially asymmetric shapes; that is, to a flattening of the nucleus in one of the two directions perpendicular to the symmetry axis. Then all three radii are different.

An attempt has been made to depict several nuclear shapes in Fig. 6.10b (as well as β and γ vibrational motions to be discussed). These pictorial images, while crude and too classical, should be helpful to readers unfamiliar with the shapes involved.

The systematics of β values (effectively, quadrupole moments) for the rare earth region is shown in Fig. 6.11(top). The qualitative behavior is easily understood in terms of a generalization of the seniority argument of Chapter 5 (see Eq. 5.17). Early in a major shell, when softness to deformation first appears, the individual j orbits are still nearly empty; hence the quadrupole moments for the nucleons in these orbits are positive. Thus a large positive $Q(\beta)$ builds up rapidly. As the shell fills, however, the contribution of successive j shells to the total quadrupole moment decreases, vanishes, and ultimately turns negative (see Fig. 5.4). On account of these negative contributions, the summation over the individual quadrupole moments steadily decreases and may even go negative (as in Pt, not shown in Fig. 6.11) near the end of the shell.

The two important quantities to describe a quadrupole deformed nucleus are the moment of inertia and the quadrupole moment of the ellipsoidal shape. Both can be written in terms of β for axially symmetric nuclei. For an ellipsoid, the so-called rigid body moment of inertia is $I = 2/5 \, Mr^2$. Integrating the radius over the nuclear surface gives (to first order in β)

$$I = \frac{2}{5} AMR_0^2(1 + 0.31\beta) \tag{6.10}$$

The intrinsic quadrupole moment is given by

$$Q_0 = \frac{3}{\sqrt{5\pi}} ZR_0^2\beta(1 + 0.16\beta) \tag{6.11}$$

to second order in β.

Note that since $R_0 \propto A^{1/3}$, $I \approx A^{5/3}$ and is a linear function of β. Q_0 is directly proportional to β in leading order. For the β values typical of actual deformed nuclei, $\beta \approx 0.3$, the higher-order terms are rather small.

[In the last two decades a new phenomenon, superdeformation, has been discussed in rapidly rotating nuclei which take on equilibrium deformations close to $\beta \sim 0.6$. Clearly, for such cases, Eqs. 6.10 and 6.11 need to be taken to higher order. We will discuss such phenomena later in this chapter.]

It is possible to calculate values of I microscopically for a rigidly rotating deformed nuclear ellipsoid by using a method called "cranking" in which one explicitly takes into account the effects of rotation on the Hamiltonian of the system. The derivation of the resulting formula is given in standard textbooks (see, for example, Ring, 1980, pp. 126–133), but would distract us here. Suffice it to give the result, namely, the so-called Inglis formula:

$$I = 2\hbar^2 \sum_{ik} \frac{|\langle i| J_x |k\rangle|^2}{\varepsilon_i - \varepsilon_k} \tag{6.12}$$

where the states i, k are single particle basis states with energies ε_i and ε_k, and where J_x is the angular momentum operator in the x-direction (i.e., perpendicular to the symmetry axis). The sum is over all occupied levels.

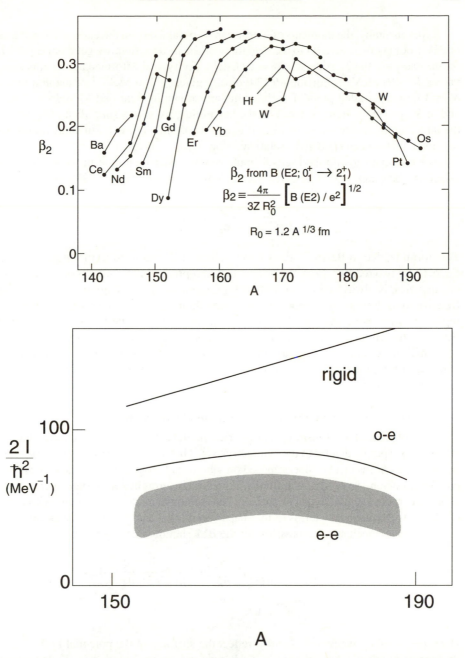

FIG. 6.11. (top) Empirical systematics of quadrupole deformation parameters β in the rare earth region. (bottom) Empirical trends of moments of inertia in the same region for even–even (e–e) and odd A (o–e) nuclei. The curve representing the latter is only a rough indication of a range of typical magnitudes. The rigid body moments obtained with the Inglis formula, Eq. 6.12, are also shown.

Experimentally, the moments of inertia can be obtained from the energies of rotating nuclei. Looking forward a bit to the beginning of the next section, we get from Eq. 6.17, for the energy of the 2_1^+ state, $2I/\hbar^2 = 6/E(2_1^+)$. For typical deformed nuclei, this gives values $I \sim 60$ MeV^{-1}. Equation 6.12 typically gives ~ 150 MeV^{-1} or about a factor of 2–3 larger. Detailed plots of such results are shown in Bohr and Mottelson (1969) and in Ring (1980). Here, to focus on the key trends, we give a simplified overview of the data and the results of Eq. 6.12 in Fig. 6.11 (bottom). The experimental values are intermediate between rigid and irrotational flow. ($I_{\text{irrot}} \sim 0.1 I_{\text{rig}}$)

The discrepancy with the Inglis formula can be understood in terms of pairing correlations (see Chapter 5). A re-derivation including pairing gives

$$I = 2\hbar^2 \sum_{ik} \frac{|\langle i| J_X |k\rangle|^2}{E_i + E_k} (U_i V_k - U_k V_i)^2 \tag{6.12a}$$

This formula, due to Belyaev, gives lower values of I compared to the Inglis formula Eq. 6.12 for two reasons. First, the pairing factor, representing the effects of partial level occupancies, is always less than unity. More importantly, the single particle energies (relative to the bottom of the potential well) and their *difference* in the Inglis formula is replaced by a *sum* of quasi-particle energies taken relative to the Fermi surface. From Eq. 5.24, the minimum value of the denominator is 2Δ, whereas, clearly, terms with very small denominator values are likely to dominate the Inglis formula. Calculations with Eq. 6.12a give results very close to the experimental values and trends shown in Fig. 6.11 (bottom).

6.4   Rotations and vibrations of axially symmetric deformed nuclei

The most obvious characteristic of nonspherical nuclei is that they can undergo rotations about an axis perpendicular to the symmetry axis. They can, of course, also vibrate and, moreover, rotations can be superimposed on vibrational motion. In order to discuss this coupled situation, we need a Hamiltonian expressed in terms of appropriate nuclear variables, incorporating both types of motion. The collective model of Bohr and Mottelson (BM) treats the nucleus in terms of the shape variables β and γ and the Euler angles.

A quantum mechanical expression for the BM Hamiltonian is given by

$$H = T + V = -\frac{\hbar^2}{2m}\left[\frac{1}{\beta^4}\frac{\partial}{\partial\beta}\left[\beta^4\frac{\partial}{\partial\beta}\right] + \frac{1}{\beta^2}\frac{1}{\sin3\gamma}\frac{\partial}{\partial\gamma}\left[\sin3\gamma\frac{\partial}{\partial\gamma}\right]\right]$$
$$+ \frac{\hbar^2}{2I}\mathbf{R}^2 + V \tag{6.13}$$

Here the partial derivatives in β and γ reflect the stiffness of the potential in these two degrees of freedom. The last term in T allows for rotational motion in the laboratory system if the rotational angular momentum $\mathbf{R} \neq 0$. A rigorous solution to this Hamiltonian for a general potential is complicated. However, for certain simple forms of the potential solutions can be obtained easily. Examples are a potential completely rigid in γ, centered at $\beta = 0$ and $\gamma = 0°$, which gives a harmonic vibrator spectrum, and a potential $V = C(\beta - \beta_0)^2$, whose solutions are those of a deformed harmonic oscillator centered at β_0, with superposed rotations.

A general form of the potential, based on an expansion in powers of β and $\cos 3\gamma$, was proposed in 1972 by Gneuss, Greiner, and co-workers. It allows a variety of shapes appropriate to actual nuclei. Including terms up to second order in the kinetic energy, the Gneuss–Greiner Hamiltonian can be written

$$H = T + V = \frac{1}{B_2} [\pi \times \pi]^{[0]} + \frac{P_3}{3} \left[[\pi \times \alpha]^{[2]} \times \pi \right]^{[0]} + C_2 \frac{1}{\sqrt{5}} \beta^2$$

$$- C_3 \sqrt{\frac{2}{35}} \beta^3 \cos 3\gamma + C_4 \frac{1}{5} \beta^4 - C_5 \sqrt{\frac{2}{175}} \beta^5 \cos 3\gamma$$

$$+ C_6 \frac{2}{35} \beta^6 \cos^2 3\gamma + D_6 \frac{1}{5\sqrt{5}} \beta^6 \qquad (6.14)$$

where the α are position coordinates and π are the conjugate momentum coordinates. In this form it is known as the Geometric Collective Model, or GCM. It contains 8 parameters: B_2 and P_3 for the kinetic energy and $C_2, C_3, C_4, C_5, C_6,$ and D_6 for the potential.

Historically, there have only been a few applications of the GCM, primarily because of the large number of parameters and the fact that most applications have utilized 8-parameter least squares fits of the parameters to the data for a given nucleus. Unfortunately, this often lands one in local minima rather than the global minimum, and fits for neighboring, and often quite similar, nuclei can display wildly varying parameter values. Recently, a simplification of the model has been proposed, which employs one kinetic energy term (in B_2) and the three potential energy terms in C_2, C_3, C_4. It is capable of reproducing all the standard limits of structure for even–even nuclei: spherical vibrator, deformed axially symmetric rotor, and γ-soft rotor (Wilets–Jean model—see below), as well as transition regions between these limits.

Although it might be logical to treat this approach here, it will make more sense to the reader if we postpone such a discussion until we have described these structural paradigms separately and, in fact, until after we have discussed the IBA model as well. We therefore return to this topic in section 6.7 below.

In order to discuss both rotational and vibrational motion, and later, their superposition and interactions, we need to discuss the deformed wave functions. Without knowing the detailed solutions to the BM Hamiltonian we will see that one can still specify the *form* of the wave functions and even delineate some of their key properties.

These wave functions incorporate the two aspects of intrinsic excitations and rotational motion. The latter is specified in terms of the well-known rotational D matrices, the former in terms of the wave functions χ_K. The adiabatic assumption of the separability of rotational and intrinsic motions leads to a product wave function in D and χ.

In a state ψ_J in a spherical nucleus, all magnetic substates are degenerate since there can be no distinction in energy as a function of angular orientation of the motion. This, of course, is not true for deformed nuclei whose energies depend on the orientation of the wave function with respect to the symmetry axis. However, whether we consider the motion of a single nucleon in a deformed field (as we shall do when we discuss the Nilsson model in Chapter 8) or the motion of some collective "wave" around the nucleus, there remains a twofold degeneracy, corresponding to clockwise and counter-clockwise motions, which persists even in the deformed field. These two motions can be distinguished by the projection, K, of their angular momenta on the symmetry axis. States with projections K and $-K$ will still be degenerate. The nuclear wave function must reflect this and thus one has the symmetrized product form for wave functions in rotational nuclei:

$$\psi_{JM} = \left(\frac{2J+1}{16\pi^2}\right)^{\frac{1}{2}} \left[D_{JMK}\chi_K + (-1)^{J-K} D_{JM-K}\chi_{-K} \right] \qquad (6.15)$$

Note that for $K = 0$, only even J values are allowed, so the wave function collapses to a single term

$$\psi_{JM} = \left(\frac{2J+1}{8\pi^2}\right)^{\frac{1}{2}} D_{J0}\chi_0$$

With these wave functions in hand, we consider axially symmetric nuclei in which the rotation has equal frequencies around the x or y axes. The rotational Hamiltonian is simply

$$H = \frac{\hbar^2}{2I}\mathbf{R}^2 \qquad (6.16)$$

where I is the moment of inertia and $\mathbf{R}$ is the rotational angular momentum operator. If we assume that the ground state is $J^\pi = 0^+$, $K = 0$, and if all the angular momentum can be ascribed to rotation (as is normally true for the low-lying, low-spin, positive parity states in deformed even–even nuclei) then the total angular momentum $\mathbf{J} = \mathbf{R}$ and we obtain the famous symmetric top rotational energy expression

$$E_{\text{rot}}(J) = \frac{\hbar^2}{2I}J(J+1) \qquad (6.17)$$

where only even J are allowed.

It is useful to comment briefly on the assumption of a 0^+ ground state. For spherical nuclei, we have seen that a δ-interaction between two identical particles in a j shell produces a 0^+ ground state. For the case of multiparticle configurations of identical nucleons in the same j shell, the seniority $v = 0$ 0^+ state also emerges as the lowest energy configuration. This behavior does not result from the special character of the interaction, but is characteristic of any short-range interaction. The same result applies when one has both valence neutrons and protons. It also occurs if j is not a good quantum number, as when there is single-nucleon configuration mixing. (As we shall see, such mixing characterizes Nilsson model wave functions, which can be written as coherent sums over several j values.) The reason is that a short-range interaction favors the lowering of the 0^+ states in each j-configuration. The presence of many j values actually serves to further lower the 0^+ state because of a build up of coherence. We have already seen this coherence effect in our discussion of pairing, in which the lowering of the 0^+ state is related to the size of the energy gap Δ, which in turn depends (see Eq. 5.22) on the number of partially filled orbits near the ground state and the strength parameter G. Thus the phenomenon of a 0^+ ground state persists in deformed even–even nuclei as well.

The symmetric top expression for rotational energies gives the values:

$$E_{2_1^+} = 6\frac{\hbar^2}{2I}$$

$$E_{4_1^+} = 20\frac{\hbar^2}{2I}$$

$$E_{6_1^+} = 42\frac{\hbar^2}{2I}$$

and so on. Thus, the energy ratio $E_{4_1^+}/E_{2_1^+} = 3.33$. This simple formula is one of the most famous results of the rotational model and still remains one of the best signatures for rotational motion and deformation. We have already seen examples of nuclei that behave according to this relation in Figs. 2.10 and 2.15.

Combining Eqs. 6.10 and 6.17 gives two characteristic features of transitional and deformed nuclei. For a *given* mass region ($A \approx$ const), $\hbar^2/2I$ decreases as β increases, leading to smaller and smaller rotational spacings as a deformed region is entered. This behavior is one of the signatures of nuclear transition regions, as we pointed out in Chapter 2 (Figs. 2.13 and 2.14). Second, since nuclear radii increase as $A^{1/3}$, $I \approx A^{5/3}$ and $\hbar^2/2I \approx A^{-5/3}$, for constant β. Rotational spacings should therefore decrease for heavier nuclei. Figure 2.12 illustrated this behavior.

As we stated earlier, in a geometrical picture (that is, a macroscopic one in which we do not worry explicitly about the Pauli principle), there is no reason why rotational motion cannot be superposed on intrinsic excitations, whether of collective vibrational or two-quasi-particle character. Now consider such an intrinsic excitation in a deformed nucleus.

Each intrinsic excitation carries intrinsic angular momentum J_0. It can be partially characterized by the projection of that angular momentum onto the symmetry (z) axis. Since for axially symmetric nuclei, any rotation of the nucleus as a whole must be about an axis perpendicular to the z axis, such rotation has vanishing projection along the z axis. Therefore, the projection of the *total* angular momentum J along the z axis, denoted K, is the same as that of the intrinsic excitation. One sometimes sees the notation Ω for the projection of the *intrinsic* angular momentum. However, the assumption of axial symmetry is generally a good approximation and K is often used interchangeably for both the projection of the intrinsic and total angular momenta. We shall follow this simplified notation here.

When nuclear rotational motion is superimposed on an intrinsic excitation characterized by projection K, the total angular momentum can take on values $J = K, K +$

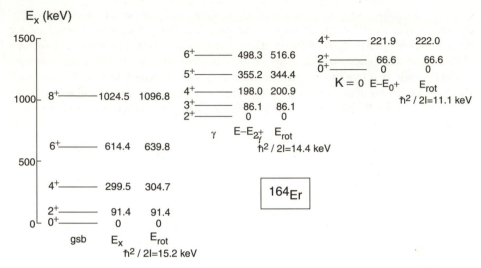

FIG. 6.12. Ground, γ, and $K = 0$ band levels of a typical deformed nucleus ^{164}Er. For each band the symmetric top rotational energy predictions (Eq. 6.18) are shown.

1, $K+2$, ..., except when $K = 0$, in which case only even spins J are allowed. Thus, for the *rotational* energies, relative to the "base" energy of the intrinsic excitation, Eq. 6.17 becomes

$$E_{\text{rot}}(J) = \frac{\hbar^2}{2I}[J(J+1) - K(K+1)] \qquad (6.18)$$

The energy expressions, Eqs. 6.17 and 6.18, are quite accurate for low spin states in deformed nuclei, thus affirming the basic validity of the rotational concept. An example is shown in Fig. 6.12 for a typical deformed nucleus, ^{164}Er (the nature of the intrinsic excitations indicated in the figure will be discussed momentarily). The energies for each band are normalized to the bandheads in order to isolate the rotational behavior. The predictions from Eq. 6.18 are reasonable, but there are also clear deviations as J increases. Also, note the changes in the inertial parameter, $\hbar^2/2I$, from band to band. Apparently the deformation is not completely constant.

Since we have just seen that the moment of inertia need not be constant from band to band, it is interesting to consider more carefully what happens when an excited band has a higher moment of inertia than the ground band. Clearly, the energies in the excited band will increase with J more slowly. At sufficiently high spin they can occur lower in energy than levels of the same spin of the ground band: that is, lowest states of a given spin (these are called "yrast" states) become those of an excited intrinsic configuration, as illustrated in Fig. 6.13 (top). This phenomenon is called "bandcrossing" although there is no real crossing in the quantum mechanical sense but rather an interaction region leading to an apparent crossing and an interchange of intrinsic structure before and after the "crossing". (See Fig. 1.9 and the inset in Fig. 6.13 (top).)

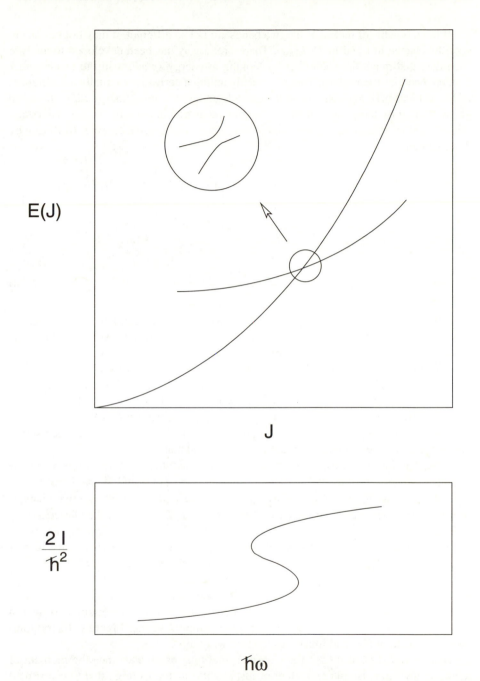

FIG. 6.13. Illustration of two "crossing" bands (with an expanded view of the crossing region (top), and a $2I/\hbar^2$ vs. $\hbar\omega$ backbending plot (bottom).

If the moments of inertia of the two bands are not so different, it may not be easy to spot the change in trend in $E(J_{yrast})$. Therefore, a way has been developed to analyze such data that magnifies the effect by looking at energy *spacings* in the bands rather than the energies themselves. This is basically taking a derivative (actually a difference) of Eq. 6.17 or 6.18 and has the additional advantage that level energy *differences* are given directly by the γ ray transition energies measured in experiment. To understand this method we start with Eq. 6.17 and obtain, for two successive energies (differing by 2 units in J):

$$E_\gamma(J \rightarrow J - 2) = E(J) - E(J - 2) = \frac{\hbar^2}{2I}[J(J + 1) - (J - 2)(J - 1)]$$
$$= \frac{\hbar^2}{2I}(4J - 2)$$
(6.19)

or

$$\frac{2I}{\hbar^2} = \frac{4J - 2}{E_\gamma}$$
(6.20)

Therefore, the transition energy E_γ (we neglect minuscule recoil effects), is linear in J. From this it follows that successive transition energy *differences*, dE_γ/dJ, should be constant at a value $2\hbar^2/I$. A spectrum of γ ray transitions representing a (say, $\Delta J = 2$) cascade down a rotational band should therefore show a set of uniformly spaced peaks. With the advent of modern arrays of high resolution detectors, the literature is filled with beautiful examples of such data. Figure 6.14 is one. (Actually, as noted in the figure, this spectrum is also an example of the decay of a superdeformed rotational band to which we shall return below.) Indeed, these regular "picket fence" sequences are themselves one of the best experimental signatures of a rotational band.

Note that this result, Eq. 6.20, is nothing more than the quantum mechanical consequence of familiar classical equations. From $E = \frac{1}{2}I\omega^2$ and $J = I\omega$ we have $E = J^2/2I$. These give $dE/dJ = E_\gamma/\Delta J = J/I = \omega$. For a transition that changes spin by 2 units, we then have $\omega = E_\gamma/2$ and $2I = 4J/E_\gamma$ which is the classical form of Eq. 6.20.

To test the constancy of the moment of inertia it has become customary to plot $2I/\hbar^2 \equiv (4J - 2)/E_\gamma$ against the rotational frequency $\hbar\omega(J)$ at spin J, where $\hbar\omega \equiv E_\gamma/2$. In earlier work on high spin phenomena, $2I/\hbar^2$ was often plotted against $(\hbar\omega)^2$ instead of $\hbar\omega$ because the (then current) Harris model expressed the moment of inertia as an expansion in even powers of $(\hbar\omega)$ and hence, in lowest order, the moment of inertia was linear in $(\hbar\omega)^2$. Recently, most such plots simply use $\hbar\omega$. Figure 6.13 (bottom) shows this kind of plot and illustrates the following discussion.

Since $E_\gamma(J)$ is linear in J for a normal rotational band, and since the moment of inertia at low spins is either constant or just slightly increases (e.g., due to centrifugal stretching), a plot of $2I/\hbar^2$ against $\hbar\omega$ is flat or slightly increasing for low $\hbar\omega$. However, at the bandcrossing, there is a more radical change. Inspection of the insert depicting the crossing region in Fig. 6.13 shows that γ-ray transition energies between yrast states

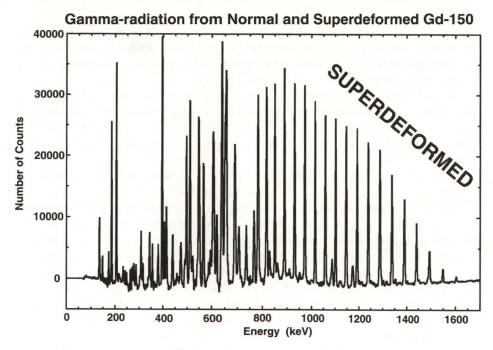

FIG. 6.14. γ-ray spectrum in ^{150}Gd showing irregularly spaced transitions at low energy and a "picket fence" of transitions within a superdeformed band on the right (see text). (Figure courtesy of C. Beausang. Based on the thesis work of S. Erturk.)

can actually decrease for a narrow range of spins. Since $\hbar\omega = E_\gamma/2$, the plot of $2I/\hbar^2$ against $\hbar\omega$ will then move backwards (to the left) in the crossing spin region. At J values above the crossing the level energies again approach a $J(J+1)$ dependence and hence E_γ (or $\hbar\omega$) increases again with spin. However, the yrast states are now the continuation of the crossing band, which has a larger value of $2I/\hbar^2$. Hence the curve of $2I/\hbar^2$ against $\hbar\omega$ again moves to the right but at a higher level. Such a plot is called a backbending plot because of the shape of the resulting curve, as seen in Fig. 6.13 (bottom).

We will return to backbending plots in Chapter 9 where we discuss a particular mechanism for them in terms of the Coriolis interaction in odd A nuclei.

We have also noted, in connection with Fig. 6.12, that the energies within a band do not exactly follow the rotor formula. An understanding of the physics involved in these deviations is simple, yet intimately connected with a number of subtle effects involving rotation-vibration coupling, bandmixing, axial asymmetry, and γ-softness. These concepts, and the various models emphasizing different aspects of them (e.g., the Davydov and Wilets–Jean pictures), are so interrelated that a logically ordered pedagogical treatment is difficult. We have chosen to first discuss the basic "Bohr–Mottelson" idea of an axially-symmetric, deformed nucleus susceptible to quadrupole vibrations of so-called β and γ type. This will allow us to proceed to consider the interactions of these excitations with the ground state band and with each other (that is, bandmixing or rotation–vibration coupling). This, in turn, will provide us with a refinement of Eqs. 6.17 and 6.18. With this in hand, we will turn to approaches that provide closely related viewpoints. In the end, we hope that both the different starting points and the intimate relationships of these models will be clear.

We turn first therefore to vibrational (phonon) excitations of deformed nuclei. The most common distortion of spherical nuclei is quadrupole in nature. Likewise, the most common low-lying vibrational excitations in deformed nuclei are quadrupole vibrations. Clearly such modes, which carry two units of angular momentum, can be of two types with $K = 0$ and $K = 2$.

The former are known as β vibrations; since $K = 0$, the vibration is aligned along the symmetry axis and therefore preserves axial symmetry. The $K = 2$ mode is called a γ vibration and represents a dynamic time-dependent excursion from axial symmetry.

These names stem from the fact that the β vibration corresponds to fluctuations in the quadrupole deformation β, while the γ vibration corresponds to oscillations in γ. A qualitative depiction of the β and γ modes is included in Fig. 6.10. The γ vibration is the more difficult to visualize on a 2-dimensional page. It may be viewed as an alternate "squashing" of an American football in two directions 90° to the symmetry or major axis. Note that, although the γ vibration involves a dynamic fluctuation in γ and has an average value of $\gamma_{ave} = 0°$, the rms value of γ is finite and can be quite large. However, the motion is more complex than a mere flattening of a prolate shape because such a vibration in γ alone would not give $K = 2$ which implies angular momentum about the symmetry axis. For an axially symmetric nucleus, no rotation can occur about the symmetry axis (z axis), but a nucleus exercising γ vibrational oscillations instantaneously *breaks* axial symmetry and therefore rotation about the z-axis can occur. This rotational angular momentum points in the z direction and has $K = 2$. The superposition of the vibrational and rotational motion is therefore a kind of oscillating tumbling motion about the z-axis, as we attempt to illustrate in Fig. 6.15*.

$K = 2$ intrinsic excitations appear in essentially all deformed nuclei, usually around 1 MeV for $A > 100$. The fact that the $K = 2$ mode occurs well below the pairing gap, $2\Delta \sim 1.5 - 2$ MeV, is not a violation of the concept of pairing correlations but an affirmation of the collective character of these excitations (as we shall see explicitly in Chapter 10). In recent years, the nature of the lowest $K = 0$ excitations in deformed nuclei has come into question (see below). The 0_2^+ state (0_1^+ is the ground state) may in fact not be a β vibration. Regardless of the structure of $K = 0$ wave functions such excitations do appear at low energies in most deformed nuclei. An example of a typical deformed nucleus with a $K = 0$ excitation and a γ vibration was shown in Fig. 6.12. Figure 2.17 showed the systematics of the lowest-lying intrinsic $K = 0$ and $K = 2$ excitations in deformed rare earth nuclei. Note that the γ vibrational energies exhibit a smoothly varying systematics while the data for the $K = 0_2^+$ excitation are more erratic.

Perhaps the most telling and interesting properties of the β and γ vibrations center on their electromagnetic decay properties. The basic E2 selection rule here is identical to, and arises from the same arguments as in, the phonon case. Microscopically, as derived

---

*I am grateful to R. V. Jolos for discussions of this point.

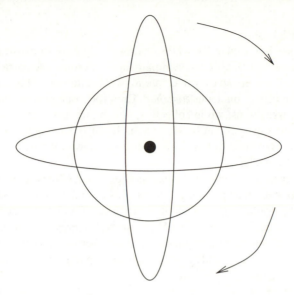

γ – vibrational motion

FIG. 6.15. Schematic illustration of the motion in a γ vibration. The z-axis points out of the page (black dot). The circle represents an end-on view of the axially symmetric ground state shape. The ellipses are the extremes of squashing during the vibrational excursions away from axial asymmetry. The arrows represent the rotational motion about the z-axis which occurs when the nucleus takes on axially asymmetric shapes.

in Chapter 10, collective vibrations can be written as linear combinations of two-quasi-particle excitations (or, in non-pairing terminology, particle–hole excitations) across the Fermi surface. Therefore, an electromagnetic transition can create or destroy at most one such vibration or phonon. A β or γ vibration can decay, therefore, by E2 radiation to the ground state band, but transitions *between* β and γ vibrational bands are forbidden since they involve the simultaneous destruction of one vibration and creation of another.

We shall see in Chapter 10 that the "collectivity" of β and γ vibrations is such that their wave functions typically involve a small handful of orbits comprising a subset of the valence space. Since collectivity in electromagnetic transitions arises from coherence in the wave functions, we can expect that $\gamma \rightarrow g$ or $\beta \rightarrow g$ E2 transitions should be collective (the matrix elements will be much larger than single-particle matrix elements), but that they will be much weaker than transitions occurring *within* a given rotational band since a change in rotational structure involves the whole nucleus (or at least all of the valence nucleons). We saw in Chapter 2 (Fig. 2.18) that rotational transitions in even–even deformed nuclei, typified by huge B(E2 : $2_1^+ \rightarrow 0_1^+$) values, can easily reach several hundred single-particle units. Vibrational transitions such as $\gamma \rightarrow g$ are typically

10–30 single-particle units.

As we shall see in Chapter 10, while it is relatively easy to construct $K = 2$ two-quasi-particle states by breaking nucleon pairs and elevating one particle to an excited quasi-particle level, it is not so easy to create $K = 0$ excitations. This normally involves the excitation of a pair of nucleons *together*. Therefore, one expects that B(E2) values for $K = 0 \rightarrow g$ transitions, while collective, will be weaker than $\gamma \rightarrow g$ transitions, typically a few single-particle units. This hierarchy of B(E2) values [Intra-band; $\gamma \rightarrow g$; $K = 0_2 \rightarrow g$] is evident in Fig. 2.18 where we showed the transition rate data for deformed rare earth nuclei.

Figures 2.17 and 2.18 allow us to assess another aspect of these excitation modes and illustrate a fundamental feature of any collective mode, namely a smooth behavior with N and Z since the available particle–hole excitations of which it is constructed should change gradually with N and Z as the Fermi surface increases with nucleon number. The resulting energies and B(E2) values should also behave smoothly. The behavior of both observables in Figs. 2.17 and 2.18 for the $K = 2$ mode surely points to a collective, slowly evolving structure.

In contrast, the $K = 0_2^+ \rightarrow g$ transitions are much weaker and are also erratic. Both features would seem *not* to point to a collective (β) vibrational mode. In fact, there is little if any solid experimental evidence for the general existence of a low-lying 1-phonon β vibration, at least in the majority of the rare earth deformed nuclei. In one or two cases (e.g., ^{154}Sm) the 0_2^+ state does have a large and collective transition to the 2_1^+ level: these particular 0_2^+ states may well be true (albeit isolated) β vibrations. In at least one nucleus, ^{166}Er, the 0_4^+ level has a collective transition [B(E2 : $0_4^+ \rightarrow 2_1^+$)] to the ground band suggesting that this higher lying 0^+ excitation could be a β vibration (see Garrett, 1997a).

The question of whether and where the β vibration exists is pointedly raised by actual collective model calculations. For typical parameter values describing actual nuclei, the GCM (briefly described above, and treated in detail below), which is an embodiment of the traditional collective model of Bohr and Mottelson, itself predicts that the low-lying $K = 0^+$ excitations have stronger B(E2) values to the γ band than to the ground band and that the latter are quite weak. Apparently the β mode is stiffer than normally assumed, and therefore higher lying or perhaps non-existent as a true collective mode. This result, which is at variance with *qualitative* arguments based on the collective picture of β and γ vibrations, is in fact consistent with a microscopic view of the respective structures of β and γ modes. As we will discuss in Chapter 10, E2 matrix elements of the operator $r^2 Y_{20}$ between two-quasi-particle Nilsson states vanish within a major shell. Any substantial B(E2) value between $K = 0^+$ states would proceed only by the mixing effects described in Chapter 10 which are due to 2-state interactions and should be rather weak and erratic.

Owing to the uncertainty in the nature of the lowest $K = 0^+$ excitation and of the collectivity of the β vibrational mode, it is safer to label the lowest $K = 0^+$ excitation neutrally by using the notation $K = 0_2^+$, and to reserve the notation β for cases where we are explicitly referring to such a vibrational excitation. Using a neutral notation is also helpful since we will discuss multi-phonon modes just below, where we will see that a double γ vibration can also have $K = 0$.

As we have noted, each intrinsic excitation can have rotational motion superposed, as in Fig. 6.12 where the ground, $K = 0$ and γ excitations have rotational bands built on each. Two-quasi-particle excitations, each also with its own rotational band, can appear near or above the "pairing gap"; that is, typically, above about 1.5 MeV. (Negative parity (octupole) vibrations may likewise occur low in energy and will be briefly mentioned later.) Extensive multi-band level schemes such as that shown later (Fig. 6.22) highlight the richness of the spectra of deformed nuclei.

In deformed nuclei the elementary collective modes might be expected to produce multi-phonon states. Since K values are projections of the angular momentum and not themselves vectors, the K values for multiple phonon excitations are obtained by simple algebraic sums and differences of the component K values. Thus, the double β vibration would have $K = 0$, the $\beta\gamma$ vibration has $K = 2$, and the double γ vibration exists in two forms with $K = 0$ and $K = 4$, but not $K = 2$. However, in contrast to spherical nuclei where 2- and even 3- and higher-phonon levels are sometimes known, albeit with strong anharmonic distortions, the identification of deformed multi-phonon states is the exception rather than the rule.

We have seen that the γ vibration, with $K = 2$, is a universal fixture in the low energy spectrum of even–even deformed nuclei. And we have noted that the double γ vibration can exist in two modes, $K = 4$ and 0, according as the projections of the angular momenta of the two γ vibrations couple parallel or anti-parallel. In the harmonic limit $K = 0$ and 4 excitations should occur at $E_{\gamma\gamma}^K = 2E_\gamma$. Given that the γ vibration typically occurs at about 1 MeV and that the pairing gap in heavy deformed nuclei is about 1.5 MeV, two-phonon modes should be found amidst a high level density of other states near 2 MeV. They are likely to mix with these non-collective excitations and be fragmented. Indeed, even aside from the issue of mixing with nearby levels, one expects greater fragmentation of collectivity in multi-phonon states in deformed nuclei compared to spherical nuclei. In spherical nuclei, a given single-particle level such as $f_{7/2}$ can contain up to eight particles, and the Pauli principle will not play a large role if the level is less than half filled. In deformed nuclei, however, each single particle excitation (Nilsson orbit with given K value) is only two-fold degenerate (see Chapter 8): if a given excitation is important in the one-phonon state, there can be a substantial "blocking" effect due to the Pauli principle in a two-phonon vibration.

Despite extensive recent discussion, the issue of the collectivity or fragmentation of multi-phonon excitations in deformed nuclei is not yet settled, either experimentally or theoretically. Calculations testing the possibility of such excitations involve large bases that incorporate both quasi-particle and collective degrees of freedom, and in most cases, they must be simplified by truncating the space. Different truncations schemes yield different results. Until recently, the most articulated microscopic model of collective states, the Quasiparticle Nuclear Model (QPNM) of Soloviev and collaborators, had predicted almost complete fragmentation: less than 1% of the 2 γ-phonon strength was predicted to exist in any given low lying state.

Experimentally, the search for evidence of multi-phonon states in deformed nuclei has been difficult. One reason has to do with the specific character of their most obvious empirical signature. For multi-phonon states the electromagnetic selection rules for collective states discussed above allow $\gamma\gamma \rightarrow \gamma$, $\beta\beta \rightarrow \beta$, and $\beta\gamma \rightarrow \gamma$ or β transitions and, indeed, such transitions are among the key signatures used in searching for such excitations.

However, the difficulty in exploiting this signature lies in the nature of electromagnetic transition rates as discussed above for transitions between spherical phonon levels. E2 transition rates go as $E_\gamma^5 B(E2)$ where here momentarily E_γ means the γ-ray energy (not the γ vibrational energy). Since transitions from states (that are candidates for double phonon vibrations), to, say, the γ-vibration, are much lower in energy than those to the ground state band, the allowed transitions are hindered by this phase space effect (which has nothing to do with structure per se) by factors typically ranging from ~ 30–1000. Thus, even though the signature transition may in fact be collective, it is not easy to observe due to the E_γ^5 factor. It is usually only with the most sensitive detectors that a few of these transitions have been seen; that is, either with high efficiency Compton suppressed Ge detectors or with the remarkable GAMS crystal spectrometer at the Institut Laue–Langevin in Grenoble.

Nevertheless, a few examples have recently been identified and, at the time of this writing, there is intense activity in searching for further examples. The first definitive evidence (Börner, 1991) for a multi-phonon excitation in a deformed nucleus resulted from a GRID study at the ILL in Grenoble on ^{168}Er: As shown in Fig. 6.16, the B(E2) value for the decay of the $K = 4$ excitation at 2055 keV to the γ-band was found to have almost half the full 2-phonon collectivity. The predicted ratio $R_{\gamma g}^4 \equiv$ B(E2 : $4_{\gamma\gamma}^+ \rightarrow 2_\gamma^+)/$B(E2 : $2_\gamma^+ \rightarrow 0_g^+)$ is 2.78, which arises from the factor of 2 representing the proportionality of the B(E2) value to the phonon number, corrected by the squared Clebsch–Gordan ratio $(\langle 442 - 2|22\rangle/\langle 222 - 2|00\rangle)^2$, and by a factor of 2 from the difference in the normalization of the $K = 0$ and 2 wave functions. The experimental ratio from the GRID measurement gives a range of values, $0.52 < R_{\gamma g}^4 < 1.61$, or about half the value predicted by the harmonic model. Subsequent experiments with different techniques have confirmed the GRID result. Other collective models which incorporate anharmonicities give values ranging from 0.5 to 1.9, which are almost identical to the experimental range. These results have led to significant revisions in the QPNM which now predicts relatively large 2-phonon $K = 4$ amplitudes in several deformed rare earth nuclei.

Evidence for additional $K = 4$ $\gamma\gamma$ modes has since been found in ^{164}Dy (Cormin-boeuf, 1997), ^{166}Er (Fahlander, 1996), ^{232}Th (Korten, 1993), and perhaps in ^{170}Er. In the Os nuclei, a $K = 4^+$ excitation appears low in the spectrum (at ~ 1 MeV) and decays preferentially to the γ band. However, these nuclei are soft both to fluctuations in the γ direction and to hexadecapole vibrations and it is possible that these $K = 4^+$ states are mixtures of $\gamma\gamma$ and hexadecapole modes.

Not knowing whether even the single phonon β vibration exists at low energy in deformed nuclei, it is, of course, difficult to identify two phonon β modes or to understand, generally, as we have said, the nature of $K = 0$ excitations. However, while no firm answers are in yet, this subject is under active scrutiny. Most likely, there is no single answer. It appears that the first excited 0^+ excitation ($K = 0_2^+$) may sometimes be a $\gamma\gamma$

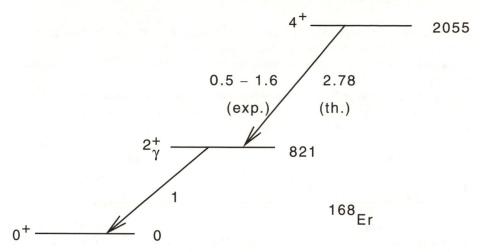

FIG. 6.16. Intrinsic states and B(E2) values in ^{168}Er relating to the $\gamma\gamma$-phonon character of the $K = 4$ excitation at 2055 keV. (B(E2) values in units of the B(E2 : $2^+_\gamma \to 0^+_1$) value).

mode. In one nucleus (^{168}Er) a moderately collective B(E2 : $J = 2, K = 0^+_2 \to 2^+_\gamma$) $\approx$ 6 W.u. has been measured (Lehmann, 1998). (In ^{172}Yb, where $E(0^+_2) < E(2^+_\gamma)$, a large B(E2 : $2^+_\gamma \to 0^+_2$) is observed: the idea that a $\gamma\gamma$ vibration could lie below the single phonon γ vibration is unconventional but such anharmonicities are not impossible.)

In a larger group of nuclei (e.g., $^{160-164}$Dy, ^{168}Er, ^{184,186}W) large values of (easier-to-measure) branching ratios of the type B(E2 : $K = 0^+_2 \to \gamma$)/B(E2 : $K = 0^+_2 \to g$) have been observed. (Some of these involve higher spin rotational band members, not the band heads of these excitations.) We stress that the criterion that these branching ratios are large is only a necessary but not a sufficient condition for assigning 2-phonon character to a $K = 0^+$ excitation. This condition merely establishes a decay *preference*, not necessarily a collective relation. It is also necessary that the B(E2 : $0^+_2 \to 2^+_\gamma$) value be collective (large relative to single particle estimates). For this, lifetime measurements (or Coulomb excitation studies) are needed to establish the absolute transition rates. Indeed, in ^{164}Dy where the branching ratio is large, the B(E2 : $0^+_2 \to 2^+_\gamma$) value is only ~ 1 W.u. (Lehmann, 1998). These $K = 0^+$ excitations may be 2-quasiparticle states that can occasionally appear below the pairing gap. In most nuclei, however, the existing data are not sensitive enough to set useful limits on the low energy transitions corresponding to B(E2 : $K = 0^+_2 \to \gamma$) values.

One can ask similar questions about *higher* lying $K = 0^+$ bands. Analysis of the existing data shows a complex pattern. In some nuclei no $K = 0^+$ band is known which shows a decay preference to the γ band. In some, as we just saw, the lowest $K = 0^+$ band does. In others (e.g., ^{160}Gd, ^{164}Dy, $^{160-170}$Er, ^{172}Yb) a higher band has this behavior. In one of these nuclei, ^{166}Er, the absolute B(E2 : $0^+_5 \to 2^+_\gamma$) = 21(6) W.u. is known and is strongly collective (see Garrett, 1997b). This would be the first definitely established

$K = 0$ double-γ phonon state. In some nuclei, several $K = 0^+$ bands decay primarily to the γ band. Finally, there has also been a suggestion (Just, 1972) of multi-phonon β vibrational excitations in the second minimum of the potential in ^{240}Pu and neighboring nuclides.

The overall situation for $K = 0$ modes is murky at best. Perhaps new data will allow a more systematic picture to emerge. Here, the advent of new generations of highly efficient arrays of γ ray detectors (such as the existing instruments GAMMASPHERE and EUROBALL), or proposed developments such as the multi-segmented tracking arrays like GRETA, or compact arrays of a small number of detectors covering large solid angles, may increase the sensitivity sufficiently for a renewed experimental attack on this intriguing question of the nature of $K = 0^+$ modes.

Lastly, we mention one other important point. Even if a given B(E2 : $0_i^+ \to 2_\gamma^+$) value is large [comparable to or larger than B(E2 : $2_\gamma^+ \to 0_g^+$], 2 phonon character is still not definitively proved. One must verify that such a matrix element cannot result from other effects, such as bandmixing if the $K = 0^+$ and γ bands are close lying. ^{158}Gd provides a nice example of this latter effect where the B(E2) values connecting the lowest $K = 0^+$ band to the γ band are quite large but where the two bands lie within ~ 60 keV of each other: the decay has been fully explained with a band mixing calculation (Greenwood, 1978).

The entire question of multi-phonon states is thus still open and actively debated. As it relates to the basic nature of the most important elementary collective excitations of deformed nuclei and to the role of the Pauli principle (the interplay of single particle and collective motion) it is also an issue of the highest importance.

Having discussed the properties of deformed vibrational states, and the possibility of multi-phonon states, we now turn to a more detailed study of the structure and interactions of the single phonon excitations themselves, focusing again on their γ decay properties since these decay matrix elements give us plentiful, detailed, and quite subtle structural information. We have already introduced some ideas concerning electromagnetic (especially E2) transitions in nuclei in general and in deformed nuclei in particular.

To further consider γ-ray transitions, both within rotational bands (intraband transitions) and between intrinsic excitations (interband transitions), we must be a bit more formal. The basic results are extremely simple to derive. Taking the wave function of Eq. 6.15 for deformed nuclei, the E2 transition matrix element (up to constant factors) is:

$$\langle \psi_f || E2 || \psi_i \rangle = \langle D_{J_f M K_f} \chi_{K_f} + (-1)^{J-K_f} D_{J_f M-K_f} \chi_{-K_f} || E2 ||$$

$$D_{J_i M K_i} \chi_{K_i} + (-1)^{J-K_i} D_{J_i M-K_i} \chi_{-K_i} \rangle$$

$$= \langle D_{J_f M K_f} || E2 || D_{J_i M K_i} \rangle \langle \chi_{K_f} || E2 || \chi_{K_i} \rangle \qquad (6.21)$$

where we have neglected the cross terms of the form $\langle K_f || E2 || - K_i \rangle$, which normally vanish by conservation of angular momentum, and have utilized the fact that the $K_f \to K_i$ and $-K_f \to -K_i$ matrix elements are identical. The separation of the wave function into rotational and vibrational components thus gives a separation of the matrix element into an intrinsic part (second factor on the right in Eq. 6.21) dependent only on

χ and a "rotational" part, which in turn depends only on the angular momenta involved and is proportional to the Clebsch–Gordan coefficient $\langle J_i K_i 2\Delta K | J_f K_f \rangle$.

The diagonal matrix elements with $\chi_i = \chi_f$, $J_i = J_f$ give the intrinsic quadrupole moments Q_0 of the excitation involved. Note that these quadrupole moments are given in the intrinsic body-fixed frame. The *observed* quadrupole moments, that is, the so-called spectroscopic quadrupole moments, involve a transformation to the laboratory frame, giving the well-known result

$$Q = Q_0 \left(\frac{3K^2 - J(J+1)}{(J+1)(2J+3)} \right) \tag{6.22}$$

The dependence on K and J reflects the fact that the *perceived* shape of a rotating nucleus is not the same as the shape in the intrinsic frame. This is easy to visualize. When a prolate deformed nucleus rotates about an axis perpendicular to the symmetry axis, the time averaged shape looks more like a disc (an oblate nucleus), which would have a quadrupole moment of the opposite sign. This effect is exacerbated for higher rotational velocities and, indeed, for $J(J+1) > 3K^2$, the *spectroscopic* quadrupole moment does have a sign opposite to the *intrinsic* quadrupole moment. In fact, for $K = 0$ this is always the case. Note that for $J = 0$ (which implies $K = 0$ since $K \leq J$), $Q = 0$: a state of zero angular momentum can have no preferred direction of the time averaged distribution in space and therefore no quadrupole moment.

For the important case of matrix elements diagonal in χ but not in J (transitions *within* a band), we have nearly the same result except for a Clebsch–Gordan coefficient connecting J_i and J_f. Thus

$$\text{B(E2} : J_i \rightarrow J_f) = \frac{5}{16\pi} e^2 Q_0^2 \langle J_i K 20 | J_f K \rangle^2 \tag{6.23}$$

For $0_1^+ \rightarrow 2_1^+$ transitions (both states have $K = 0$), this gives

$$\text{B(E2} : 0_1^+ \rightarrow 2_1^+) = \frac{5}{16\pi} e^2 Q_0^2 \tag{6.24}$$

Since the intrinsic quadrupole moment $Q_0 \propto \beta(1 + 0.16\beta)$, $\text{B(E2} : 0_1^+ \rightarrow 2_1^+) \sim \beta^2$. The large β values (of about 0.3) that characterize deformed nuclei can lead to a one to two order of magnitude increases in this B(E2) value above that of near spherical nuclei ($\beta \sim 0.05$). This explains the systematics we showed in Fig. 2.16, which provides the most obvious evidence of deformed collective behavior in nuclei.

Frequently one can extract very sensitive and critical information on structure effects and rotation–vibration interactions from B(E2) values for a pair of transitions connecting the same two intrinsic states. These transitions can be either both intraband or both interband with the same initial and final bands. Then the *intrinsic* matrix element will clearly be identical for both transitions and will cancel in their ratio. Such branching ratios depend only on the squares of Clebsch–Gordan coefficients, and are therefore model independent in the sense that they do not depend on the microscopic structure of the excitations involved. They depend on the assumption of the separability of rotational and vibrational motions. They are known as Alaga rules.

Specifically, we have

$$\frac{B(E2 : J_i \rightarrow J_f)}{B(E2 : J_i \rightarrow J'_f)} = \frac{\langle J_i K_i 2\Delta K \mid J_f K_f \rangle^2}{\langle J_i K_i 2\Delta K \mid J'_f K_f \rangle^2} \qquad (6.25)$$

Note that, since the intrinsic structure has canceled out in such ratios, they are equally valid for transitions involving any intrinsic states (e.g., two-quasi-particle states) as well as for those involving vibrational excitations.

As examples of these ratios, we have

$$\frac{B(E2 : 2^+_\gamma \rightarrow 0^+_g)}{B(E2 : 2^+_\gamma \rightarrow 2^+_g)} = 0.7$$

$$\frac{B(E2 : 2^+_\gamma \rightarrow 4^+_g)}{B(E2 : 2^+_\gamma \rightarrow 2^+_g)} = 0.05$$

Equally simple but numerically different results are obtained for other transitions. There are two important uses of such ratios. First, since they depend on the K values of the initial and final states, they can sometimes be empirically used to assign K quantum numbers to different intrinsic excitations. Secondly, as we shall see momentarily, small admixtures of different intrinsic excitations (bandmixing effects) can induce enormous changes in these branching ratios, so the empirical ratios can provide very sensitive tests of small details of the nuclear wave functions.

It is easy to look up or calculate values of the Clebsch–Gordan coefficients involved in these branching ratios. However, transitions involving $K = 0$ and $K = 2$ bands are so important and so common that it is useful to collect the results here. Table 6.7 shows the relative B(E2) values for transitions involving low-spin states in $K = 0$ and $K = 2$ bands.

These Clebsch–Gordan coefficients are often thought to be rather annoying and opaque factors that obscure otherwise simple ideas. However, in fact, they reflect these ideas and often help in understanding the physics. An example illustrates this. Consider the transition in Table 6.7 from a $K = 0$ band to a $K = 2\,\gamma$ band (the $0 \rightarrow 2$ column). Notice that all the spin-*decreasing* values are small. This is especially true for transitions which lower the spin by 2 units, such as $4 \rightarrow 2_\gamma$ and $6 \rightarrow 4_\gamma$. The reason is simple and reflects the geometry and the rotational motion. Consider the contribution of rotational angular momentum to total angular momentum. For a state in a $K = 0$ band all the angular momentum is rotational. For the $K = 2$ band, with two units of intrinsic angular momentum, the $2^+_{K=2}$ state can have components with rotational angular momentum, $\mathbf{R} = 0, 2,$ or 4 with 0 dominating. Thus, the transition $4^+_{K=0} \rightarrow 2^+_{K=2}$ is largely a $\Delta R = 4$ change in rotational angular momentum. Crudely speaking, the $4^+_{K=0}$ state is 4 units of angular momentum further removed from its bandhead than is the $2^+_{K=2}$ state. Hence the transition should be weak: an E2 transition can change $\mathbf{R}$ by 2 units at most. It proceeds only through the weaker $\mathbf{R} = 2$ and 4 components in the $2^+_{K=2}$ state, as is

Table 6.7 *Some useful Alaga rules for E2 transitions in deformed nuclei*[*]

J_i	J_f	$\langle J_i K_i 2\Delta K \vert J_f K_f \rangle^2$ $K_i \to K_f$			
		$0 \to 0$	$2 \to 0$	$0 \to 2$	$2 \to 2$
0	2	1.0	–	1.0	–
2	0	0.200	0.200	–	–
	2	0.286	0.286	0.286	0.286
	3	–	–	0.500	0.500
	4	0.515	0.014	0.215	0.215
3	2	–	0.358	–	0.358
	3	–	–	–	0
	4	–	0.143	–	0.343
	5	–	–	–	0.300
4	2	0.286	0.120	0.008	0.120
	3	–	–	0.112	0.267
	4	0.260	0.351	0.351	0.042
	5	–	–	0.389	0.234
	6	0.455	0.031	0.142	0.340
5	3	–	–	–	0.191
	4	–	0.319	–	0.191
	5	–	–	–	0.093
	6	–	0.182	–	0.167
	7	–	–	–	0.360
6	4	0.315	0.098	0.021	0.235
	5	–	–	0.154	0.141
	6	0.255	0.364	0.364	0.130
	7	–	–	0.347	0.124
	8	0.431	0.039	0.116	0.371

[*]The entries are the squares of the Clebsch–Gordan coefficients for each indicated transition. Relative B(E2) values connecting states J_i, J_f in bands with K_i, K_f are B(E2 : $J_i K_i \to J_f K_f$) $\propto \langle J_i K_i 2\Delta K \vert J_f K_f \rangle^2$.

reflected by the small Clebsch–Gordan coefficient. Following similar arguments, it is easy to see why spin-*increasing* $\gamma \to g$ band transitions are also weak.

It is interesting to compare the Alaga rules with the data for deformed nuclei. To this end, we show comparisons of $\gamma \to g$ transitions for three nuclei in Table 6.8, of which one, [154]Gd, is situated at the beginning of the deformed region while the other two, [168]Er and [178]Hf, are near midshell. The table shows a number of very interesting features:

1. The general agreement is remarkably good, indicating that these simple expressions are a reasonable leading-order approximation. Despite the deviations to be

Table 6.8 *Comparison of some relative* B(E2: $\gamma \rightarrow g$) *values in deformed rare earth nuclei with the Alaga rules* *

J_i	J_f	Alaga	Relative B(E2 : $J_i \rightarrow J_f$) ^{154}Gd	^{168}Er	^{178}Hf
2	0	70	43	54	88
	2	100	100	100	100
	4	5	14	6.8	5.8
3	2	100	100	100	100
	4	40	105	65	52
4	2	34	16	20	18
	4	100	100	100	100
	6	9	—	14	—
5	4	100	—	100	100
	6	57	—	123	107
6	4	27	—	12	18
	6	100	—	100	100
	8	11	—	37	—

* One transition is normalized to 100 for each initial state. The Alaga rule entries are relative values from Table 6.7 for the $K = 2 \rightarrow K = 0$ case.

discussed at considerable length next, it is important to stress that the approximate validity of the Alaga rules is one of the strongest arguments for axially deformed nuclei and for the concept of separable rotational motion. Nevertheless, there are substantial deviations from them and their study greatly deepens our understanding of deformed nuclei.

2. The deviations increase substantially with increasing spin.

3. Transitions in which the spin increases ($J_f > J_i$) are nearly always empirically larger than the Alaga rules, while spin decreasing transitions ($J_f < J_i$) are nearly always smaller.

4. The deviations can become quite large, leading to factors of three or four discrepancies from the predictions.

5. The deviations are, on average, larger in Gd than in Er.

Combined with all the evidence from rotational energy sequences, measurements of quadrupole moments, and the like, point 1 provides a vast body of evidence that supports the idea of a superposition of rotational and intrinsic motion and the approximate separability of the two. Point 5 suggests that this separability is most applicable in mid-shell and least just after the transition region from spherical to deformed nuclei. This is reasonable, of course, since the energy scale of rotational motion decreases systematically toward mid-shell, and therefore, the distinction in energy between rotational and vibrational behavior is larger there than closer to the vibrational regions at the beginning and end of major shells. We shall soon see more dramatic evidence of this point in terms of a systematic measure of the rotation–vibration coupling. First, however, in order to understand points 2, 3, and 4, and in particular why point 4 does not indicate a serious breakdown of the rotational description, we must introduce the concept of bandmixing and discuss a quantitative formalism to treat it in a simple way.

6.4.1 Bandmixing and rotation–vibration coupling*

Bandmixing is a widespread phenomenon in even–even nuclei. We shall limit our discussion to its simplest and most common manifestations, namely $\Delta K = 2$ mixing between the γ and the ground bands: the same physical concepts and formalism also apply to other cases but the extent and systematic behavior of $\gamma–g$ mixing makes it the most interesting and informative to study. The basic scheme is simply an example of two-state mixing. For $\gamma–g$ mixing, the ground and γ band wave functions can be written as

$$\psi_g = \phi_g - \varepsilon'_\gamma(J)\phi_\gamma$$
$$\psi_\gamma = \varepsilon'_\gamma(J)\phi_g + \phi_\gamma \tag{6.26}$$

where ε'_γ is the *small* mixing amplitude of each band in the other. It is convenient to separate the spin dependent and spin independent parts of the mixing by writing

$$\varepsilon'_\gamma = \sqrt{2}\,\varepsilon_\gamma f_\gamma(J) \tag{6.27}$$

In order to derive the spin dependence, we need to anticipate a result from Chapter 9 on Coriolis mixing in the Nilsson model. Coriolis mixing is a well-known effect in any rotating system and arises from the transition from a body-fixed (nuclear) frame of reference to the laboratory. We will show in Chapter 9 that the Coriolis effect in nuclei mixes intrinsic states differing by $\Delta K = \pm 1$. The dependence of the Coriolis mixing matrix element on the total angular momentum J is contained in the mixing operator J_+, given by (see Eq. 9.2)

$$\langle K|J_+|K+1\rangle = \sqrt{(J-K)(J+K+1)}$$

where K is the lower K value. We now interpret $\gamma \to g\,\Delta K = 2$ mixing as proceeding via a two-step process. For weak mixing, this can be viewed as a sequence of two separate two-step mixing effects.

We know from the discussion in Chapter 1 of weak two-state mixing that if three states ϕ_1, ϕ_2, ϕ_3, mutually mix, the mixing of states ϕ_2 and ϕ_3 gives

$$\phi'_2 = \alpha\phi_2 + \beta_{23}\phi_3$$

Then, if the already mixed state ϕ'_2 mixes with state ϕ_1, we have

---

*I am indebted to J.Kern and C. Günther for many discussions on bandmixing and for corrections they gave me to the first edition.

$$\psi_I = \alpha_1\phi_1 + \beta_{12}\phi_2'$$
$$= \alpha_1\phi_1 + \beta_{12}\alpha\phi_2 + \beta_{12}\beta_{23}\phi_3$$

or, since α and $\alpha_1 \sim 1$,

$$\psi_I = \phi_1 + \beta_{12}\phi_2 + \beta_{12}\beta_{23}\phi_3$$

Thus, the overall mixing amplitude of state ϕ_3 in state ϕ_1 is simply given by the product of the individual two-state mixing amplitudes β_{12} and β_{23}.

Applying this to the present case, we have the mixing sequence $(K = 0) \rightarrow (K = 1) \rightarrow (K = 2)$. Hence the spin dependence of the $\Delta K = 2$ mixing amplitude, $f_\gamma(J)$ is

$$f_\gamma(J) = \sqrt{J(J+1)}\sqrt{(J-1)(J+2)}$$
$$= \sqrt{(J-1)J(J+1)(J+2)} \qquad (6.28)$$

Similarly, for mixing between the $K = 0$ and ground bands, the mixing sequence is $(K = 0) \rightarrow (K = 1) \rightarrow (K = 0)$. Then $f_0(J)$ is given by $f_0(J) = J(J+1)$. Note that both $f_0(J)$ and $f_\gamma(J) \rightarrow J^2$ for large J: the band mixing increases rapidly for high spin.

The spin dependence of ε_γ' explains point 2 concerning the increase of the deviations from the Alaga rules with increasing spin. However, we have yet to explain why these deviations can be so large without implying a corresponding destruction of the entire rotational picture on which the Alaga rules and the present formalism are based.

We can now calculate the interband E2 matrix elements very simply using the admixed wave functions of Eq. 6.26.

$$\langle\psi_g||E2||\psi_\gamma\rangle = \langle\phi_g||E2||\phi_\gamma\rangle - \varepsilon_\gamma'(J_g)\langle\phi_\gamma||E2||\phi_\gamma\rangle + \varepsilon_\gamma'(J_\gamma)\langle\phi_g||E2||\phi_g\rangle \quad (6.29)$$

The first term in Eq. 6.29 is the direct matrix element in the absence of mixing. Thus, the perturbed matrix element can be written as a sum of a direct term plus contributions proportional to ε_γ'. In deriving this expression we have dropped terms in $\varepsilon_\gamma'^2$ since the mixing is assumed to be small. Each of the two terms multiplying ε_γ' is proportional to a Clebsch–Gordan coefficient multiplied by the intrinsic quadrupole moment of the γ or ground band. Therefore, even if we assume these intrinsic moments to be equal (as is commonly done since the deformation does not differ much from band to band), the K dependence of the Clebsch–Gordan coefficients prevents these two terms from canceling.

In the case of $0_2^+ \rightarrow g$ mixing, exactly the same formalism applies with a substitution of f_0 for f_γ. One interesting result for the special case of transitions that do not change spin $(J_0 = J_g)$ follows immediately. Then the mixing amplitudes that appear in the expression analogous to Eq. 6.29 for $0_2^+ \rightarrow g$ mixing are $\varepsilon_0'(J_0)$ and $\varepsilon_0'(J_g)$ which are identical and hence factor out. For identical quadrupole moments, the two terms multiplying ε_0' are also identical and vanish: $0_2^+ \rightarrow g$ bandmixing has no effect on transitions which do not change the spin. This is a special case of the result derived in Eq. 1.17.

Incorporating the spin dependence of ε' and expressions for the Clebsch–Gordan coefficients implicit in Eq. 6.29 leads to a general form for the effect of bandmixing on interband B(E2) values. We obtain in this way the well-known expressions:

$$B(E2 : J_\gamma \rightarrow J_g) = B_0(E2)[1 + Z_\gamma F_\gamma(J_\gamma, J_g)]^2 \tag{6.30}$$

$$B(E2 : J_0 \rightarrow J_g) = B_0(E2)[1 + Z_0 F_0(J_0, J_g)]^2 \tag{6.31}$$

Here $B_0(E2)$ is the unperturbed value and Z_γ or Z_0 is a bandmixing parameter proportional to ε_γ or ε_0. For the γ-g case, B(E2)$_0$ is defined by

$$B_0(E2 : J_\gamma \rightarrow J_g) = 2\langle J_\gamma 02 - 2|J_g 2\rangle^2 \langle \phi_g |\mathbf{E2}| \phi_\gamma \rangle^2$$

and Z_γ is given by

$$Z_\gamma = -\left(\frac{15}{2\pi}\right)^{\frac{1}{2}} \frac{eQ_0\varepsilon_\gamma}{\langle \phi_g |\mathbf{E2}| \phi_\gamma \rangle} \tag{6.32}$$

where $\langle \phi_g |\mathbf{E2}| \phi_\gamma \rangle$ is the intrinsic matrix element (usually expressed in eb with the quadrupole moment in barns). In typical deformed nuclei $\langle \phi_g |\mathbf{E2}| \phi_\gamma \rangle$ is a small fraction of an eb and $Q_0 \sim 5$–10 b. Hence, roughly, $Z_\gamma \sim -40\varepsilon_\gamma$. [We will see a quantitative example momentarily.] Note the factor of 2 in the definition of the unperturbed B(E2) value which stems from the normalization of the $K = 0$ and $K = 2$ wave functions. Treatments of bandmixing differ in this definition. We follow Bohr and Mottelson (their Eq. 4–92). If the B(E2 : $J_\gamma \rightarrow J_g$) is defined without the factor of 2, the formalism and the results for the $F_\gamma(J_\gamma, J_g)$ functions (Table 6.9) as well as the Mikhailov formalism below are unchanged as long as ε'_γ is then also defined without the factor of $\sqrt{2}$ (i.e., as long as $\varepsilon'_\gamma \equiv \varepsilon_\gamma f(J_\gamma)$]. The functions $F_\gamma(J_\gamma, J_g)$ and $F_0(J_0, J_g)$ are given in Table 6.9 for the three possible cases of $\Delta J = 0, \pm 1, \pm 2$. Clearly, the case $\Delta J = \pm 1$ does not apply to the $K = 0^+ \rightarrow g$ transitions, and the result obtained earlier for $\Delta J = 0$, $K = 0^+ \rightarrow g$ transitions is reflected by the value of unity.

One can analyze experimental data in terms of this formalism in several different ways. One is to extract a Z_γ value from each branching ratio between a pair of bands and then test for a consistent value. An example of this is shown in Table 6.10, where the Z_γ values for ^{154}Gd are given. In this case, $Z_\gamma \sim 0.08$ provides reasonable agreement with the data.

Another approach exploits a particularly useful form of Eq. 6.30 by combining the spin dependence of the f_γ functions with that of the Clebsch–Gordan coefficients. We obtain

$$B(E2 : J_\gamma \rightarrow J_g) = 2(J_\gamma 22 - 2|J_g 0)^2 \left[M_1 - M_2[J_f(J_f + 1) - J_i(J_i + 1)]\right]^2 \tag{6.33}$$

M_1 and M_2 are defined by

$$M_1 = \langle \phi_g |\mathbf{E2}| \phi_\gamma \rangle - 4M_2$$

Table 6.9 *Correction factors* $[1 + ZF]$ *for* $\gamma \rightarrow g$ *and* $0^+ \rightarrow g$ *reduced* $E2$ *matrix elements due to* $\gamma \rightarrow g$ *and* $0^+ \rightarrow g$ *bandmixing.*

J_i	J_f	Correction factor	
		$\gamma \rightarrow g$	$0 \rightarrow g$
$J_f - 2$	J_f	$1 + (2J_f + 1)Z_\gamma$	$1 + 2(2J_f - 1)Z_0$
$J_f - 1$	J_f	$1 + (J_f + 2)Z_\gamma$	–
J_f	J_f	$1 + 2Z_\gamma$	1
$J_f + 1$	J_f	$1 - (J_f - 1)Z_\gamma$	–
$J_f + 2$	J_f	$1 - (2J_f + 1)Z_\gamma$	$1 - 2(2J_f + 3)Z_0$

* Riedinger, 1969.

Table 6.10 Z_γ *values for* ^{154}Gd*

Branching ratios	Experiment	Alaga	Z_γ $(\times 10^2)$
$\dfrac{4_\gamma \rightarrow 4_g}{4_\gamma \rightarrow 2_g}$	6.4(10)	2.94	5.1(9)
$\dfrac{3_\gamma \rightarrow 4_g}{3_\gamma \rightarrow 2_g}$	1.05(6)	0.4	8.2(5)
$\dfrac{2_\gamma \rightarrow 2_g}{2_\gamma \rightarrow 4_g}$	7.34(10)	20	11.4(23)
$\dfrac{2_\gamma \rightarrow 2_g}{2_\gamma \rightarrow 0_g}$	2.30(12)	1.43	8.3(9)
$\dfrac{2_\gamma \rightarrow 0_g}{2_\gamma \rightarrow 4_g}$	3.71(42)	14.0	9.8(11)

*Errors given on last digit. Riedinger, 1969.

$$M_2 = \left(\frac{15}{8\pi}\right)^{\frac{1}{2}} e Q_0 \varepsilon_\gamma \tag{6.34}$$

and are related to Z_γ by

$$Z_\gamma = -\frac{2M_2}{(M_1 + 4M_2)} = -\frac{2M_2}{\langle \phi_\gamma | \mathbf{E2} | \phi_g \rangle}$$

For $K = 0^+ \rightarrow g$ transitions, the formalism is similar if $M_1 = \langle \phi_g | \mathbf{E2} | \phi_0 \rangle$ and $M_2 = (5/16\pi)^{1/2} e Q_0 \varepsilon_0$.

Thus, M_1 is essentially the direct intrinsic $\Delta K = 2$ matrix element (the correction term, $-4M_2$ for $\gamma - g$ mixing, is normally very small) and M_2 is proportional to the spin independent mixing amplitude ε_γ. The advantage of Eq. 6.33 is that it can be rewritten for the $\gamma \rightarrow g$ case as

$$\frac{\sqrt{\mathrm{B}(E2 : J_\gamma \rightarrow J_g)}}{\sqrt{2}\langle J_\gamma 22 - 2 | J_g 0 \rangle} = M_1 - M_2[J_g(J_g + 1) - J_\gamma(J_\gamma + 1)] \tag{6.35}$$

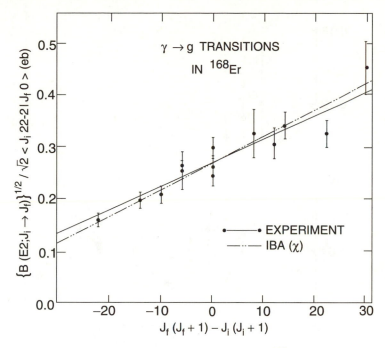

FIG. 6.17. Mikhailov plot for $\gamma \to g$ transitions in ^{168}Er (Warner, 1982).

A plot of the left side against the spin function on the right is a straight line with intercept M_1 at $J_\gamma = J_g$ and slope M_2. From such a plot, called a *Mikhailov plot*, one can extract directly from the empirical results both the direct *intrinsic unperturbed* $\Delta K = 2$ *matrix element* and the *mixing amplitude* $\varepsilon_\gamma (\propto M_2)$, provided the data can be fit by a straight line. Deviations from a straight line can arise from several sources: unequal quadrupole moments of the bands, more than two-bandmixing, undetected M1 components in the interband transitions, or two-bandmixing that follows a different spin dependence than that given by the $f_\gamma (J)$ functions.

An example of a Mikhailov plot, for ^{168}Er, is shown in Fig. 6.17 (the dashed line labeled IBA will be discussed later). It is clear that the data points are very well approximated by a straight line, thus validating the use of the Mikhailov formalism. Such data are typical of deformed nuclei.

The use of the bandmixing formalism in either the Z_γ or Mikhailov forms is a powerful tool for analyzing deviations of relative E2 transitions from the rotational (Alaga) values and for studying rotation–vibration interactions. Empirically, it is invariably found that $M_2 << M_1$ and negative (by convention, M_1 is positive): that is, ε'_γ is negative and therefore Z_γ is positive. In this way, Z_γ values have been extracted for a number of rare earth nuclei. The results are summarized in Fig. 6.18.

These systematics exhibit a parabolic behavior that minimizes at midshell. This smooth pattern highlights, indirectly, the collective structure of the unperturbed states involved: if the mixing were with single-particle excitations, it would surely be more

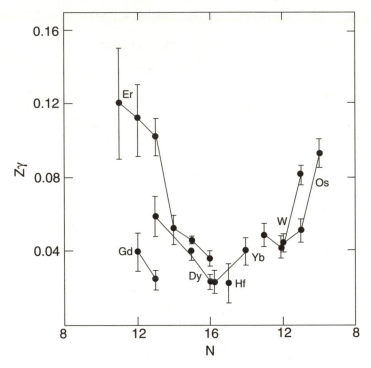

FIG. 6.18. Empirical systematics of Z_γ in the rare earth region (Casten, 1983). N is half the number of valence nucleons.

erratic. As we shall see in Chapter 10, the microscopic structure of collective vibrations changes smoothly and slowly with N and Z. The minimum in Z_γ at midshell, and the generally small values of Z_γ, demonstrate the overall validity of the separation of rotational and vibrational degrees of freedom and show that, as expected, this separability is best at midshell.

With this bandmixing formalism and the empirical results in hand, we can now address points 2, 3 and 4 mentioned earlier. First of all, Eq. 6.35 shows that, regardless of the sign of M_2, the effect of bandmixing (on $\Delta J \neq 0$ transitions at least) must increase with increasing spin as we anticipated earlier from the spin dependence of the mixing amplitudes ε_γ and ε_0. Second, in the case of $\gamma - g$ mixing, the negative values of M_2, that is, the positive slopes in a Mikhailov plot, and the fact that the abscissa is positive for $J_f > J_i$, implies that the B(E2) values are *increased* for *spin-increasing* transitions and *decreased* for *spin-decreasing* transitions, as we observed in the examples given in Table 6.8.

This point is also clear from inspection of the analytic formulas in Table 6.9. Since M_2 has the opposite sign to Z_γ, the data in Figs. 6.17 and similar data for other deformed regions show that Z_γ is always positive. Then, we see from Table 6.9 that, for *spin increasing* transitions, the B(E2) correction factor always has the form

$$\left[1 + g(J)Z_\gamma\right]^2 \quad \text{(spin increasing)}$$

while for *spin decreasing* transitions, we have

$$\left[1 - h(J)Z_\gamma\right]^2 \quad \text{(spin decreasing)}$$

where $g(J)$ and $h(J)$ are positive functions of the final spin J.

It is worth working through an explicit example to see how the bandmixing technique is used. We take the case of γ – ground mixing in ^{168}Er and use both the analytic approach with Table 6.9 and the Mikhailov formalism.

According to Tables 6.7 and 6.9,

$$\frac{B\left(E2 : 2_\gamma^+ \to 2_1^+\right)}{B\left(E2 : 2_\gamma^+ \to 0_1^+\right)} = 1.43 \left[\frac{1 + 2Z_\gamma}{1 - Z_\gamma}\right]^2 \tag{6.36}$$

where the first factor is the unperturbed (Alaga) ratio. From the experimental value of 1.85, we obtain $Z_\gamma = 0.044$. Similar values are obtained from other transitions. A good average value is $Z_\gamma \sim 0.038$. The small magnitude of Z_γ confirms the adequacy of the two-state bandmixing approach.

Turning to the Mikhailov approach, which generally is easier and yields interesting physics more directly, Fig. 6.17 gives

$$M_1 = 0.268\,\text{eb} \qquad M_2 = -0.0045\,\text{eb}$$

Thus, we can immediately deduce the direct, unperturbed $\gamma \to g$ E2 matrix element

$$\langle \phi_g | \text{E2} | \phi_\gamma \rangle = 0.250\,\text{eb}$$

Using $Q_0 = 7.61$ b, the spin independent part of the mixing amplitude is

$$\varepsilon_\gamma = \left(\frac{8\pi}{15}\right)^{\frac{1}{2}} \frac{M_2}{e Q_0} = -0.00077$$

The full mixing amplitudes $\varepsilon_\gamma'(J) = \sqrt{2}\,\varepsilon_\gamma\, f_\gamma(J)$ are then $-0.0053\,(2^+)$, -0.012 (3^+), and $-0.021\,(4^+)$.

We can now calculate the actual mixing matrix element since for such small mixing we have, from Eq. 1.10

$$\varepsilon_\gamma'(J) \sim \frac{\langle \phi_g(J) | V_{\Delta K=2} | \phi_\gamma(J) \rangle}{\left[E_\gamma(J) - E_g(J)\right]_{\text{unp}}}$$

Neglecting the tiny difference between perturbed and unperturbed spacings, and taking $E_{2_\gamma^+} - E_{2_g^+} = 741$ keV we get

$$\langle \phi_g(2^+) | V_{\Delta K=2} | \phi_\gamma(2^+) \rangle = \varepsilon_\gamma'(741\,\text{keV}) = 3.9\,\text{keV}$$

In a way, it is remarkable how much detailed information, including a mixing ampli-tude, an interaction matrix element, and even an absolute unperturbed transition matrix element, can be obtained in this simple way from the measurement of relative interband B(E2) values.

Finally, we now see that the rather large changes in the interband B(E2) values result from extremely small residual interactions, on the order of a few keV, and mixing amplitudes $\sim 10^{-2} - 10^{-3}$. Returning to our earlier point 4, we see that the rotational description of the wave functions is still an excellent approximation although certain observables deviate substantially from their rotational predictions. Later we shall see other cases, such as Coriolis mixing, where small disturbances of pure wave functions grossly affect certain observables and, conversely, where the measurement of those observables provides very sensitive probes of specific wave function components.

The fundamental reason that small interactions such as the one we are considering can lead to such large effects is obvious from Eq. 6.29: the mixing introduces an effectively *intraband* contribution to the originally *interband* transition. The $4_\gamma^+ \rightarrow 2_g^+$ transition, for example, contains small amplitudes for the very large $4_\gamma^+ \rightarrow 2_\gamma^+$ and $4_g^+ \rightarrow 2_g^+$ *rotational* matrix elements. From this, we can immediately appreciate the well-known empirical fact that contrary to *interband* transitions, *intraband* transitions are virtually unaffected by bandmixing because the effect is reversed, namely, adding a small *inter*band amplitude to a much larger *intra*band amplitude.

By formalizing this argument we can deduce some interesting results. We consider the set of bands shown in Fig. 6.19, where we illustrate states of spin J, $J + 1$, $J + 2$ occurring somewhere in these bands. We do *not* restrict ourselves to small mixing. First, let us isolate the two-band system of intrinsic states 1 and 2 and actually calculate the relevant B(E2) value for the transition between states of spin J' and J. If the mixing is large, we might expect transitions *between* bands to be comparable to those *within* a band. However, we shall see that under one rather reasonable assumption, this is not the case.

We explicitly write the B(E2) value, using an obvious notation for the initial and final wave functions analogous to the notation in Eq. 1.7 except that it distinguishes amplitudes α, α', β, β' for the spins J and J'. We have, for an *intra*band transition in band 2,

$$\text{B(E2} : (J' \rightarrow J)_2) = \frac{1}{2J_2' + 1} \left[\langle (\alpha'\phi_1 + \beta'\phi_2)_{J'} \| \mathbf{E2} \| (\alpha\phi_1 + \beta\phi_2)_J \rangle \right]^2$$

$$= \frac{1}{2J_2' + 1} \left[\alpha\alpha' \langle \phi_1(J') \| \mathbf{E2} \| \phi_1(J) \rangle + \beta\beta' \langle \phi_2(J') \| \mathbf{E2} \| \phi_2(J) \rangle \right.$$

$$\left. + \alpha'\beta \langle \phi_1(J') \| \mathbf{E2} \| \phi_2(J) \rangle + \alpha\beta' \langle \phi_2(J') \| \mathbf{E2} \| \phi_1(J) \rangle \right]^2$$

But, here, the interband matrix elements connecting unperturbed states in bands 1 and 2 are negligible compared to the intraband rotational matrix elements. We assume for simplicity that the intrinsic matrix elements are band independent and obtain

$$\text{B(E2} : (J' \rightarrow J)_2) \sim \frac{1}{2J_2' + 1} \left[(\alpha\alpha' + \beta\beta') \langle \phi_1(J') \| \mathbf{E2} \| \phi_1(J) \rangle \right]^2 \qquad (6.37)$$

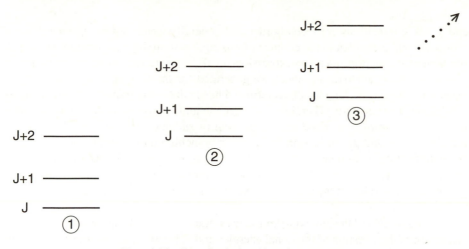

FIG. 6.19. Set of admixed bands (see text).

If we now assume that the mixing interaction, though possibly large, is not very spin dependent, then the composition of the mixed wave functions will also not depend much on spin and therefore, $\alpha \sim \alpha'$ and $\beta \sim \beta'$. But, then, the factor $(\alpha\alpha' + \beta\beta') \sim 1$ by orthonormality and the intraband transition has an identical B(E2) value as in the unmixed case.

Extending this argument to multiband mixing, the factor $(\alpha\alpha' + \beta\beta')$ in Eq. 6.37 will simply be replaced by $(\alpha\alpha' + \beta\beta' + \gamma\gamma' + \delta\delta' + \ldots)$. If, again, the primed and unprimed mixing amplitudes are approximately equal, this is just the orthogonality sum, which is unity. (For *inter*band transitions, the amplitude sum is $(\alpha_1\alpha_2' + \beta_1\beta_2' + \ldots)$, and in this case nearly vanishes by the same orthogonality argument.) Thus we see that, although the mixing is large, *intra*band transitions are barely affected and retain their normal rotational strengths.

This has many repercussions, two of which are worth citing briefly. It means, for example, as we argued already, that Alaga rules for *intra*band transitions are essentially unaffected by mixing. Thus, observed deviations from the Alaga rules can be ascribed to other mechanisms (e.g., M1 components) and can be used to estimate these. Second, consider heavy ion reactions that bring large amounts of angular momentum into the nucleus, which then decays by a series of cascade transitions. It has been observed that these cascades flow through many rotational bands, but that the population within a band tends to remain intact as J decreases, even though these relatively high-lying quasi-particle excitations are expected to mix considerably. The preceding derivation provides a simple explanation: the mixing can indeed be strong, but as long as it does not change rapidly with J, the *intra*band transitions are only slightly affected and remain dominant.

In closing this section, we note that extensions of the formalism to include $0 - \gamma$ mixing have also been developed and are available in the literature. One point that will be useful in our later discussion of the IBA can be deduced immediately without a formal development of the mixing expressions. The effects of $0 - \gamma$ bandmixing on $0 \rightarrow g$

and $\gamma \rightarrow g$ transitions are second order and generally weak; however, since $0 \rightarrow \gamma$ transitions are forbidden in the absence of mixing, such mixing can strongly break this fundamental selection rule. The expression for $0 \rightarrow \gamma$ transitions in the presence of $0 - \gamma$ mixing is analogous to those we generated for the $0 \rightarrow g$ and $\gamma \rightarrow g$ cases, except that there is no longer a direct term and hence the *entire* transition strength arises *solely* from a mixing term. Therefore, although the finite transition matrix elements arise from mixing, *branching ratios* are *independent* of $\varepsilon_{0\gamma}$ and are given only by ratios of functions of J_0 and J_γ. At least in ^{168}Er, these branching ratios disagree with the data. We note for future reference that in the IBA model, $0 \rightarrow \gamma$ transitions are, in contrast, allowed for deformed nuclei but their branching ratios depend on the detailed structure (in effect on the value of the asymmetry parameter γ). We will discuss this further later in this chapter.

Having discussed the low-lying, intrinsic excitations of axially symmetric nuclei, we can return to the question of rotational energies and corrections to the simple first-order expressions in Eqs. 6.17 and 6.18. It was useful to discuss these intrinsic excitations, in particular, the bandmixing between them, first, because the corrections to the symmetric top formula are intimately connected with excursions from axial symmetry and rotation–vibration coupling. Indeed, the first order rotational expression makes several implicit assumptions, the most important of which are that there is no coupling between rotational and intrinsic degrees of freedom and that the mixing is independent of J. These two assumptions are, in fact, related. As the nucleus rotates, it experiences a centrifugal force that tends to increase the deformation and moment of inertia and decrease the rotational spacings, and leads to an enhanced coupling to vibrational modes. There are several ways of incorporating these effects into a rotational energy expression. One of the first and most common is simply to expand the rotational energy in powers of $J(J + 1)$ and keep the second term.

One then has

$$E = AJ(J + 1) + BJ^2(J + 1)^2. \tag{6.38}$$

where $A = \hbar^2/2I$. (We will derive this formula in a moment.) From our earlier comments, we know that empirical values of B are negative. If they are also small ($B/A << 1$), the expansion converges rapidly and Eq. 6.38 will be a significant improvement.

In some cases, still higher-order terms such as $CJ^3(J + 1)^3$ are necessary to produce adequate fits for higher J values. Rather than explore this, we shall turn shortly to an alternate expression that automatically includes Eq. 6.38 and all higher-order terms. First, we show an example in Fig. 6.20 of the ground state rotational band of ^{168}Yb compared with the energies calculated from Eqs. 6.17 and 6.38, as well as other expressions to be discussed. Evidently, the first-order expression (Eq. 6.17) is reasonable only for very low-spin states. Equation 6.38 (AB in the figure) is an improvement for higher spins, although it too encounters serious difficulties for still larger J. A fit with the $CJ^3(J+1)^3$ terms (ABC) included further improves the predictions, but is also inadequate for large J: the opposite signs empirically deduced for B and C tend to produce wild oscillations in predicted energies (compressions of levels, even spin inversions) at high enough J values.

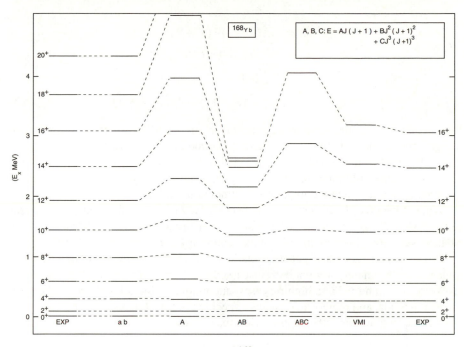

FIG. 6.20. Empirical ground band levels of ^{168}Yb compared with various models. The labels ab and A, AB, ABC refer to the coefficients in Eq. 6.40 and in the expansion of rotational energies in powers of $J(J+1)$ (see Eq. 6.40 and following discussion).

An alternate approach to incorporating rotation–vibration or centrifugal effects into the rotational energy expression is to make the moment of inertia spin dependent. This approach is known as the Variable Moment of Inertia (VMI) model and has enjoyed considerable success. In general, its predictions are better than those of Eq. 6.38, and it is not limited to the realm of strongly deformed nuclei. Figure 6.20 includes VMI predictions and shows their advantages. We shall not dwell on this approach, as it has been extensively covered in other literature.

Interestingly, it is easy to see how both effects (a change in the moment of inertia and the addition of a higher order term) result immediately from the effects of $\gamma - g$ bandmixing.

We have seen that the mixing is generally small so we can use the approximation of Eqs. 1.12 to write the energy shift (lowering) of the ground state band as $\Delta E_s^{gsb}(J) = V^2/\Delta E_{unp}^2$. But, from Eq. 1.10, the mixing amplitude $\varepsilon'_\gamma = \sqrt{2}\varepsilon_\gamma f_\gamma(J) = V/\Delta E_{unp}$. So, $\Delta E_s^{gsb} = 2\varepsilon_\gamma^2 f_\gamma^2(J)$. Hence, from Eq. 6.28,

$$\Delta E_s^{gsb}(J) = 2\varepsilon_\gamma^2 (J-1)J(J+1)(J+2)$$
$$= 2\varepsilon_\gamma^2 J(J+1)\left[(J-1)(J+2)\right]$$
$$= 2\varepsilon_\gamma^2 J(J+1)\left[J(J+1)-2\right]$$
$$= 2\varepsilon_\gamma^2 J^2(J+1)^2 - 4\varepsilon_\gamma^2 J(J+1) \qquad (6.39)$$

The second term is the promised correction to the standard rotational formula, and can give the variation with J of the inertial parameter $\hbar^2/2I$, while the first gives the required second-order correction term. From this derivation it is clear that Eq. 6.38 is, as we implied earlier, ultimately connected with the concept of rotation–vibration coupling (bandmixing) and also that it implicitly assumes small mixing.

When J becomes large enough such that $|\varepsilon_\gamma| f_\gamma(J) \sim 1$ we must anticipate a breakdown of Eq. 6.38 and thus a need for many higher-order terms or an alternate formula. We have seen this effect empirically in the failure of Eq. 6.38 for $J \geq 14$ in Fig. 6.20.

However, there is a much superior rotational expression that is valid for even higher spins that unfortunately has not been discussed much in the literature. It automatically gives Eqs. 6.17 and 6.38 as limiting cases, automatically includes all the higher order correction terms, and moreover, contains a specific relationship between the coefficients of each successive term. One simply writes the two-parameter formula

$$E(J) = a\left[\sqrt{1 + bJ(J+1)} - 1\right] \qquad (6.40)$$

where a and b are parameters.

This expression can be derived in the Bohr–Mottelson picture by including small deviations from axial symmetry. A trivial rationale for this was presented by Lipas many years ago. Suppose that we make the ansatz that we can write the moment of inertia I as a function of excitation energy (this would therefore also contain an implicit dependence on J as in the VMI):

$$I = \alpha + \beta E(J)$$

Then, substituting this into Eq. 6.17 gives

$$E(J) = \frac{\hbar^2}{2\left[\alpha + \beta E(J)\right]} J(J+1)$$

or

$$2\beta E^2(J) + 2\alpha E - \hbar^2 J(J+1) = 0$$

Hence

$$E(J) = \frac{-2\alpha \pm \sqrt{4\alpha^2 + 8\beta\hbar^2 J(J+1)}}{4\beta}$$

or, taking positive energies,

$$E(J) = \frac{\alpha}{2\beta} \left[\sqrt{1 + \frac{2\beta\hbar^2}{\alpha^2} J(J+1)} - 1 \right] \tag{6.41}$$

which is simply Eq. 6.40 with $a = \alpha/2\beta$ and $b = 2\hbar^2\beta/\alpha^2$. Note that energy ratios, such as $E_J/E_{2_1^+}$, depend only on the single parameter b in Eq. 6.40. Nevertheless, this formula is far more accurate than any of the expressions we have considered, as shown in Fig. 6.20 for ^{168}Yb where the predictions are compared with one-, two-, and three-term expansions in $J(J+1)$ and with the VMI model. Its success extends to softer (transitional) nuclei (e.g., ^{152}Sm, ^{184}Pt). Since the expression works so well for higher J, we anticipate a later discussion to caution that it is only applicable below any "backbend" that may be present.

Aside from its empirical success, Eq. 6.40 is interesting because, for relatively low spins such that $bJ(J+1) << 1$, expansion of the square root naturally recovers the second (and higher) order terms in the rotational formula of Eq. 6.38:

$$\left[\sqrt{1 + bJ(J+1)} - 1 \right] = \frac{b}{2} J(J+1) - \frac{b^2}{8} J^2(J+1)^2 + \dots \tag{6.42}$$

Here, however, the coefficients of each power of $J(J+1)$ are interrelated, whereas in Eq. 6.38, they are arbitrary parameters. It is remarkable that this constrained version of the expansion in powers of the angular momentum produces such an excellent fit.

For large J such that $bJ(J+1) >> 1$ (this limit may or may not be reached in practice depending on the value of b), Eq. 6.40 is almost linear in J, reflecting the enormous compression of the ground band due to both bandmixing effects and to centrifugal stretching. Thus, Eq. 6.40 incorporates the limit of small mixing (Eq. 6.38), but also extends into spin regions where the mixing is large.

We will show additional data for high J values shortly but first a brief digression on the experimental means of accessing high spin states is perhaps useful.

A projectile incident upon a target nucleus carries an orbital angular momentum of $\mathbf{l} = \mathbf{r} \times \mathbf{p}$. The study of high-spin states, which necessarily involves the transfer of large amounts of angular momentum to a target nucleus, will clearly benefit from the use of heavy ion projectiles. Early examples of this approach were the $(\alpha, xn\gamma)$ studies of Lark and Morinaga in the 1960s. They were able to populate the ground state rotational bands of many even–even deformed nuclei up to the 6^+ and 8^+ levels. These early studies provided important information on rotational structure, supplying the impetus, for example, for the study of the higher-order terms in the rotational energy expansions we have just discussed. A natural outgrowth of these studies was the use of true heavy ion projectiles; ^{12}C, ^{16}O at first, and subsequently nearly all nuclear species up to and including ^{208}Pb and U. Nowadays, evidence for discrete levels up to spins of $50\hbar$ and beyond is routine and has led to the discovery of phenomena such as superdeformation and identical bands (see below).

Figure 6.21 illustrates the basic population ideas behind a (H.I., $xn\gamma$) reaction. An incident heavy ion impinges on and is captured by a target nucleus, forming a compound nucleus which is almost always neutron deficient because of the curvature of the valley of

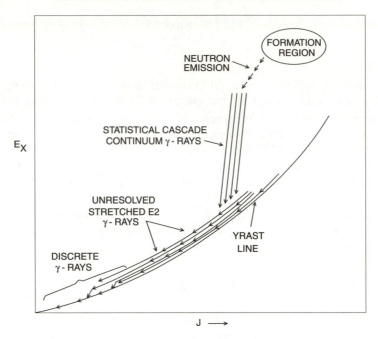

FIG. 6.21. Schematic idea of the flow of population and the de-excitation mechanisms in (heavy ion, $xn\gamma$) reactions.

stability: heavy nuclei near stability have a greater neutron excess than the lighter stable projectiles and targets whose fusion produces them. The compound nucleus is formed at extremely high spin and excitation energy. It is invariably particle unstable. The first step in the return to a quiescent state is the emission of a number of evaporation nucleons, most often neutrons. This process is favored over γ-ray emission both because it involves the strong rather than the electromagnetic interaction and also because the more massive neutrons provide a somewhat more efficient way of removing angular momentum for a given energy emission. (Recall our discussion in Chapter 1 of the hindrance inherent in high multipolarity γ-ray emission.) The additional evaporation of several neutrons from the initial compound nuclear state increases the neutron deficiency. Thus, (H.I., $xn\gamma$) reactions are useful both for studying high-spin states and gaining access to proton rich nuclei not otherwise observable.

Nucleon emission, which proceeds along a path slanted slightly towards the origin in Fig. 6.21, terminates when the nucleus eventually reaches an excitation energy where γ decay can compete with particle emission. The γ decay tends to proceed along nearly vertical de-excitation paths. Given the high complexity of the compound state just prior to γ-ray emission, there are a myriad of possible routes and the decay is characterized by a nearly statistical "rain cloud" pattern. Ultimately, states relatively near the yrast levels are reached. (The term yrast state refers to the lowest lying state of each spin.) At this point the de-excitation path closely tracks the yrast line. The de-excitation usually

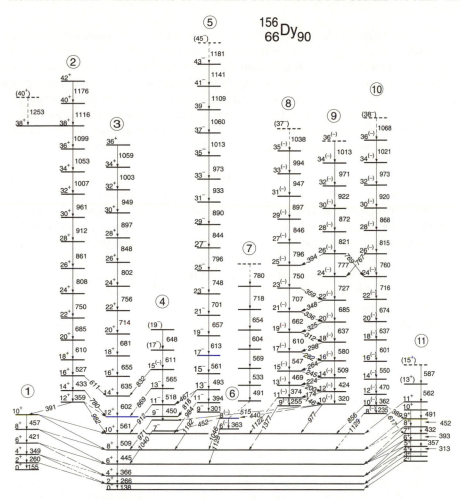

FIG. 6.22. Typical (H.I., $xn\gamma$)-based level scheme (Riley, 1988).

results in the emission of a sequence of E2 γ-ray transitions feeding down any of several nearly parallel rotational bands. (In the newly discovered phenomenon called magnetic rotation (Baldsiefen, 1994), or shears bands, the cascade consists of M1 transitions.) Eventually, the de-excitation of these "side bands" terminates at or near their bandheads and the intensity feeds into the yrast sequence itself. From here on all the decay proceeds simply by stretched E2 transitions in the ground state band.

A typical level scheme observed in a (H.I., $xn\gamma$) reaction is shown in Fig. 6.22. One notes the separation of the levels into sequences of quasi-rotational bands extending over extremely large ranges of spin. Even though these bands may mix with one another, the decay, as we have discussed just above, is preferentially down the rotational band sequences rather than to high spin yrast states.

Given the large number of bands, there will be many γ-ray transitions of nearly comparable energies. This poses a serious experimental problem. It is only with the development of multifold coincidence techniques using large arrays of detectors in the last decade or so that we have reached the experimental sophistication required to produce level schemes such as those in Fig. 6.22. The typical experimental arrangement consists of a target surrounded by a large array of γ-ray detectors, usually Ge detectors flanked by anti-Compton shields, designed to enhance the photopeak efficiency and discriminate against γ ray Compton scattering events. The latter lead to the deposition of a continuum of energies, below the γ ray transition (photopeak) energy, that forms an unwanted background. Such arrays are commonly called "balls" and may involve upwards of 50 to 100 detectors, or smaller numbers of higher efficiency Ge detectors such as "clover" detectors. Gammasphere is an example of the former class of instruments, while YRAST Ball at Yale exemplifies a clover-detector-based array. An excellent review and discussion of this important development and its applications may be found in Beausang (1996). With fusion evaporation reactions and modern arrays of γ ray detectors, a wealth of fascinating phenomena have been discovered, among them backbending and bandcrossing, rotation alignment (both discussed later in Chapter 9), shears bands, superdeformation and identical bands (see below).

Such phenomena are nearly always revealed by an analysis of rotational energies and we therefore now return to the behavior of rotational spectra at high spin. Indeed, access to higher and higher spin regions with heavy ion fusion evaporation reactions in the last two decades has led to the alternative way of presenting rotational energies that we discussed earlier when we introduced the ideas of bandcrossing and backbending.

It is implicit in Eqs. 6.17, 6.18 that the moment of inertia is a constant. We have already seen, however, in Figs. 6.12, 6.13 that this is not generally the case. A number of effects make the moment of inertia spin dependent: centrifugal stretching, changes in the pairing strength, Coriolis mixing, mixing with other intrinsic excitations (see the discussion of bandcrossing and backbending above and in Chapter 9). Therefore, in the last two decades the preferred approach has been to define a *spin*-dependent moment of inertia (this is not unlike the energy dependent moment of inertia just discussed, which embodies some of the same physics motivation). Therefore, for each J value, we can *define* a "local" moment of inertia, $I(J)$, still defined by Eq. 6.20, using $E_\gamma(J \rightarrow J-2)$. Moreover, to focus on the *changes* in the moment of inertia it has become common practice to define two moments of inertia, $I^{(1)}$ and $I^{(2)}$, the "kinematic" and "dynamic" moments, respectively. Their definitions are

$$I^{(1)} = \frac{J}{\omega}$$

and   (J in units of $\hbar$)

$$I^{(2)} = \frac{\mathrm{d}J}{\mathrm{d}\omega}$$

$I^{(1)}$ is the moment of inertia we have been discussing and is just the slope of J against ω. If the rotation is rigid $I^{(1)}$ is constant and $I^{(2)} = \mathrm{d}J/\mathrm{d}\omega = I^{(1)} + \omega \mathrm{d}I^{(1)}/\mathrm{d}\omega = I^{(1)}$.

Non-constancy of $I^{(1)}$, and differences between $I^{(1)}$ and $I^{(2)}$, signal dynamic frequency effects on the rotation.

Figure 6.23 shows an example of yrast energies $E(J)$ as a function of J, for ^{168}Yb, as well as $I^{(1)}$ and $I^{(2)}$ against ω for the same data. We see that, at low J (low $\hbar\omega$), $I^{(1)}$ is nearly constant but increases slightly, due perhaps to centrifugal stretching increasing the deformation or to a reduction of pairing correlations. At some J value, or frequency $\hbar\omega$, $I^{(1)}$ turns sharply upward and $I^{(2)}$ shows wilder gyrations. Note that, for the same spins, $E(J)$ still appears rather smooth: hence the advantage of the quantities $I^{(1)}$ and $I^{(2)}$ in magnifying effects on rotational energies. The behavior of $I^{(1)}$ and $I^{(2)}$ signals a rather drastic change in structure. We have already seen that such effects, often called "upbending" or "backbending", can result from the ground state band "crossing" a band built on a different intrinsic state with higher moment of inertia. In Chater 9, we will see that they can also arise from strong effects of the Coriolis force.

One of the most exciting discoveries in heavy ion reactions has been that of superdeformation (see Twin, 1986); that is, nuclear states with deformation $\beta \sim 0.5 - 0.6$. Superdeformation was first observed in 1986 in ^{152}Dy. Since then numerous superdeformed bands have been found, in several mass regions, and in both even and odd mass nuclei. At first, only higher spin levels were associated with superdeformation but superdeformed states in the Hg region down to spins $J \sim 10$ are now known. Typically, the deformation is deduced from the rotational spacings. From Eq. 6.19, γ ray transition energies are proportional to $1/I$ which, in turn, via Eq. 6.10, gives the deformation β. Closely spaced γ ray energies imply large I and hence large β (but see below for further discussion of rotational energies in superdeformed bands).

A spectrum showing a typical cascade of γ rays in a superdeformed band was shown in Fig. 6.14. This spectrum is also a superb example of shape and structure coexistence. The nucleus ^{150}Gd (with $Z = 64$) is nearly spherical in its ground and low lying states. There is no simple pattern to the γ ray de-excitations of the spherical excited states. However, the decay of the superdeformed levels in the *same* nucleus is a cascade of $E2$ transitions down the band. Hence, the transition energies are regular, and, indeed, evenly spaced: From Eq. 6.19 $\Delta E_\gamma = 2\hbar^2/I$ which is constant if the moment of inertia does not change. Figure 6.14 beautifully illustrates this. A second regularity stems from the feeding pattern. Since most of the population of any given state in the superdeformed band results from feeding from the next higher band member, the *intensities* of successive γ rays should change smoothly (down to a spin where the superdeformed band finally decays out to the normal states).

The idea behind superdeformation is that the potential energy surface of a nucleus has a "normal" spherical or deformed minimum and then, instead of a monotonic increase of the potential with β, there is a second minimum at large (superdeformed) β values. Such a double-minimum in the potential usually but not always occurs at high spin. [Of course, at still higher angular momenta, the nucleus becomes unstable to centrifugal forces and flies apart in the fission process. In that case, the potential rises with β at first but then tops out and descends to the fission point.]

Interpretations of the origin of the second (superdeformed) minimum in the potential generally involve the concept of energy gaps in the Nilsson diagram (see Chapters 8, 9)

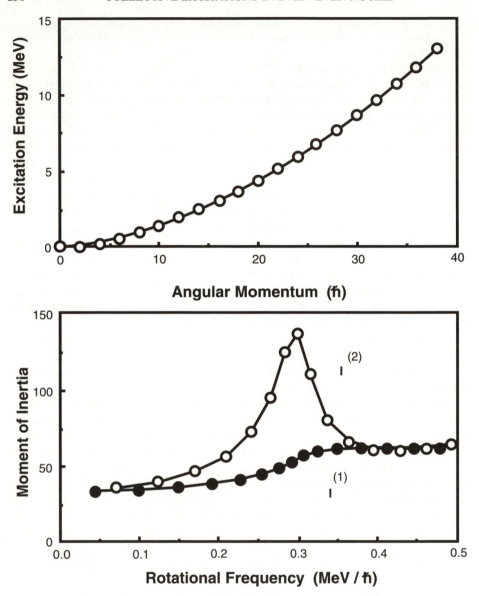

FIG. 6.23. Energies, kinematic, and dynamic moments of inertia in ^{168}Yb. (Figure courtesy of C. Beausang)

at large deformations associated with the presence of particles in orbits (Nilsson states) that strongly decrease in energy with ellipsoidal elongation of the nucleus. These orbits originate in higher oscillator shells. These deformed "shell gaps" act in a similar way to spherical shell gaps and lend greater stability to superdeformed configurations in those nuclei whose Fermi surface is near the gap. The strongly downsloping states from higher oscillator shells drive the nucleus towards larger deformation. These states are not unlike the intruder states encountered earlier in our discussion of spherical nuclei. They are "deformed intruders". The detailed characterization of superdeformed levels has involved the development of sophisticated "cranked" shell model calculations that properly account for the effects of extremely fast rotational motion, along with the incorporation of specific corrections for shell structure not included by simply specifying the deformation parameters.

The decay-out of a superdeformed band is a fascinating, and only recently discovered, process. Indeed, until this discovery, superdeformed bands were "floating" at unknown excitation energies and even with unknown bandhead spins since they were disconnected from the rest of the level scheme. The decay-out process can occur in two ways. If (as seems to be less common and certainly less interesting) the superdeformed band reaches its bandhead, further deexcitation has no other choice but to proceed to a different intrinsic configuration. The other decay-out process occurs at higher excitation energies where the level density of the normally deformed levels is high and a superdeformed state occurs very close. Then the nucleus may tunnel from one minimum in the potential well (the superdeformed) to the normal well. Re-phrased, the normal and superdeformed states may mix slightly. Since the normal and superdeformed states are so different in structure, the mixing matrix elements are weak and, typically, this process only occurs if the normal and superdeformed levels are within a few 10s of eV of each other. However, in fact, not much mixing is required because the decay-out transition is of dipole multipolarity and higher energy than the intraband transition and therefore inherently favored. The weak mixing amplitudes allow the two decay routes to compete. This process has been observed and this interpretation developed in the Hg–Pb region (Krücken, 1996, 1997).

To be historically accurate we should note that superdeformed states (though not called that) were actually first found in the 1960s at low spin as "fission isomers" in the actinide nuclei. These nuclei have a particularly complicated potential energy surface with multiple minima (even at low spin) as a function of deformation. States near the bottom of the second minimum, typically at a couple of MeV in excitation energy, have large deformations and have relatively long lifetimes prior to tunneling the fission barrier (hence the name fission isomers).

Superdeformation at high spin has been intensely studied for over a decade now and has revealed much about the interplay of rotation, single particle structure, and the role of pairing. For example, as we shall see in Chapter 9, Coriolis effects in rapidly rotating nuclei can often make it energetically favorable to break $J = 0$-coupled pairs of nucleons and to recouple their spins to $J_{max} = 2j - 1$ (e.g., $J_{max} = 12$ for two identical nucleons in $i_{13/2}$ orbits). Hence the generation of higher and higher spins in rotating nuclei is often accompanied by a successive reduction in pairing leading, eventually, to its disappearance at very high spin values. This phenomenon is sometimes referred to as pairing collapse. Since one of the effects of pairing is a reduction of the moment of inertia below that which would apply to a rigid rotor of the same shape (compare the Inglis and Belyaev formulas, Eqs. 6.12 and 6.12a), the rotational spacings in superdeformed states where pairing is weak are even smaller than would be suggested by the increase in deformation.

One of the most intriguing aspects of high spin studies and (often) of superdeformed states is the phenomenon of identical bands (see Byrski, 1990). These are pairs (or sets) of rotational bands, usually in different, adjacent (even and odd A) nuclei that have almost identical energy spacings between corresponding levels. This phenomenon shows up directly in the γ ray spectra following fusion evaporation reactions since the γ ray energies directly give the level spacings. The identity of the level spacings is remarkable, on the order of 0.1% (e.g., typically the corresponding γ ray energies, and hence the level spacings in the two bands are equal to within a keV or less for spacings ranging from several hundred keV to one MeV).

It is easy to understand the surprise that the identical band phenomenon presented to nuclear physicists by recalling that, for a rotating ellipsoid, the moment of inertia $I \propto A^{5/3}$ (see Eq. 6.10) and hence rotational energies should scale as $A^{-5/3}$. For adjacent nuclei (e.g., ^{152}Dy — ^{151}Tb) in the $A \sim 150$ mass region $\Delta(A^{-5/3})$ is about 1%. Moreover, due to "blocking" and pairing effects one expects the moment of inertia parameter to differ by considerably more than 1% in adjacent even and odd mass nuclei: In an odd A nucleus, one orbit (that occupied by the last nucleon) is blocked in the sum in Eq. 6.12a. Hence, the moment of inertia is slightly less in the odd A nucleus and rotational spacings should be larger. Therefore, in identical bands, there must be some other mechanism(s) which, remarkably, act to almost exactly cancel these effects.

Following their initial discovery, many other examples of identical bands were found– in other pairs of even and adjacent odd A nuclei, in pairs of adjacent even A nuclei, and also in non-adjacent nuclei, and at low spin as well. The phenomenon has developed from its initial status as an isolated curiosity to a rather widely observed feature of rotational spectra.

No completely satisfactory explanation exists, though not for lack of proposed interpretations. These interpretations range from cancellation of mass and deformation effects on the moment of inertia (e.g., the increase of I with A is canceled by a slightly smaller deformation), to the interplay of pairing and deformation, to more exotic explanations in terms of Bose–Fermi symmetry or supersymmetry, or to a simple consequence of the p–n residual interaction. No explanation fully accounts for the widespread nature of the phenomenon.

6.5 Axially asymmetric nuclei

The models we have discussed thus far incorporate excursions from axial symmetry that are both *small* and dynamic. Certainly, such an approach accounts reasonably well for the deviations of most well-deformed nuclei from the properties of the pure axial rotor. However, there have long been indications that larger and possibly permanent (static) asymmetries also occur. Naturally, this would lead to more radical departures from the energy and transition rate expressions we have considered. In fact, in certain limiting cases of large asymmetry, new selection rules appear. In another sense, however, such models for larger asymmetry are extensions of the small excursions from axiality dealt with so far, and their predictions go over into the latter as $\gamma \rightarrow 0°$. It also turns out that many predictions of models for large, fixed asymmetries γ are identical, or nearly so, to models incorporating dynamic fluctuations in γ so long as γ_{rigid} in one equals γ_{rms} in the other.

The best known model of fixed stable asymmetry (*triaxiality*) is that of Davydov and co-workers developed around 1960. Here, the potential $V(\gamma)$ is envisioned to have a steep, deep minimum at a particular value of γ so that the nucleus takes on a rigid shape with that asymmetry.

We have seen that, if the rotational and vibrational motions are not completely decoupled, and there is an interaction (mixing) between the γ and ground bands, the latter will acquire a finite γ_{rms} and K will no longer be a good quantum number. Therefore, it is not surprising that in the Davydov model K is not a good quantum number either. Here, however, since γ can be large, the K admixtures can reach levels far beyond those we have encountered.

The relation between the Davydov model and models with axially symmetric but γ-soft potentials runs deeper than this. In a nucleus with such a potential, the greater the softness the lower the γ vibration will lie, and the larger γ_{rms} will be in the ground state. In the Davydov model there is no distinction in intrinsic structure between what is normally called the ground state rotational band and the γ vibrational states. The levels of these two bands simply become the so-called normal and anomalous levels of a new ground state band whose energies depend explicitly on γ, which can take on values from $0° \rightarrow 30°$ (prolate symmetric $\rightarrow$ maximum asymmetry: $30° \leq \gamma \leq 60°$ corresponds to the "oblate" region of asymmetry). Figure 6.24 shows the lowest levels as a function of γ and clearly illustrates the descent of the γ vibrational levels. Indeed, for $\gamma \geq 25°$, $E_{2^+_\gamma} < E_{4^+_1}$. In contrast, as $\gamma \rightarrow 0°$, the normal levels $0^+_1, 2^+_1, 4^+_1, 6^+_1$, which are rather insensitive to γ, go over into those of an axially symmetric ground state band, while the "γ band" energies increase rapidly. An important feature of the anomalous levels is their energy "staggering": They tend to be grouped into couplets as $(2^+, 3^+), (4^+, 5^+)\ldots$

The behavior in Fig. 6.24 is easy to understand. As γ increases, the nucleus becomes increasingly flattened. Therefore, states whose wave functions are predominantly aligned in the direction of the flattening attain lower energies, since the nuclear force is attractive and they are, on average, closer to the bulk of the nuclear matter. This is exactly the case for the $K = 2$ (and higher K) levels which, therefore, rapidly decrease in energy with increasing γ. They also mix with the normal ground state band levels (yrast states) and, as we have seen, K is no longer a good quantum number. The leveling off of the energies of the anomalous levels for $\gamma \geq 25°$ is easily understood in terms of that mixing. In Table 6.11 we give a number of interesting quantities relating to the Davydov model, including the amplitude for $K = 0$ in the 4^+_1 and 4^+_2 states as a function of γ. For γ between $25°$ and $30°$, the major amplitudes in the 4^+_2 state actually interchange so that the $K = 0$ amplitude is larger than the $K = 2$ one. (The 4^+_1 amplitudes do not quite "cross" since there is actually substantial three-state mixing involving the 4^+_3 state.) The interchange of amplitudes in the 4^+_2 level occurs near the energy inflection point in Fig. 6.24. The decreasing trend of the quasi-γ-band, or anomalous level energies, would have caused these energies to cross the normal levels at this point. Instead, the interaction

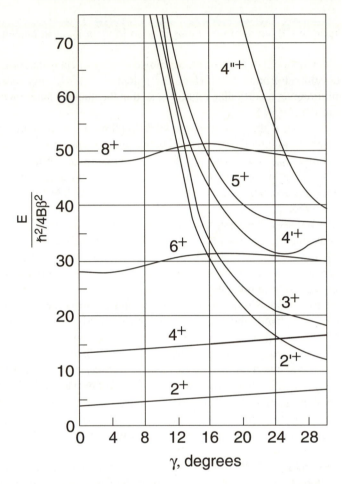

FIG. 6.24. Normal and anomalous levels of the triaxial rotor (Preston, 1975).

causes a repulsion. This is a nice example of this two-state mixing effect discussed in Chapter 1.

In practical applications of the Davydov model, one usually extracts γ from the energy ratio $E_{2_2^+}/E_{2_1^+}$ of the first two 2^+ states. This ratio is given in Table 6.11 for several values of γ and is plotted in Fig. 6.25 (along with two B(E2) ratios). It can be calculated for any γ value from the expression

$$\frac{E_{2_2^+}}{E_{2_1^+}} = \frac{[1 + X]}{[1 - X]} \tag{6.43}$$

where

$$X = \sqrt{1 - \frac{8}{9} \sin^2(3\gamma)}$$

The individual energies are equal to the respective numerators and denominators multiplied by $(9/\sin^2 3\gamma)$. Note that $X \to 1$ for $\gamma \to 0°$. Thus, $E_{2_2^+}/E_{2_1^+} \to \infty$ for $\gamma \to 0°$ as seen in the table and figure. Also, $X \to 1/3$ for $\gamma \to 30°$. Table 6.12 gives the empirical values of $E_{2_2^+}/E_{2_1^+}$ for a number of heavy rare earth nuclei as well as the associated γ values. If, on account of the anomalously low $E_{2_2^+}/E_{2_1^+}$ ratios, these nuclei are considered to have large rigid triaxiality (this is the term usually used for the concept of fixed asymmetries), then these γ values are the only input needed to make Davydov model predictions for other observables. Such predictions are included in Table 6.11 for several γ values. The branching ratio,

$$R_2 = \frac{B\left(E2 : 2_2^+ \to 2_1^+\right)}{B\left(E2 : 2_2^+ \to 0_1^+\right)}$$

is also a useful indicator, as shown in the same table and in Fig. 6.25. R_2 can be written analytically and calculated for any γ from

$$R_2 = \frac{\dfrac{10}{7}\left[\dfrac{\sin^2(3\gamma)}{9X^2}\right]}{\dfrac{1}{2}\left[1 - \dfrac{3 - 2\sin^2(3\gamma)}{3X}\right]} \tag{6.44}$$

where the numerator and the denominator are the individual B(E2) values. Note that both B(E2) values in R_2 vanish for $\gamma = 0°$, yet they have a finite ratio that is the Alaga rule: the vanishing is reasonable since for $\gamma = 0°$, $E_{2_2^+}/E_{2_1^+} \to \infty$, corresponding to infinite rigidity in the γ direction and to vanishing vibrational amplitude. R_2 increases rapidly with γ and $R_2 \to \infty$ for $\gamma = 30°$. This latter result is identical to the selection rule for an alternate model of axial asymmetry that we will soon discuss, the γ flat or γ-unstable model of Wilets and Jean in which the $2_2^+ \to 0_1^+$ transition is forbidden.

Some other B(E2) values and branching ratios are given in Table 6.11. Those corresponding to rotational transitions in either the γ or ground bands, are nearly γ independent. Others vanish at both $\gamma = 0°$ and 30° but attain small, finite values for intermediate γ values. These are transitions *between* normal and anomalous levels. The $2_2^+ \to 2_1^+$ and $3_1^+ \to 4_1^+$ transitions form a third category: small at $\gamma = 0°$ and rising rapidly toward $\gamma = 30°$. This behavior is easily understandable if we note that, in the $\gamma = 30°$ case, the Davydov model has the same selection rules as both the quadrupole vibrator model and the Wilets–Jean γ-unstable model. For example, at $\gamma = 0°$, the $2_2^+ \to 2_1^+$ and $3_1^+ \to 4_1^+$ transitions are interband ($\gamma \to g$): given the built-in stiffness in γ, they must be forbidden. At $\gamma = 30°$, the 3_1^+ state is analogous to a three-phonon level, the 4_1^+ level to a two-phonon excitation, so the transition becomes allowed. Similarly the $2_2^+ \to 2_1^+$ transition is analogous to a $2 \to 1$-phonon transition.

Since finite γ values correspond to mixed K values, one might expect a close relation between B(E2) values for finite γ in the Davydov model and in the bandmixing

Table 6.11 *Some useful predictions of the asymmetric rotor (Davydov) model**

$\gamma \rightarrow$	$0°$	$5°$	$10°$	$15°$	$20°$	$25°$	$27.5°$	$30°$
$E_{2_2^+}/E_{2_1^+}$	∞	65.2	15.9	6.85	3.73	2.41	2.10	2.00
$\dfrac{B\left(E2:2_2^+ \rightarrow 2_1^+\right)}{B\left(E2:2_2^+ \rightarrow 0_1^+\right)}$	1.43	1.52	1.85	2.71	5.33	20.4	83	∞
$\psi_{4_1^+}$ $\quad K=0$	1	1	0.999	0.993	0.955	0.852	0.792	0.739
$\quad\quad K=2$	0	0.003	0.030	0.114	0.296	0.522	0.605	0.661
$\psi_{4_2^+}$ $\quad K=0$	0	-0.003	-0.030	-0.114	-0.296	-0.523	-0.602	0.559
$\quad\quad K=2$	1	1	0.999	0.993	0.954	0.842	0.754	0.500
$B\left(E2:2_1^+ \rightarrow 0_1^+\right)$	1	0.993	0.972	0.947	0.933	0.955	0.985	1
$B\left(E2:2_2^+ \rightarrow 0_1^+\right)$	0	0.0074	0.028	0.053	0.067	0.0425	0.015	0
$B\left(E2:2_2^+ \rightarrow 2_1^+\right)$	0	0.011	0.051	0.143	0.357	0.865	1.23	1.43
$\dfrac{B\left(E2:2_2^+ \rightarrow 0_1^+\right)}{B\left(E2:2_1^+ \rightarrow 0_1^+\right)}$	0	0.0075	0.0283	0.056	0.072	0.0444	0.015	0
$B\left(E2:4_2^+ \rightarrow 2_1^+\right)$	0	0.0004	0.011	0.008	0.0004	0.021	0.018	0
$B\left(E2:4_1^+ \rightarrow 2_2^+\right)$	0	0	0.0023	0.010	0.033	0.039	0.016	0
$B\left(E2:4_2^+ \rightarrow 4_1^+\right)$	0	0.0138	0.0624	0.167	0.313	0.311	0.271	0.273
$B\left(E2:4_1^+ \rightarrow 2_1^+\right)$	1.429	1.418	1.395	1.377	1.372	1.365	1.378	1.389
$B\left(E2:3_1^+ \rightarrow 4_1^+\right)$	0	0.006	0.034	0.130	0.406	0.821	0.955	1
$B\left(E2:3_1^+ \rightarrow 2_1^+\right)$	0	0.0132	0.0492	0.095	0.12	0.079	0.027	0
$B\left(E2:3_1^+ \rightarrow 2_2^+\right)$	1.78	1.77	1.74	1.69	1.67	1.70	1.76	1.78

*$B(E2)$ values are in units of $e^2 Q_0^2/16\pi$.

formalism. This is indeed so. Consider, as an example, the case we worked out earlier of ^{168}Er. We found that the Mikhailov plot analysis gave a full mixing amplitude in the 4^+ states of $\varepsilon_\gamma'(4^+) = -0.021$. In the Davydov model, the experimental ratio $R_2 = 1.85$ yields $\gamma \sim 11°$ (see Table 6.11). This, in turn, corresponds to a $\gamma \rightarrow g$ mixing amplitude of -0.03. The agreement is not exact since the comparison is not quite on an equal footing. In the bandmixing case, the spin independent mixing amplitude was deduced from a Mikhailov plot, which gives an overall average value for all transitions, while the Davydov mixing value was deduced from R_2 alone. In any case the essential point is that both the bandmixing formalism and the Davydov model lead to K mixtures in γ and ground bands, and give comparable mixing amplitudes and B(E2) values for small γ. (The bandmixing formalism is a first-order perturbation treatment and is therefore inapplicable for large γ.) Though the physcial pictures are different, predictions for many observables are nearly identical.

When we turn to the large γ extreme, one might think that the extremely low-lying

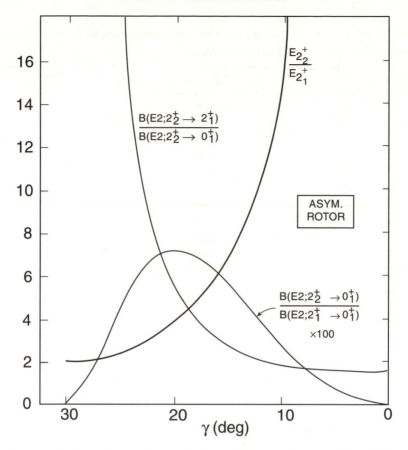

FIG. 6.25. Dependence of several observables on γ (compare Fig. 6.48).

γ band levels of the Davydov model would be an appropriate signature for stable axially asymmetric shapes. However, we can also envision large γ values as dynamic quantities by picturing a deformed nucleus that is totally free to vibrate in the γ degree of freedom. Such a nucleus is completely γ-soft: it is a "γ-unstable" rotor. This corresponds to a nuclear potential centered at a finite β but completely flat in the γ degree of freedom. The nucleus oscillates smoothly from $\gamma = 0° - 60°$ and has $\gamma_{rms} = 30°$. In the extreme limit of complete γ instability, known as the *Wilets–Jean* model, the "rotational" energies are given by

$$E(\Lambda)_{W-J} = \vartheta \Lambda(\Lambda + 3) \tag{6.45}$$

where ϑ is a constant analogous to $\hbar^2/2I$ and the levels are now classified according to the quantum number Λ. This classification scheme is given in Fig. 6.26. The yrast levels have $J = 2\Lambda$. Note that each Λ value (for $\Lambda > 1$) corresponds to more than one level

Table 6.12 *Values of $E_{2_2^+}/E_{2_1^+}$ for some deformed and transitional nuclei and the corresponding γ values (rounded to nearest degree).*

Nucleus	$E_{2_2^+}/E_{2_1^+}$	γ
^{152}Sm	8.9	13°
^{160}Dy	11.1	12°
^{168}Er	10.3	13°
^{172}Yb	18.6	9°
^{176}Hf	15.2	10°
^{182}W	12.2	12°
^{184}Os	7.9	14°
^{188}Os	4.1	19°
^{192}Os	2.4	25°
^{196}Pt	1.94	30°

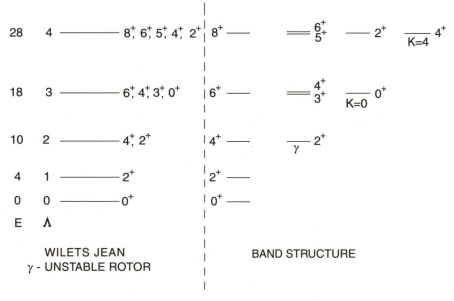

FIG. 6.26. Ground band levels in the γ-unstable or Wilets–Jean model. (Left) In terms of Λ multiplets; (Right) Displayed in analogy to the quasi-band structures of a normal rotor.

and that the Λ values $2, 3, 4, \ldots$ include a low-lying set of levels analogous to the γ vibrational band and to the anomalous levels of the Davydov models for large γ.

Now that we have the three basic extreme geometric models, the harmonic vibrator, the axially symmetric deformed rotor, and the γ-soft axially asymmetric deformed rotor,

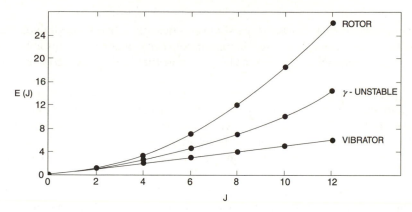

FIG. 6.27. Dependence of ground band energies on spin for different models. An identical set of curves is obtained in the U(5), O(6) and SU(3) symmetries of the IBA (see Eq. 6.72).

it is interesting to compare their rotational energies. The results are shown in Fig. 6.27. The spacing of the normal rotational levels of the Wilets–Jean rotor are quite different than in the symmetric rotor. This is easy to see if one notes that for these levels (the yrast states) $\Lambda = J/2$, so that $E_{W-J}(J\text{-yrast}) = (\vartheta/4)J(J+6)$, which increases with J considerably slower than the $J(J+1)$ law for a symmetric rotor. For example, $E_{4_1^+}/E_{2_1^+} = 2.5$ compared to 3.33 for the rotor and $E_{6_1^+}/E_{4_1^+} = 4.5$ instead of 7. We note for later reference that the γ unstable limit of Wilets–Jean is very closely related to the O(6) limit of the IBA Model.

Another characteristic of the Wilets–Jean model concerns E2 selection rules. The allowed transitions must satisfy $\Delta\Lambda = \pm 1$. Comparison with Table 6.11 shows that the Davydov model goes over to the same selection rules as the Wilets–Jean model for $\gamma = 30°$, again highlighting the similarity of a γ-flat model with $\gamma_{rms} = 30°$ to a rigid asymmetric model with fixed $\gamma = 30°$. The Wilets–Jean picture also resembles the phonon model in its E2 selection rule and gives some identical predictions. For example, the $2_2^+ \rightarrow 2_1^+$ transition ($\Lambda = 2 \rightarrow \Lambda = 1$) is allowed, as is $4_1^+ \rightarrow 2_1^+$, but the crossover $2_2^+ \rightarrow 0_1^+$ transition is forbidden ($\Delta\Lambda = 2$), as is the $4_1^+ \rightarrow 2_2^+(\Delta\Lambda = 0)$. These are the same results one would have if the 0_1^+, 2_1^+ and $(4_1^+, 2_2^+)$ doublet were treated as the zero-, one-, and two-phonon vibrational states. This is not surprising since the potential for the spherical vibrator, while parabolic in β, is independent of γ, so it is trivially γ-unstable as well. The real difference between the vibrator and Wilets–Jean limits is that in one $\beta_{ave} = 0$, whereas the other has a deformed minimum. Another difference between the two models (with the vibrator now considered in its harmonic, degenerate multiplet limit) is that the vibrator has a two-phonon triplet of levels $0^+, 2^+, 4^+$ while the Wilets–Jean scheme has only a $2^+, 4^+$, doublet, the first excited 0^+ state having $\Lambda = 3$. Thus, not only is it higher in energy but its allowed decay is to the *second* 2^+ level (2_2^+) rather than the first as in the vibrator; more precisely, $B(E2 : 0_2^+ \rightarrow 2_2^+)/B(E2 : 0_2^+ \rightarrow 2_1^+) \rightarrow \infty$ in the γ-unstable limit but is zero in the vibrator. The reader is warned, however, that these differences do not *necessarily* persist in a sufficiently anharmonic vibrator. We will encounter this point, and this close relationship between vibrator and γ-unstable models, again, in our discussion of the U(5) and O(6) symmetries of the IBA model later in this chapter.

Despite the obvious similarities between the Davydov and Wilets–Jean models, there is one outstanding difference by which to distinguish them empirically: the energy staggering in the low-spin anomalous or γ vibrational levels is *opposite* in the two models. In the γ unstable case, these levels group according to (2^+), $(3^+, 4^+)$, $(5^+, 6^+) \ldots$ as seen in Fig. 6.26: in the simplest version of this model, the levels within each couplet are degenerate.

It is interesting to use a quantitative measure of the asymmetry as a distinguishing signature. To this end, we define the quantity $\Delta E_J = E_J - E_{J-1}$, where the energies refer to the γ-band, and then define the energy staggering, $E_s(J)$, as

$$E_s(J) = \Delta E_J - \Delta E_{J-1} \tag{6.46}$$

For example,

$$E_s(4) = (E_{4_\gamma^+} - E_{3_\gamma^+}) - (E_{3_\gamma^+} - E_{2_\gamma^+}) \tag{6.47}$$

These double energy diferences involve three levels with spins J, $J - 1$, and $J - 2$, and we use the convention that the level of spin J (the starting level) is always of even spin. The usefulness of E_s in comparing different models is evident if we relate it to the energy of the first 2^+ level, $E_{2_1^+}$. Although the harmonic vibrator is not a rotational scheme, the levels of the "quasi-γ band" in that scheme also display a staggering similar to the γ-unstable model and include the same degenerate couplets. We can therefore include this model in this intercomparison as well. For the four cases of the symmetric rotor, the triaxial rotor/ Davydov model with $\gamma = 30°$, the γ-unstable or Wilets–Jean model, and the harmonic vibrator, the following analytical expressions result:

$$E_s(4) = \frac{\hbar^2}{2I}\{[J(J + 1) - (J - 1)J] - [(J - 1)(J) - (J - 2)(J - 1)]\}$$

$$= \frac{\hbar^2}{2I}2 = \frac{E_{2_1^+}}{3} \qquad \text{(symmetric rotor)}$$

$$E_s(4) = A[(34 - 18) - (18 - 12)] = 10A = \frac{5}{3}E_{2_1^+} \qquad (\gamma \text{ rigid}, \gamma = 30°)$$

$$E_s(4) = A\{0 - [3(6) - 2(5)]\} = -8A = -2E_{2_1^+} \qquad (\gamma \text{ unstable})$$

$$E_s(4) = [0 - (3E_{2_1^+} - 2E_{2_1^+})] = -E_{2_1^+} \qquad \text{(vibrator)}$$

$$\tag{6.48}$$

Generally, if $E_s(4) < E_{2_1^+}/3$ (γ-soft case), the even spin states are depressed relative to the odd spin; if $E_s(4)) > E_{2_1^+}/3$ (γ-rigid), the odd spin levels are depressed relative to the even spin ones.

Before inspecting empirical values of $E_s(4)$, we note another interesting feature of the vibrator limit. As we have seen, empirical values of the energy ratio $R \equiv E_{2ph}/E_{1ph}$ for vibrational nuclei are typically ~ 2.2 rather than the strict harmonic limit of 2.0. Table 6.4 gave the energy levels of the three-phonon states, which include the 3^+ and 4^+ levels of the "quasi-γ band" in terms of the two-phonon levels. $E_s(4)$ cannot be defined analytically in this case since the unknown energies of these 3^+ and 4^+ levels implicitly involve the energy of the 4_1^+ level. However, if we make the approximation that the two-phonon 2_2^+ and 4_1^+ levels are degenerate (i.e., $\varepsilon_2 = \varepsilon_4$ in the notation of Table 6.4), then the 3^+ and 4^+ quasi-γ band states will still be degenerate at an energy $3E_{2_1^+}(R-1)$. The energy anharmonicity of the two-phonon states is tripled in the three-phonon states. Simple manipulations then give $E_s(4) = -E_{2_1^+}(2R - 3)$. Of course, in the harmonic limit $R = 2$, and this goes to $E_s(4) = -E_{2_1^+}$ as in Eq. 6.48. For the more typical case of $R > 2$, however, $E_s(4)$ becomes increasingly negative compared to the harmonic value.

We see that E_s nicely distinguishes different models ranging from vibrational to rotational to axially asymmetric, and is also a useful signature of transitional regions between these ideal limiting cases. We show in Fig. 6.28 the empirical $E_s(4)$ values for rare earth nuclei. Near the $N = 82$ closed shell, the nuclei are close to vibrational with only a small component of β softness. $E_s(4)$ is negative and, in units of $E_{2_1^+}$, ranges from $-0.8 \rightarrow -0.4$. As the neutron number increases, the β softness rapidly increases and $E_s(4)$ approaches the rotational limit of $+E_{2_1^+}/3$. The fact that most of the deformed rare earth nuclei have $E_s(4)$ values slightly less than $E_{2_1^+}/3$ reflects both their predominantly axially symmetric rotational character and the presence of a small amount of rotation-vibration coupling. Towards the end of the $N = 82$–126 shell, the behavior becomes more erratic, with the Pt isotopes displaying tendencies toward extreme γ softness.

Despite years of popularity of the Davydov model because of its simplicity and analytic formulas, there is virtually no evidence for rigid triaxial behavior. Axial asymmetry at low energy in nuclei seems associated with γ softness instead.

In a later section of this chapter we shall see that such γ softness also characterizes the O(6) symmetry of the IBA and nuclei in O(6) $\rightarrow$ rotor or SU(3) transitional regions. To anticipate that discussion briefly, it is interesting to note that the classic O(6) nuclei, $^{192-196}$Pt, have larger $E_s(4)$ values than some of the other Pt isotopes, even though they are supposedly completely γ-soft. We can already guess from Fig. 6.28, however, that nuclei with nearly γ-independent potentials but with shallow minima at $30°$ (that is, with the addition of a small component of triaxiality at $\gamma = 30°$) might have $E_s(4)$ values higher than the extreme γ-unstable limit. This is in fact the case, as will be commented later in the discussion of the IBA.

The γ-rigid/γ-soft ambiguity is not the only one that can obfuscate an interpretation of the structure. Figure 6.28 also shows the difficulty in distinguishing near harmonic vibrational structure from γ softness. There is one useful, albeit qualitative, indicator stemming from the systematic behavior of certain absolute energies that can sometimes clarify whether a *transitional* region is vibrational $\rightarrow$ symmetric rotor or axially asymmetric $\rightarrow$ symmetric rotor in character. This indicator is the relation between the energy of the 2_γ^+ (normally 2_2^+) level and the 4_1^+ state of the quasi-ground state band.

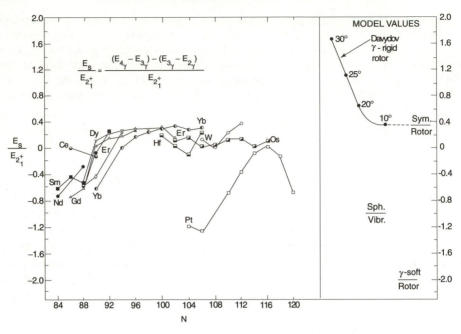

FIG. 6.28. Systematics of $E_S(4)/E_{2_1^+}$ in the rare earth region. Predicted values for several models are given at the right.

In a transition from vibrator to symmetric rotor, the 2_γ^+ level starts off initially degenerate with the 4_1^+ level at rather high energies (since $E_{2_1^+}$ is then also rather high). As the phase transition develops, the 4_1^+ level drops rapidly (as the 2_1^+ energy drops) to become a member of the ground state rotational band, while the 2_γ^+ level remains rather high in energy (although it may drop slightly). In a transition from an axial asymmetric rotor (either soft or rigid) to symmetric rotor, the 2_γ^+ and 4_1^+ levels start out nearly degenerate (for large γ) and rather low-lying, but the former rises rapidly as $\gamma \to 0°$ while the 4_1^+ level drops slightly as the deformation increases. These contrasting systematics are shown for the Sm and Os nuclei in Fig. 6.29. This discussion somewhat over simplifies the situation, however, as other effects are at work (see Chapter 7). Moreover, this signature, while valid, cannot be quantified as well as the energy staggering since the two levels involved have intrinsically different structure in the rotational limit: the 4_1^+ state belongs to the ground state band while the 2_γ^+ state is an intrinsic excitation whose energy depends on details of its microscopic collective wave function (see the discussion of the RPA in Chapter 10). In any case, it is evident that the Ba–Gd region represents a vibrational-rotational transition, whereas the Pt–Os region is axially asymmetric $\to$ rotational.

We will discuss the Sm–Gd transition region again, rather extensively, in Chapter 7, where we consider the evolution of structure in nuclei, and where other facets of structure in this region such as the effects of the p–n interaction, the relation of structural evolution

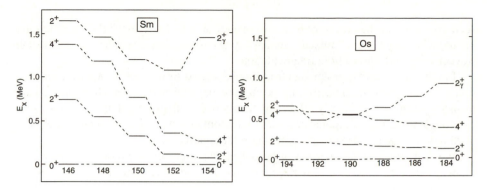

FIG. 6.29. Systematics of low-lying levels in the Sm and Os nuclei. The symmetric rotor limit is at the right in each case.

here to that involving intruder states in nuclei such as Hg–Pt and Cd, and the possibility of phase transitional behavior in finite nuclei, will all be discussed.

This completes our brief survey of some of the essential features of geometric models for deformed even–even nuclei. We will return to study detailed predictions of these models in the context of the Geometric Collective Model (GCM) of Gneuss, Greiner, and colleagues at the end of this chapter after we discuss the IBA model next. These geometric models have been enormously successful, especially when the first order energy and B(E2) predictions are modified by incorporating the higher order terms that reflect changes in shape and adiabaticity of the rotational motion as J increases. These models are all truly phenomenological in that they are applied to real nuclei by inspecting empirical properties in order to *assign* the appropriate shapes (β, γ, or higher-order moments).

In closing this section, it is important to stress that we have hardly exhausted the collective vibrational modes in even–even nuclei. For example, we have not touched on the most basic low-lying negative-parity excitations—octupole vibrations—nor have we considered hexadecapole excitations. Just as quadrupole vibrations in deformed nuclei can have $K = 0$ or $K = 2$, octupole excitations exist in $K = 0, 1, 2,$ and 3 forms. In many deformed nuclei, several of these have been identified, primarily via strong E1 and/or E3 transitions to the ground state band. The ordering of these octupole K values changes systematically through a shell: near the beginning, the low K modes are lowest, while they are highest toward the end. Though these excitations account for most of the known negative parity states below the pairing gap in deformed nuclei, they have been less well studied than their positive parity quadrupole counterparts. Hexadecapole vibrations are much less well known: perhaps their most likely manifestation is in the low-lying $K = 4$ bands in the Os isotopes.

Another phonon mode is related to the isovector M1 scissors mode mixed proton–neutron symmetry states discussed in Section 6.2. Just as an isovector dipole mode yields a 1^+ state, an isovector quadrupole mode gives a 2^+ mixed symmetry state. This state is expected to decay to the 2_1^+ level with a strong $B(M1)$ value. A candidate for such an excitation has recently been discovered at 2.13 MeV in ^{136}Ba (see Pietralla, 1998a). Contrary to the 1^+ mixed symmetry states in neighboring nuclei, the quadrupole isovector strength shows little fragmentation.

Finally, a collective proton–neutron mode, related to magnetic rotation (See Section 6.2) and called chiral bands, may occur in some deformed nuclei where protons and neutrons occupy unique parity orbits with nearly perpendicular angular momenta. These can couple to the core rotational angular momentum in either a left- or right-handed way, leading to two nearly degenerate, identical rotational bands.

We will return to the question of hexadecapole after our discussion of the Nilsson model in Chapter 9 and to the microscopic structure and systematics of octupole excitations in Chapter 10.

6.6 The interacting boson model

Up until the mid-1970s the two principal strains of nuclear structure theory were embodied in the shell model, which emphasized the single-particle aspects of nuclear structure along with a careful accounting of the effects of the Pauli principle, and the collective model pioneered by Bohr and Mottelson and generalized by Gneuss and Greiner, Kumar, and numerous others. Each of these models, of course, had numerous offshoots: the shell model for spherical nuclei led to the development of the Nilsson model for deformed shapes and the collective models developed refinements that incorporated axial asymmetry, higher moments in the shape such as hexadecapole components, and the like. The link between these models was provided in the early 1960s by the development of microscopic approaches to collective motion utilizing such techniques as the TDA and RPA, which we will discuss in Chapter 10.

In 1974 the *Interacting Boson Approximation* (IBA) model was proposed, by Arima and Iachello. It is based on a third approach that is group theoretical or algebraic and that recalled methodology developed in the 1950s by Elliott and co-workers for light deformed nuclei. The IBA has been extraordinarily successful, and has also generated its own family of offshoots inspiring alternate, sometimes competitive, algebraic approaches such as the pseudo-SU(3) and symplectic group studies of Draayer and co-workers.

The IBA is a model for collective behavior. It has become customary to refer to collective models of the Bohr–Mottelson type as "geometric" models and those of the IBA or other group theory-based approaches as "algebraic" models. Today one has a situation in which there is a triad of models—shell, geometric, and algebraic—with which one can attack the basic problems of nuclear structure. These models are not generally incompatible, although there are differences in certain important details, but rather reflect three approaches to nuclear structure that emphasize different complementary aspects of that structure.

As we have commented repeatedly, the shell model rapidly becomes intractable far from closed shells. In order to circumvent this, two basic alternatives have been tried. In one, that of geometric models, the whole microscopic approach is abandoned and replaced by a macroscopic one involving an assumed or deduced nuclear shape, with rotations and vibrations about that shape. The other, of which the IBA is an example, seeks to effectively truncate the shell model space: the practical utility of such an approach depends on the *extent* of the truncation, while its success depends on the *appropriateness* of the truncation in isolating the key configurations involved (at least in the low-lying states).

The truncation inherent in the IBA is shockingly extreme. For example, it reduces the 3×10^{14} 2^+ shell model basis states in ^{154}Sm to 26! It is a wonder that such a scheme can work at all, much less have the extensive and repeated success it has enjoyed.

The basic idea of the IBA is to assume that the valence fermions couple in pairs *only* to angular momenta 0 and 2 and that the low-lying collective excitations of medium and heavy nuclei can be described in terms of the energies and interactions of such pairs. These fermion pairs, having integer spin, are treated as bosons (called s and d bosons for obvious reasons).

More formally, the model is founded on and embodies the following assumptions and ideas:

- Closed shells of either protons or neutrons, and excitations out of them, are neglected.

- The low-lying excitations of even–even nuclei depend only on the valence space.

- The valence nucleons are treated in pairs, as s and d bosons, with angular momenta 0 and 2. In the IBA-1 no distinction is made between proton and neutron bosons. The number of bosons is half the number of valence protons and neutrons, both of which are always counted to the nearest proton and neutron closed shells. This counting rule generates a finite, fixed, number of s and d bosons. This finite number has profound effects, and leads to numerous predictions that are different from those of the geometric analogues of IBA structures.

- The states of this boson system result from the distribution of the fermions in s and d pairs, and thus depend only on the s and d boson energies and on interactions *between* bosons. These interactions are assumed to be simple (at most two-body).

A fundamental feature of the IBA that results from these assumptions is its group theoretical structure. Since an s boson ($J = 0$) has only one magnetic substate and a d boson ($J = 2$) has five, the s–d boson system can be looked at mathematically as a six dimensional space. The basis states span that space. It turns out that such a system can be decribed in terms of the algebraic group structure U(6). As we shall discuss at length, such a "parent" group has various subgroups and different "decompositions" of U(6) into sequences of subgroups that lead to different symmetries (dynamical symmetries, to be exact). There are three of these symmetries that are physically interesting, known by the labels U(5), SU(3), and O(6). Each has specific, characteristic properties and a definite geometric analogue. Actual examples of nuclei manifesting these symmetries have been identified.

This symmetry structure is central and critical to the IBA. Many predictions can be obtained analytically by powerful group theoretical (algebraic) methods, rather than by tedious numerical diagonalizations. Moreover, the group structure keeps the underlying physical picture close at hand. Even when analytic results are not obtainable (for nuclei with structures "intermediate" between two symmetries) the symmetries act as benchmarks or touchstones that provide a physical backdrop and a simpler starting point for detailed calculations.

As Broglia and others have pointed out, the twin aspects of the grounding of the IBA in fermion pairs, and its symmetry structure with geometrical analogues, confers on the model a "Janus-like" character. On the one hand, it looks to the shell model for its microscopic justification (is it a reasonable truncation to ignore all configurations except those corresponding to $J = 0$ and $J = 2$ valence fermion pairs?) and for the ultimate derivation of its parameters and their systematics. On the other, it leads to a picture of the nucleus, and specific predictions, very closely allied to macroscopic geometrical collective models.

To summarize, the two essential distinguishing features of the IBA are its symmetry structure resulting from the s–d boson truncation and its emphasis on the valence space, with explicit recognition of the finite number of valence nucleons. The first leads to its algebraic formulation and to the dynamical symmetries so intimately associated with the model, while the latter leads to many key predictions, often different from otherwise closely related geometrical models. As we shall see, it confers a microscopic aspect on an otherwise basically phenomenological model.

We now present a simplified outline of some key elements of the IBA-1 model (protons and neutrons treated together). We first discuss the bosons and the basis states that can be constructed from them, and then a suitable IBA-1 Hamiltonian. We then turn to a discussion of the group theory of the IBA and its symmetry structure. Finally, we consider realistic (nonalgebraic) calculations for actual nuclei and a simplified approach to many of these, the so-called consistent Q formalism. Throughout, we give a number of concrete examples of IBA predictions and stress the relationship to the geometrical models discussed earlier in this chapter.

The basic entities of the IBA are $s(J = 0)$ and $d(J = 2)$ bosons, which are assigned energies ε_s and ε_d. (Note that it is conventional in the IBA literature to use L for angular momentum both for the individual bosons, s and d, and for the total spin of a state. Here, we keep to the convention of this book and use J for these quantities.) A given nucleus with $N_p + N_n$ valence protons and neutrons (each counted to the nearest closed shell) has $N = (N_p + N_n)/2$ s and d bosons. For example, ^{152}Sm has $N = 6 + 4 = 10$ and both ^{144}Ba and ^{196}Pt have $N = 3 + 3 = 6$. No distinction is made whether the valence nucleons are particles or holes. Ground and excited states are formed by distributing the bosons in different ways among s and d states and coupling them to different total J. The level structures that result depend on these distributions and couplings.

The simplest situation is to imagine all N bosons in s boson states. By convention, $\varepsilon_s = 0$, the absolute ground state. The lowest excited state will have $(N - 1)s$ bosons, one d boson, and an energy $E = \varepsilon_d$. The next states, in this simplest case, will be a group with two d bosons ($n_d = 2$). Clearly, as in the phonon model, the two d bosons can couple to $J = 0, 2, 4$. This triplet will be degenerate. Higher d boson multiplets will also occur up to $n_d = N$. This is a purely harmonic spectrum identical to the harmonic vibrator except for the limitation due to finite boson number.

Since the IBA is configured explicitly in terms of s and d bosons, most of the formalism is phrased in terms of creation and destruction operators for these entities,

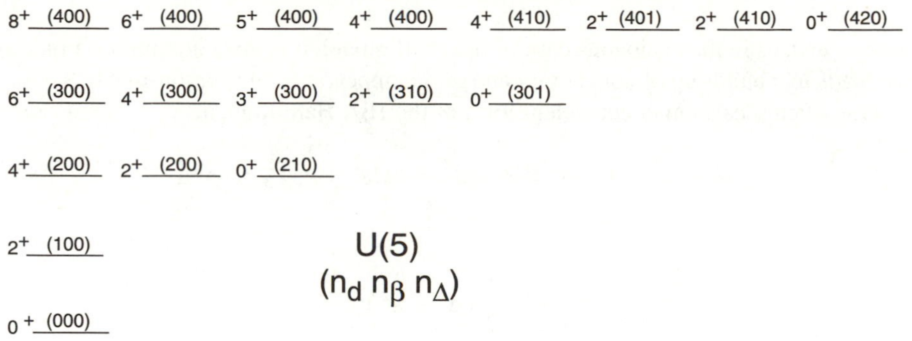

8^+ _(400)_ 6^+ _(400)_ 5^+ _(400)_ 4^+ _(400)_ 4^+ _(410)_ 2^+ _(401)_ 2^+ _(410)_ 0^+ _(420)_

6^+ _(300)_ 4^+ _(300)_ 3^+ _(300)_ 2^+ _(310)_ 0^+ _(301)_

4^+ _(200)_ 2^+ _(200)_ 0^+ _(210)_

2^+ _(100)_

$$\mathsf{U(5)}$$
$$(n_d \; n_\beta \; n_\Delta)$$

0^+ _(000)_

FIG. 6.30. Basis states of the IBA. The U(5) limit (Casten, 1988a).

$s^\dagger$, s, $d^\dagger$, $\tilde{d}$, and combinations thereof. The basic rules for operating with these are the same as for the phonon operators b, $b^\dagger$ used earlier in this chapter (Eqs. 6.5–6.7). The Hamiltonian for the harmonic system just described is simply

$$H = \varepsilon_d (d^\dagger \tilde{d})^0 = \varepsilon n_d \tag{6.49}$$

that is, the energies are $E = \varepsilon n_d$, where for simplicity here and henceforth we drop the subscript "d" on ε.

Different states in a multiplet can be distinguished by their angular momentum J and by the number $n_\beta (n_\Delta)$ of d bosons coupled pairwise (tripletwise) to $J = 0$. Sometimes one specifies not n_β, but the number of d bosons *not* coupled to $J = 0$, and denotes this "boson seniority" quantum number by v. These states form a convenient basis set for the IBA and are illustrated in Fig. 6.30. Note that for $n_d \geq 4$, more than one state of a given J can occur.

Having defined the basis states in this way, we can now consider more general IBA Hamiltonians composed of creation and destruction operators for s and d bosons limited to a maximum of two-body (boson) interactions. If we keep only those terms relevant to excitation energies (i.e., if we ignore terms contributing to binding energies), we can write

$$H = \varepsilon' n_d + \frac{1}{2} \sum_J C_J (d^\dagger d^\dagger)^{(J)} \bullet (\tilde{d}\tilde{d})^{(J)}$$
$$+ \frac{v_2}{\sqrt{10}} \left[(d^\dagger d^\dagger)^{(2)} \bullet \tilde{d}s + H.c. \right] + \frac{v_0}{2\sqrt{5}} (d^{\dagger 2} s^2 + H.c.) \tag{6.50}$$

where ε', C_J, v_2 and v_0 are six free parameters. As we have discussed, the first term simply counts the number of d bosons and multiplies it by a d boson energy. This gives the unperturbed energy of a state with n_d noninteracting d bosons. The second group of three terms introduces interactions between *pairs* of d bosons that depend on the angular momentum to which they are coupled but that do not change the relative numbers of s and d bosons nor mix the basis states. The other terms have the property of changing the number of d bosons by $\Delta n_d = \pm 1, \pm 2$. These terms *mix* different basis states of a given J and, as in the analogous case of the shell model, it is this configuration mixing that leads to a build- up of collectivity and to the appearance of rotation-like behavior.

One often sees another equivalent form of the IBA Hamiltonian,

$$H = \varepsilon'' \mathbf{n}_d + a_0 \mathbf{P}^\dagger \mathbf{P} + a_1 \mathbf{J}^2 + a_2 \mathbf{Q}^2 + a_3 \mathbf{T}_3^2 + a_4 \mathbf{T}_4^2 \tag{6.51}$$

where

$$\mathbf{P} = \frac{1}{2} \left(\tilde{\mathbf{d}}^2 - \mathbf{s}^2 \right)$$

$$\mathbf{T}_J = \left(\mathbf{d}^\dagger \tilde{\mathbf{d}} \right)^{(J)}, \quad J = 0, 1, 2, 3, 4$$

$$\mathbf{Q} = \left(\mathbf{d}^\dagger \mathbf{s} + \mathbf{s}^\dagger \tilde{\mathbf{d}} \right) - \frac{\sqrt{7}}{2} \left(\mathbf{d}^\dagger \tilde{\mathbf{d}} \right)^{(2)}$$

$$\mathbf{n}_d = \sqrt{5} \mathbf{T}_0, \quad \mathbf{J} = \sqrt{10} \, \mathbf{T}_1 \tag{6.52}$$

The operators in Eqs. 6.51 and 6.52 are convenient combinations of those in Eq. 6.50 that have simple physical interpretations in terms of, for example, boson pairing and quadrupole operators. The most important point to note in Eq. 6.51 is the Δn_d character of the various terms; those in $\mathbf{n}_d$, $\mathbf{J}^2$, $\mathbf{T}_3^2$ and $\mathbf{T}_4^2$ have $\Delta n_d = 0$, $\mathbf{P}^\dagger \mathbf{P}$ has $\Delta n_d = 0, \pm 2$ contributions, while $\mathbf{Q}^2$ has $\Delta n_d = 0, \pm 1, \pm 2$ parts.

An important aspect of IBA predictions focuses on E2 transitions. The relevant operator, $\mathbf{T}(E2)$, is simply related to $\mathbf{Q}$ in Eq. 6.52, by

$$\mathbf{T}(E2) = e_B \mathbf{Q} = e_B \left[\mathbf{s}^\dagger \tilde{\mathbf{d}} + \mathbf{d}^\dagger \mathbf{s} + \chi \left(\mathbf{d}^\dagger \tilde{\mathbf{d}} \right)^2 \right] \tag{6.53}$$

where e_B is a boson charge similar to the effective charge for fermions and is often treated as a free parameter. In the original IBA formalism, the parameter χ in $\mathbf{Q}$ is fixed at $\chi = -\sqrt{7}/2$ in the Hamiltonian and treated as a free parameter in $\mathbf{T}(E2)$. An alternate formalism, the *consistent Q formalism* (CQF), uses the same χ in both H and $T(E2)$, which leads to certain simplifications and to a clearer physical picture of this model. This formalism will be discussed shortly.

We have stated that the s and d boson structure of the IBA leads to a six-dimensional space and hence to a description in terms of the unitary group U(6). We shall not delve into the group theory of the IBA in any detail, but a few ideas are useful to understand how the symmetries so characteristic of this model arise. Much of the following discussion is based on a review by the author and D. D. Warner to which the reader is referred for additional material on the IBA and its literature.

The basic concept underlying the group theory of the IBA is that of the "generators" of a group. These are sets of operators that "close on commutation" (i.e., the commutator of any pair $[\mathbf{A}, \mathbf{B}] = \mathbf{AB} - \mathbf{BA}$ either vanishes or is proportional to another member of the group, or a linear combination thereof). For the IBA, the 36 operators $\mathbf{s}^\dagger \mathbf{s}$, $\mathbf{s}^\dagger \tilde{\mathbf{d}}_\mu$, $\tilde{\mathbf{d}}_\mu^\dagger \mathbf{s}$ and $(\mathbf{d}_\mu^\dagger \tilde{\mathbf{d}}_\mu)^J$ where $J = 0, 1, 2, 3, 4$ and $|\mu| \leq J$ satisfy this condition and are the generators of U(6). As an example, we show this closure for the particular pair $\mathbf{d}^\dagger \mathbf{s}$ and $\mathbf{s}^\dagger \mathbf{s}$.

Using Eqs. 6.5–6.7 we have

$$[\mathbf{d}^\dagger \mathbf{s}, \mathbf{s}^\dagger \mathbf{s}]|n_d n_s\rangle = (\mathbf{d}^\dagger \mathbf{s} \mathbf{s}^\dagger \mathbf{s} - \mathbf{s}^\dagger \mathbf{s} \mathbf{d}^\dagger \mathbf{s})|n_d n_s\rangle$$
$$= \mathbf{d}^\dagger \mathbf{s} n_s |n_d n_s\rangle - \mathbf{s}^\dagger \mathbf{s} \mathbf{d}^\dagger \mathbf{s}|n_d n_s\rangle$$

or, since n_s is just a number and is factorable

$$= (n_s - \mathbf{s}^\dagger \mathbf{s})\mathbf{d}^\dagger \mathbf{s}|n_d n_s\rangle$$
$$= (n_s - \mathbf{s}^\dagger \mathbf{s})\sqrt{n_d + 1}\sqrt{n_s}|n_d + 1, n_s - 1\rangle$$
$$= \sqrt{n_d + 1}\sqrt{n_s}[n_s - (n_s - 1)]|n_d + 1, n_s - 1\rangle$$
$$= \sqrt{n_d + 1}\sqrt{n_s}|n_d + 1, n_s - 1\rangle$$
$$= \mathbf{d}^\dagger \mathbf{s}|n_d, n_s\rangle$$
$$\left[\mathbf{d}^\dagger \mathbf{s}, \mathbf{s}^\dagger \mathbf{s}\right] = \mathbf{d}^\dagger \mathbf{s} \qquad (6.54)$$

The other commutators can be similarly evaluated and indeed close on commutation. This set of 36 generators of the group of transformations of U(6) is said to form a Lie algebra.

Another key concept is that of a *Casimir operator* of a group. This is an operator that commutes with *all* of the generators of the group. Such operators can be composed of linear or higher order combinations of the generators and are appropriately called linear, quadratic,... Casimir operators.

The linear Casimir operator of U(6), which commutes with all 36 generators, is the total boson number operator $\mathbf{N} \equiv \mathbf{d}^\dagger \mathbf{d} + \mathbf{s}^\dagger \mathbf{s}$ whose eigenvalue is N. This result follows trivially from the fact that all 36 combinations of the $\mathbf{s}$ and $\tilde{\mathbf{d}}$ operators must conserve the total boson number. For example,

$$\left[\mathbf{N}, \mathbf{s}^\dagger \tilde{\mathbf{d}}\right]\psi = \mathbf{N}(\mathbf{s}^\dagger \tilde{\mathbf{d}})\psi - (\mathbf{s}^\dagger \tilde{\mathbf{d}}\mathbf{N})\psi$$
$$= N(\mathbf{s}^\dagger \tilde{\mathbf{d}})\psi - N(\mathbf{s}^\dagger \tilde{\mathbf{d}})\psi = 0$$

Suppose now that some smaller set of operators also closes on itself under commutation. This set forms the generators of a smaller subgroup of U(6). It will have linear and/or quadratic Casimir operators associated with it that commute with all the generators of the subgroup. There are several subgroups for U(6), so the reduction process continues until the rotational subgroup O(3) is reached.

It is now necessary to find the quantum numbers that label the states. In general, the generators of a group may change some quantum numbers (e.g., n_d) but there will be one (or more) that are *not* changed by *any* of the generators. For U(6), the 36 generators always conserve N. The set of basis states that have a particular fixed value of an unchanged quantum number (or numbers) is called an *irreducible representation* of the group.

Since the generators of a given group cannot connect different irreducible representations, the Casimir operator(s) of a group that commute with all the generators by definition must be diagonal and therefore must conserve all quantum numbers, including those of the subgroups. Indeed, each Casimir operator has eigenvalues that are functions only of the conserved quantum numbers of the particular subgroup. Thus we have the central result that a Hamiltonian consisting of Casimir operators of a group and subgroups cannot mix different representations of any of the groups involved. Furthermore, its eigenvalues are simple linear combinations of its component Casimir operator eigenvalues and are functions of the quantum numbers characterizing each group and subgroup. Since the quantum numbers characterizing a subgroup are constant for all the states of the particular representation it defines, all the states of that representation must be degenerate. This degeneracy is broken only by the next step in the chain, which subclassifies the levels according to another quantum number (for a subsequent subgroup). This whole process is illustrated for one of the group chains (the so-called O(6) limit) of the IBA in Fig. 6.31. The precise meanings of the quantum numbers and eigenvalue terms will be clarified shortly. The key point here is the successive degeneracy breaking and the classification of sets of states (representations) of each subgroup by specific quantum numbers. Another important result is now also obvious: a transition operator consisting of generators of a given group or subgroup cannot connect states in different representations of that group. This leads to many essential selection rules.

A central task in developing any group chain or group reduction scheme is to identify the quantum numbers that label the irreducible representations of each subgroup. This is the basic procedure followed in the algebraic treatment of the IBA. Group chains are constructed starting from U(6), where all the states are degenerate for a given value of N, and ending with O(3). A Hamiltonian for any such chain is written as a sum over the Casimir operators of the subgroups of the specific chain, and is therefore diagonal in a basis defined by the corresponding representation labels. Each step in the chain reduction introduces one or more free parameters (coefficients of terms in H) into the eigenvalue expression and requires one or more quantum numbers to distinguish the representations of the particular subgroup; it also breaks a previous degeneracy. Thus, the solution of the eigenvalue problem for such a chain reduces to that of the (known) eigenvalues of each of the Casimir operators.

The structure defined by such a Hamiltonian is referred to as a *dynamical symmetry*. One of the elegant aspects of these symmetries is that the excitation energy spectrum can be written down immediately and each state can be labeled by appropriate quantum numbers even though these symmetries may correspond to a complex physical situation and, in terms of Eq. 6.51, to a complex Hamiltonian. Since transition operators can often be written in terms of the group generators, transition selection rules appear naturally, and the rates for allowed transitions can be written analytically. Moreover, many ratios of transition rates depend only on general characteristics of the symmetry (group chain) and are parameter free. This should not be surprising: the Alaga rules for E2 branching ratios in deformed nuclei are a familiar geometrical analogue.

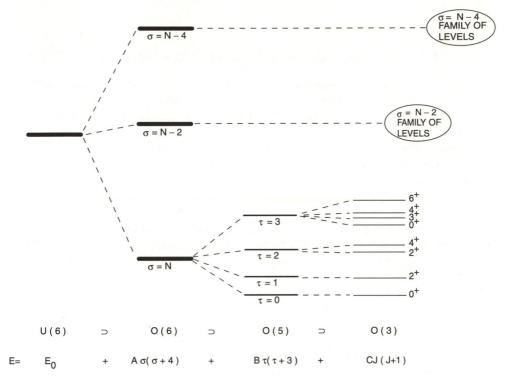

FIG. 6.31. Illustration (using the O(6) symmetry) of the successive degeneracy breaking in a dynamical symmetry group chain (Casten, 1988a).

Returning now to the basic problem in the IBA, there are exactly three group chains of U(6) that end in O(3), which is the rotation group. This group is a necessary subgroup in any physical chain because it provides for rotational invariance. The three group chains can be written, along with their relevant quantum numbers (see discussion to follow) as:

$$
\begin{array}{llllll}
\text{I.} & U(6) \supset & U(5) \supset & O(5) \supset & O(3) & U(5) \\
& N & n_d & \nu & n_\Delta J & \quad(6.55) \\
\text{II.} & U(6) \supset & SU(3) \supset & O(3) & & SU(3) \\
& N & (\lambda, \mu) & K J & & \quad(6.56) \\
\text{III.} & U(6) \supset & O(6) \supset & O(5) \supset & O(3) & O(6) \\
& N & \sigma & \tau & \nu_\Delta J & \quad(6.57)
\end{array}
$$

We now discuss each of these symmetries in turn.

I. U(5)

The U(5) symmetry is the IBA version of a vibrator. Its representation labels were already introduced, since this limit provides the basis states used in most treatments of the general IBA Hamiltonian.

The eigenvalues of U(5) are

$$E(n_d, v, J) = \alpha n_d + \beta n_d(n_d + 4) + 2\gamma v(v + 4) + 2\delta J(J + 1) \qquad (6.58)$$

where α, β, γ, and δ are parameters. A *harmonic* version of U(5) was illustrated in Fig. 6.30. Note, however, that U(5) is a very rich symmetry and allows much anharmonicity. The degenerate multiplets with a given value of n_d include levels with different values of v, J and the energies can depend on these quantum numbers. The U(5) wave functions, of course, are trivial. Since they are themselves the normally-used IBA basis states, each wave function has but a single term. Even highly anharmonic U(5) spectra maintain the same simple wave functions: the anharmonicity is a diagonal effect on the energies and does not lead to mixing of the basis states.

An interesting result concerning the anharmonicity follows from the Hamiltonian for U(5) written in the form of Eq. 6.50. Here U(5) includes all terms with $\Delta n_d = 0$. Although the interactions that break the degeneracies may appear to be complex, they never involve higher order interactions in **d** and **s** boson operators (i.e., operators such as $(\mathbf{d}^\dagger \mathbf{d}^\dagger \mathbf{d}^\dagger)(\tilde{\mathbf{d}}\tilde{\mathbf{d}}\tilde{\mathbf{d}})$). Thus the anharmonicities, *whatever* they may be, arise from two-body interactions. This is exactly the ansatz used to discuss the relationship between three-phonon and two-phonon energies in the anharmonic vibrator model earlier. In neither case are the results dependent on the details of the interactions, but only on the *number* of the interacting entities. Thus, these same relationships, listed in Table 6.4, characterize the U(5) limit. No choices of U(5) parameters can violate them.

In each symmetry that we shall discuss, key signatures and tests are provided by E2 transitions. So it is useful to summarize some typical results for each.

The operator $\mathbf{T}(E2)$ has a term that changes n_d by ± 1, and a term with $\Delta n_d = 0$. If $\mathbf{T}(E2)$ is chosen to be a generator of U(5), only the latter term is used. (Since n_d is a good quantum number in U(5), a generator of U(5) cannot change n_d.) The predicted E2 matrix elements would then vanish between states differing by ± 1 d boson, while they would yield nonzero diagonal contributions (quadrupole moments). This situation is essentially the inverse of what is expected and observed for vibrational nuclei, making it customary to use the first term of the E2 operator in the U(5) limit, as this term produces results similar to those of the geometric vibrational picture.

One could also use both terms in T(E2). Then one gets finite quadrupole moments for U(5) nuclei, as is indeed observed in nuclei such as ^{114}Cd. However, one would also obtain transitions *within* a phonon multiplet (that is, transitions with $\Delta n_d = 0$). It is an open question whether such transitions are observed.

Regardless of the value of χ, one obtains the general result

$$\sum_{J'} \mathrm{B}(E2 : J, n_d + 1 \to J', n_d) = e_B^2 (n_d + 1)(N - n_d) \qquad (6.59)$$

The sum on the left side of Eq. 6.59 accounts for the distribution of strength from a given initial state if the angular momentum selection rules allow decay to more than one level of the next lower multiplet. This sum contains more than one term only for decay of $n_d \geq 3$ states, and is identical in origin to the sum in the phonon model expression.

The factor $(n_d + 1)$ in Eq. 6.59 is analogous to the phonon model result proportional to $(N_{ph} + 1)$. The factor $(N - n_d)$ in the IBA case has no analogue in the phonon model and arises specifically from the finite boson number. Its origin can easily be seen. The matrix element $\langle n_d, n_s | s^\dagger \tilde{d} | n_d + 1, n_s - 1 \rangle$ can be calculated as:

$$\left\langle n_d, n_s | s^\dagger \tilde{d} | n_d + 1, n_s - 1 \right\rangle = \sqrt{(n_d + 1)} \sqrt{n_s} \langle n_d, n_s | n_d, n_s \rangle$$

$$= \sqrt{(n_d + 1)} \sqrt{n_s} = \sqrt{(n_d + 1)} \sqrt{N - n_d} \qquad (6.60)$$

In terms of the number of quadrupole excitations only (i.e., d bosons or quadrupole phonons) the U(5) limit and the geometrical vibrator are identical. The 2_1^+ level has one such excitation, the levels of the $0^+, 2^+, 4^+$ triplet have two, and so on. The difference is that, in such excitations in the IBA, the restriction to a finite fixed N imposes another complementary constraint on the number of s bosons, which gives rise to the second factor on the right in Eqs. 6.59 and 6.60. In the phonon model, a creation (or destruction) of a phonon takes place in isolation: the transition rate is related to the number of phonons available. However, in going from one U(5) representation to another in the IBA, the creation (or destruction) of a d boson *must* involve the destruction (or creation) of an s boson to conserve N. As n_d grows, there are fewer available s bosons, and the s boson factor n_s (or $N - n_d$) decreases. In an $(n_d + 1) \rightarrow n_d$ transition in the IBA, larger n_d values facilitate the transition (there is more freedom in choosing a particular d boson to destroy) but the smaller number of s bosons hinders the transition. These two counterbalancing aspects are reflected in the two factors in Eqs. 6.59 and 6.60.

Equation 6.59 gives, for the transitions between the lowest levels,

$$B(E2 : 2_1^+ \rightarrow 0_1^+) = e_B^2 N \qquad (6.61)$$

and

$$B(E2 : 2_2^+ \rightarrow 2_1^+) = 2e_B^2 (N - 1) \qquad (6.62)$$

The ratio of these two equations gives the useful result

$$R = \left[\frac{B(E2 : 2_2^+ \rightarrow 2_1^+)}{B(E2 : 2_1^+ \rightarrow 0_1^+)} \right]_{U(5)} = \left[\frac{N - 1}{N} \right] \left[\frac{B(E2 : 2_2^+ \rightarrow 2_1^+)}{B(E2 : 2_1^+ \rightarrow 0_1^+)} \right]_{Phonon} \qquad (6.63)$$

Since U(5) is usually relevant only near closed shells where N is rather small, differences with the geometric model can be significant. For example, Eq.6.63 gives $R = 1.6$ for $N = 5$, compared to $R = 2.0$ for the geometric picture. Finally, when the initial state is the fully aligned $J = 2N$ excitation, the factor $(N - n_d)$ is reduced to unity. This is an example of the well-known cutoff effect in B(E2) values involving high spin states, which is another characteristic distinction of the IBA from geometric models.

Whenever some model predicts a symmetry, it is always a critical test to search for empirical examples. This is particularly true for the IBA since it is so intimately connected with the concept of dynamical symmetry. Searches for U(5)-like nuclei naturally focus on those regions where the geometric vibrational model is also appropriate. The nucleus ^{118}Cd has been proposed as a near-harmonic empirical manifestation of U(5). We have also discussed earlier a multiphonon interpretation of ^{114}Cd extending up to 5-phonon states. The $E2$ transitions are in excellent agreement with U(5) although the energies are highly anharmonic. Alternate interpretations of ^{114}Cd involving vibrational and co-existing intruder states (see discussion surrounding Figs. 6.6–6.8 and Table 6.5) also reproduce the data.

II. SU(3)

This symmetry is the IBA version of a deformed rotor, but with special characteristics that distinguish it from its geometric analog. The SU(3) limit is obtained when the $\mathbf{Q}^2$ term dominates in Eq. 6.51 (a $\mathbf{J}^2$ term may also be present). Thus

$$H = a_2 \mathbf{Q}^2 + a_1 \mathbf{J}^2 \tag{6.64}$$

Here, for $\mathbf{Q}$ to be a Casimir operator of SU(3), χ must equal $-\sqrt{7}/2$. We saw before that $\mathbf{Q}^2$ strongly mixes U(5) basis states with $\Delta n_d = 0, \pm 1, \pm 2$. Therefore, SU(3) wave functions are no longer simple in terms of an expansion in U(5). On the contrary, they are rather complex, and certainly not very physically transparent, combinations of many U(5) states. The simplicity of SU(3), or any IBA symmetry, results from its geometrical structure, from the analytic nature of many results, and from simple selection rules, despite the fact that the wave functions, when expressed in the basis of *another* symmetry, may be complex. Such complexity signals only that the symmetries are different from each other, not that one is more complicated than any of the others. Nevertheless, one gets an insight into the symmetry structure by explicitly showing some wave functions in the same basis. This is done for all three symmetries in a U(5) basis in Table 6.13.

On account of the mixing of basis states with different n_d values in SU(3), the expectation values of the operator $(\mathbf{d}^\dagger \mathbf{d})^0 = n_d$, are very different in SU(3) and U(5). They are shown for the ground band states in Fig. 6.32 along with the values for O(6). In U(5), $\langle n_d \rangle = 0, 1, 2, \ldots$ etc., up to $\langle n_d \rangle = N$. In SU(3), $\langle n_d \rangle$ is already substantial in the ground state. This has three important effects that we can see without detailed calculation. First, any effects of finite boson number will be relatively larger in the SU(3) ground state than in U(5). Second, since in both cases $\langle n_d \rangle_{\max} = N$, the expectation value of n_d must increase *slower* with J in SU(3) than in U(5). This will have important consequences for certain B(E2) values that, we shall see, will increase more slowly with J in SU (3) than in U(5). Third, in U(5), $\langle n_d \rangle$ for a given state is independent of N, while in SU(3), it is roughly proportional to N. This has enormously important effects on collective E2 transitions.

The SU(3) energies are given in terms of quantum numbers of the group chain II of Eq. 6.56.

$$E(\lambda, \mu, J) = \frac{a_2}{2}\left(\lambda^2 + \mu^2 + \lambda\mu + 3(\lambda + \mu)\right) + \left[a_1 - \frac{3}{8}a_2\right] J(J+1) \tag{6.65}$$

Table 6.13 *Wave functions expressed in the U(5) basis for the first three 0^+ states in each limit of the IBA. For N = 6.*

State*	Limit	(000)	(210)	(301)	(420)	(511)	(602)	(630)
				Basis States $(n_d n_\beta n_\Delta)$				
	U(5)	1	0	0	0	0	0	0
0_1^+	O(6)	−.43	−.75	0	−.491	0	0	−.095
	SU(3)	.134	.463	−.404	.606	−.422	−.78	.233
	U(5)	0	1	0	0	0	0	0
0_2^+	O(6)	.685	.079	0	−.673	0	0	−.269
	SU(3)	.385	.600	−.204	−.175	.456	.146	−.437
	U(5)	0	0	1	0	0	0	0
0_3^+	O(6)	0	0	−.866	0	−.463	0	0
	SU(3)	−.524	−.181	−.554	.030	−.114	−.068	−.606

* The states are ordered for pedagogical clarity and not necessarily in the order of increasing energy: indeed, the $\tau = 3\ 0^+$ state in O(6) (here labeled 0_3^+) is usually the 0_2^+ state.

where a_1, a_2 are the coefficients of the multipole form of the Hamiltonian of Eq. 6.51. Each set of (λ, μ) values defines a representation of the subgroup of SU(3) and corresponds to a set of one or more rotational bands. Each band is characterized by a quantum number, sometimes denoted K', which is almost identical to the usual K projection quantum number of geometrical shape models. (Technically, some SU(3) states contain small admixtures of other K values; this has notable effects on certain B(E2) values since it is a bandmixing effect, but as far as the wave functions are concerned, it is an excellent (and useful) approximation to ignore these mixtures and use the usual notation K.) The rule that determines the K values that occur in a given (λ, μ) representation is $K = 0, 2, \ldots \min(\lambda, \mu)$, K even. For typical values of $a_2 < 0$, the ground state band has $(\lambda, \mu) = (2N, 0)$. The next representations are $(\lambda, \mu) = (2N − 4, 2)$ with $K = 0, 2$ bands, $(2N − 8, 4)$ with $K = 0, 2, 4$ bands; and $(2N − 6, 0)$ with a single $K = 0$ band. These states are illustrated in Fig. 6.33. The similarities to a deformed rotor are clear: we see sequences of states resembling a ground state band, $K = 0$ and γ vibrational bands, and bands that can be characterized as the double $K = 0$, $K = 0 \otimes K = 2$, and $\gamma\gamma (K = 0, 4)$ two-phonon intrinsic excitations. However, we note two specific features, exemplified by the $K = 0$ and γ bands, that distinguish SU(3) from a general deformed rotor and act as characteristic signatures for the symmetry. They are schematically shown in Fig. 6.34. Since the $K = 0$ and γ bands appear in the same representation, states of the same spin of these two bands must have the same energies. Thus, SU(3) is a *special case* of a deformed rotor with degenerate $K = 0$ and γ bands. We stated that a transition operator consisting of the Casimir operators of a subgroup cannot connect different

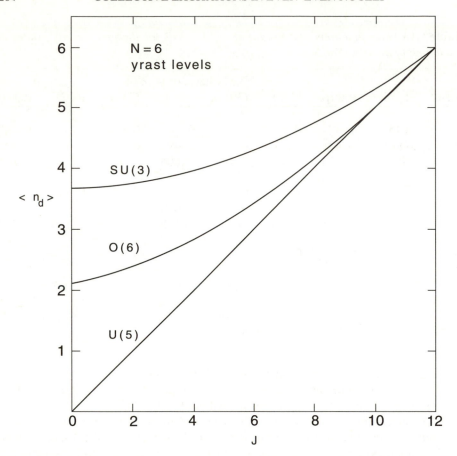

FIG. 6.32. Expectation values of n_d in the ground band in the three limits of the IBA (Casten, 1988a).

representations. Therefore, in the SU(3) limit, the E2 operator with $\chi = -\sqrt{7}/2$ cannot lead to transitions from either the $K = 0$ or γ bands to the ground band! This is in direct contrast to the usual picture of harmonic collective $K = 0$ and γ vibrations in deformed nuclei. Moreover, since these bands are in the same representation, the E2 operator leads to *allowed*, collective $\gamma \rightarrow K = 0$ E2 transitions, again violating the $\Delta N_{ph} = \pm 1$ selection rule of geometrical models. Given decades of success of the latter model, this would appear to be an argument against the IBA. However, neither the issue nor the data is as simple as at first appears, and both approaches can be made more or less mutually consistent. We will discuss this shortly to help us to better understand the IBA.

There are three more distinguishing characteristics of the SU(3) symmetry shown in Fig. 6.34. Since there is no connection between representations, there can be no γ–g or $K = 0 - g$ bandmixing. $Z_\gamma(Z_{K=0})$ are effectively zero in SU(3). And, although both $K = 0 \rightarrow g$ and $\gamma \rightarrow g$ E2 transitions vanish in SU(3), they have a finite ratio.

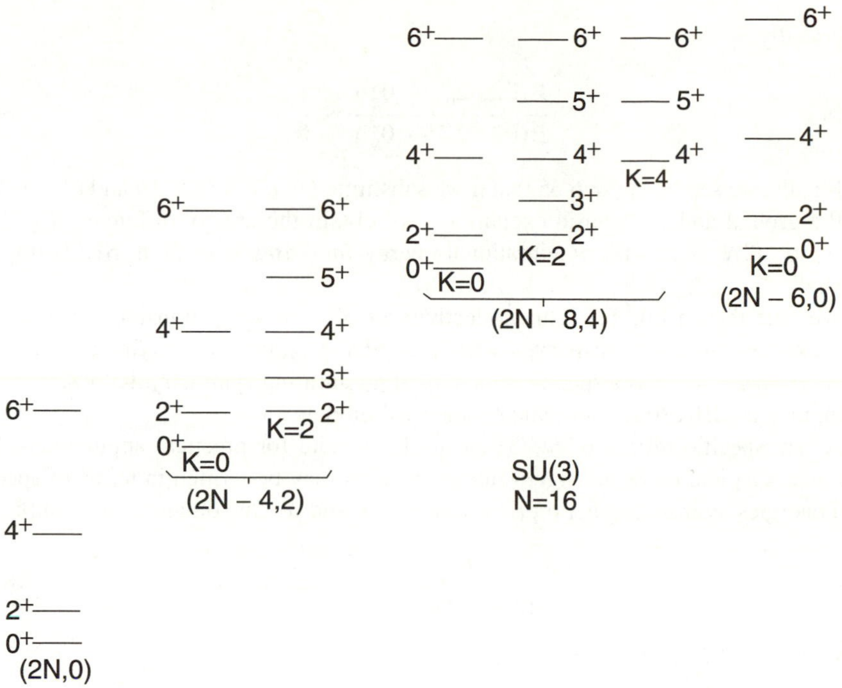

6^+—   —6^+ —6^+ — 6^+

—5^+ —5^+

4^+—   —4^+ —4^+ —4^+
K=4

6^+—   —6^+ —3^+
2^+—   —2^+ —2^+
0^+\ K=0   K=2
—5^+ 0^+\ K=0 $\overline{K=0}\ 0^+$
(2N – 6,0)

4^+—   —4^+ (2N – 8,4)

—3^+
6^+— 2^+—   —2^+
0^+\ K=0   K=2

(2N – 4,2)   SU(3)
N=16

4^+—

2^+—
0^+—
(2N,0)

FIG. 6.33. Characteristic spectrum of SU(3) (Casten, 1988a).

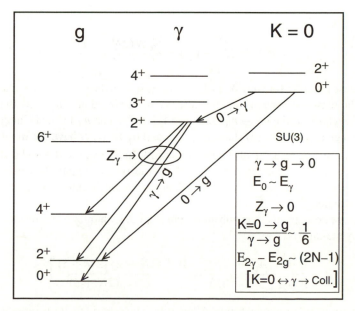

FIG. 6.34. Characteristic signatures of SU(3) (Casten, 1985b).

Specifically,

$$\frac{B(E2 : 2_0^+ \rightarrow 0_g^+)}{B(E2 : 2_\gamma^+ \rightarrow 0_g^+)} \sim \frac{1}{6}$$

Finally, we see from Eq. 6.65 that if we substitute $(\lambda, \mu) = (2N, 0)$ and $(2N - 4, 2)$ for the ground and γ intrinsic excitations, we obtain the energy difference $E_\gamma(J) - E_g(J) \propto (2N - 1)$—the γ vibrational energy *increases* with N in SU(3) towards midshell.

We note that in Fig. 6.34, the collectivity of $K = 0 \rightarrow \gamma$ transitions is enclosed in brackets because it persists even with large SU(3) symmetry breaking and, as such, cannot properly serve as a specific signature of the limiting symmetry. It does, however, distinguish the IBA from harmonic geometrical models.

A few specific results of SU(3) are useful to cite for practical applications. The parameters a_2 and a_1 of the eigenvalue expression may be written in terms of specific level energies by inserting appropriate values of λ and μ. One obtains, for example,

$$a_2 = -\frac{E_{2_\gamma^+} - E_{2_1^+}}{3(2N - 1)}, \quad a_1 = \frac{E_{2_1^+}}{6} + \frac{3}{8}a_2 \tag{6.66}$$

Ground band B(E2) values are given by

$$B(E2 : J + 2 \rightarrow J) = e_B^2 \frac{3}{4}\left[\frac{(J + 2)(J + 1)}{(2J + 3)(2J + 5)}\right](2N - J)(2N + J + 3) \tag{6.67}$$

Hence,

$$B(E2 : 2_1^+ \rightarrow 0_1^+) = e_B^2 \frac{N(2N + 3)}{5} \tag{6.68}$$

Note that these go as N^2 for large N, in direct contrast with the linearity in N characteristic of U(5). The reason is obvious and has already been hinted at. The U(5) B(E2) values scale as N because of the $(N - n_d)$ factor in Eq. 6.59. The n_d factor is independent of N because a given pair of U(5) states (e.g., 2_1^+ and 0_1^+) always have the *same* pair of n_d values (e.g., 1 and 0, respectively) regardless of N. In SU(3), the $\tilde{\mathbf{d}}^\dagger\mathbf{s}$ and $\mathbf{s}^\dagger\tilde{\mathbf{d}}$ operators in T(E2) give factors involving N from *both* operators in each pair, because, as we just saw, both n_d and n_s increase with N. So, a dependence on N enters twice, leading to the $\sim N^2$ dependence.

Finally, we note, from direct subsitution in Eq. 6.67, an interesting result that we can illustrate by the ratio

$$\frac{B\left(E2 : 4_1^+ \rightarrow 2_1^+\right)}{B\left(E2 : 2_1^+ \rightarrow 0_1^+\right)} = \frac{10}{7}\left[\frac{(2N - 2)(2N + 5)}{(2N)(2N + 3)}\right] \tag{6.69}$$

The first factor is the rotational model Alaga rule. The second factor is (another example of) an N-dependent finite boson number effect, which means that *even* in the strict SU(3) limit, B(E2) ratios deviate from the Alaga rules. Note that the second factor goes to unity as $N \rightarrow \infty$. That its predictions go over into those of the usual geometrical model for large N is a characteristic feature of the IBA. Many of the unique aspects of the model (such as allowed, collective $K = 0 \rightarrow \gamma$ E2 transitions) stem directly from the explicit incorporation of finite N, which in turn, reflects the model's emphasis on the valence space.

Since SU(3) is such a specific type of deformed rotor, we already recognize that it does not characterize most deformed nuclei since such nuclei exhibit nondegenerate $K = 0$ and γ bands, collective $K = 0 \rightarrow g$ and (especially) $\gamma \rightarrow g$ transitions, and finite Z_γ. Moreover, in the first half of the deformed rare earth region $E_\gamma(J) - E_g(J)$ actually decreases rather than displaying a proportionality to $(2N - 1)$. We will consider shortly how the IBA can treat such nuclei. First we ask if there are *any* nuclei that do display the limiting characteristics of SU(3). The answer is (probably) yes, the rare earth isotopes of Yb and Hf near neutron number $N = 104$. The empirical evidence is displayed in Figs. 6.35 and 6.36, where each of the signatures of SU(3) is approached in the same general N, Z region (and in no other). At the same time, it is clear that no single nucleus in the $N = 104$ region displays all the SU(3) characteristics. Moreover, some of these same empirical features (high intrinsic $K = 2$ energies, weak B(E2 : $\gamma \rightarrow g$) values, small bandmixing) also characterize high-lying noncollective two-quasi-particle excitations. Thus the evidence is ambiguous, although the author feels that it points toward an *underlying* SU(3) character that may be mixed with noncollective degrees of freedom; the coincidence of signatures is too significant to be dismissed as fortuitous.

III. O(6)

The O(6) symmetry is the least familiar geometrically, although it is now recognized as corresponding to a deformed, axially asymmetric but γ-soft rotor, the Wilets-Jean model. The O(6) Hamiltonian is

$$H = a_0 \mathbf{P}^\dagger \mathbf{P} + a_1 \mathbf{J}^2 + a_3 \mathbf{T}_3^2 \tag{6.70}$$

and the eigenvalue equation is

$$E(\sigma, \tau, J) = A(N - \sigma)(N + \sigma + 4) + B\tau(\tau + 3) + CJ(J + 1) \tag{6.71}$$

where $A = a_0/4$, $B = a_3/2$ and $C = a_1 - a_3/10$ (one sometimes encounters a notation with coefficients $A/4$ and $B/6$). The characteristic quantum numbers are σ for the O(6) group and τ for the O(5) subgroup. A typical O(6) spectrum was used to illustrate the idea of a group chain in Fig. 6.31, and is shown more completely (for $N = 6$) in Fig. 6.37. The lowest levels (for A, $B > 0$) have $\sigma = N$, and $\tau = 0, 1, 2, \ldots$. For each value of $\tau \geq 2$, there is a multiplet of states whose degeneracy is broken by the $J(J + 1)$ term. For $\tau = 2$, there are only 2^+ and 4^+ levels, and no triplet as in U(5). Major families of O(6) levels are grouped and characterized according to σ, and within each family, by τ (and J, of course). Note the characteristic behavior within a σ family: energies of states with the same τ are monotonic in J (usually decreasing with J since $C > 0$) and splittings increase rapidly with τ. The ground band or yrast levels increase

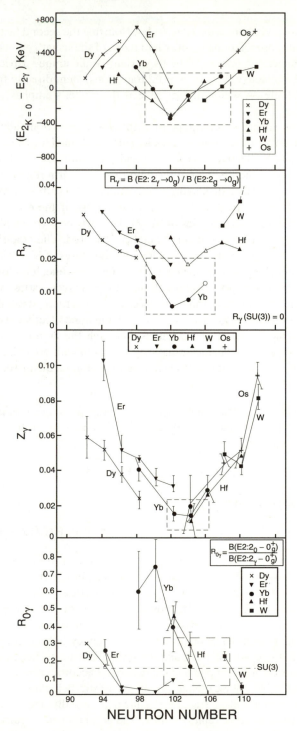

FIG. 6.35. Empirical evidence relative to four of the SU(3) signatures near $N = 104$ (Casten, 1985b).

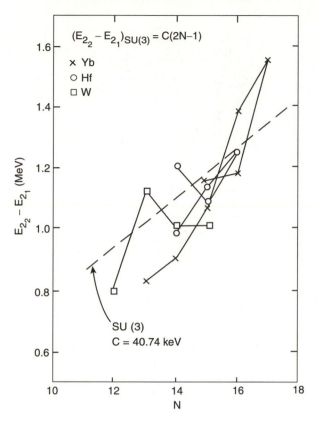

FIG. 6.36. Comparison of empirical and SU(3) values for $E_{2_\gamma^+} - E_{2_1^+}$ in the rare earth region (Casten, 1988a). N is the boson number.

as $\tau(\tau+3) = (J/2)(J/2+3) \propto J(J+6)$ as in the Wilets–Jean model. As noted for the corresponding geometric models, this increase is faster than in U(5) where $E(J) \propto J$, and slower than in SU(3) where $E(J) \propto J(J+1)$. These relative ground band energies may be summarized

$$E_2 : E_4 : E_6 : E_8 := \begin{cases} E_{n_d=1} : E_{n_d=2} : E_{n_d=3} : E_{n_d=4} = 1 : 2 : 3 : 4 & \text{U(5)} \\ E_{\tau=1} : E_{\tau=2} : E_{\tau=3} : E_{\tau=4} = 1 : 2.5 : 4.5 : 7 & \text{O(6)} \\ E_{J=2} : E_{J=4} : E_{J=6} : E_{J=8} = 1 : 3.33 : 7 : 12 & \text{SU(3)} \end{cases}$$

(6.72)

These expressions are identical to those of the vibrator, Wilets–Jean, and rotor models shown in Fig. 6.27. Despite the apparent differences in ground band energies for each of the symmetries, it is important to recall that the curves in Fig. 6.27 are defined by the characteristic quantum number n_d, τ, and J for U(5), O(6), and SU(3), respectively. For U(5) and O(6), however, there is also a separate $J(J+1)$ term in the Hamiltonian and,

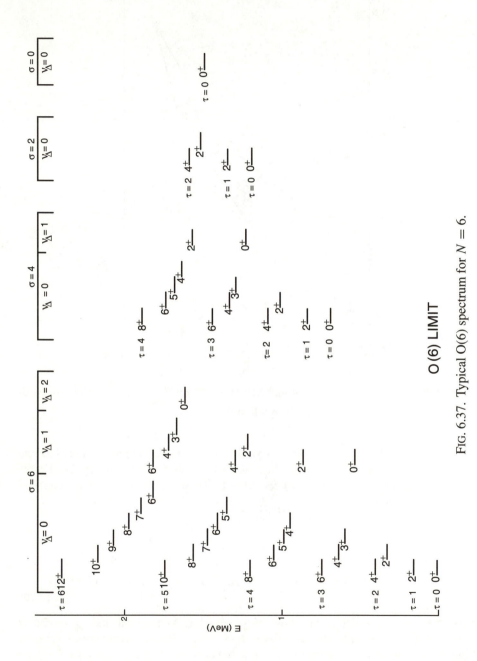

FIG. 6.37. Typical O(6) spectrum for $N = 6$.

depending on the strength of its coefficient, the actual ground band energies in these two symmetries can be made to resemble each other or SU(3) itself.

In terms of the U(5) basis states, the nondiagonal term in $H_{O(6)}$ is $\mathbf{P}^\dagger\mathbf{P}$, which has $\Delta n_d = 0, \pm 2$ matrix elements. Thus the wave functions are mixtures in a U(5) basis, but are not as complex as in SU(3). Table 6.13 illustrates this, showing that in O(6) the finite amplitudes always differ from each other by the addition of a zero coupled pair of d bosons for a given state. This implies that the number of *unpaired d* bosons is constant for a given state. It is zero for the 0_1^+ and 0_2^+ states, and 3 for 0_3^+. But note that this quantity is just the boson seniority v, and in fact, τ and v are identical. The use of different notations has historical origins only and no physical content. Indeed, v or τ arise in U(5) and 0(6) because both chains involve the same subgroup O(5). This has been the source of some confusion since the common occurrence of this subgroup means that many predictions of the two symmetries are identical. Differences between them do exist, but reside principally in transition rates, which depend on the detailed d boson structure and occur among higher-lying states belonging to higher n_d multiplets in U(5) and to $\sigma < N$ representations in O(6). (We shall return to this point momentarily).

The fact that $\mathbf{P}^\dagger\mathbf{P}$ has only $\Delta n_d = 0, \pm 2$ terms suggests that one might be able to construct O(6) wave functions in another way. We pointed out that the $\mathbf{Q}^2$ term has $\Delta n_d = 0, \pm 1, \pm 2$ components. However, if $\chi = 0$, it has only $\Delta n_d = 0, \pm 2$ terms. Therefore, the Hamiltonian

$$H = a_2\mathbf{Q}^2 + a_1\mathbf{J}^2 \quad \text{(with } \chi = 0\text{)} \tag{6.73}$$

also produces O(6) wave functions and spectra. This Hamiltonian has only two terms, so it cannot give three independent terms in the eigenvalue equation. This way of producing O(6) thus leads to a special case of Eq. 6.71 where $A = B$. Although this is only one of an infinite set of possible $A : B$ relations, it turns out to be the one empirically observed in O(6) nuclei, suggesting the usefulness of this alternate form for $H_{O(6)}$. We will see that this alternative is in fact one limiting case of the CQF, referred to below Eq. 6.53, that offers a simplified approach to many IBA problems. First, however, we need to delineate a few additional O(6) predictions. The E(2) transition selection rules are clear from the form of the E2 operator that is a generator of O(6), namely

$$\mathbf{T}(E2) = e_B\left(s^\dagger\tilde{d} + d^\dagger s\right)$$

The allowed transitions must satisfy $\Delta\sigma = 0$ and $\Delta\tau = \pm 1$. The first rule is a direct consequence of the fact that a generator of a group cannot connect different representations. It means that there are no allowed transitions from one σ family to another. Eventually, excited levels with $\sigma < N$ must decay, but only by violations of the strict symmetry. This selection rule provides the most telling contrast with U(5). (See the following.) The $\Delta\tau = \pm 1$ rule is similar to the ΔN_{ph} or $\Delta n_d = \pm 1$ rules for the geometrical vibrator and U(5) limits. Naturally, B(E2) values between yrast states are allowed, and given by

$$B(E2 : J + 2 \rightarrow J) = B\,(E2 : 2(\tau + 2) \rightarrow 2\tau)$$

$$= e_B^2 \left(N - \frac{J}{2} \right) \left(N + \frac{J}{2} + 4 \right) \left(\frac{\frac{J}{2} + 1}{J + 5} \right) \qquad (6.74)$$

from which we get,

$$B(E2 : 2_1^+ \rightarrow 0_1^+) = e_B^2 \frac{N(N + 4)}{5} \qquad (6.75)$$

As in SU(3), B(E2) values in O(6) scale approximately as N^2 for large N. The yrast B(E2) values are illustrated in Fig. 6.38.

There has recently been much discussion of the differences and similarities between O(6) and U(5). The level schemes in Figs. 6.30 and 6.37 appear very different. Specifically, U(5) has the well-known two-phonon triplet, while O(6) lacks the 0^+ state. On account of this difference, the first excited 0^+ state in U(5) decays to the 2_1^+ level, while the first excited 0^+ state in O(6) has $\tau = 3$ and therefore decays to the 2_2^+ state. These distinctions have been used as evidence both for or against each of these symmetries. This issue is more subtle, however. Each figure embodies a specific choice of parameters ($\alpha, \beta, \gamma, \delta$ for U(5) and A, B, C for O(6)); indeed, the U(5) scheme is the harmonic limit. Actually, both symmetries permit a rich variety of level scheme configurations by appropriate choices of parameters. Since both group chains contain O(5) and O(3) subgroups, the only real structural differences center on the O(6) and U(5) parent groups. Thus, it turns out that the energies of the entire lowest representation of O(6), with $\sigma = N$, can be exactly replicated in U(5). To do so, the 0_2^+ (two-phonon or $n_d = 2$) U(5) level must be forced up in energy above the $n_d = 3\,0_3^+$ state in order to reproduce the O(6) decay pattern in which the first excited 0^+ state decays to the 2_2^+ level. Such an O(6)-like U(5) spectrum would be enormously anharmonic but still valid within the context of this dynamical symmetry.

The real difference between O(6) and U(5) occurs in two other realms: absolute transition rates and higher-lying levels. Table 6.13 shows that, though their energies may be identical, the U(5) and O(6) wave functions are completely different. Thus, B(E2) values will be different; in particular, the different expectation values of n_d imply different finite N effects. We have seen an example of this in the yrast B(E2) values in Fig. 6.38. In U(5), the yrast band experiences greater changes in $\langle n_d \rangle$ from state to state (see Fig. 6.32), and hence the B(E2) values change more rapidly with spin.

High-lying levels in U(5) have high n_d and decay to $(n_d - 1)$ levels. There are many such $(n_d - 1)$ states and thus many allowed E2 transitions. In O(6), high-lying levels often belong to low τ states of $\sigma < N$ representations. These typically have only one allowed deexcitation transition—that permitted by the $\Delta\tau = \pm 1$ selection rule.

O(6) nuclei are now well known in two regions: the Pt isotopes, especially ^{196}Pt, and the Xe–Ba nuclei near $A = 130$. A comparison of the ^{196}Pt level scheme with the O(6) limit is shown in Fig. 6.39. The agreement with all the E2 selection rules is impressive. All allowed transitions are observed and dominate the decays of their respective levels; all forbidden transitions are either weak or unobserved. At the same time, there are at least

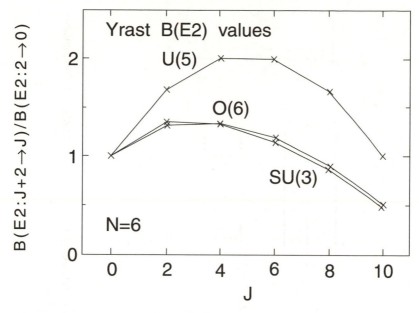

FIG. 6.38. Dependence of ground band B(E2) values on J in the three limits of the IBA.

three important discrepancies in energies: the $\tau = 3\,0^+$ state is *not* below the 3_1^+ level as it should be, the splitting among the high τ states (e.g., the $0^+ - 2^+ - 2^+, \tau = 3 - 4 - 5$ states) is much less than predicted, and the energies in the γ band are less staggered than predicted.

The data for Xe and Ba is comparable to Pt: only $\sigma = N$ and $N - 2$ levels are known, but many $\tau = 4, 5, 6$ states have been assigned and the O(6) character extends over a large range of nuclei. There are some striking analogies between Pt and Xe–Ba. Besides their common manifestation of O(6)-like characteristic, fits of the O(6) eigenvalue equation in each region show nearly identical ratios $A/B \sim 0.9$. Interestingly, this common value is very close to the special case $A/B = 1$, which corresponds to the Hamiltonian of Eq. 6.73. Moreover, exactly the same discrepancies with O(6) that occur in ^{196}Pt are repeated in Xe–Ba. One of these, the weaker γ band energy staggering, provides a useful clue to the nature of the responsible symmetry breaking. Recall from our earlier discussion of asymmetric rotor models that the rigid triaxial rotor of Davydov is characterized by staggering exactly opposite in phase to that of the Wilets–Jean γ unstable model. The O(6) limit corresponds to the latter, that is, to a completely γ independent potential and to a nucleus whose shape fluctuates uniformly over a range of γ values from $\gamma = 0° - 60°$ such that $\gamma_{rms} = 30°$. The staggering data suggests that a realistic potential for Pt and Xe–Ba might contain some small γ dependence that would shift the characteristic γ band energy staggering pattern slightly toward the opposite couplings of the Davydov model. It turns out that the introduction of a very small (3–4%) γ dependence in the potential with a minimum at $30°$ corrects not only the γ band energy staggering, but the other discrepancies as well. This idea is in line with our earlier discussion in the

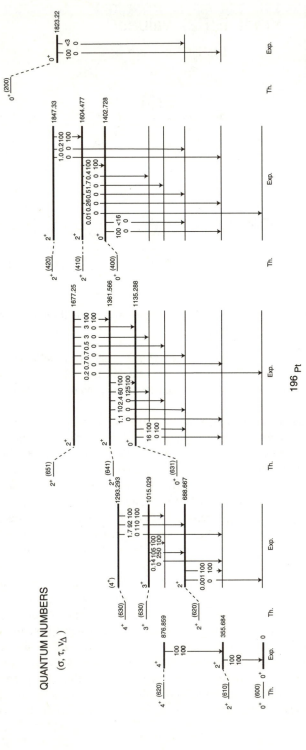

Fig. 6.39. Comparison of empirical energies and E2 branching ratios in ^{196}Pt with the O(6) limit. The lower numbers on the transition arrows are the predictions, the upper are the measured values (Cizewski, 1978).

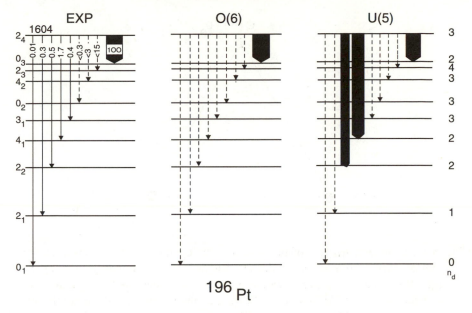

FIG. 6.40. Comparison of the decay of the 1604 keV level in ^{196}Pt with O(6) and U(5) symmetries (Cizewski, 1978).

context of Fig. 6.28 in which we suggested that the $E_s/E_{2_1^+}$ data could be accounted for by a potential that was predominantly γ independent but with a small minimum near $\gamma = 30°$.

We commented earlier that a strict distinction between O(6) and U(5) can not rely on the energies or branching ratios of low-lying levels alone, but requires absolute B(E2) values or branching ratios for $\sigma < N$ states. Both of these types of data are abundant for Pt and for Xe–Ba. As an example, Fig. 6.40 shows the decay of the most telling $\sigma = (N-2) = 4$ level for ^{196}Pt and compares this with O(6) and U(5) predictions. The high n_d U(5) levels have multiple allowed decay routes while the O(6) $\sigma < N$ levels often have only one such route. The data clearly support the O(6) interpretation.

Further support comes from the E2 decay of the $\sigma = 4\,0^+$ state at 1402 keV in ^{196}Pt which is forbidden by the $\Delta\sigma = 0$ selection rule. The transition to the 2_1^+ state breaks only this selection rule while that to the 2_2^+ level also breaks the τ selection rule. Therefore, one expects that the E2 branching ratio will favor the 2_1^+ level as is observed experimentally. However, a critical test of O(6) is whether this B(E2) value is also weak on an absolute scale, as predicted. A collective B(E2 : $0_3^+ \to 2_1^+$) transition would be in serious violation of O(6). A lower limit on the lifetime of this level has been measured (Börner, 1990), giving B(E2 : $0_3^+ \to 2_1^+$) < 5 W. u. (compared to 205 W.u. for the $0_1^+ \to 2_1^+$ transition) confirming the predicted forbiddenness of this transition, establishing the validity of the σ quantum number, and cinching an O(6) description for ^{196}Pt.

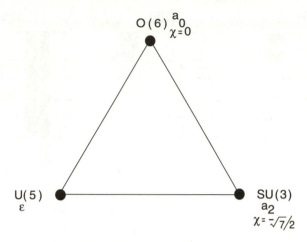

FIG. 6.41. Symmetry triangle of the IBA with the coefficients giving each dynamical symmetry. For U(5) most calculations to date use $\chi = 0$ although $\chi = -(7)^{1/2}/2$ is suggested by microscopic analyses and gives the finite quadrupole moments observed in nuclei such as the Cd isotopes. Following common practice (albeit not theoretical rigor) we will take $\chi = 0$ for U(5). The value $\chi = 0$ for O(6) refers to the CQF Hamiltonian of Eq. 6.73 where O(6) is produced by the $\mathbf{Q} \cdot \mathbf{Q}$ term rather than the $\mathbf{P}^{\dagger}\mathbf{P}$ term.

As we have suggested in each of the preceding discussions, it appears that good examples of all three IBA symmetries exist. This provides support for the general structure of the IBA model and, moreover, these limiting cases offer convenient benchmarks to assess the structure of nuclei similar in character to the symmetries but exhibiting some degree of symmetry breaking.

Of course, there is no a priori reason why a particular nucleus should satisfy a given symmetry or, indeed, why any nuclei should satisfy the particular constraints of one of these symmetries. Most nuclei do not, yet the IBA is appealing precisely because it provides a simple way to treat such nuclei. Such treatments also reveal new aspects of collective behavior not heretofore encountered or expected.

To study some of these ideas, we now turn to transitional nuclei between IBA symmetries. A convenient way to picture the symmetry structure of the IBA is in terms of the symmetry triangle shown in Fig. 6.41. The three symmetries mark the vertices. In terms of the Hamiltonian of Eq. 6.51, the characteristic wave functions of each symmetry are generated by a specific term whose coefficient is labeled at the appropriate vertex: ε for U(5), a_2 for SU(3), and a_0 for O(6). We have also seen that for $\mathbf{Q}$ to be a generator of SU(3) one must set $\chi = -\sqrt{7}/2$.

Indeed, in early applications of the IBA, it was customary to keep χ in the quadrupole operator in the Hamiltonian fixed at the SU(3) value of $-\sqrt{7}/2$, while treating χ in the E2 operator as a free parameter. This is certainly permissible, even though one might feel slightly uncomfortable using different forms of the quadrupole operator since it is

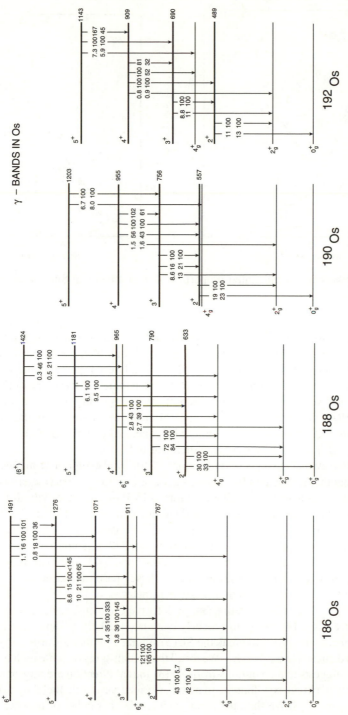

FIG. 6.42. Empirical (upper numbers) and calculated relative B(E2) values for the Os 0(6) → rotor transitional region (Casten, 1978).

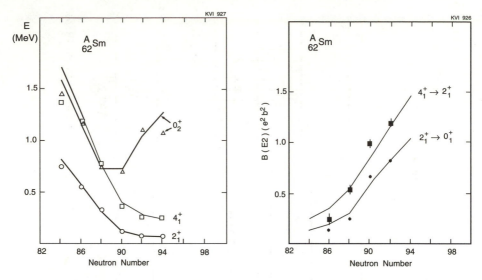

FIG. 6.43. Empirical and calculated properties of the transitional Sm nuclei (Scholten, 1978).

same quadrupole force that produces deformed nuclei and collective E2 transitions. This discomfort initiated interest in pursuing an approach in which consistent forms of Q are used throughout. This approach, the CQF, is widely used, perhaps due more to its simplicity than to any philosophical preference. One additional advantage of the CQF is that it involves fewer free parameters, which makes it easier to establish a relationship between the IBA and corresponding geometrical analogues. As noted in the discussion of Eq. 6.73, the CQF allows the O(6) limit to be generated with the $\mathbf{Q} \cdot \mathbf{Q}$ terms, simply by setting $\chi = 0$. These χ labels are also included in Fig. 6.41.

Treating transitional nuclei is trivially simple. Structural evolution from one limiting symmetry to another merely involves changing the ratio of the two parameters associated with the two vertices from 0 to ∞. For example, a U(5) $\rightarrow$ SU(3), or vibrator $\rightarrow$ rotor, transition, such as occurs in the Nd, Sm, and Gd nuclei near $A = 150$, is effectuated by varying ε/a_2 and an O(6) $\rightarrow$ rotor transition, as in the Os isotopes, by changing a_0/a_2 or, setting $a_0 = 0$, a_2 finite, by varying χ from zero to $-\sqrt{7}/2$. Examples of such structural evolution are shown for these two cases in Figs. 6.42 and 6.43. The extremely complex changes in level schemes and transition rates in these phase transitions are rather well reproduced in terms of a variation of a single parameter.

While the SU(3) symmetry represents a deformed nucleus, few deformed nuclei satisfy its strict rules. Therefore, to properly calculate the bulk of deformed rare earth and actinide nuclei, one needs to break the SU(3) symmetry either towards U(5) or O(6). Most deformed nuclei have $E_0 > E_\gamma$. We recall that a high-lying 0_2^+ state is characteristic of O(6). This suggests that deviations from SU(3) in that direction are appropriate. It is indeed possible to obtain excellent fits to most deformed rare earth nuclei by using finite ratios of a_0/a_2 near -4. We shall not show such results, but rather turn to the alternate method, the consistent Q formalism (CQF), which gives an even simpler way of dealing with these nuclei.

In the CQF, the $O(6) \rightarrow SU(3)$ transition leg in the symmetry triangle is accomplished by varying χ from 0 to $-\sqrt{7}/2$ in *both* H and T(E2). Now, H has the form shown in Eq. 6.73

$$H = a_2 \mathbf{Q}^2 + a_1 \mathbf{J}^2 \quad \left[\text{with} -\frac{\sqrt{7}}{2} \leq \chi \leq 0 \right] \tag{6.76}$$

Since $\mathbf{J}^2$ is diagonal and has the same effect on any states of the same spin, it plays no structural role. Thus, if we write

$$H = a_2 \left[\mathbf{Q}^2 + \frac{a_1}{a_2} \mathbf{J}^2 \right] \tag{6.77}$$

we see that a_2 is just a scale factor on energies and has no structural influence. Thus the wave functions, relative B(E2) values, and relative energies of states of the same spin are determined *solely* by χ and N. Whereas in the traditional form of H in Eq. 6.51 the structure and relative energies in an $O(6) \rightarrow SU(3)$ region depend on a_0, a_2, a_3, and E2 transitions depend separately on χ, now a *single* parameter determines all. The only loss of generality is that the $O(6)$ symmetry approached and obtained in this way is a special case, with $A/B = 1$. However, this does not seem to be a deficiency, since it is this special case that is experimentally observed.

Since predictions of branching ratios and of relative energies (of states of the same J) depend only on χ and N in the CQF, it is possible construct *universal* plots for a given N, or *universal* contour plots against χ and N. Some examples are shown in Figs. 6.44 and 6.45. (In the energy ratio plotted, $E_{2_1^+}$ is always subtracted to remove the structurally inconsequential effects of the $\mathbf{J}^2$ term.) Figure 6.44 (top) shows the behavior of the $K = 0$ vibration relative to the γ band energy. These excitations belong to the representation $(N-4, 2)$ in SU(3). As $O(6)$ is approached, the $K = 0$ vibration increases rapidly relative to the γ band. The reason is that the γ band goes over into a quasi-γ band sequence starting with the $\tau = 2_2^+$ level of the $\sigma = N$ family, while the $K = 0$ bandhead becomes the *yrast* 0^+ level of the higher-lying $\sigma = N - 2$ family. Figure 6.44 also shows that for essentially all χ values, and especially for those values typical of deformed nuclei (cross hatched box), the $K = 0$ band lies well above the γ band. This is in excellent agreement with the data, but not unexpected given our discussion. What is more impressive is that the B(E2) ratio in the middle panel shows that the CQF *automatically* predicts that $B(E2 : K = 0 \rightarrow g) \ll B(E2 : \gamma \rightarrow g)$.

In Fig. 2.18 we saw that this feature was one of the most characteristic empirical properties of deformed nuclei. It is particularly interesting that the IBA predictions for realistic deformed nuclei come out this way since *both* $K = 0 \rightarrow g$ and $\gamma \rightarrow g$ transitions are forbidden in SU(3). However, in the transition towards $O(6)$, the γ band levels go over into the quasi-γ band (in Davydov language, into the anomalous levels of the ground band) with collective transitions to the latter, while the $K = 0$ band goes

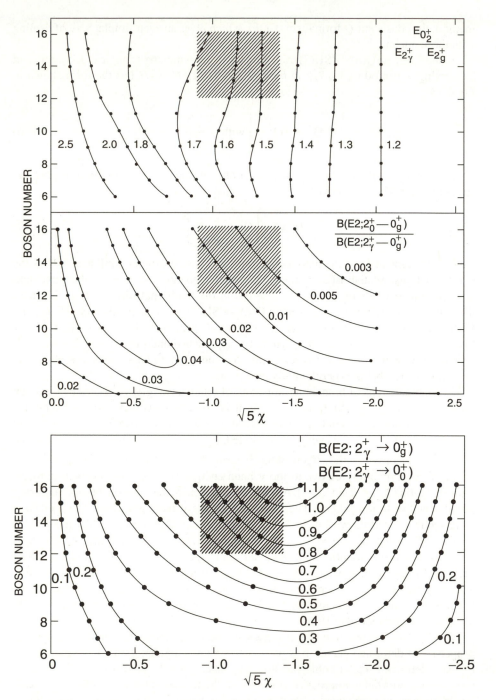

FIG. 6.44. Three contour plots in the CQF formalism of the IBA (Warner, 1983).

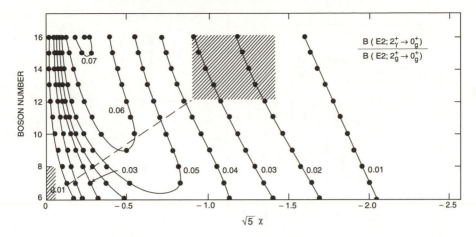

FIG. 6.45. Another CQF contour plot (Warner, 1983).

over into the low-lying states of the $\sigma = N - 2$ family and the $\Delta\sigma = 0$ selection rule forbids their decay to the ground band. Thus they remain weak throughout the transition leg while the $\gamma \rightarrow g$ transitions become collective.

The lowest contour plot in Fig. 6.44 shows one of the most remarkable and surprising predictions of the IBA, that of *collective* $K = 0 \leftrightarrow \gamma$ transitions that are comparable in strength to $\gamma \rightarrow g$ transitions. This is completely contrary to normal expectations and to the traditional perception of the experimental situation. When this prediction of the IBA was initially realized, it was thought to contradict a wide body of empirical evidence showing systematically collective $\gamma \rightarrow g$ and weakly collective $K = 0_2 \rightarrow g$ transitions but no $K = 0_2 \leftrightarrow \gamma$ transitions. However, given the typical closeness in energy of E_0 and E_γ and the fact that E2 transition rates $T(E2) \propto B(E2) \, E_\gamma^5$, such transitions would be extremely weak in *intensity* even if the *matrix elements* are large. Then, extremely sensitive experiments, mostly utilizing the (n, γ) reaction and powerful γ-ray spectrometers installed at an Institute Laue–Langevin in Grenoble, showed that collective $K = 0_2 \leftrightarrow \gamma$ transitions do indeed exist and often appear whenever the appropriate experiments have been carried out.

It is evident in Fig. 6.44 that the $\gamma \rightarrow g/K = 0 \leftrightarrow \gamma$ matrix elements generally increase with N, and in the large N limit, $K = 0 \leftrightarrow \gamma$ matrix elements become negligible relative to either $\gamma \rightarrow g$ or $g \rightarrow g$ intraband transitions, thus recovering the geometrical picture. Of course, geometrical models can be modified (perturbed) so as to produce $K = 0 \leftrightarrow \gamma$ matrix elements of collective strength simply by introducing mixing *between* $K = 0$ and γ bands. However, as we pointed out in the bandmixing discussion earlier in this chapter, the B(E2) values so produced result not as a correction to an unperturbed value (since the latter vanishes), but *solely* from the mixing. Thus, all $K = 0 \rightarrow \gamma$ B(E2) values are proportional to the same mixing parameter $\varepsilon_{0\gamma}$ and hence, in branching ratios, $\varepsilon_{0\gamma}$ cancels out: the branching ratios depend only on known functions of spin. In [168]Er the branching ratios obtained by assuming the total transition

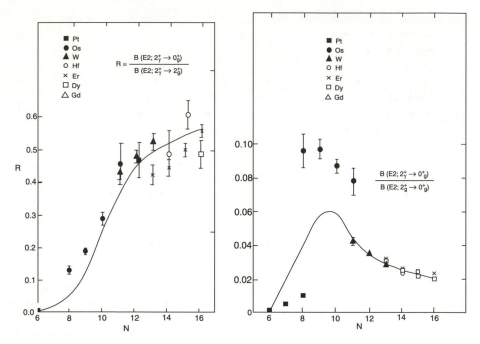

FIG. 6.46. Comparison of the data with predictions of the CQF for two observables. In each case a straight-line trajectory in χ (see dashed line in Fig. 6.45) was used (Warner, 1983).

strength arises from mixing disagree with the data.

Analysis of the bandmixing Mikhailov plots for ^{168}Er indeed shows that most of the $K = 0 \rightarrow \gamma$ transition strength in fact comes via a direct $\Delta K = 2$ matrix element (albeit not as large as the lowest contour plot in Fig. 6.44 would suggest).

The last contour plot, that of Fig. 6.45 is also interesting. It shows a branching ratio that vanishes in both SU(3) and O(6) limits but is finite in between. There is no possible path between these limits that bypasses finite values. Thus, in a totally parameter-free manner, the IBA automatically predicts that such a branching ratio (and, indeed, many others) will peak in transitional regions.

When carrying out actual calculations, one generally chooses χ to reproduce a specific energy ratio or branching ratio and then inspects other predictions. However, a more generic viewpoint is obtained by taking the simplest possible trajectory in χ (shown as the straight dashed line in Fig. 6.45) between O(6) and rotor nuclei. Figure 6.46 shows predictions for such a trajectory for the branching ratio of Fig. 6.45 and for a branching ratio involving the same pair of intrinsic states, and compares them with the data for rare earth nuclei. One does not expect to get exact agreement in such a simplified approach, but the general pattern of the predictions is in remarkably good accord with the data.

It is useful to present a more detailed set of predictions for a typical deformed nucleus situated between SU(3) and O(6) but closer in structure to the former: ^{168}Er has become,

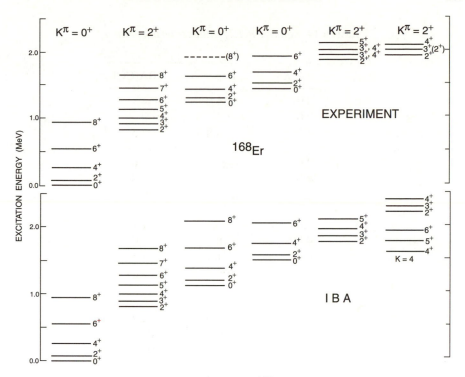

FIG. 6.47. Observed and calculated energies in ^{168}Er. The $K = 4$ band predicted near 1.6 MeV is observed near 2 MeV (Warner, 1980).

by virtue of the extensive data available for it, the standard case. Calculated and empirical energies are compared in Fig. 6.47. The empirical B(E2 : $\gamma \to g$) values were shown earlier in a Mikhailov snapshot in Fig. 6.17. The dashed line is the IBA prediction in the CQF. Table 6.14 also presents this comparison. The agreement is outstanding and demonstrates the *automatic* incorporation of bandmixing in the IBA.

An interesting feature of this bandmixing is that it decreases with increasing N. This can be discerned on the left in Fig. 6.46 from the flattening of the predictions and their asymptotic approach to the Alaga rule value of 0.7 as N increases. Thus, the IBA automatically predicts a parabolic behavior for Z_γ against N, minimizing at midshell, that is nearly identical to the empirical pattern shown earlier in Fig. 6.18.

Of course, the agreement in energy for the higher bands in Fig. 6.47 is probably at least partially fortuitous since the empirical excitations seem to have some two-quasi-particle character. Also, the predicted $K = 4$ band is not observed below 2 MeV. (It may be necessary to include g bosons in the IBA to account for this discrepancy.)

From all these results we hope that the power of the IBA as a simple yet general phenomenological model becomes apparent. In other approaches to collective nuclear structure, one must invoke different models (vibrator, rotor, etc.), in an ad hoc way to accommodate different structures. In the IBA, a single framework, embodied in the simple Hamiltonian

Table 6.14 *Relative B(E2) values from the γ band in* ^{168}Er

J_i	J_f, K_f	Exp	IBA(CQF)
2	0,0	54.0	54
	2,0	100	100
	4,0	6.8	7.6
3	2,0	2.6	2.6
	4,0	1.7	1.8
	2,2	100	100
4	2,0	1.6	1.7
	4,0	8.1	9.6
	6,0	1.1	1.5
	2,2	100	100
5	4,0	2.9	3.5
	6,0	3.6	4.4
	3,2	100	100
	4,2	122	95
6	4,0	0.44	0.44
	6,0	3.8	4.9
	8,0	1.4	1.0
	4,2	100	100
	5,2	69	57
7	6,0	0.7	1.9
	5,2	100	100
	6,2	59	36

Based on Warner, 1982.

of Eq. 6.51, encompasses all three limiting symmetries, and most intermediate situations as well. It does so simply through appropriate relative magnitudes of the coefficients ε, a_2, a_0, or χ, and by diagonalization of a small set of basis states. The "adhocness" is still there, of course, now appearing in the choice of parameters, but the simplicity of the *framework* greatly facilitates calculations and helps us to understand transitional cases in a unified context.

Having dealt at some length with the IBA predictions in the CQF, it is useful to comment on the physical significance of χ. Since the CQF has only one significant parameter, it is possible to compare IBA predictions as a function of χ with geometrical model predictions as a function of β or γ. One can relate β and χ by the equation

$$\beta_{\text{IBA}} = \frac{1}{2}\beta_0 \left[-\sqrt{\frac{2}{7}}\chi \pm \sqrt{\left[\frac{2}{7}\right]\chi^2 + 4} \right] \tag{6.78}$$

where β_0 is a normalization factor because the scale of β_{IBA} is undetermined. Thus we see that, as χ varies from $-\sqrt{7}/2 \rightarrow 0$, β_{IBA} goes from $\sqrt{2}\beta_0 \rightarrow \beta_0$; that is, both SU(3) and O(6) are deformed rotors, the latter slightly less so. The $\beta - \chi$ correlation, however, is relatively minor in importance. The essential structural evolution in the IBA is one of χ with γ. This is easy to see by carefully comparing Figs. 6.25 and 6.48 in which the same three observables are plotted against χ for the IBA and γ for the Davydov model. (The use of the latter as a point of comparison is convenient but not valid in the strict sense since the O(6) limit is γ soft, not rigid; however, the reader will recall our statement that most predictions of the Davydov model for fixed γ are nearly identical to those of a γ soft model with the same γ_{rms}.) These two figures show that each observable passes through the *same* set of values via a similar path. Simply by equating values of each observable, one can assign an effective γ to each χ value. Correlations for two of these observables are shown in Fig. 6.49, along with yet another that was not shown in Figs. 6.25 and 6.48. The fact that all three observables show the same correlation supports the validity of associating each χ value with an asymmetry γ. The picture provided by this correlation is simple: increasing deviations of χ from SU(3) toward O(6) correspond to increasing axial asymmetry and to larger and larger values of γ_{rms}. Since the IBA never introduces a minimum in $V(\gamma)$, this increase in γ_{rms} can *only* arise if the potential becomes increasingly flatter in γ as $\chi \rightarrow 0$. Figure 6.50 confirms this by showing the effective potential $V(\gamma)$ for several χ values. This figure shows one other interesting point. The reader may have noted in Fig. 6.49 that $\gamma \neq 0°$ for SU(3), even though this limit is supposed to be that of a symmetric rotor. Figure 6.50 shows the reason. Although the minimum in $V(\gamma)$ occurs at $\gamma = 0°$, the potential is not infinitely steep (for finite N), and zero point motion leads to a finite γ_{rms}.

The body of research relating to the IBA in the last decade is enormous. We have only summarized a few highlights, emphasizing the symmetries of the IBA, transition regions, the role of finite boson number, some experimental tests of the model, and a geometrical understanding of it. We have completely ignored important topics such as the intrinsic state formalism, which allows many IBA results to be obtained analytically; the extension of the CQF to the SU(3) $\rightarrow$ U(5) transition region by adding an εn_d term to Eq. 6.76; the use of effective boson numbers, especially in regions where important subshell gaps occur that might alter the proper counting of valence nucleons; the $N_p N_n$ parameterization of the IBA; and numerous extensions to the model.

The two most important of these latter are the IBA-2, in which protons and neutrons are distinguished and treated separately, and the IBFM, which incorporates a single fermion coupled to the boson core so that odd A nuclei can be calculated. The IBA-2 has its own symmetries and has led to the recognition of an important new collective mode, the so-called M1 scissors mode, in which protons and neutrons oscillate (in angle) relative to each other (see Fig. 6.1). This mode, frequently discussed by Richter and collaborators in the rare earth region, is now known to be widespread near 3 MeV excitation energy in deformed nuclei. Its existence and properties are closely connected with a new quantum

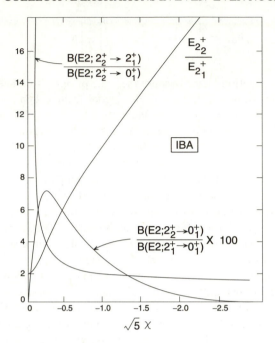

FIG. 6.48. Three observables, relating to axial asymmetry, as a function of χ (for $N = 16$) in the IBA (compare Fig. 6.19).

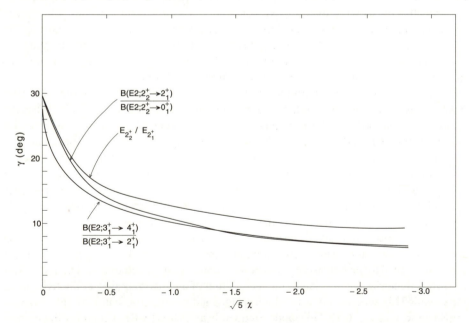

FIG. 6.49. Relation between the geometrical asymmetry γ and the IBA parameter χ (Casten, 1984). For $N = 16$ (the dependence on N is weak).

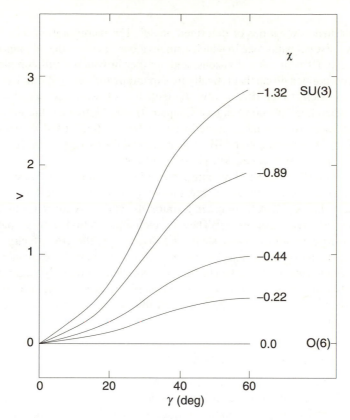

FIG. 6.50. Dependence of the IBA potential on γ for several χ values (Casten, 1984, Ginocchio, 1980).

number that arises in the IBA-2, F-spin, which describes the degree of proton–neutron symmetry. For many IBA-2 Hamiltonians, F-spin is a good quantum number. This classification leads to the concept of an F-spin multiplet—a group of nuclei, widely dispersed throughout a mass region but displaying similar level schemes. F-spin also facilitates the projection of complex IBA-2 calculations into simpler, equivalent IBA-1 cases. Finally, the development of boson models for both even and odd-mass nuclei raises the possibility of treating both in a single unified framework. This leads to the concept of Bose–Fermi symmetries or even supersymmetries (often called $SUSY$'s). Very recently, new data on [196]Au (Metz, 1999) has extensively supplemented earlier results (Warner, 1982a) and seems to support the existence of SUSY in a nuclear context.

   These are all extensive topics that merit their own discussion but that are well outside the scope of this book. Also beyond that ken are important studies extending the basic model to incorporate g bosons. These efforts aim both to reduce the impact of "boson cutoffs" that lead to reduced B(E2) values between high-spin states and to account for classes of excitations (e.g., $K = 3$ and $K = 4$ bands) that regularly appear among the low-lying intrinsic excitations of deformed nuclei. The incorporation of g bosons has followed two distinct paths, one in which a single g boson is introduced numerically and allowed to interact with the s–d bosons, and another in which a full complement of g bosons is incorporated group theoretically by expanding the parent group U(6) to U(15).

Other extensions to the IBA involve the inclusion of two-quasi-particle excitations so that "backbending" phenomena (see Chapter 9) may be treated, the introduction of higher-order terms such as those cubic in s or d bosons (e.g., $(\mathbf{d}^\dagger \mathbf{d}^\dagger \mathbf{d}^\dagger)^j (\tilde{\mathbf{d}}\tilde{\mathbf{d}}\tilde{\mathbf{d}})^J$) that incorporate a γ dependence in the IBA potential, and the expansion of the model basis with p and f bosons so that negative parity states appear.

Finally, a significant facet of algebraic modelling in nuclear structure is the development of sophisticated approaches that invoke symmetries in the fermions directly. Some of these, like the IBA, emphasize fermion dynamical symmetries in the valence space. Others are even more general. They use techniques founded in symplectic group theory to incorporate all oscillator shells simultaneously, thereby affecting important renormalizations and offering the possibility to describe giant E2 resonances and low-lying collectivity without effective charges, in a single coherent algebraic framework. Though far beyond the scope of this book to describe, these more microscopic models are nonetheless current research areas. For all this work the reader is referred to the recent literature.

6.7 Geometric collective model (GCM)

Earlier (Section 6.4) we introduced the Bohr–Mottelson Hamiltonian and a particular realization of it, the Geometric Collective Model (GCM) of Gneuss and Greiner. We have delayed discussion of this approach until we had introduced the basic structural classification of collective nuclei and the IBA model.

Until recently, the GCM has, unfortunately, seen rather little use, primarily because of the complexity arising from the 8 parameter Hamiltonian. The problem is that there are only a few truly independent observables that give the key structural characteristics of a nucleus. Other observables generally introduce only some fine tuning.

For example, in a rotational nucleus, as in Fig. 6.51, the bandhead energies, one or two rotational energies [e.g., $E(2_1^+)$ and $E(4_1^+)$], a couple more in the case of the γ-band (to define both the odd and even spin sequences), one intraband transition strength [e.g., $B(E2 : 2_1^+ \rightarrow 0_1^+)$], and a few interband $B(E2)$ values (one connecting each pair of intrinsic structures) basically determine the structure. In many nuclei not all of these are known (e.g., $B(E2 : 0_2^+ \rightarrow 2_\gamma^+)$ is seldom known, as we have discussed). Moreover, even some of the above quantities are themselves correlated (e.g., $E(2_1^+)$ and the $B(E2 : 2_1^+ \rightarrow 0_1^+)$ value are both related to the moment of inertia and the deformation; staggering of γ band energies and $E(2_\gamma^+)$ are both related to the γ-softness of the potential). While deviations of higher band energies from the rotational formula may give clues to the stiffness of the potential, and deviations of $B(E2)$ values from the Alaga rules allow us to discuss band mixing, such subtleties generally do not alter the basic structural classification of a nucleus. Hence, a handful of observables gives the basic structural signatures and an 8-parameter model is usually significantly under-

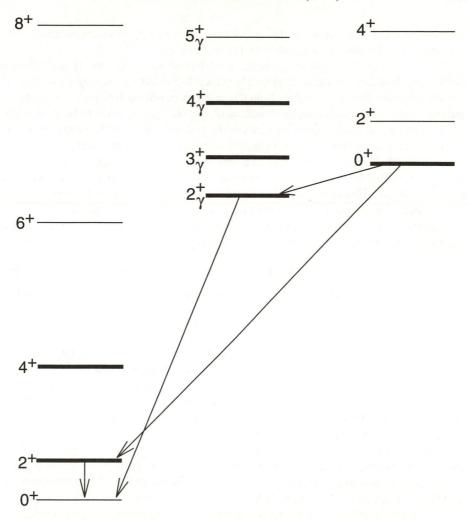

FIG. 6.51. Typical key observables (thicker energy levels, and the E2 transitions indicated) in a deformed even–even nucleus that give essential information on the basic structure. An analogous figure would apply to near closed shell, vibrational or transitional nuclei. The $R_{4/2} \equiv E(4_1^+)/E(2_1^+)$ energy ratio gives directly the basic structure of the nucleus: near closed shell for $R_{4/2} < 2.0$, vibrational for $R_{4/2} \sim 2.2$, transitional for $R_{4/2} \sim 3.0$, rotational for $R_{4/2} \sim 3.33$. The 2_1^+ energy and the B(E2 : $2_1^+ \rightarrow 0_1^+$) give the overall scale of collectivity and, for deformed nuclei, the deformation β. Also in deformed nuclei the energies of the $2^+(K = 2^+)$ and $0^+(K = 0^+)$ states give the most elementary intrinsic excitation modes of the system. In spherical vibrators these two states and the 4_1^+ state comprise the 2-phonon triplet and give the spin dependent anharmonicities. The inter-excitation B(E2) values give the collective relationships of these excitations. Finally, the energy staggering in the γ-band for deformed nuclei gives a measure of the γ-softness. Besides these spectral properties, the mass (separation energies) and, for unstable nuclei, the β-decay half-life, are important.

determined. Fits for adjacent, similar, nuclei may result in very different parameter sets and one can easily land in local rather than global minima.

To circumvent this problem a simpler model prescription for the Hamiltonian is needed and has recently been developed in which the standard existing paradigms of nuclear structure discussed in this chapter, as well as transition regions between them, are obtained with a much simpler treatment of the GCM. The price to be paid is the loss of some higher order anharmonicities in the potential and kinetic energy terms but the advantages are a tremendous gain in simplicity and intuitive physical content and a much lower likelihood of falling into local minima in a least squares fitting procedure. With the truncated Hamiltonian most fits to data can be done rather easily. In any case, having achieved a reasonable reproduction of a given nucleus with such a Hamiltonian, one can then always add in the neglected terms to fine tune the calculations.

The simplified GCM Hamiltonian is given by the B_2 term in Eq. 6.14 for the kinetic energy and the potential is restricted to three terms.

$$V_{\text{GCM}} = (C_2/\sqrt{5})\beta^2 + (C_4/5)\beta^4 - (C_3\sqrt{2/35})\beta^3 \cos 3\gamma \qquad (6.79)$$
$$= C_2'\beta^2 + C_4'\beta^4 - C_3'\beta^3 \cos 3\gamma \qquad (6.80)$$

where we have defined the parameters C_2', C_4', and C_3' to incorporate the awkward numerical constants in Eq. 6.79.

It is convenient to discuss the GCM potential by analogy with the IBA which is why we have postponed the discussion till here. We now discuss idealized limits for the GCM Hamiltonian of Eq. 6.80 analogous to those of the IBA. In Fig. 6.52 we show a GCM structural triangle along with the IBA symmetry triangle repeated from Fig. 6.41. Reference to the GCM level spectra and B(E2) values illustrated in Figs. 6.53 and 6.54 will be helpful in the following discussion.

Suppose only the C_2' term contributes, that is, $V = C_2'\beta^2$ with C_2' positive. Then the potential corresponds to a spherical harmonic vibrator and the levels will be equally spaced phonon multiplets. The Hamiltonian is also similar to the IBA in the harmonic $U(5)$ limit $[H_{\text{IBA}} = \varepsilon n_d]$, as in Fig. 6.30. The GCM results for $C_2' = 44.8$ MeV (that is, $C_2 = 100$ MeV) are shown on the far left in Fig. 6.53. The phonon energy $\hbar\omega = E(2_1^+)$ and $R_{4/2} \equiv E(4_1^+)/E(2_1^+) = 2.0$. Anharmonities can be introduced with the $C_4'\beta^4$ term, although certain degeneracies such as 2_2^+, 4_1^+, which inherently reflect γ softness, cannot be broken without introducing the C_3' term.

To obtain a deformed shape, we need to displace the minimum in V to finite β. A little thought shows that this can be obtained with $C_2' < 0$ and $C_4' > 0$, that is $V = C_2'\beta^2 + C_4'\beta^4$. This potential, however, has no γ-dependence so it is a γ-flat deformed oscillator: it corresponds to the Wilets–Jean Model and to the O(6) limit of the IBA. As with these models, the yrast energies follow a $\Lambda(\Lambda+3)$ or $\tau(\tau+3)$ dependence where the correspondence of quantum numbers is $\Lambda = \tau = J/2$. $R_{4/2}$ is close to 2.50.

A typical spectrum with this H_{GCM} is shown on the far left in Fig. 6.54. The E2 selection rules and branching ratios are analogous to those of the O(6) limit of the IBA. In particular, using the IBA quantum number τ to characterize the GCM levels (this is pedagogic rather than rigorous as the IBA $\rightarrow$ GCM only for $N_B \rightarrow \infty$), we have, for E2

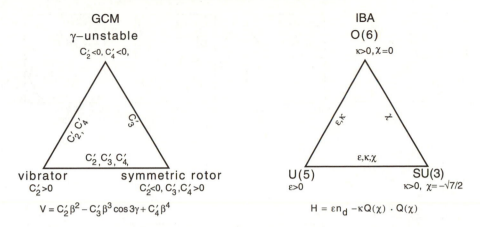

FIG. 6.52. Structural triangle for the GCM, in analogy with that for the IBA. The vertices represent structural paradigms and the legs are transition regions. The parameters needed to obtain each type of structure are indicated (based on Zhang, 1997).

transitions, $\Delta\tau = \pm 1$, giving the result that the first excited 0^+ state (0_2^+) has allowed transitions to the 2_2^+ but not to the 2_1^+ state. The minimum in this GCM potential occurs at a deformation value of $\beta = 1.06(-C_2/C_4)^{1/2} = 1.06[(C_2'/((5)^{1/2}C_4')]^{1/2}$, where $V_{min} = -C_2^2/(2(5)^{1/2}C_4) = -C_2'^2/(2(5)^{1/2}C_4')$.

To obtain a prolate deformed, axially symmetric nucleus, that is, a nucleus with the minimum in the potential at finite positive β and $\gamma = 0°$, we use the potential $V = C_2'\beta^2 + C_4'\beta^4 - C_3'\beta^3 \cos 3\gamma$, with $C_2' < 0, C_4' > 0$, and $C_3' > 0$. This potential produces a deformed rotor with ground state rotational band [$E \sim J(J+1)$] and with higher excited intrinsic excitations such as the γ-vibration with $K = 2$. Each such excitation also sports a rotational band. The spectra on the right sides of Figs. 6.53 and 6.54 show this limit, for two choices of parameter values. They show typical spectra for deformed nuclei with the γ vibration at ~ 800 keV and a $K = 0$ excitation a little above 1 MeV. The minimum in this potential is at $\beta \sim (-C_2/C_4)^{1/2} = [(C_2'/((5)^{1/2}C_4')]^{1/2}$ as long as $C_3 << |C_2|, C_4$. [An oblate rotor with V_{min} at $\beta < 0, \gamma = 0°$ is obtained with $C_3' < 0$.]

An interesting feature of the deformed rotor limit of the GCM is that it automatically incorporates a rotational-vibration interaction (i.e., bandmixing). For example, the ratio $R_{\gamma g} \equiv B(E2 : 2_\gamma^+ \rightarrow 0_g^+)/B(E2 : 2_\gamma^+ \rightarrow 2_g^+) = 0.37$ with the parameters of Fig. 6.53, compared with the Alaga rule of 0.7. Also $R_{4/2} = 3.25$, instead of the limiting rotor value of 3.33.

Another fascinating result is that, for a wide variety of GCM parameters that give a deformed rotor spectrum, the 0_2^+ state is calculated to preferentially decay to the γ-band rather than the ground band. This is contrary to the expected result in the Bohr Mottelson picture of β and γ vibrations but is identical to the situation found in the IBA.

This result is important for two reasons. First, the dichotomy in the decay properties of the 0_2^+ intrinsic state in the IBA and the traditional BM geometric picture has long

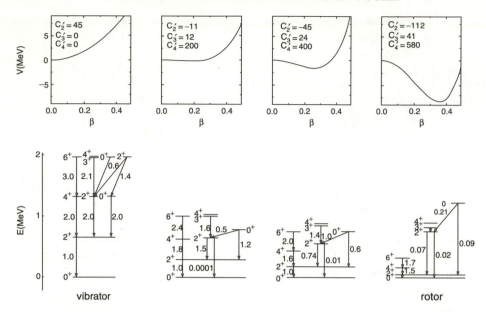

FIG. 6.53. Potential energy surface (in the $\gamma = 0°$ plane) and level schemes in the GCM for a vibrator (left), a rotor (right), and the transitional region between them (middle). The parameter B_2 in Eq. 6.14 is 100×10^{-42} MeV.s^2. This corresponds to $B_2 = 100$ in the input to the GCM code in Langanke (1991) (based on Zhang, 1997).

been puzzling. Secondly, in actual nuclei, with perhaps an occasional exception such as ^{154}Sm or ^{166}Er, the empirical 0^+ states do *not* have collective B(E2) values to the ground state band. Rather, as seen in Fig. 2.18 for the 0_2^+ excitations, these B(E2) values are often an order of magnitude or more weaker than $\gamma \rightarrow g$ B(E2) values. In fact, although it has been traditional in nuclear physics for 40 years or more to refer to the lowest quadrupole excitations (with $K = 2$ and 0) of a deformed nucleus as γ and β vibrations we have seen that there is actually very little evidence that the lowest $K = 0$ excitations are fluctuations in β. As we have discussed, it is at least as likely is that they contain significant amplitudes for 2-phonon γ vibrational excitations. They may also correspond to some kind of pairing or even 2-quasiparticle modes. Pending a better understanding, it is probably best to drop the almost reflex tendency to label the lowest $K = 0$ band as a β vibrational band. Apparently, in the data, in the IBA, and even in the geometric model, the β mode, if it exists intact at all, occurs at higher energies in most nuclei.

As we saw just above, a variety of rotor spectra can be produced with different choices of C_2', C_4', and C_3'. For example, a large C_3' makes the potential stiffer in γ and raises the γ-band energy. Naturally, this also decreases the B(E2 : $\gamma \rightarrow g$) values, raises $R_{4/2}$ toward 3.33 and $R_{\gamma g}$ towards the Alaga rule value of 0.7. The latter two results correspond to less rotation–vibration mixing (larger energy separation of rotational and vibrational energy scales). For example, with $C_3' = 167$ MeV and the other parameters as in Fig. 6.53, $E(2_\gamma^+)$ rises to ~ 1.7 MeV, $R_{4/2}$ goes to ~ 3.32 and $R_{\gamma g}$ is ~ 0.55.

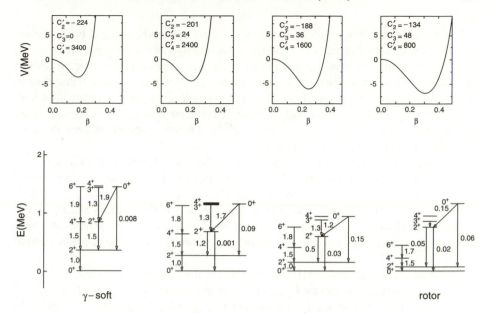

FIG. 6.54. Similar to Fig. 6.53, for a γ-soft to rotor transition region. The same B_2 is used as in Fig. 6.53. Note that the rotor spectrum on the right in this figure and Fig. 6.53 are slightly different in order to exhibit the flexibility inherent in the model (based on Zhang. 1997).

We can summarize the three structural paradigms as follows:

$$V = C_2'\beta^2 \qquad\qquad C_2' > 0 \qquad\qquad \text{Spherical vibrator}$$

$$V = C_2'\beta^2 + C_4'\beta^4 \qquad\qquad C_2' < 0, C_4' > 0 \qquad\qquad \gamma\text{-unstable deformed rotor}$$

$$V = C_2'\beta^2 + C_4'\beta^4 - C_3'\beta^3 \cos 3\gamma \qquad\qquad C_2' < 0, C_3', C_4' > 0 \qquad\qquad \text{Deformed axially symmetric rotor}$$

As noted, level sequences and transition matrix elements calculated with these potentials, with particular values of B_2, C_2', C_3', C_4' chosen to produce typical energy and B(E2) values of actual nuclei that have these three types of structure, are shown on the far left and right sides of Figs. 6.53, 6.54. Inspection of these, and comparison to Figs. 6.2, 6.12, 6.30, 6.33, and 6.37 will show the similarity to the ideal limits of collective nuclear structure already discussed in this chapter.

The simplification of V_{GCM} thus introduced has another appealing feature. It allows almost trivially simple treatments of shape transition regions between these limits. The obvious analogy is to the transition regions between the dynamical symmetries of the IBA discussed earlier in this chapter. Indeed, the deep relation of these approaches is made even more apparent by comparison of the "structural" triangle for the GCM with the symmetry triangle of the IBA.

The simplest case to treat (as also in the IBA) is the γ-unstable to symmetric rotor transition region. This requires only a change in C_3' from 0 to a finite positive value (while maintaining finite values $C_2' < 0$ and $C_4' > 0$). [Note: the paragraphs just below focus on the *essential* changes in parameters needed to achieve each transition leg: in actual calculations, one should expect to vary other parameters as well in order to achieve detailed fits to the characteristic magnitudes of energies and B(E2) values appropriate to actual realizations of each type of structure. For example, changing C_3', as we just noted, causes a change from γ-soft to axially symmetric rotor, but changing *only* C_3' might lead to unrealistic rotor spacings or yrast B(E2) values. Therefore one might need to adjust C_2' and C_4' to fine tune a fit to actual data.]

The kind of structural transition induced by a finite C_3' (and adjusting C_2' and C_4' to achieve realistic spectra) is illustrated in Fig. 6.54, along with a projection of the potential along the β axis (i.e., for $\gamma = 0°$). We see that C_3' plays exactly the role of χ in the IBA (compare Fig. 6.41). [Recall: $\chi = 0$ gives O(6), $\chi = -(7)^{1/2}/2$ gives SU(3), for the Hamiltonian $H = -\kappa Q \bullet Q$.] Since χ is intimately related to the rms axial asymmetry, this association is completely reasonable since C_3' multiplies the GCM potential term in $\cos 3\gamma$.

A vibrator to rotor transition [U(5) $\rightarrow$ SU(3) in the IBA] is effected by lowering C_2' to negative values and increasing C_4' and C_3' from zero to positive values. The structural changes are illustrated for typical parameters in Fig. 6.53. The parameter changes in C_2', C_4', and C_3' are analogous to the changes in ε, κ, and χ, respectively, in the IBA, as sketched in the structural triangle in Fig. 6.52. Finally, the vibrator to γ-unstable rotor is achieved by changing C_2' and C_4', just as the U(5) to O(6) transition in the IBA requires changing ε and κ.

Of course, the IBA and GCM are not identical models. One is algebraic, the other geometric. One explicitly involves the number of valence nucleons and its predictions are nucleon number dependent while the other is independent of nucleon number (except implicitly in that the potential may depend on N and Z). In general one expects the IBA predictions to go over to those of the GCM as $N_B \rightarrow \infty$.

Very recently, an alternate expression for the GCM potential has been developed (by Mark Caprio, a graduate student at Yale) in terms of physically intuitive parameters, denoted d, e, and f, namely,

$$V(\beta, \gamma) = f \left[\frac{9}{112} d \left[\frac{\beta}{e} \right]^2 - \sqrt{\frac{2}{35}} \left[\frac{\beta}{e} \right]^3 \cos 3\gamma + \frac{1}{5} \left[\frac{\beta}{e} \right]^4 \right] \qquad (6.81)$$

The relations of the d-e-f and C parameters are as follows:

$$d = \frac{112}{9\sqrt{2}} \frac{C_2 C_4}{C_3^2} \qquad e = \frac{C_3}{C_4} \qquad f = \frac{C_3^4}{C_4^3}$$

$$C_2 = \frac{9\sqrt{5}}{112} \frac{fd}{e^2} \qquad C_3 = \frac{f}{e^3} \qquad C_4 = \frac{f}{e^4} \qquad (6.82)$$

Figure 6.55 shows a selection of potentials, for different d values, to exhibit the variety of structures obtainable with the GCM. [The parameters e and f are adjusted in each case to give potentials of reasonable and similar radial extent and depth.]

Shapes of the GCM potential

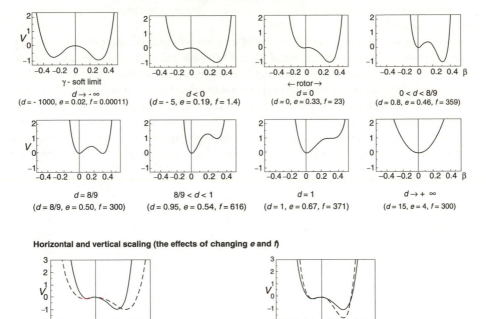

Horizontal and vertical scaling (the effects of changing e and f)

FIG. 6.55. Illustration of the effect of the parameter d (see text and Eqs. 6.81, 82) on the potential for the GCM. The bottom two panels show the scaling effects of the parameters e and f. The parameter f and the potential V are in MeV. (Figure courtesy of M. Caprio.)

The advantage of the (d, e, f) formulation is that these parameters modify the potential in easily visualizable ways: e expands the potential in the radial direction, f factors the depth of the potential, and d describes the shape. (We have highlighted three letters above that may offer mnemonic help in recalling the effects of the three parameters.)

The specific ranges of d have the following structural meanings:

$d < 0$ corresponds to a potential with a prolate minimum and oblate saddle pont;

$d = 0$ is a potential with a saddle point at $\beta = 0$ and a prolate minimum;

d between 0 and 8/9 gives a potential with a spherical minimum and a deeper prolate minimum;

d between 8/9 and unity also gives two minima but in this case the spherical minimum is deeper;

$d > 1$ has a single minimum in the potential centered at $\beta = 0$.

The effects of the parameters e and f as quasi-scaling variables for the potential are also shown (at the bottom) in Fig. 6.55.

It should be clear that the GCM model can accommodate a rich phenomenology. Note in particular the existence, for $d \leq 1$, of double minima or saddle points in the potential. The GCM, even in this alternate formulation, is therefore capable of dealing with shape coexistence phenomena (e.g., prolate and oblate or spherical and deformed shapes). The model, for example, has recently been used to discuss phase coexistence of spherical and deformed phases in ^{152}Sm. Given that one has some reasonable feeling (whether based on theory, intuition, or data) for the structure of a specific nucleus, it is now fairly straightforward to estimate initial values for d, e, and f. Fine tuning will usually produce a reasonable fit.

7

EVOLUTION OF NUCLEAR STRUCTURE

Until now in this book we have tended to study the excitations of a given nucleus. This can be called a "vertical" approach to structure and has been the empirical cornerstone of nuclear physics and the inspiration for countless nuclear models. With the accumulation of data over several decades, however, we have come to a situation where we can study in a systematic way *how* structure evolves with the changing numbers of nucleons. This is a "horizontal" approach where we study changes in various observables across different regions of the nuclear chart (or even across the whole chart).

It is our purpose to outline in this chapter some of the elements of a horizontal approach that focuses on structural *evolution*. We will see that this offers important insights into nuclear structure, and into phase transitions in finite nuclei, that it provides a powerful means of predicting the properties of newly discovered nuclei, that it provides new, more efficient, signatures of structure, and that it can even give clues to the underlying shell structure. We will focus on two types of approaches, known as Valence Correlation Schemes (VCS) and Correlations of Collective Observables.

Many of these ideas are especially applicable to the new exotic nuclei that are becoming available with the advent of radioactive beams. This highly exciting new field which, in many ways, represents the principal future direction for nuclear structure, will be discussed in some detail in Chapter 11, where we will also return to some of the ideas to be presented here to show their special relevance to understanding nuclear structure, shell structure, and the evolution of collectivity in regions of extremes of proton to neutron composition.

7.1 Overview of structural evolution

The behavior of magic nuclei is fairly simple. Doubly magic nuclei (e.g., ^{48}Ca, ^{208}Pb) are especially rigid to excitation. The lowest lying states are often of negative parity, representing particle–hole excitations across major shells. Singly magic nuclei are likewise simple. They often reflect the properties of the seniority scheme, as seen in Figs. 2.3, 2.6, 5.2–3 which show the constant excitation energies and inverted parabolic behavior of B(E2: $2_1^+ \rightarrow 0_1^+$) values typical of this scheme. The behavior of other key observables—quadrupole and dipole moments—in $|j^n\rangle$ configurations, as a function of n, were also discussed in Chapter 5 and summarized in Fig. 5.4 which has rather broad implications.

As soon as we depart from magic nuclei, spectra begin to change rapidly when *both* valence protons *and* valence neutrons are present. This was dramatically evident in Figs. 2.6–2.8, which show the behavior of $E(2_1^+)$ in a set of singly magic nuclei, the Sn isotopes, and in nuclei with both valence protons and neutrons. We have commented that the much lower values in the latter strongly suggest that it is the *valence proton–neutron interaction* that leads to softness. Earlier, we formalized this idea slightly by giving some qualitative arguments why the $T = 0$ component of the p–n interaction can induce single nucleon configuration mixing more readily than its $T = 1$ component and why such mixing is tantamount to the development of nonspherical nuclear shapes.

Collectivity and a softness toward deformation go hand in hand as valence nucleons are added beyond closed shells. In midshell regions of medium and heavy nuclei, one invariably encounters a large concentration of deformed nuclei exhibiting rotational behavior and low-lying (often collective) excitations of γ, $K = 0$, and octupole type.

A schematic view of such a structural evolution, typical in broad brush strokes of many regions, is shown in Fig. 7.1a. The basic trend reflects the systematics of the Sm nuclei shown in Fig. 6.29. As long as neither type of nucleon is magic, $E(2_1^+)$ drops and $E(4_1^+)/E(2_1^+)$ ranges from $< 2 \rightarrow 3.33$ as nucleons of either type are added. A typical structural sequence is shell model (rigid spherical) $\rightarrow$ vibrator $\rightarrow$ transitional $\rightarrow$ rotor. In the transitional region, $R_{4/2} \equiv E(4_1^+)/E(2_1^+)$ usually jumps rapidly from < 2.5 to > 3.0. We have discussed two kinds of transitions, one spherical-deformed, the other spherical-γ-soft-deformed. $E(4_1^+)/E(2_1^+)$ values near ~ 2.5 can occur in either case, but in one case they imply softness to quadrupole deformation; in the other an extreme softness to axial asymmetry.

Although $E(2_1^+)$ and $E(4_1^+)/E(2_1^+)$ are often thought to reflect similar aspects of structural evolution, they actually access slightly different physics and provide complementary insights. We will discuss this point below since it is, surprisingly, not well known.

One can view structural evolution either in terms of changes to the nuclear potential (in the β and γ degrees of freedom—we are not referring here to the single particle potential of the spherical shell model), or microscopically in terms of configuration mixing. In terms of a potential, imagine a sequence of nuclei starting with a roughly harmonic oscillator potential centered on a spherical shape ($\beta = 0$). As valence nucleons are added this potential will generally soften; that is, widen and become less steep. Our discussion of potentials and zero point motion in Chapter 3 tells us that the levels (e.g., $E(2_1^+)$) characterizing the equilibrium (i.e., ground state) structure will decrease in energy. At some point, the nuclei will go deformed. The 2_1^+ energy is then primarily determined by the moment of inertia, and hence it continues to decrease as β increases.

Microscopically, this evolutionary process results from configuration mixing and the lowering of the 2_1^+ (and other) levels is due to the kind of multistate mixing discussed in Chapter 1. We recall that, if N degenerate states all mutually mix with equal interactions, $-V$, one level is lowered by an energy $(N - 1)V$ and its mixed wave function is $\psi = \frac{1}{\sqrt{N}}(\phi_1 + \phi_2 + \phi_3 + \ldots + \phi_N)$. Energy lowering and coherence in turn leads to enhanced transition rates (e.g., increases in the B(E2 : $2_1^+ \rightarrow 0_1^+$) value—as seen in the data for mid-shell nuclei in Chapter 2). Indeed, the Grodzins rule relates the B(E2 : $2_1^+ \rightarrow 0_1^+$) value to $1/E(2_1^+)$. Hence, microscopically, the lowering of the 2_1^+ level is a signature of increasing coherence in the wave functions and collectivity in structure.

The $E(4_1^+)/E(2_1^+)$ ratio samples this process a bit differently. First, we consider

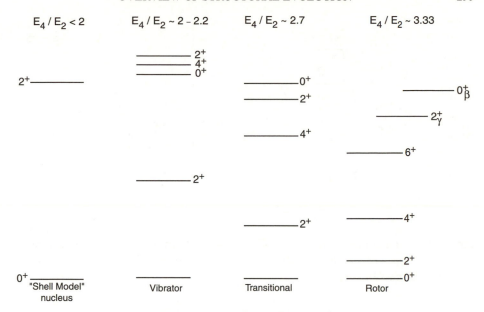

SCHEMATIC EVOLUTION OF STRUCTURE
NEAR CLOSED – SHELL → MID SHELL

FIG. 7.1a. An imaginary, but typical, sequence of level schemes and nuclear structure
types from a near closed shell region to a well deformed midshell nucleus.

nuclei with axial symmetry ($\gamma = 0°$). As we have just noted, the usual, qualitative view
is that this ratio takes on values < 2.0 for $|j^n\rangle$ configurations near-closed shell nuclei,
$\sim 2 – 2.3$ for typical near-harmonic vibrational nuclei, moves towards 3.0 in transitional
nuclei and converges on 3.33 for deformed nuclei. However, we recall that the last value,
3.33, is the standard result for a quantum mechanical rotor. It does *not* depend on the
magnitude of the deformation of the rotor, but only on the *fact* that the nucleus rotates
(i.e., that its shape, becoming deformed, allows a direction in space to be specified).

Consider, therefore, a sequence of nuclei starting with one having a potential centered
at $\beta = 0$ while the next is centered and localized at small but finite β (i.e., V_{min} is at
$\beta \neq 0$, and $V(\beta = 0)$ is substantially higher than V_{min} so that the nucleus spends
little time near $\beta = 0$). Then the 2_1^+ energy will drop a small amount but $R_{4/2}$ will
jump quickly toward 3.33. If potentials for successive nuclei are localized at larger and
larger β, $E(2_1^+)$ will continue to drop steadily while $R_{4/2}$ will remain near 3.33. The Th
isotopes illustrate this scenario as shown in Fig. 7.1b.

Imagine instead a different scenario, namely that several successive nuclei have
potentials that remain centered at $\beta = 0$, while softening steadily (becoming less steep
in β) until a point at which the equilibrium deformation suddenly jumps to a large,
nearly saturated β value (e.g., $\beta \sim 0.25$). Then, $E(2_1^+)$ will remain fairly large and $R_{4/2}$
small ($\sim 2 – 2.5$) until the transition point where both will jump [$E(2_1^+)$ down, $R_{4/2}$ up]

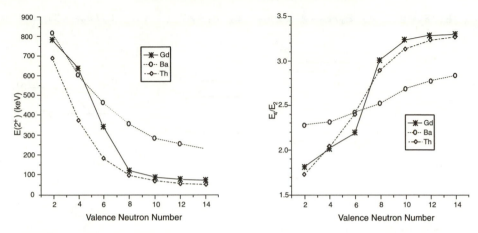

FIG. 7.1b. Empirical phenomenology of $E(2_1^+)$ and $R_{4/2}$ for selected nuclei exhibiting different types of structural evolution in spherical-deformed transition regions. See text for discussion.

towards saturation values. The Sm–Gd nuclei exhibit this trend (see Fig. 7.1b).

All of the above has invoked axial symmetry. However, we recall from Chapter 6 that deformed nuclei can exhibit axial asymmetry. In that case, $R_{4/2}$ can vary from ~ 2.5 for $\gamma = 30°$ to ~ 3.33 if $\gamma \sim 0°$. Hence, in a transitional region that passes from spherical through an axially asymmetric shape towards a deformed symmetric rotor, $E(2_1^+)$ will drop steadily and $R_{4/2}$ will increase steadily. The Ba ($N < 82$) and Os nuclei typify this kind of spherical-axially asymmetric-deformed trajectory (see Fig. 7.1b).

Despite the apparent simplicity of Fig. 7.1 and the concepts behind it, nuclear systematics over a *region* of nuclei is usually much more complex. This was exemplified by Fig. 2.13, which showed the behavior of $E(2_1^+)$ in the $A = 100$ region, and by Fig. 7.2 (left), which shows similar data for the energy ratio $E(4_1^+)/E(2_1^+)$ in the $A = 150$ region. While the general pattern of increased collectivity toward midshell can be discerned, these plots provide few obvious clues to a simple understanding of the structural behavior. Hence, we wish next to illustrate and discuss a phenomenological approach that greatly simplifies the systematics of nuclear transition regions and provides some insights into the operation of the p–n interaction and to the development of collectivity and deformation.

7.2 Valence correlation schemes: the $N_p N_n$ scheme

The approach is called the $N_p N_n$ scheme, and is an attempt to parameterize nuclear data in such a way as to explicitly emphasize the valence p–n interaction. Suppose we make the assumption that the onset of collectivity, configuration mixing, and deformation in nuclei is solely due to the p–n interaction and, moreover, that this interaction is fairly long range, orbit independent, and relevant only for the *valence* protons and neutrons. These assumptions, especially the orbit independence of the p–n interaction, may seem extreme but we will see that, averaged over many valence nucleons, they are reasonable

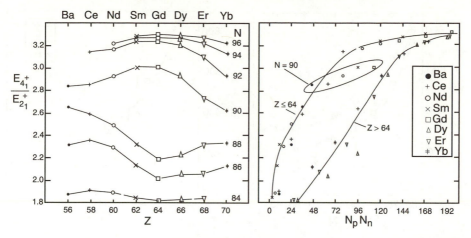

FIG. 7.2. Normal (left) and $N_p N_n$ (right) plots of $E_{4_1^+}/E_{2_1^+}$ for the $A = 150$ region (Casten, 1985a).

approximations. Under these assumptions, the total valence nucleon integrated p–n interaction strength will scale with the product, $N_p N_n$, of the number of valence protons times the number of valence neutrons. N_p and N_n are always counted to the nearest closed shells, whether the valence nucleons are particles or holes. It is important to understand that the valence p–n interaction embodied in the $N_p N_n$ scheme is not the total p–n interaction, but the collectivity- and deformation-driving part of it (primarily the $T = 0$ component). The $T = 1$ component must be identical to the p–p and n–n $T = 1$ forces, and as we have emphasized repeatedly, aside from the pairing interaction, these are repulsive on average and do not lead to as much configuration mixing, collectivity, and deformation. (The Sn nuclei in Fig. 2.6 are a classic example of this.)

To illustrate the use of the $N_p N_n$ scheme, the data shown for the $A = 130$ region in Fig. 2.14 are replotted, in Fig. 7.3, both against N (left) and in terms of $N_p N_n$ (right). Figure 7.4 compares normal and $N_p N_n$ plots for $E(4_1^+)/E(2_1^+)$ in the same region.

It is evident that the $N_p N_n$ plot substantially simplifies the systematics: the data that fell on several distinct curves before now coalesce so that they can be described by a single curve for a given mass region.

The $N_p N_n$ scheme is an example of a Valence Correlation Scheme (VCS). A VCS is a parametrization of some observable in terms of an intrinsic quantity based on the valence space. Others have been discussed and are often equally helpful. The P-factor (see below) is one. We will also briefly discuss one that is applicable to two nucleon separation energies.

If the simplification seen in Figs. 7.2–7.4 with the $N_p N_n$ scheme is general, this scheme offers a powerful phenomenological tool to simplify and unify the treatment of nuclear evolutionary behavior, and one which also has a simple underlying microscopic basis. However, applying the $N_p N_n$ scheme to other regions (e.g., the $A = 150$ region in Fig. 7.2) can be slightly less straightforward although sometimes more revealing (as

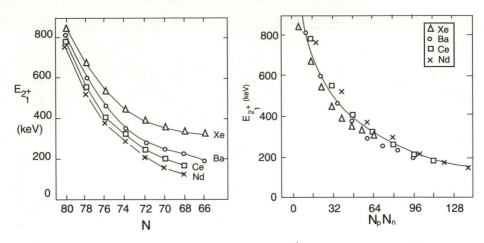

FIG. 7.3. Normal (left) and $N_p N_n$ (right) plots of $E(2_1^+)$ for the $A = 130$ region (Casten, 1985a).

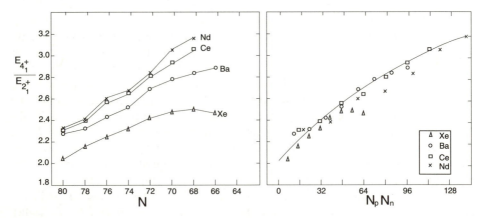

FIG. 7.4. Normal (left) and $N_p N_n$ (right) plots of $E_{4_1^+}/E_{2_1^+}$ for the $A = 130$ region (Casten, 1985a).

we now discuss).

To proceed further, we need to digress for a moment and return to a topic discussed in Chapter 3, namely the effects of residual interactions on single particle energies (and, hence, as we shall see, on shell and subshell gaps). First, we note that, if there are substantial subshell gaps in the single-particle level energies, the counting of N_p and N_n may be ambiguous. Moreover, the gaps themselves may also evolve, partly as a consequence of the p–n interaction itself.

If the p–n interaction is expanded in multipoles, the monopole and quadrupole components generally dominate. The monopole component $P_0(\cos\theta)$ is obviously constant as a function of the angles between the proton and neutron orbits, and is therefore inde-

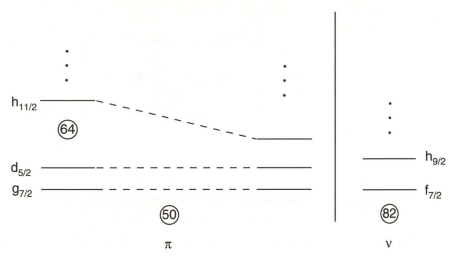

FIG. 7.5. Illustration of the eradication of a proton subshell gap at $Z = 64$ as a function of neutron number due to the monopole p–n interaction.

pendent of the total angular momentum J to which a pair of protons and neutrons are coupled. The monopole p–n interaction does depend on the relative radial behavior of the proton and neutron orbits, and therefore on which orbits are filling in a given mass region. Its effect is to shift the effective proton and neutron single-particle energies. As one proceeds through a pair of proton and neutron major shells, the specific behavior of the integrated monopole p–n interaction induces distinct patterns of shifts in the single-particle energies that can and do cause the appearance and disappearance of subshell gaps. The monopole p–n interaction thus effectively modifies the *valence space* for the protons and neutrons. This in turn affects the *numbers* of protons and neutrons on which the quadrupole p–n interaction may act. Hence, if one defines N_p and N_n based on the known evolution of these subshell gaps, $N_p N_n$ may still be a realistic estimate of the integrated quadrupole p–n interaction strength.

Since the radial integral of the monopole interaction is largest for orbits with similar shell model quantum numbers (see Fig. 3.10), the largest effects in heavy nuclei generally occur when spin-orbit partner orbits, such as $g_{9/2p}$ and $g_{7/2n}$, are filling. In such a situation, the single-particle energies of both these orbits are significantly lowered relative to their neighbors. The consequences can be rather dramatic and are now thought to be the underlying reason for the very sudden onsets of deformation near $A = 100$ and $A = 150$. The idea is illustrated in Fig. 7.5 for the latter region. As neutrons begin to fill the $h_{9/2}$ orbit near $N = 90$, the strong $1h_{11/2p} - 1h_{9/2n}$ interaction causes the effective single-particle energy of the $1h_{11/2}$ proton orbit to decrease and to obliterate the gap at $Z = 64$. Thus, for $N < 90$, the effective proton shell in this region is $Z = 50–64$, while for $N \geq 90$, the effective shell is the normal major shell $Z = 50–82$.

There is some very simple and beautiful empirical evidence for this concept from the energies of the first 2^+ states in the $N = 88$ and $N = 90$ isotones. These data are

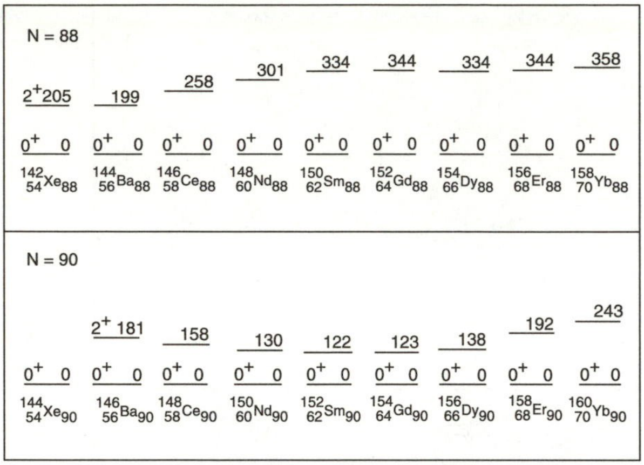

FIG. 7.6. Contrasting behavior of $E_{2_1^+}$ for $N = 88$ and $N = 90$ isotones.

shown in Fig. 7.6. If the effective proton shell were the normal one from $Z = 50$–82, one would expect $E_{2_1^+}$ to decrease as the number of protons were increased past $Z = 50$ until the near midshell region at $Z = 66$, after which $E_{2_1^+}$ should increase once again. This is precisely what happens for $N = 90$. Exactly the opposite behavior, however, characterizes the $N = 88$ isotones. (The $N = 84$, 86 isotones are similar to $N = 88$.) Without the concept of a significant subshell gap at $Z = 64$ for these neutron numbers, such behavior would be completely incomprehensible. But if one assumes an effective proton shell $Z = 50$–64 for $N = 88$, the midshell point is $Z = 57$, and one would now expect $E_{2_1^+}$ to increase for Z between 57 and 64. This is exactly the behavior observed.

Of course, the idea of an instantaneous disappearance of the $Z = 64$ gap at $N = 90$ is an unrealistically simple scenario. More likely, the dissipation of the gap is more gradual. Nuclear g factors for 2_1^+ states can be used to extract *effective* valence proton and neutron numbers. The details of this technique are beyond the scope of this book, but the results for N_p^{eff} and N_n^{eff} are shown in Fig. 7.7. They support the idea of a rapid but not instantaneous dissipation of a proton subshell gap as manifested by an increase in the effective value of N_p with neutron number for the Ce, Nd, and Sm nuclei. The solid lines in the figure show the abrupt dissipation scenario at $N = 90$.

A similar story characterizes the $A = 100$ region. Indeed, it originated in that region with the pioneering microscopic calculations of Federman and Pittel; the concept involved is sometimes known as the Federman–Pittel mechanism. The filling of the $1g_{7/2n}$ orbit lowers the $1g_{9/2p}$ orbit, obliterating the shell gap at $Z = 38$ (or 40) and suddenly increasing the size of the proton shell from $Z = 38$–50 to $Z = 28$–50. Nuclei such as Zr effectively go from $N_p = 2$ to $N_p = 10$. Deformation promptly ensues. It is only because of a mechanism such as this that one can explain the precipitous drop in $E_{2_1^+}$ for Zr (and Sr) at $N = 60$ in Fig. 2.13. (The effect comes full circle when the proton and neutron shells are filled, at 50 and 82, where a lowering of the $1g_{7/2}$ neutron orbit (Fig. 3.9) is caused by the filling of the $1g_{9/2}$ proton orbit between Zr and Sn.)

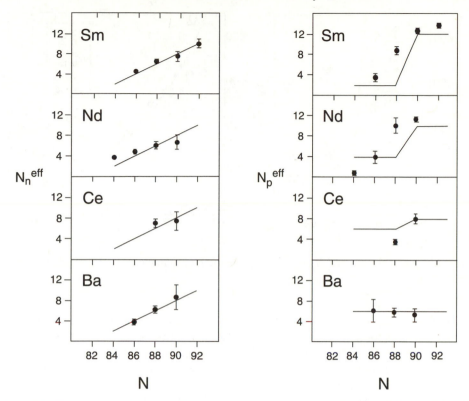

FIG. 7.7. Effective numbers of valence protons and neutrons in the $A = 150$ transitional region empirically extracted from $g(2_1^+)$ factors. The solid lines give the normal dependence of N_n against N and the changes in N_p against N, assuming the $Z = 64$ gap disappears suddenly between $N = 88$ and $N = 90$ (Wolf, 1987).

This discussion provides a way to link the origin of deformation in terms of subshell evolution to the structure of intruder states discussed in the last chapter. There, we commented that intruder states appear as low-lying excitations in some nuclei (e.g., Sn, Pb) because they effectively correspond to an increase in the number of valence nucleons, and therefore, to an increase in the p–n interaction. On account of this, they are also more deformed than the normal states. How low the deformed intruder states go depends on two factors: the size of the energy gap to be overcome and the strength of the p–n interaction in the particular intruder orbits involved. Intruder excitations of the type we considered in the Cd nuclei, or in Pb, correspond to a larger gap ($Z = 50, 82$) and a moderate interaction. Suppose, however, that the gap is smaller and/or the interaction stronger. Then it becomes possible for the intruder energies to drop below those of the "normal" states and become the (now deformed) new ground state. In this view, the presence of excited intruders in some regions and the sudden onset of deformation in others are just two facets of a common mechanism. Such a model, discussed extensively

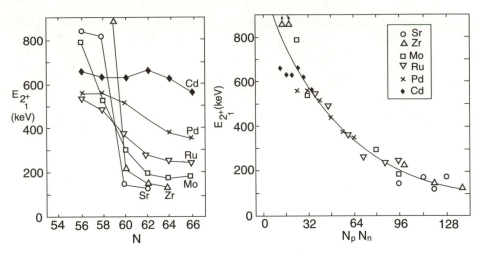

FIG. 7.8. Normal (left) and $N_p N_n$ (right) plots for $E(2_1^+)$ in the $A = 100$ region (Casten, 1985a).

by Heyde and co-workers in recent years, works and can account for the Pb intruders. It also accounts for the spherical-deformed phase transitions near $A = 100$ (Sr, Zr) and $A = 150$ (Ce, Nd, Sm, Gd), where the $Z = 40, 64$ subshell gaps come into play.

Figure 7.8 demonstrates this for $E(2_1^+)$ in the $A = 100$ region. The left panel repeats Fig. 2.13 while the right side gives the $N_p N_n$ plot of the same data assuming that the proton shell is taken as $Z = 38$–50 for $N < 60$ and 28–50 for $N \geq 60$. The simplification of the trends is remarkable. An extremely compact trajectory results. For the $A = 150$ region, Fig. 7.2 includes, on the right, the $N_p N_n$ plot where we have taken $Z = 50$–64 for $N < 90$ and $Z = 50$–82 for $N \geq 90$ for the nuclei with $Z \leq 64$. [We assume normal shells for $Z > 64$.] The effects are again dramatic: the extremely complex systematics seen in normal plots against N or Z are instantly simplified and the data coalesce into single smooth curves in $N_p N_n$ plots. The existence of two curves on the right in Fig. 7.2 is not surprising; they correspond to the two halves of the proton shell.

The circled points at $N = 90$, which deviate from the smooth curve, demonstrate the imperfectness of the assumption of an instantaneous change in proton shell structure at $N = 90$. Indeed, it was actually simply by manually shifting these $N = 90$ data points onto the smooth curves that the first extractions of effective N_p values were carried out.

The $N_p N_n$ scheme thus provides a simple, yet powerful, guide to understanding and predicting the systematic behavior of nuclear properties. It can also be used as a more reliable way of estimating the properties of unknown nuclei. The reason is that, whereas in traditional plots against N or Z such nuclei constitute *extrapolations* beyond the known ranges, the $N_p N_n$ values of many unknown nuclei far off stability are actually *smaller* than those for known nuclei in the same region. The $N_p N_n$ scheme converts the normal process of *extra*polation into one of *inter*polation, which is inherently much more reliable. Figure 7.9 illustrates this, giving as well one more illustration of the

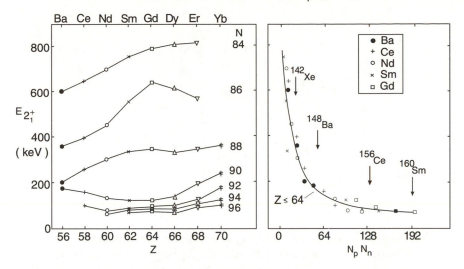

FIG. 7.9. Normal (left) and N_pN_n (right) plots of $E_{2_1^+}$ for the $A = 150$ region. In the N_pN_n plot, only the $Z \leq 64$ points are shown. The arrows point to the N_pN_n values for the neutron rich nuclei indicated. Predictions for $E_{2_1^+}$ (or other observables) for these nuclei are obtainable simply by reading off the appropriate curve at these N_pN_n values (Casten, 1986).

simplification achieved by the N_pN_n scheme. The vertical arrows point to the locations of four unknown neutron-rich nuclei in the rare earth region whose N_pN_n values are considerably less than those for known nuclei. Since this plot was constructed, ^{148}Ba and ^{142}Xe have been studied and the predictions verified.

The ^{142}Xe results are worth a somewhat closer look, as shown in Fig. 7.10. On the left, we have indicated by a ? symbol the uncertainty in estimating $E(2_1^+)$ for ^{142}Xe in a normal plot. One might choose to follow the $N = 88$ trend downward to a low $E(2_1^+)$ value. Or, one might realize that Xe is close to $Z = 50$ and expect the 2_1^+ energy to rise as the closed shell is approached. But, how much should it rise? There is no simple or reliable way to estimate this with the phenomenology on the left in Fig. 7.10. One concludes, dispiritedly, that we can only guess that $E(2_1^+)$ for ^{142}Xe is somewhere between ~ 160 and ~ 400 keV.

On the N_pN_n plot, though, there is little uncertainty (perhaps ± 20 keV due to the scatter in the points). One deduces, by *interpolation*, a value closely in agreement with the experimental result $E(2_1^+) = 287$ keV.

This process is an example of the value of paradigms in physics. This is an appropriate point to reflect a little on this idea. The establishment of a standard, a new benchmark, a simple behavior, permits a *paradigm shift* in which the approach to a problem is shifted to a new and better starting point. We have already encountered a number of examples. The emphasis in the shell model on the valence space, ignoring nucleons in closed shells, is a classic case; instead of dealing with, say, 150 nucleons, one has to deal with the energies

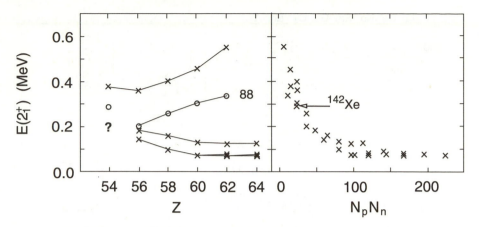

FIG. 7.10. Similar to Fig. 7.9 but including $Z = 54$ and focusing on the $A = 140$–160 region. The question mark on the left indicates the large uncertainty in extrapolating an estimate of $E(2_1^+)$ for ^{142}Xe: on the right the ^{142}Xe data point lies right amidst the compact trend with $N_p N_n$.

and interactions of maybe 10–20 valence nucleons. The IBA makes the same paradigm shift.

In the next three chapters we will see a paradigm shift of a similar sort, to the Fermi surface. Even the valence nucleons are more or less neglected (except for one at a time) in most applications of the deformed shell model or Nilsson model for odd-A nuclei, in microscopic (e.g., RPA) treatments of collective vibrations, and in the pairing formalism where all the complex correlations of the wave functions are embedded in the paradigm shift to the ground state as starting point. One considers only particle–hole (or 2 quasi-particle) excitations *relative* to the Fermi surface. Finally, the treatment of deformed nuclei included another paradigm shift in first going from the laboratory system to the body-fixed frame where rotation could be ignored.

A most elegant illustration of a paradigm shift is provided by the rotational formula $E \sim J(J + 1)$ discussed in Chapter 6. It is worth illustrating this. In Fig. 7.11, we show on the left an imaginary set of yrast levels. The question is how to understand them. The understanding one would have had in the time periods before and after 1952 (when Bohr and Mottelson introduced the idea of rotational bands into nuclear physics) is shown on the right: Before 1952 — utter confusion and no clue as to what was happening; Post 1952 — one instantly recognizes two aspects of the structure. First, a nearly perfect rotational band with energies very close to $\sim J(J + 1)$. Thus we are dealing with a deformed nucleus. Secondly, the $J(J + 1)$ dependence now provides a new paradigm and one exploits the paradigm shift to look at the small *deviations* from this new standard. In this way, as we have done in Chapter 6, we instantly see evidence for centrifugal stretching.

In the same vein, one can use the $N_p N_n$ scheme to parameterize collective model calculations. Normally, in phenomenological models such as geometric collective models or the IBA, the Hamiltonian contains a number of terms, each incorporating a free strength

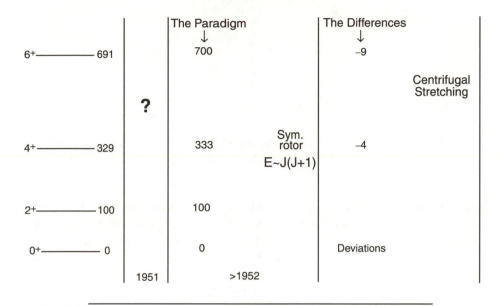

FIG. 7.11. Illustration of the value and use of a paradigm. Imaginary set of levels of an even–even nucleus along with an indication of how they would have been interpreted pre- and post-1952.

parameter. These must be estimated for each individual nucleus, and therefore, the calculation of extensive sets of nuclei can involve an enormous proliferation of parameters. If instead, the parameters are written as *functions* of N_pN_n, it is often possible to calculate sets of 50–100 nuclei with only 4–6 parameters. Moreover, for reasons that are obvious from the previous discussion, this process automatically give parameter estimates for unknown nuclei further from stability, so their predictions become interpolative.

Finally, the N_pN_n scheme has been applied to odd A nuclei by Bucurescu, Zamfir, and colleagues (Bucurescu, 1989) with intriguing results. Energy ratios based on different initial single-particle configurations probe the p–n interaction for different orbit combinations. Moreover, N_pN_n plots in odd nuclei provide a sensitive signature of the evolution of strongly coupled and decoupled band structures (see discussion of these in Chapter 9).

Thus far, we have discussed the N_pN_n scheme for a given region and have noted its simplifying power. Further advantages of this approach are evident when one compares different regions. As we saw in Chapter 2, different regions appear to behave in entirely different fashion vis a vis the development of collectivity and deformation: one needs only to glance at the left sides of Figs. 7.3, 7.8, 7.9 for a dramatic illustration of this. In contrast, N_pN_n plots are similar for essentially all regions of medium and heavy mass

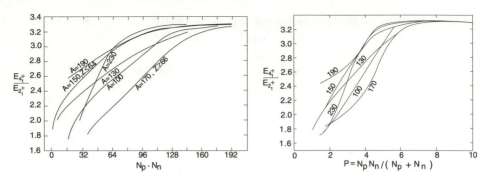

FIG. 7.12. $N_p N_n$ (left) and P-factor (right) plots summarizing the systematics of $E_{4_1^+} / E_{2_1^+}$ in six regions of medium and heavy nuclei (based on Casten, 1987).

nuclei. Figure 7.12 (left) collects the smooth curves, drawn through $N_p N_n$ plots, of the ratio $E_{4_1^+} / E_{2_1^+}$ for six mass regions in heavy nuclei. What appeared earlier to be radically different behavior from one region to another becomes a set of nearly parallel curves in $N_p N_n$. The similarity in *structure* of these curves gives confidence that the $N_p N_n$ scheme is a meaningful indicator. It also provides some confidence in extrapolating the curves for a new region once a few nuclei in that region are studied.

The other feature in Fig. 7.12 is that the curves for different regions are rather widely displaced: *changes* in structure are correlated with *changes* in $N_p N_n$ throughout medium and heavy nuclei but the *absolute* value of $N_p N_n$ provides little information. However, a slight modification of the $N_p N_n$ scheme puts it on an absolute scale and provides further physical insight. To see this we consider the parameter

$$P = \frac{N_p N_n}{(N_p + N_n)} \tag{7.1}$$

P can be viewed in several ways. It is simply a normalized value of $N_p N_n$. It is also a measure of the integrated strength of the valence p–n interaction compared to the valence pairing interaction that scales as the total number of valence nucleons. Finally, it is the average number of p–n interactions *per* valence nucleon. For this reason, P has been called a nucleonic "promiscuity factor." If we now plot the same data used to obtain Fig. 7.12 (left) against P, we obtain the results shown on the right of that figure. Individually, the six curves show exactly the same behavior as their counterparts in the $N_p N_n$ plot, but the different regions are coalesced into a narrow envelope, providing a unifying framework for understanding the systematic behavior of nearly all medium and heavy mass nuclei and highlighting, even more than $N_p N_n$, the correlation between the p–n interaction and collectivity.

A slightly more detailed glance at the right-hand side of Fig. 7.12 reveals a fascinating point directly correlated with our understanding of the residual pairing and p–n interactions. If we take the value $E_{4_1^+} / E_{2_1^+} \sim 3.0$ as a measure of the "transition point" from spherical and vibrational to deformed, we see that all the regions pass through this value in the narrow range of P values between 4 and 5. This value, denoted P_{crit}, gives a kind of "critical" value that acts as a signature for deformation. One can formulate a rule: nuclei with $P < P_{crit} = 4$–5 will not be deformed, unlike nuclei with $P > P_{crit}$. Simple consideration of the formula for P refines this idea: a necessary condition for $P > 4$ is that *both* $N_p, N_n \geq 4$, while a sufficient condition for $P \geq 5$ is that *both* $N_p, N_n \geq 10$. Thus, no nucleus can be deformed unless there are at least four valence protons *and* four valence neutrons and a nucleus *must* be deformed if there are at least 10 valence nucleons of each type. Apparent exceptions to this rule such as the light Hg ($Z = 80$) nuclei, which have two proton holes relative to $Z = 82$, are indeed only apparent since the reason deformation sets in these nuclei is another example of the movement of single-particle energies of one type of nucleon as a function of the number of the other: as N decreases, the $h_{9/2p}$ orbit from above $Z = 82$ descends across the $Z = 82$ gap and enters the shell below. The counting of N_p should therefore be based on some effective Z value between 82 and 92.

The value of $P_{crit} \sim 4$ to 5 is interesting in itself. We know from the energy gap, 2Δ, in even–even nuclei that typical like-nucleon pairing interactions have strengths of $V_{pair} \sim 1$ MeV. Similarly, p–n interactions are on the order of 200–300 keV (see Chapter 4). Thus, P, which gives the ratio of the number of p–n interactions to pairing interactions, equals 4–5 at precisely the point at which the *integrated* p–n interaction strength begins to dominate the pairing strength. This provides an appealing physical picture that simply states that softness to deformation and the phase transition to deformed shapes occur just when the deformation-driving p–n interaction begins to dominate the spherical-driving like-nucleon pairing interaction.

The $N_p N_n$ scheme is based on the rather crude assumption that the p–n interaction is orbit independent. Since it is not, one might expect situations in which $N_p N_n$ is not the best scaling parameter. In fact, we have already seen evidence for this: most observables are smooth against $N_p N_n$ but they are not *linear* in $N_p N_n$. The valence p–n interaction increases with $N_p N_n$, but is not necessarily proportional to it. The most dramatic evidence for this occurs in the empirical behavior of B(E2: $0_1^+ \rightarrow 2_1^+$) values in deformed nuclei. These increase from the beginning of a shell through the transition region. But, instead of continuing this increase unabated until midshell, they saturate. This was illustrated earlier for rare earth nuclei in Figs. 2.16 and 5.3. The reader will recall from our discussion of the IBA that this B(E2: $0_1^+ \rightarrow 2_1^+$) $\sim N^2$ both in SU(3) and in realistic symmetry breaking calculations for deformed nuclei. Thus, assuming constant boson effective charges e_B, such IBA calculations must disagree significantly with the data near midshell. These and other data suggest that, although some simplicity will be lost, it might be useful to have a more refined estimate of p–n strength than $N_p N_n$. This can be obtained very easily by a simple, explicit calculation of the integrated quadrupole p–n interaction among the valence nucleons. The result, called $|S_{pn}|$, is shown in Fig. 7.13. Clearly, instead of $N_p N_n$, one can use these calculated $|S_{pn}|$ values to define effective $(N_p N_n)_{eff}$ products and therefore effective values of N_p and N_n themselves. If these are then used to recalculate quantities such as B(E2: $0_1^+ \rightarrow 2_1^+$) values, the observed saturation is excellently reproduced.

If we anticipate our discussion of the Nilsson model in the next chapter, we can easily explain the behavior of $|S_{pn}|$. (Readers unfamiliar with the Nilsson model, please

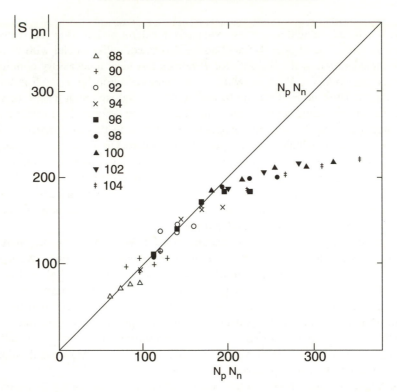

FIG. 7.13. Calculated $|S_{pn}|$ values for the deformed rare earth nuclei, illustrating the saturation in quadrupole collectivity. Compare the behavior of empirical B(E2) values in Fig. 5.3 (Casten, 1988b). The legend gives the neutron number corresponding to each symbol.

forgive this short digression: better yet, return to it after reading the next chapter.) $|S_{pn}|$ is obtained simply by integrating (summing) the products of the individual proton and neutron quadrupole moments for each Nilsson orbit over all filled orbits up to a given N, Z. Downward sloping lines in the Nilsson diagram correspond to equatorial orbits, while flat or upward sloping lines represent more polar orbits. Thus, at the beginning of a shell, when both protons and neutrons are filling the equatorial downward sloping orbits, there will be high overlap and significant contributions to the integrated quadrupole p–n interaction. Near midshell, however, some neutrons will enter flat or upward sloping orbits that will have lower overlap with the downward sloping proton orbits, and vice versa. These, and proton–neutron pairs in which both particles are in flat Nilsson orbits with zero quadrupole moments, will contribute little or nothing to further increases in the integrated quadrupole p–n interaction strength. This strength should therefore increase linearly with $N_p N_n$ at the beginning of a shell and then saturate toward midshell.

One last point is relevant before ending this discussion: the concept of the p–n interaction scaling approximately with $N_p N_n$, at least at the beginning of a shell, was

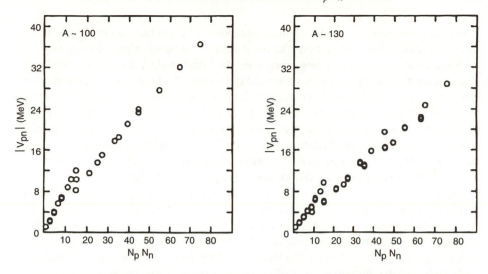

FIG. 7.14. Empirical values of V_{pn} (based on Zhang, 1989).

originally proposed as a convenient ansatz in order to provide a simple phenomenological approach to the systematics of certain collective observables. However, Zhang and co-workers have *empirically* extracted actual p–n interaction energies for the last proton and last neutron in a given nucleus (see further discussion in Section 11.3). By carrying out appropriate sums over the valence nucleons they have obtained the *total* valence p–n interactions (V_{pn}). Results for two shells are shown in Fig. 7.14. The plots show exactly the initial linearity with $N_p N_n$ (and subsequent slower growth) that we have been discussing. That they do not completely saturate (in contrast to Fig. 7.13) is because S_{pn} is just the quadrupole interaction: The total p–n interaction includes the (always attractive) monopole component whose strength is monotonic throughout a shell and so provides a continuously increasing "base" to the total $T = 0$ p–n strength.

We have seen that the $N_p N_n$ scheme, and its siblings P and $|S_{pn}|$ plots, provide a simple yet powerful way of correlating a vast amount of systematic data on the development of collectivity, phase transitions, and deformation in medium and heavy nuclei. These concepts have an appealing microscopic foundation in the deformation-driving $T = 0$ component of the valence p–n interaction and its competition with its opposite number, the spherical-driving like-nucleon pairing interaction. In this way, these ideas, albeit phenomenological, bring together a number of threads running throughout this book. Through this phenomenology, we are beginning to develop a unified, coherent view of the evolution of nuclear structure. This view emphasizes the importance of the p–n interaction, and its role both in modifying the underlying shell structure and in inducing correlations, configuration mixing, and deformation. What is needed now is to graduate from phenomenology to a real microscopic theory of nuclear structure and its evolution that embodies these ideas.

Another key issue is to understand the relation between the obvious centrality of the p–n interaction and the equally obvious successes of collective models that make no *explicit* mention of this interaction. The resolution of this seeming paradox appears to be twofold. First, the p–n interaction determines the "mean field"; that is, the correlations and deformation of the ground state (or the base state of a family of states such as the intruders), upon which the collective excitations are then constructed. In models such as the GCM or the IBA-1, where no distinction is made between protons and neutrons—that is, both are assumed to exhibit the same overall shape and structure—there is, implicitly, a large p–n interaction to enforce the congruency. The same is true of microscopic models (e.g., the Nilsson model and various equilibrium potential energy surface models that utilize it) that assume the same deformations for protons and neutrons.

Secondly, the p–n interaction, through its monopole component, affects the detailed distribution and energies of the underlying single-particle states. As we have seen, this is critical to the evolution of subshell gaps and therefore to the content of the valence space on which the quadrupole component acts. Moreover, these single-particle states are the fodder with which (see Chapter 10) the detailed microscopic structure (e.g., energies, collectivity) of the vibrational excitations is constructed. Thus, the p–n interaction implicitly enters collective models both in the equilibrium shapes they present and in the single-particle energies used in obtaining their predictions.

To close this section we give another example of a VCS. Figure 7.15 (left) shows single nucleon separation energies for the $A \sim 150$ region. They show regularities but are likewise complex. However, inspection of the regularities, and reference to concepts we have already discussed, lead to a new VCS. We note in Fig. 7.15 (left) that the S_n values for constant Z decrease with increasing neutron number but increase with increasing proton number. The underlying physics is easily understandable. The non-pairing part of the like nucleon interaction is repulsive, as we have noted in Chapter 4. In contrast the p–n interaction is attractive. Therefore the last neutron should become less bound as neutrons are added (smaller S_n) but more bound with added protons. These trends are easily accommodated if we plot the same data as in Fig. 7.15 (left) against the valence nucleon number function $\alpha N_p - N_n$. Fitting α to a given region is easy and Fig. 7.15 (right) shows the same data as Fig. 7.15 (left) plotted against $2.6 N_p - N_n$. Now a compact, nearly perfectly linear behavior emerges. Similar $\alpha N_p - N_n$ correlations result for both S_n and S_p in all mass regions. As with the $N_p N_n$ scheme, they often permit predictions for new nuclei by interpolation.

7.3 Correlations of collective observables

The $N_p N_n$ scheme, P-factor and other VCSs are powerful aids in tracking the evolution of structure. We will see in Chapter 11 that they are also very useful in highlighting special nuclei that deviate from the behavior of the majority of their siblings.

There are other correlation schemes, equally powerful, which are based on a different approach, namely correlations *between* collective observables. These approaches give different physics insights and are complementary to the VCS approach. The best and, to date, most exploited example is the $E(4_1^+) - E(2_1^+)$ correlation between the energy of the first 4^+ level and the first 2^+ level in even–even nuclei. This correlation has been generalized to higher spin states, to non-yrast excitations (e.g., the γ band energies),

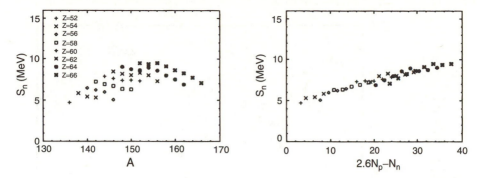

FIG. 7.15. Empirical S_n values $Z = 52$–66 plotted against A (left) and $2.6\,N_p - N_n$ (right) (based on Streletz, 1996).

to B(E2) values, and to similar observables in both odd-A and odd–odd nuclei. These correlations give insights into the structure of long sequences of nuclei, the evolution of structure from spherical to deformed shapes, and the concept of phase transitions in nuclei.

We start by showing the behavior of $E(4_1^+)$ against neutron number for all known even–even nuclei with $R_{4/2} > 2.05$ in the $Z = 38$–82 region in Fig. 7.16. The most obvious aspects of Fig. 7.16 are that the 4_1^+ energies change substantially across the plot, suggesting corresponding structural changes of some sort, and that there is large scatter in the points, even for a given neutron number. An important point to note for the subsequent discussion is that the data points are discrete, both vertically and horizontally. The horizontal discreteness simply comes from the fact that nuclei contain integer numbers of nucleons. The vertical discreteness reflects the changes in properties that occur with changes in the numbers of protons, even for a given neutron number.

Suppose that, now, instead of plotting these data against N (or Z or A) we plot the 4_1^+ energies against another collective observable, namely the 2_1^+ energy. Then, Fig. 7.17 results. This figure comprises about a third of the nuclear chart and yet an extraordinarily compact phenomenology prevails. There are two key features. First, along most of the trajectory the data for each specific element in Fig. 7.17 lie along a *straight line*. Secondly, the data for *each* element lie, to good accuracy, along the *same* trajectory. This trajectory is a straight line from high 2_1^+ energies down to the region around 100–150 keV, where there is a turnover to a steeper trajectory. As a result the data for the entire region all lie along a single, simple, compact trajectory.

This trajectory is clearly bi-linear with a very narrow transition region between the two linear segments. We show two solid lines in the figure, which are least squares fits to the slopes of the data in the two segments. Not only do the data unexpectedly follow a simple path, but the slopes of the two linear segments, 2.00 and 3.33, have simple physical interpretations. To understand the implications of this we recall the vibrator and rotor models of Chapter 6. The data in the linear segment with slope 2.00(2) in Fig. 7.17 satisfy the equation

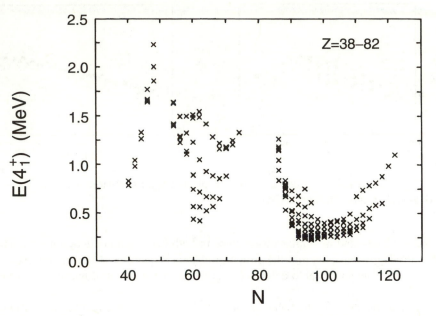

FIG. 7.16. $E(4_1^+)$ values for even–even nuclei with $R_{4/2} > 2.05$ from $Z = 38$–82.

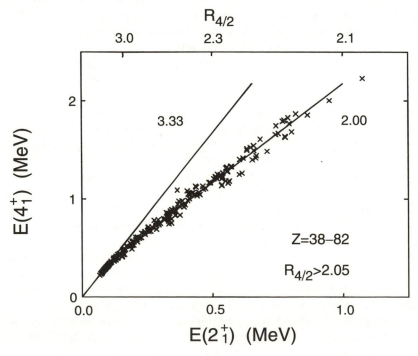

FIG. 7.17. Same data as in Fig. 7.16, here plotted against $E(2_1^+)$ (based on Casten, 1993).

$$E(4_1^+) = 2E(2_1^+) + \varepsilon_4 \tag{7.2}$$

where ε_4 is the intercept and takes on the best fit value $\varepsilon_4 = 161(8)$ keV. This is the equation of an Anharmonic Vibrator (AHV) where ε_4 is the anharmonity, which can be thought of as the phonon–phonon interaction energy. Obviously, Eq. 7.2 can be satisfied for any individual nucleus simply by choosing the needed value of ε_4. The absolutely remarkable feature, however, is the constancy of the anharmonity for such a wide span of nuclei–nuclei with 4^+ energies ranging from several hundred keV to nearly 2 MeV. Clearly these nuclei are not all spherical anharmonic vibrators (see Chapter 6). They span a variety of structures ranging from spherical vibrators to intermediate nuclei, γ-soft nuclei, and transitional nuclei near the edge of the deformed region. The $R_{4/2}$ values range from ~ 2.05 to 3.0 or slightly larger. Thus, although the internal microscopic structure of the phonon must be changing drastically across this segment of Fig. 7.17, somehow the interactions of the phonons remain unchanged.

We can test this hypothesis by writing the more general n-phonon AHV energy expression for the state of highest J for a given number of phonons:

$$E(n - \text{phonon}) = E(J = 2n) = nE(2_1^+) + \frac{n(n-1)}{2}\varepsilon_4 \tag{7.3}$$

The interpretation of ε_4 as a phonon–phonon interaction is now clear: an n-phonon system has $n(n-1)/2$ pairwise interactions of identical phonons. Eq. 7.3 for the 6^+ state gives

$$E(6_1^+) = 3E(2_1^+) + 3\varepsilon_4 \tag{7.4}$$

The corresponding data are shown in Fig. 7.18. Again a straight line results. And, indeed, a line of slope 3.00 and intercept $3\varepsilon_4 = 483$ keV fits the main trend quite well. With the addition of the next order anharmonic term to Eq. 7.3, the AHV gives an excellent reproduction of the data up to the highest spins of the ground band intrinsic configuration.

As we follow $E(2_1^+)$ to lower values in Fig. 7.17, we encounter an abrupt change in slope to the rotor value of 3.33. The slope of 3.33 for small $E(2_1^+)$ values is not in itself surprising: we have seen that these highly collective nuclei are good rotors. What is surprising is the rapidity of the change in structure from AHV to rotor.

How does one interpret these results? The most important point is that they challenge the traditional, and heretofore accepted, idea that structure changes smoothly with increasing valence nucleon number. Certainly there *is* a smooth change in structure (i.e., in phonon structure) as demonstrated by the changing 2_1^+ energies and the changing value of $R_{4/2}$ (see upper scale in Fig. 7.17). But, at the same time, a certain underlying constancy remains. The meaning of this is not yet fully understood. Yet, models such as the IBA and GCM naturally reproduce the behavior in Figs. 7.17, 7.18 (see below). At this point in time, it seems that the explanation lies in the idea that the minimum in the potential as a function of deformation β does not evolve gradually with valence nucleon number from zero to some finite value, but, at some point, changes abruptly from a potential corresponding to a spherical nucleus to one with stable deformed character. We will return to this point below and in Section 7.4.

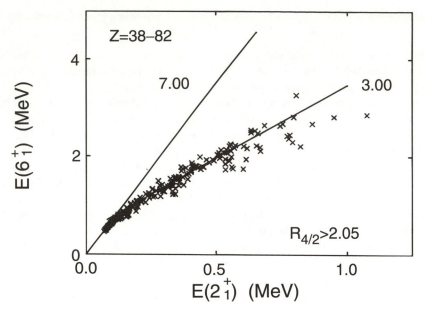

FIG. 7.18. Data for $E(6_1^+)$ for $Z = 38$–82 (based on Zamfir, 1994).

First, though, it is useful to discuss two other points, namely, the results of theoretical approaches to Fig. 7.17 and exploitation of the compact phenomenology, in its own right, to gain further insights. We do the latter first by again invoking the concepts of paradigm and paradigm shifts. With the smooth and compact trajectory of Fig. 7.17, we can shift our point of view from the zero of excitation energy to the line $E(4_1^+) = 2E(2_1^+)$; that is, the harmonic vibrator and plot, instead of $E(4_1^+)$, the *deviation* of $E(4_1^+)$ from $2E(2_1^+)$. That is, we plot the anharmonities ε_4 themselves. In Fig. 7.19, we show this for the entire plot in Fig. 7.17, not only the AHV section, and see immediately the advantages of such a paradigm. The data points now lie along a compact envelope, of width about ± 40 keV about the mean. This envelope is essentially flat in the AHV region and curves downward in the rotor region with decreasing $E(2_1^+)$.

This downturn itself is not surprising. One can transform the rotor equation, $E_{\mathrm{rotor}} \sim J(J+1)$, into the form of the AHV expression, namely, for $E(4_1^+)$,

$$E(4_1^+)_{\mathrm{rotor}} = 3.33E(2_1^+) = 2E(2_1^+) + (4/3)E(2_1^+) \qquad (7.5)$$

This is the same as Eq. 7.2 if we equate ε_4 and $(4/3)\,E(2_1^+)$. The difference with the results discussed in the context of Eq. 7.2 is that ε_4 *varies* with $E(2_1^+)$ *in the rotor region*. This corresponds to the downturn in Fig. 7.19 for small $E(2_1^+)$ values.

We have noted that one of the features of a paradigm is that it magnifies deviant behavior. Figure 7.19 is an ideal case in point. Over 90% of the data lie within the compact envelope of points sketched but a few nuclei clearly stand out. As it happens, these are not random nuclei but fall into three distinct classes. The points below the

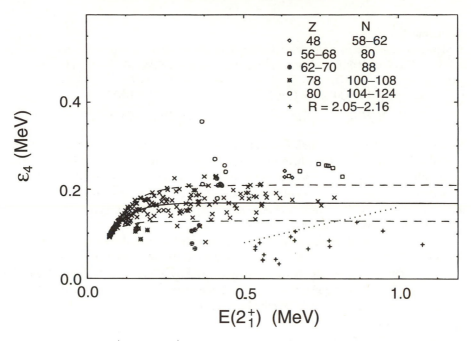

FIG. 7.19. $\varepsilon_4 = E(4_1^+) - 2E(2_1^+)$ for the nuclei of Fig. 7.17. The legend identifies the nuclei that fall outside the compact envelope given by the dashed lines (based on Jolos, 1996).

envelope at higher $E(2_1^+)$ are nearly harmonic vibrators with $2.05 \leq R_{4/2} \leq 2.16$. The points below the envelope at low $E(2_1^+)$ are nuclei in spherical-deformed transition regions ($A \sim 150$ and 190). The data points above the envelope all correspond to nuclei with two holes relative to a closed shell (e.g., Cd ($Z = 48$), Hg ($Z = 80$), $N = 80$, etc.)

Thus, from the simplest-to-obtain data, a plot like Fig. 7.19 can be used to identify (or at least suggest) nuclei in such categories. Clearly, in newly accessible region of exotic nuclei (those becoming accessible with radioactive beams) such clues will be invaluable. As we will see in Chapter 11, exotic nuclei far from the valley of stability may no longer be characterized by the same shell structure and magic numbers as we have discussed in Chapter 3. Magicity will turn out to be fragile, and not the robust concept long considered sacrosanct in nuclear physics. Hence, exploitation of plots such as Fig. 7.19 can give valuable clues to the underlying shell structure even in the absence of more extensive data.

It is worth noting explicitly the sensitivity at work here. Consider the Cd data points in Fig. 7.19 lying above the envelope in comparison to Fig. 6.6. Few if any readers would have noted anything unusual if we had mistakenly plotted the 4_1^+ energies in Cd 30–40 keV lower in Fig. 6.6. Yet, that would have been sufficient to bring the high Cd points in Fig. 7.19 within the envelope. Without the sensitivity enhancement brought about by the paradigm shift produced by the AHV, one would never have been able to spot this anomalous behavior.

The next question to address is whether standard models can reproduce the results in Fig. 7.17. To discuss this we consider both the IBA and GCM. We use the Hamiltonians

$$H = \varepsilon \mathbf{n}_d - \kappa Q \bullet Q \qquad \text{(IBA)} \qquad (7.6)$$

$$H = T + C_2' \beta^2 + C_4' \beta^4 + C_3' \beta^3 \cos 3\gamma \qquad \text{(GCM)} \qquad (7.7)$$

We are interested not in whether one can choose specific parameters to produce points along the line in Fig. 7.17 but whether, in some sense, these models unavoidably mandate the correlation in Fig. 7.17 as somehow natural. We will see that they do this, and more.

To show this we need to calculate a large variety of nuclei, spanning a full range of collective structures. We therefore choose random parameter combinations subject only to some very general constraints.

For the IBA, the parameters spanning the symmetry triangle are ε, κ, and χ. We also allow the boson number N_B to vary from 4–16. It turns out that, for constant κ, the IBA produces a set of data points that cluster tightly along a line of slope 2.0 in the $E(4_1^+) - E(2_1^+)$ plot. The quantity κ determines the intercept of the line. We therefore fix κ at an appropriate value and take hundreds of combinations of ε, χ, and N_B subject only to avoidance of the region very close to 0(6) [i.e., $\varepsilon = \chi = 0$]. Some of these random parameter sets will be appropriate to real nuclei, others not. Nevertheless, the results are remarkable, as shown on the left in Fig. 7.20. The IBA calculations all lie along a straight line of slope almost exactly 2.00 down to some 2_1^+ energy and then abruptly curve over to a line of slope 3.33. The agreement with the data is nearly perfect.

Note the rather strong and robust statement that this makes. Not only does the IBA reproduce the compact empirical trajectory but it *cannot do* otherwise. The IBA produces a slope (in the AHV region) of 2.0 and no other. Were the data different (e.g., were the slope, say, 1.6 or 2.5, or were there large fluctuations in the data) the IBA could not be forced to reproduce it without changing κ for each nucleus.

For the GCM we follow a similar procedure. We choose random values of C_2', C_3', and C_4' subject only to the limitation that $C_2' < -100$ MeV and that C_3' and C_4' take on values similar to the broad ranges shown in Figs. 6.53, 6.54. The results are shown in the right panel of Fig. 7.20 and show the same features as the data and the IBA—a slope of almost exactly 2.0 and a finite intercept. Once again, were the data very different, the GCM would be hard pressed to fit it.

We conclude that the observed phenomenology has a deep basis in the nature of collective structure, although its actual origin is not (yet) at all clear. We know that rather general collective models reproduce the empirical behavior: we do not know why. For the IBA, for example, it is not clear whether the key ingredient is the truncation of the shell model basis space or the specific types of interactions. The generality of the IBA results just discussed suggests that it may be the former.

In any case, further insight may be gained by considering the potential energy surface associated with the IBA (via the intrinsic state formalism), at least for a vibrator-to-rotor [U(5) towards SU(3)] transition region. We illustrate this in Fig. 7.21 as a function of β and ξ, where ξ is defined by re-writing Eq. 7.6 (up to an overall scale factor)

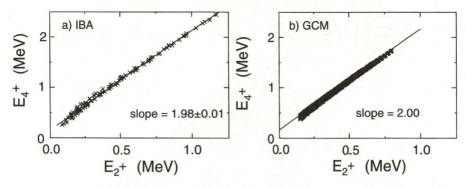

FIG. 7.20. IBA and GCM calculations of $E(4_1^+)$ against $E(2_1^+)$. Compare to the data in Fig. 7.17 (based on Zamfir, 1994 and 1999).

$$H = (1 - \xi)\mathbf{n}_d - \xi Q \bullet Q \qquad (7.8)$$

That is,

$$\varepsilon/\kappa = (1 - \xi)/\xi \qquad (7.9)$$

so that $\xi = 0$ for U(5) and $\xi = 1$ for SU(3). We see that the potential shows a minimum at $\beta = 0$ for a range of small ξ values and then a minimum at finite β for larger ξ values. There is *no* gradual migration of $\beta_{\min}$ from zero to larger values but a sudden jump at a critical value of ξ. This suggests a sudden jump in structure from vibrator to rotor, corresponding to the sharp kink in Fig. 7.17. Figure 7.21 was calculated for $N = 10$ and $\chi = -\sqrt{7}/2$; similar behavior results for other choices. We have actually seen this type of behavior in the Gd isotopes illustrated in Fig. 7.1b. We will return to this point in the next Section.

Finally, we note that the $E(4_1^+) - E(2_1^+)$ correlation can be extended (and generalized) to odd A nuclei by inspecting energy *differences*, $\Delta E_j(2)$ and $\Delta E_j(4)$, between the "bandhead" and the energies of states 2 and 4 units of angular momentum above that of the bandhead:

$$\Delta E_j(2) \equiv E(j + 2) - E(j) \qquad (7.10)$$
$$\Delta E_j(4) \equiv E(j + 4) - E(j) \qquad (7.11)$$

An example for non-rotational states of unique parity orbits is shown in Fig. 7.22. Again, a remarkably compact correlation, with slope 2.0, is observed. Studies of normal parity bands in odd mass nuclei, of excited bands, and of odd–odd nuclei all show exactly the analogous behavior. Again, this supports a basic origin in fundamental aspects of nuclear collectivity and structural evolution which is not yet understood.

7.4 Phase transitions in finite nuclei

In this section we exploit the correlations just discussed to address the questions of phase transitions in nuclei. [We refer here to phase transitions involving abrupt structural

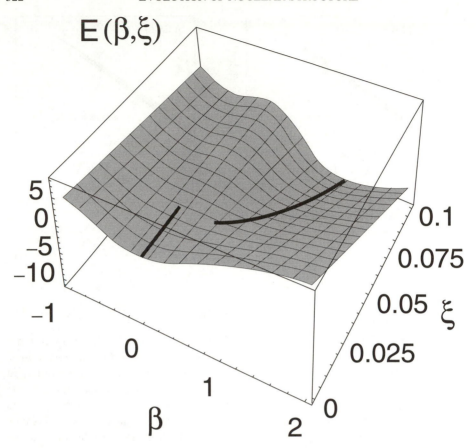

FIG. 7.21. The potential energy surface in the IBA as a function of ξ and β. The thick
    line denotes the β value at the minimum in the potential for each ξ (Casten, 1999).
    The author thanks his colleagues, D. Kuznezov and N. V. Zamfir for permission to
    use this figure and Fig. 7.25.

changes, such as spherical-deformed, at low energy, as a function of nucleon number.
We are not referring to phase transitions as a function of temperature or density such as
liquid-gas or confinement-deconfinement (quark gluon plasma) phase transitions.]

    We have discussed many times the changes in structure that occur with neutron and
proton number. Shape transition regions such as that near $N = 90$ in the Nd, Sm, and
Gd isotopes or the Os–Pt nuclei have been mentioned and have contributed key testing
grounds for nuclear models for four decades.

    One often encounters statements that nuclei exhibit phase transitions in these (and
other) regions. However, at the same time it is often said that finite systems cannot exhibit
real phase transitions in the sense of condensed matter systems and therefore that the
rapidity of structural changes will always be muted to a greater or lesser extent. Both

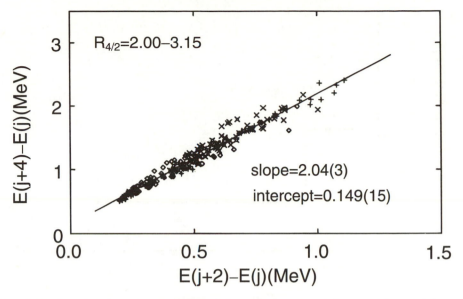

FIG. 7.22. $\Delta E_j(4)$ plotted against $\Delta E_j(2)$ for unique parity orbits in odd A nuclei (based on Bucurescu, 1996).

statements have an element of truth in them. We will see that nuclei in some regions do exhibit behavior very reminiscent of an actual phase transition but, of course, it is not completely abrupt. Indeed, the concept of abruptness itself in a region of nuclei differing by integer numbers of nucleons needs clarification.

In order to discuss these points, we need to understand the meaning of a phase transition and the conditions required for it. The essential elements, as sketched in Figure 7.23, are a control parameter (such as temperature) with a critical point (c) at which some other observable called an order parameter (e.g., the specific heat or the magnetization) changes abruptly. Of course, to identify such an abrupt change, the fluctuations in the order parameter must be small–that is, the data for the order parameter must follow a compact trajectory as a function of the control parameter. Depending on the nature of the phase transition, the derivative of the order parameter may be discontinuous. [Of course, in actual laboratory measurements with instruments of finite resolution and precision such discontinuities will be seen only as extremely sharp changes.] Also associated with a phase transition is the concept of phase coexistence (both phases of the system existing simultaneously as in an ice–water mixture), and a latent heat needed to transform the system from one phase to another.

The minimum prerequisites for a phase transition then are a suitable control parameter, minimal fluctuations in the order parameter, and (for a first order phase transition) a sharp change in the latter at some critical value of the control parameter.

In applying these ideas to nuclei we immediately encounter the problem that nuclei are composed of integer numbers of nucleons and adjacent nuclei differ by discrete changes in nucleon number. Observable properties also change discretely. This is illus-

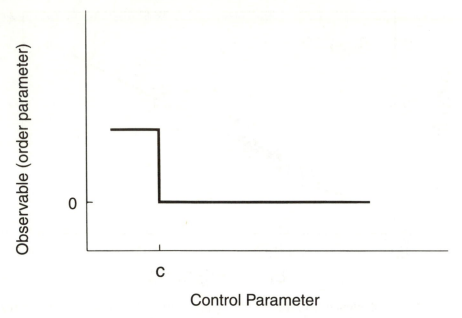

FIG. 7.23. Schematic illustration of the idea of a phase transition.

trated in the left panel of Fig. 7.24 which shows the 4^+_1 energies for the Sm isotopes against N. These nuclei undergo a well-known spherical-deformed shape transition at $N \sim 90$ and the figure clearly shows a structural change occurring with neutron number. However, the discrete nature of the abscissa makes it impossible even to talk about a phase transition, abruptness, or to define a derivative. All one is entitled to do is connect the points pair-wise by straight line segments. If we try to ameliorate this situation by combining data for an entire region as in the right panel of Fig. 7.24, the situation is worse. The abscissa, by definition, remains discrete and, in addition, the data for different elements are not superposed. Instead there are large fluctuations. These systematics offer evidence for a pattern of structural change but, even if the abscissa were continuous, the large fluctuations and the absence of any sharp change in behavior would preclude a discussion in terms of phase transitions.

There is, however, a way around this difficulty. Suppose instead of plotting the 4^+_1 energies against N, Z, or A that, inspired by the results of Section 7.3, we plot them against $E(2^+_1)$. We start on the upper left in Fig. 7.25 with the data for $Z = 56$. At first, there is no apparent improvement compared to the left panel of Fig. 7.24: The points are still at discrete abscissa values. However, the discreteness of the 2^+_1 energies is no longer a necessary consequence of integer nucleon number, but a consequence of structure. In principle, at least, any 2^+_1 energy is possible. We can see the implications of this if we again add data for other elements. To show the effects most clearly, we do this in a sequential way in Fig. 7.25, starting with $Z = 56$ in panel a), $Z = 56 + 58$ in b), 56–60 in c), and 56–62 in d). This figure should be read from left to right and then top to bottom

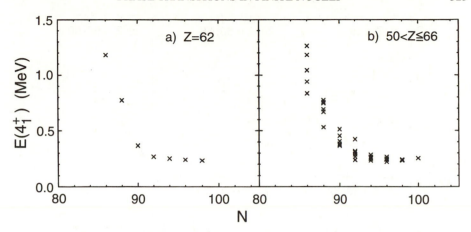

FIG. 7.24. $E(4_1^+)$ against neutron number. Left: for Sm. Right: for $Z = 50$–66, $N = 82$–126.

as the successive frames of a movie culminating in a figure like Fig. 7.17.

We see a qualitative change from the situation we encountered in Fig. 7.24. There are two key points: The data for different elements occur at *different* $E(2_1^+)$ values, thus gradually filling in a trajectory against $E(2_1^+)$, and, secondly, as in Fig. 7.17, the data for each element follow almost exactly the *same trajectory*. This means that the combined trajectory shows very small fluctuations. If we now add in further elements, namely $Z = 56$–74 in panel e) and, finally, the entire shell $Z = 50$–80 in panel f) we obtain a nearly continuous correlation with very little scatter.

By virtue of the nearly continuous nature of the plot, and the small fluctuations, we can now at least ask the "phase transition question" – is there a sharp change in $E(4_1^+)$ against $E(2_1^+)$? Of course, $E(2_1^+)$ is not really, rigorously, a control parameter; that is, a quantity like temperature that can be varied independently of the status of the system being studied. $E(2_1^+)$ and all other observables self-consistently result from the same Hamiltonian. Nevertheless, because of the smooth and compact behavior of Fig. 7.25f, we can treat $E(2_1^+)$ as an empirical proxy for a control parameter. From our earlier discussion, we already know that the 4_1^+ energies follow two linear segments, with slopes 2.00 and 3.33, linked by a very narrow transition region near $E(2_1^+) = 120$ keV. One can now fit an analytic expression to the data in Fig. 7.25f and differentiate it to obtain the *slope* of $E(4_1^+)$ against $E(2_1^+)$. The slope S is constant at 2.00 as $E(2_1^+)$ decreases and then, at a critical 2_1^+ energy about 120 keV, it suddenly rises toward the rotor value of 3.33. The derivative of the slope is as close to discontinuous at $E_{\text{crit}}(2_1^+)$ as one could hope for in finite nuclei.

In this view, the 2_1^+ energy serves the role of a control parameter, the slope is the order parameter (or, to make the order parameter vanish above the critical point, we can define it as S-2), and the transition is basically first order. Actually, a better and more physically intuitive order parameter is the deformation β. We have already seen, for example, that the IBA and the GCM fit the data in Figures 7.17 or 7.25. For the IBA, we showed in

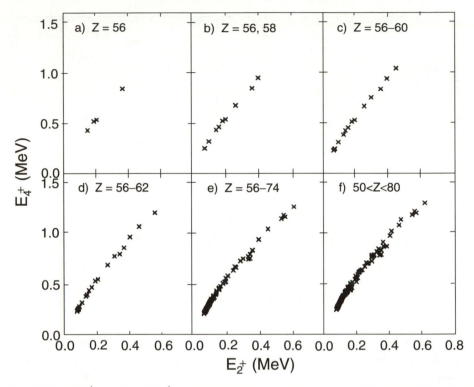

FIG. 7.25. $E(4_1^+)$ against $E(2_1^+)$ for successively larger numbers of elements. The lower right-hand panel resembles Fig. 7.17.

Fig. 7.21 that β jumps almost discontinuously in a vibrator-to-rotor transition region, as a function of the model parameters, from zero to a large finite value. [Note that the behavior is more gradual in a γ-soft to rotor transition region.] Since both $E(4_1^+)$ and the slope in Fig. 7.25f reflect the deformation, it is reasonable to think of β itself as the order parameter which jumps at $E_{\text{crit}}(2_1^+)$.

Of course, treating $E(2_1^+)$ as a quasi-control parameter is admittedly unconventional but it seems to produce an intuitive and very useful approach to the evolution of structure. Using $E(2_1^+)$ as a control parameter (instead of N, Z, or A) obviates the pesky integer nucleon number problem and enables a phase transitional analysis. The phase transition (spherical-deformed in this case) does not occur in a single nucleus, or even in a single element, but is the property of an entire region (ensemble) of nuclei. The actual critical point may or may not occur in any given nucleus.

Finally, we return to the issue of phase coexistence. If we have a phase transition, perhaps we have coexisting phases at the nuclear critical point. The best evidence for this seems to occur in the $A = 150$ region, just above $N = 82$, where the ground state configuration is spherical while an excited configuration is deformed. With increasing neutron number the energy of this deformed configuration drops until it becomes the ground state. In fact, the concept of a first order phase transition is intimately related to the idea of a level crossing of two configurations.

Relating this to the above picture, the crossing point of these two configurations should occur when $E(2_1^+)$ is at the critical point. There is no guarantee that a particular nucleus happens to exist near the critical point. As it happens, however, one does, namely ^{152}Sm, for which $E(2_1^+) = 121$ keV. The details of the arguments for coexistence are too extensive to repeat here but an indication of the key points may be enlightening. The experimental levels of ^{152}Sm are shown on the left in Fig. 7.26. One sees, in fact, a nearly deformed ground state band, with $R_{4/2} = 3.01$ and an excited sequence built on the 0_2^+ level that resembles an anharmonic vibrator, with $R_{4/2}(0_2^+) = 2.68$. Figure 7.26 shows an anharmonic vibrator fit to the 0_2^+-based levels on the right. Comparison with the data shows good empirical candidates for a one-phonon 2^+ state (2_2^+), a 2-phonon triplet ($4_2^+, 2_3^+, 0_3^+$) and for some members of higher multiplets. Many B(E2) values support this picture. Indeed, it was the near-vanishing of the $2_3^+ \rightarrow 0_2^+$ transition ($\Delta N_{ph} = 2$ in this scenario) that led to the suggestion [Iachello, 1998] of phase coexistence in the first place.

Both the IBA and the GCM can reproduce these data. We illustrate this, for the IBA, in Fig. 7.27 which is calculated with the parameters $\chi = -\sqrt{7}/2$ and $\varepsilon/\kappa = 30$. GCM calculations give almost identical predictions. We also show the pure vibrator predictions for the low lying 0_2^+-based levels in the middle panel of Fig. 7.27. While the vibrator description accounts well for many properties of ^{152}Sm there are also serious difficulties with it. Specifically, some phonon-allowed transitions are nearly an order of magnitude weaker than the vibrator predictions.

Overall, the IBA predictions are in excellent agreement with the data and, interestingly, also show strong (albeit not strong enough) reductions from the vibrator B(E2) values for transitions from the 2_3^+ and 4_3^+ levels. The IBA wave functions themselves explicitly reflect the coexistence picture. This is seen in Fig. 7.28 where we show the d-boson probability distribution of the calculated IBA wave functions. We see, for the yrast levels, $0_1^+, 2_1^+$, and 4_1^+, a broad distribution in n_d typical of deformed or SU(3) wave functions in the IBA. However, for the $0_2^+, 2_2^+, 4_2^+$ and 2_3^+ levels we see a very high probability for the particular n_d values appropriate to the vibrational picture: $n_d = 0$ for the 0_2^+ level, $n_d = 1$ for 2_2^+, and $n_d = 2$ for 4_2^+ and 2_3^+.

This kind of phase coexistence is related to the shape coexistence exhibited by intruder states that we discussed for nuclei such as the Sr, Cd, and Hg isotopes. As we discussed earlier, in both cases the proton–neutron interaction, particularly the monopole component, plays a key role in the structural evolution. Both cases involve particle–hole excitations across a shell gap in a spherical nucleus which produce an excited deformed configuration. It is the relative size of the gap and the strength of the attractive p–n interaction that determines whether the deformed states will descend far enough in energy to become the deformed equilibrium ground state configuration or only far enough to appear as low lying intruder states. In the intruder case, the excitation is cross-major-shell. In the shape transition case, the structural development takes place within a single major shell and the particle–hole excitation is across a sub-shell gap.

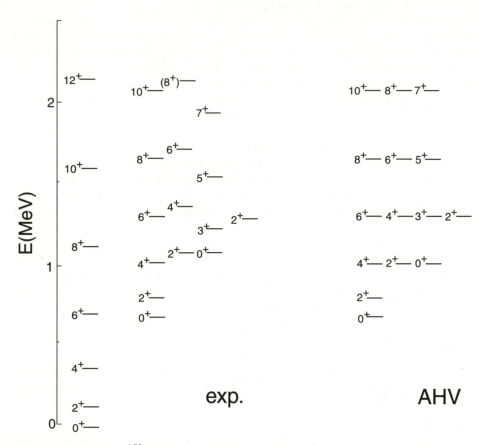

FIG. 7.26. Levels of ^{152}Sm. For the levels above the 0^+_2 state, a comparison with the anharmonic vibrator is shown on the right.

Very recently, a potentially major breakthrough has occurred which augers well for our understanding of phase transitions and critical point nuclei. To date, symmetries in nuclear structure (see Chapter 6) have pertained to idealized limits of structure, such as the vibrator and rotor (vertices in Fig. 6.41). Yet the most interesting and challenging nuclei are those in phase transitional regions where structure is changing most rapidly with N and Z and where there are competing degrees of freedom. Predictions for such transition regions have emerged traditionally only from complex numerical calculations.

Recently, however, a new class of symmetry, based on an analytic treatment of special potentials, has been developed which is applicable specifically to nuclei at the critical point of a phase transition. Two specific symmetries, E(5) and X(5), correspond to vibrator to γ-soft rotor and vibrator to axial rotor transition regions, respectively (Iachello,

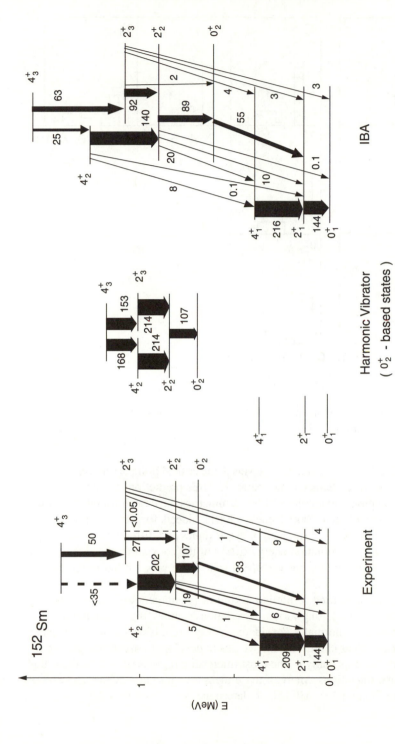

Fig. 7.27. Comparison of B(E2) values and level energies for ^{152}Sm with harmonic vibrator and IBA predictions. The numbers on the transition arrows are B(E2) values in W.u. Several result from recent measurements [Zamfir, 1999 on which this figure is based, as well as Klug, 2000] and are quite different than previous literature values (e.g., the 107 W.u. B(E2: $2_2^+ \rightarrow 0_2^+$) value replaces the previously accepted value of 520 W.u. and the B(E2: $4_2^+ \rightarrow 2_2^+$) value of 202 W.u. replaces ~ 400 W.u.). The ^{152}Sm level scheme has also recently been described in terms of a new type of symmetry, for critical points, called X(5), whose parameter-free predictions closely ressemble the data and the IBA.

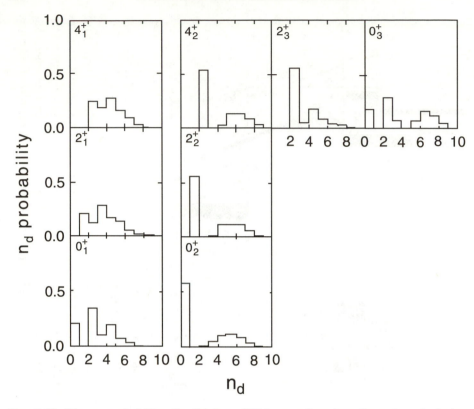

FIG. 7.28. The n_d probability distribution of IBA wave functions from the calculations shown in Fig. 7.27 (Zamfir, 1999).

2000; Iachello, 2001). In these symmetries, levels at the critical point are assigned quantum numbers, and there are characteristic, parameter free predictions for energy ratios and B(E2) values. Empirical examples of both symmetries have been found, [134]Ba for E(5) and [152]Sm for X(5) (Casten, 2000; Iachello, 2001). Indeed, for X(5), the predictions for the yrast and yrare levels of [152]Sm, including the values and differences in $R_{4/2}(0_1^+)$ and $R_{4/2}(0_2^+)$, and even the excitation energy $E(0_2^+)$ of the coexisting phase relative to $E(2_1^+)$, as well as intra- and intersequence B(E2) values, are very similar to the data in Fig. 7.27.

In this chapter, we have tried to outline a kind of phenomenological approach to the evolution of structure, exploiting various correlation schemes, that substantially simplifies our view of what is normally considered a very complex variation of nuclear properties across the nuclear chart. This allowed us to develop a unified perspective on structural evolution, and especially of the most interesting regions of nuclei, those undergoing phase/shape transitions. Many of these ideas are quite new. Time, and access to exotic nuclei (see Chapter 11), will help to determine how useful they are.

8

THE DEFORMED SHELL MODEL OR NILSSON MODEL

The purpose of this chapter is to describe the basic single-particle model applicable to nearly all deformed nuclei—the Nilsson model. This model is surely one of the most successful nuclear models ever developed. It accounts for most of the observed features of single-particle levels in hundreds of deformed nuclei and is always the first model turned to when new experimental information on such levels is obtained. It also provides a microscopic basis for the existence of rotational and vibrational collective motion that is directly linked to the spherical shell model. It is also very easy to incorporate extensions, refinements, and corrections to it. Essentially a single-particle model, the Nilsson model has enjoyed particular success in the interpretation of single nucleon transfer reactions.

Even before discussing this model, we are faced with a conceptual difficulty arising from the nonspherical shape, or the separation of the motion of an individual nucleon around the nucleus from rotations of the nucleus itself in space. These motions can be very different. Imagine a nucleus with prolate quadrupole distortion and a single nucleon orbiting in an equatorial plane, as shown by orbit K_1 in Fig. 8.1. Now imagine that this nucleus can rotate about an axis perpendicular to the symmetry axis. With rapid rotation about this axis, the time averaged shape of the core becomes oblate (disc-like). What shape the orbit of the single nucleon takes then depends on the extent to which its motion is coupled to that of the core, that is, it depends on the separation of rotational and single-particle degrees of freedom. A rigorous separation is, in general, not possible. An approximate separation can be made, however, if the frequency of the nucleonic motion is much larger than the frequency of the nuclear rotation, in which case the individual nucleon executes many orbits during a single nuclear rotation, or, alternately phrased, the nucleus is essentially stationary during a single orbit of that nucleon.

This discussion is not only of formal interest, but alerts us to the possibility that the separability of these motions may be rather poor for extremely high rotational velocities. Modern experimental techniques have approached the limit where the characteristic frequencies for rotational and single nucleon motion are not distinct. It is then necessary to explicitly incorporate the effects of the rotation on the single-particle motion. Coriolis mixing is one such effect that we shall discuss at length, but there are others (relating to the underlying core shape, the effective single-particle energies, and so on) that are beyond the scope of this book. Mathematically, the separation of single-particle and rotational motion greatly simplifies calculations and is the principal reason why in a body-fixed frame of reference, one evaluates the single nucleon motion first and later superimposes the rotational motion. (Incidentally, the same basic type of problem applies to the spherical shell model, in which the "global" motion is linear motion of the center of mass, which is more easily distinguished.)

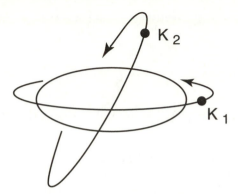

FIG. 8.1. Illustration of two single-particle orbits at different inclinations to a prolate deformed nucleus.

8.1 The Nilsson model

We will discuss in considerable detail the quantitative properties of the Nilsson model and in particular the Nilsson energies and wave functions. First, however, we wish to proceed in a rather unconventional way in order to show that it is simple to "derive," without detailed calculation, the entire Nilsson diagram and many of the prominent features of the Nilsson wave functions. The two necessary ingredients are simply a choice of deformed single-particle potential, such as a shell model potential with quadrupole deformation β, and the recognition that the nuclear force is short range and attractive.

Consider first, then, a valence nucleon in a single j orbit in such a prolate deformed potential (Fig. 8.1). It will have lower energy if its orbit lies closer to the rest of the nuclear matter than if it lies at larger distances from it. Clearly, then, the orbit labeled K_1 will be lower in energy than K_2. The energy depends on the *orientation* with respect to the nuclear symmetry axis. This is contrary to the spherical shell model where there is no preferred direction in space. One can specify this orientation by the magnetic substate of the nucleon—that is, the projection of the total angular momentum on the symmetry axis as shown in Fig. 8.2. As in our earlier discussion of even–even nuclei, this quantity is usually denoted by the symbol K. (Technically, Ω is used for the projection of the single-particle angular momentum on the symmetry axis and K for the projection of the total angular momentum. However, since the rotational angular momentum of axially symmetric nuclei is perpendicular to the symmetry axis, it contributes nothing to K and therefore $K = \Omega$ and is often substituted for it.) The low K values correspond to equatorial motion near the bulk of the nuclear matter for a prolate quadrupole distortion, and have lower energy.

One can easily go a step further. Consider the classical orbit angles corresponding to different K values. Suppose $j = 13/2$ (e.g., the $i_{13/2}$ orbit) with $K = 1/2, \dots, 13/2$. Classically, as illustrated in Fig. 8.2, we can approximate the angle of an orbital plane by $\theta = \sin^{-1}(K/j)$. These angles are given in Table 8.1. The interesting feature is that θ changes slowly for low K values and rapidly for high K values. Therefore one expects that the difference in energy between low K values is rather slight and increases rapidly

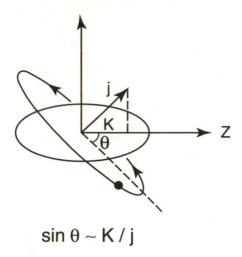

$$\sin\theta \sim K/j$$

FIG. 8.2. Diagram defining the quantities j, K, and θ in the discussion of the Nilsson model.

Table 8.1 *Classical orbit angles, relative to the nuclear equator, for $j = 13/2$.*

K	1/2		3/2		5/2		7/2		9/2		11/2		13/2
θ(deg)	4.4		13.3		22.6		32.6		43.8		57.8		90
$\Delta\theta$(deg)		8.9		9.3		10.0		11.2		14.0		32.2	

for the higher ones. With these simple considerations, we can now develop a limited region of a Nilsson diagram for a single j. This is shown in Fig. 8.3. The characteristic features are just as we have derived: for $\beta > 0$, the energy drops rapidly with β for low K values and rises rapidly for the higher K values and the separation of adjacent K values increases sharply with K. We shall see later that the energies for small deformations depend on K^2.

The only additional step needed to construct the full Nilsson diagram of deformed single-particle energies as a function of β is to combine several j values. As discussed earlier, the characteristic feature of a deformed field is single-nucleon configuration mixing. Therefore, we must now superimpose this configuration *mixing* of different j values on the K *splitting* just considered. Recalling a fundamental rule of quantum mechanics that no two levels with the same quantum numbers may cross (an infinitesimal interaction will cause them to repel when they get sufficiently close) and noting that the only remaining good quantum number for these orbits is K, it then follows that no two lines in the Nilsson diagram corresponding to the same K value (and parity) cross. As two such lines approach each other they must repel (see Fig. 1.9). Thus, it is now possible to incorporate several j values into the Nilsson diagram and to extend

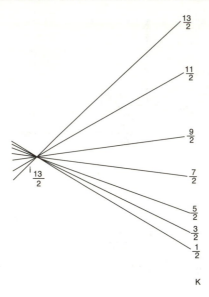

FIG. 8.3. Variation of single-particle energies of $i_{13/2}$ orbits with different projections K (orientations θ) as a function of deformation ($\beta > 0$, prolate, to the right).

it to realistic deformations where the energies of different orbits intermingle. This is shown in Fig. 8.4, which gives the Nilsson diagram for two different regions. Each line, representing a Nilsson state, starts out straight and is downward or upward sloping according to the angle of the orbit relative to the main mass of the nucleus. It only starts to curve when it approaches another level with the same K and parity. The entire structure of the diagram relies thus on only three factors: K splitting (resulting from the effects of a short-range nuclear interaction in a deformed field), level–level repulsion, and the input single-particle shell model energies.

The Nilsson wave functions are equally easy to deduce qualitatively, even though they are a complex result of a multistate diagonalization. (We ignore here the phases of the various terms in these wave functions, most of which can only be obtained by explicit diagonalization.) The interaction that leads to configuration mixing in the Nilsson model is of quadrupole form. (We will discuss other deformed shapes later.) For very small quadrupole deformations β, the nuclear wave functions must be nearly pure in j; as the deformation increases, the configuration mixing will increase.

The nondiagonal mixing matrix elements of the quadrupole interaction, not surprisingly, tend to mix configurations that differ by two units in angular momentum and in which the nucleon spin orientation is not changed. For example, in the 50–82 shell, the $d_{5/2}$ and $s_{1/2}$ orbits have large matrix elements and mix substantially, even though they are slightly further separated than the $d_{5/2}$ and $d_{3/2}$ orbits. The $g_{7/2}$ and $d_{3/2}$ mix more than $g_{7/2}$ and $d_{5/2}$ do. Likewise, in the 82–126 shell, the quadrupole matrix element between the $p_{3/2}$ and $f_{7/2}$ orbits is strong. However, the closeness of the energies of the $f_{7/2}$ and $h_{9/2}$ orbits leads to substantial mixing, even though the matrix element is not favored. Therefore, combining a regard for the energy separations of different shell model orbits and the most important quadrupole mixing matrix elements, one can estimate the Nilsson wave functions, that is, the composition of the wave functions in terms of amplitudes for different j subshells.

Consider the example of the 82–126 neutron shell shown in Fig. 8.4b. As β increases, the $f_{7/2}$ and $h_{9/2}$ orbits begin to mix. We recall that the angle of the orbital orientation depends primarily on the ratio K/j ($\theta \approx \sin^{-1} K/j \approx K/j$ for small K). Small angles can occur *either* because K is low, or for given K, because j is high. Thus, the energies of the $K = 1/2, 3/2$, and $5/2$ orbits from the $h_{9/2}$ shell decrease in energy faster with deformation than those from the $f_{7/2}$ orbit. This difference in rate of decrease of the Nilsson energies with deformation can overcome the small spherical $f_{7/2}$–$h_{9/2}$ energy separation. The low K $f_{7/2}$ and $h_{9/2}$ orbits therefore approach each other, mixing more and more. However, the two orbits cannot cross and so repel each other, leading to an inflection point at the value of β where they would have crossed. This effect is very clear for the $K = 5/2$ and $K = 7/2$ pairs of $f_{7/2}$ and $h_{9/2}$ orbits in Fig. 8.4b.

An interesting feature of the Nilsson diagram is apparent if one looks at the energies past the "pseudo crossing." Starting at large deformations and tracing back toward $\beta = 0$ the energy of the lowest $K = 5/2$ orbit is drawn *as if* it stems from the $f_{7/2}$ shell. However, one sees that it actually points directly back to the $h_{9/2}$ spherical energy. This reflects the fact that this orbit, for large deformations, is actually the continuation of the $h_{9/2}$ shell. In effect, while the energies do not cross, the wave functions do "exchange" principal wave function components near the inflection point.

In contrast to these examples, the $K = 9/2$ orbit from the $h_{9/2}$ shell is virtually straight since there is no other nearby (negative parity) j shell with a $K = 9/2$ component. This reflects the general feature that the Nilsson wave functions for the highest K values in a given shell are very pure.

The extreme example of this, in the 82–126 shell, is the orbits stemming from the $i_{13/2}$ orbit. This orbital, with $l = 6$, has positive parity and lies amidst a grouping of negative parity orbits. It has been brought down from the next major shell by the strong spin-orbit interaction in the shell model. Having opposite parity it cannot mix with any other orbits in the 82–126 shell. Therefore, the wave functions of these *unique parity* Nilsson orbits are extremely pure, consisting almost solely of $j = 13/2$ components even to rather large deformations. This special structure has many extremely important consequences. The simplicity of their wave functions makes a number of physical effects particularly simple to understand. Moreover, certain residual interactions such as Coriolis effects are both simple and particularly strong in these orbits. Finally, many of these same features make the states stemming from these orbits easily amenable to empirical study.

This discussion of the structure of the Nilsson diagram for a given major shell can be applied to any shell and the entire Nilsson diagram can be constructed. The only other point to note is that the higher the shell, the stronger the effects we have been considering will be, since a particle in a higher shell is at a larger radius, further outside the spherical nucleus, and therefore has more to gain energetically, upon deformation, if it is equatorial.

It should be evident by now that it is easy to write down not only the approximate

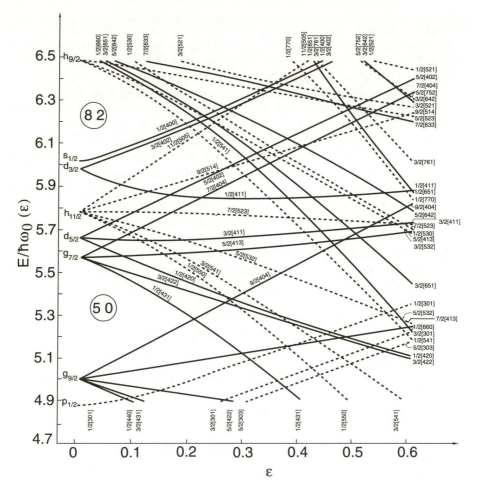

FIG. 8.4. (a) Nilsson diagram for the $Z = 50–82$ region. The abscissa is the deformation parameter ε, which is nearly the same as β. (Redrawn from Gustafson, 1967). The units for the ordinate, $\hbar\omega_0$, are given approximately by $41\,A^{-1/3}$ MeV. Hence, a value of 6, for $A = 150$, corresponds to 46.3 MeV.

energies but also estimates of the wave functions of almost any Nilsson orbit without detailed calculation. We shall give a couple of examples of this in a moment, but first it is convenient to define the Nilsson quantum numbers labeling each orbit. This is also instructive because it highlights the physical nature of the various orbits. A typical Nilsson orbit is labeled as follows:

$$K^\pi[Nn_z\Lambda]$$

The first quantum numbers give the K value and parity. Inside the brackets the three quantum numbers are N, the principal quantum number denoting the major shell; n_z,

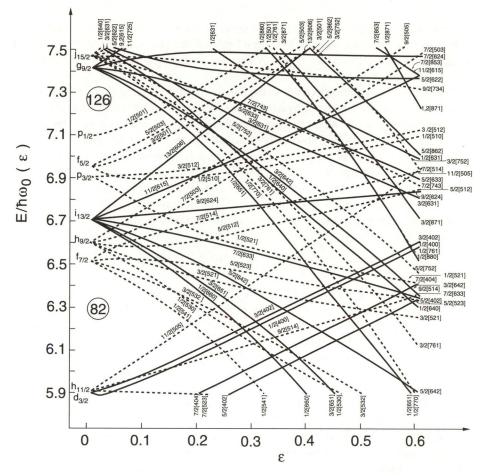

FIG. 8.4. (b) Nilsson diagram for the neutron shell 82–126. The abscissa is the deformation parameter ε, which is nearly the same as β. (Redrawn from Gustafson, 1967).

the number of nodes in the wave function in the z direction (n_z is particularly critical for
understanding the structure of the wave function); and Λ, the component of the orbital
angular momentum along the z, or symmetry, axis. By definition, $K = \Lambda + \Sigma = \Lambda \pm 1/2$,
where Σ is the projection of the intrinsic nucleon spin on the symmetry axis. Hence, one
sometimes sees an alternate notation $[Nn_z\Lambda\uparrow]$ or $[Nn_z\Lambda\downarrow]$, where the arrow replaces K
and indicates whether the spin angular momentum aligns ($\uparrow$) or antialigns ($\downarrow$) with the
orbital angular momentum. The two notations are equivalent, but using K is somewhat
more convenient and common.

For the most common case of a prolate nucleus, equatorial orbits are nearest the
nuclear matter and lie lowest. It is clear from Fig. 8.1 that these orbits are also those
in which the nucleon wave function is most extended in the z direction. Their wave

functions have the largest number of nodes in the z direction and hence the largest values of n_z. We note also that there is a relation between the permissible values of n_z and Λ such that their sum must be even if N is even (positive parity) and odd if N is odd (negative parity).

We are now in a position to label all the Nilsson orbits in a given shell. There are two ways this can be approached, one by labeling the sequence of orbits from each spherical j shell and another in which all orbits of a given K value are labeled according to increasing energy. To make the labeling absolutely clear, we will illustrate both approaches (for a prolate nucleus). We start from the fact that the lowest-lying orbit has the highest possible n_z. Clearly the maximum value of n_z for a given principal quantum number N is $n_z = N$. The lowest orbit for a given N has $K = 1/2$, since this is the most equatorial orbit. Taking $N = 5$ for illustration, noting that if $K = 1/2$, Λ can only be 0 or 1, and that $n_z + \Lambda$ must be odd since N is odd, the Nilsson quantum numbers for the lowest $N = 5$ orbit must be

$$K^{\pi}[Nn_z\Lambda] = 1/2^-[550]$$

This Nilsson labeling describes the $K = 1/2$ orbit stemming from the $h_{11/2}$ shell. Although most of the $N = 5$ orbits occur in the 82–126 major shell, the $h_{11/2}$ orbit is the one that is pushed down by the spin orbit interaction into the lower, predominantly $N = 4$, 50–82 shell. Continuing for the other orbits from the $h_{11/2}$ shell, the next has $K = 3/2$. Since its orbital orientation is slightly more inclined away from the equatorial plane, it is less extended in the z direction and has the next lower n_z value. Its Nilsson quantum numbers are $3/2^-[541]$. Again, the Λ value is fixed by the requirements that $n_z + \Lambda$ is odd and $\Lambda = K \pm 1/2$. The rest of the $h_{11/2}$-based orbits are then $5/2^-[532]$, $7/2^-[523]$, $9/2^-[514]$, $11/2^-[505]$.

The next $N = 5$ orbits are in the 82–126 shell proper, and stem from the $f_{7/2}$ parent (although their actual wave functions contain large $h_{9/2}$ amplitudes). The $K = 1/2$ orbit, having higher energy than the $K = 1/2$ orbit from the $h_{11/2}$ shell, must be less extended in the z direction and must have a lower n_z. Its Nilsson quantum numbers are then trivially, $1/2^-[541]$. The $K = 3/2$ orbit is $3/2^-[532]$. One can continue filling the entire shell in this way, and it is easy to reproduce the labels shown in Fig. 8.4.

An alternative way is to proceed not by energy for a given parent j shell but by K value. For example, the sequence of $K = 1/2$ orbits, starting with that from the $h_{11/2}$ orbit, will be (recall that $\Lambda = 0$ or 1)

$$1/2^-[550],\ 1/2^-[541],\ 1/2^-[530],\ 1/2^-[521],\ 1/2^-[510],\ \text{and}\ 1/2^-[501]$$

The unique parity orbit in the 82–126 shell is the lowest $N = 6$ orbit, and therefore its Nilsson quantum numbers must be $1/2^+[660]$, $3/2^+[651]$, $\ldots$, $13/2^+[606]$.

Let us now consider the wave functions in more detail. These wave functions can be written in many forms. Because they involve single-j configuration mixing, they can be expanded in a spherical basis. For many purposes the easiest and most physically transparent form is one that the previous discussion anticipated, an expansion in shell model orbits specified by their j values. Thus, we write

$$\psi_{\text{Nils}_i} = \sum_j C_j^i \phi_j \tag{8.1}$$

where the ϕ_j are solutions to the spherical independent particle model and the C_j^i are configuration mixing coefficients. Using this language, it is easy to make at least crude estimates of the actual Nilsson wave functions. For example, for $\beta = 0.23$ ($\varepsilon \approx 0.95\,\beta \approx 0.22$) the $5/2^-[523]$ orbit will have a wave function that its predominantly $h_{9/2}$, with the next largest component $f_{7/2}$. Crudely, we can estimate that $\psi(5/2^-[523]) \approx 0.5\phi_{7/2} + 0.8\phi_{9/2} + \ldots$. We emphasize that this is at best a *guess* as to the Nilsson wave functions and that the phases are arbitrary. We engaged in this exercise simply because it is useful to have at least *some*, albeit crude, a priori feeling for the structure of these wave functions. It will also remove some of the mystery from the actual Nilsson wave functions. To take another example, consider that the $3/2^-[532]$ orbit for a deformation of $\beta \approx 0.23$ is just past the inflection point with the $3/2^-[521]$ orbit. Its wave function should be roughly equal admixtures of $f_{7/2}$ and $h_{9/2}$ components. In contrast, the wave function for the same orbit for a deformation of $\beta \approx 0.05$ would be largely $f_{7/2}$.

The wave functions near the top of the 82–126 shell are particularly simple: $\psi(1/2^-[501])$ is dominated by a $p_{1/2}$ component. Near midshell, the situation is somewhat more complicated since a given wave function will contain components from j shells both above and below it. A particularly nice example is the $1/2^-[521]$ orbit. Careful inspection of the Nilsson diagram shows that although it starts out from the $p_{3/2}$ shell, it soon mixes with the $K = 1/2\ f_{5/2}$ orbit, undergoing a virtual crossing before $\beta \approx 0.1$. At this point its wave function is a very thorough mixture of $\phi_{3/2}$ and $\phi_{5/2}$ components. Continuing to larger deformations, at $\beta \approx 0.2$ there is another inflection point due to interaction with a combination of the $K = 1/2$ orbits from the $f_{7/2}$ and $h_{9/2}$ shells. Therefore, we might expect at $\beta \approx 0.25$ that $\psi(1/2^-[521]) \approx C_{3/2}\phi_{3/2} + C_{5/2}\phi_{5/2} + C_{7/2}\phi_{7/2} + C_{9/2}\phi_{9/2}$ with all four C_j values substantial in magnitude.

As we anticipated, the unique parity orbits are extremely pure, and increasingly so as K increases. For example, the $13/2^+[606]$ orbit *must* (assuming no N mixing) be pure $j = 13/2$, while $\psi(1/2^+[660]) \approx 0.95\,\phi_{13/2} + 0.3\,\phi_{9/2}\ldots$. (These numbers are only rough estimates but do embody the physics and reflect the actual amplitudes from real calculations.) Note that the only other significant contribution besides $i_{13/2}$ comes from the $g_{9/2}$ orbit, which differs by two units of angular momentum and has its spin and orbital components aligned in the same way as the $i_{13/2}$.

Table 8.2 gives examples of some actual Nilsson wave functions for a typical Nilsson potential ($\delta = 0.22$, $\beta \approx 0.23$). Inspection of the table shows that all our guesses as to the structure are semiquantitatively correct. Of course, fine details are beyond this discussion and some amplitudes are more difficult to intuit a priori.

It is worth pausing here to reflect on and to re-emphasize what we have done. Without any detailed calculation whatsoever, using only simple considerations of the attractive nature of the nuclear force and the nuclear shapes involved, we have essentially "derived" the entire Nilsson diagram, the Nilsson energies, and the asymptotic Nilsson quantum numbers. We have also discussed the basic structure of the wave functions.

Table 8.2 *Nilsson wave functions (C_j coefficients) for some $N = 5$ orbits*

$K^\pi[Nn_z\Lambda]$	j					
	1/2	3/2	5/2	7/2	9/2	11/2
$3/2^-[532]$		0.234	0.369	−0.560	−0.651	0.268
$5/2^-[523]$			0.237	−0.472	−0.826	−0.196
$7/2^-[514]$				0.323	0.938	0.128
$1/2^-[521]$	−0.510	0.345	0.473	0.431	0.444	0.120
$5/2^-[512]$			−0.023	0.836	−0.515	0.157
$1/2^-[510]$	0.021	−0.676	0.586	−0.343	0.277	0.067
$3/2^-[512]$		0.379	0.815	0.283	0.327	0.063
$7/2^-[503]$				0.937	−0.336	0.099
$9/2^-[505]$					0.998	0.071
$1/2^-[501]$	−0.821	−0.361	−0.411	−0.122	−0.104	−0.019

*$\delta = 0.22$, $\kappa = 0.0637$, $\mu = 0.42$.

At this point, however, we can obtain a deeper understanding of the Nilsson model and diagram and of the role of the quantum number, n_z, by a slightly more formal approach.

To begin, we consider the Nilsson Hamiltonian for a single-particle orbiting in a deformed potential and inspect two instructive limits, corresponding to small and large deformation. Actually there are many Nilsson-type Hamiltonians incorporating many variants of the single-particle deformed potential. Various authors have used deformed harmonic oscillator or modified harmonic oscillator potentials, Wood–Saxon potentials, and others. The differences reside primarily in details that do not concern us here so we will content ourselves with the modified harmonic oscillator originally used by Nilsson.

The Nilsson model is a shell model for a deformed nucleus. It provides a description of single particle motion in a nonspherical potential, $V = V_0(r) + V_2(r)P_2(\cos\theta)$. The original and basic form incorporated only quadrupole deformed axially symmetric shapes. An appropriate single-particle Hamiltonian for a nucleus with symmetry axis z is:

$$H = T + V$$
$$= \frac{\mathbf{p}^2}{2m} + \frac{1}{2}m\left[\omega_x^2\left(x^2 + y^2\right) + \omega_z^2 z^2\right] + C\mathbf{l} \bullet \mathbf{s} + D\mathbf{l}^2 \tag{8.2}$$

where ω_x, ω_y, and ω_z are one-dimensional oscillator frequencies in the x, y, and z directions. This Hamiltonian gives the eigenvalue equation $H\psi_i = E_i\psi_i$, where ψ_i is a Nilsson wave function written in the form $\psi_i = \Sigma_j C_j^i \phi_j$. The $\mathbf{l}^2$ and $\mathbf{l} \bullet \mathbf{s}$ terms ensure the proper order and energies of the single-particle levels in the spherical limit ($\beta = 0$).

Although the form of the Hamiltonian in Eq. 8.2 is useful, it is also convenient to introduce an alternative version written directly in terms of a nuclear deformation parameter $\delta \approx 3/2\sqrt{5/4\pi}\,\beta \approx 0.95\beta$. To do this, one writes the frequencies as

$$\omega_x^2 = \omega_y^2 = \omega_0^2\left(1 + \frac{2}{3}\delta\right)$$

$$\omega_z^2 = \omega_0^2\left(1 - \frac{4}{3}\delta\right) \tag{8.3}$$

where ω_0 is the oscillator frequency ($\hbar\omega_0 = 41A^{-1/3}$) in the spherical potential with $\delta = 0$. It is assumed that the nuclear volume remains constant as a function of ω_0. Therefore, one has the condition $\omega_x\omega_y\omega_z$ constant or

$$\omega_0 = \left(1 - \frac{4}{3}\delta^2 - \frac{16}{27}\delta^3\right)^{-\frac{1}{6}} = \text{constant} \tag{8.4}$$

For positive deformations δ or $\beta > 0$ (prolate shapes), ω_z decreases with increasing deformation while ω_x and ω_y increase. This is physically reasonable since an increasing prolate deformation elongates the nucleus in the z direction. This increases the "length" of a circumferential route and therefore lowers the frequency of orbiting in this direction. In contrast, the nucleus is "squeezed" in the x and y direction so the orbit frequencies can be larger for a given energy.

Inserting these definitions into Eq. 8.2 allows us to rewrite the Hamiltonian in terms of the operator $\mathbf{r}^2 Y_{20}$ as follows

$$H = \frac{\mathbf{p}^2}{2m} + \frac{1}{2}m\omega_0^2\mathbf{r}^2 - m\omega_0^2\mathbf{r}^2\delta\frac{4}{3}\sqrt{\frac{\pi}{5}}Y_{20}(\theta, \phi) + C\mathbf{l}\bullet\mathbf{s} + D\mathbf{l}^2 \tag{8.5}$$

The two equivalent versions of the Nilsson Hamiltonian in Eqs. 8.2 and 8.5 allow us to understand the structure of the model in the limits of large and small deformations, respectively. Note that, in the literature the $\mathbf{l}\bullet\mathbf{s}$ and $\mathbf{l}^2$ terms are usually expressed in terms of parameters $\kappa = C/2\hbar\omega_0$ and $\mu = 2D/C$. κ typically takes on values around 0.06 and μ varies from 0 to ≈ 0.7 for different shells.

For small deformation, j is approximately a good quantum number. Equation 8.5 consists of a Hamiltonian for an isotropic oscillator with $\mathbf{l}^2$ and $\mathbf{l}\bullet\mathbf{s}$ terms plus a perturbation proportional to $\delta\mathbf{r}^2Y_{20}$. The former part gives the spherical shell model energies and is spherically symmetric. The eigenstates of this Hamiltonian can be labeled by the quantum numbers Nlj and m of the spherical single-particle states. Treating the Y_{20} term as a perturbation, the shift in energies relative to $\delta = 0$ (i.e., to spherical energies) is

$$\Delta E(Nljm) = -\frac{4}{3}\sqrt{\frac{\pi}{5}}m\omega_0^2\delta\left\langle Nljm|\mathbf{r}^2Y_{20}(\theta, \phi)|Nljm\right\rangle \tag{8.6}$$

We can evaluate this by separating the radial and angular parts and using the relation for a harmonic oscillator potential that

$$\frac{1}{2}m\omega_0^2 \left\langle Nljm|\mathbf{r}^2|Nljm\right\rangle = \frac{1}{2}\hbar\omega_0 \left(N + \frac{3}{2}\right) \tag{8.7}$$

Evaluating the matrix element of the spherical harmonic Y_{20} gives the final result for *small* δ

$$\Delta E(NljK) = -\frac{2}{3}\hbar\omega_0 \left(N + \frac{3}{2}\right)\delta \frac{[3K^2 - j(j+1)]\left[\frac{3}{4} - j(j+1)\right]}{(2j-1)j(j+1)(2j+3)} \tag{8.8}$$

where we have replaced the projection m with K, the projection of the total angular momentum on the z axis.

This simple result has three facets that account for the structure of the Nilsson diagram for small deformations:

- There is a proportionality to δ, the quadrupole deformation.
- The shifts display a dependence on K^2.
- They depend linearly on the oscillator quantum number N.

We have seen exactly these features in our intuitive derivation and in the Nilsson diagram, especially for unique parity orbits for which j is a good quantum number out to rather large deformations.

Another direct implication of Eq. 8.8 is that for $\delta > 0$ there are more downward sloping than upward sloping orbits. For $j > 1/2$, the $[3K^2 - j(j+1)]$ term is negative, giving downward sloping orbits (since $3/4 - j(j+1)$ is negative) if

$$K < \sqrt{\frac{j(j+1)}{3}} \approx \frac{j}{1.8} = 0.65j \tag{8.9}$$

and upward sloping for $K > 0.65j$. For example, for $j = 13/2$, orbits with $K = 1/2, 3/2, 5/2, 7/2$ should be downward sloping and $K = 9/2, 11/2$, and $13/2$ upward sloping. This feature is indeed displayed by the exact numerical diagonalizations depicted in the Nilsson diagram of Fig. 8.4. Note the interesting physical correlation here. The angular orientation of an orbit to the symmetry axis is approximately given by $\sin\theta \approx K/j$ and $K/j \approx 0.65$ corresponds to $\theta \approx 40°$. Inclinations greater than these are unfavored energetically by a prolate quadrupole deformation.

The dependence on N implies that the slopes of the energy levels in a Nilsson diagram are steeper for larger N. Thus, heavier nuclei are easier to deform than lighter ones. We commented implicitly on this N effect earlier and can now explain its physical origin a bit more precisely. A nucleon in a high oscillator shell will have a larger average radius [indeed, we just utilized the fact that the expectation value $\langle \mathbf{r}^2 \rangle \propto (N+3/2)$]. Therefore, as the nucleus deforms, the nuclear matter approaches this outer orbit. Since the nuclear force is attractive, the energy of a particle in this orbit decreases. The effect is obviously less for a particle in a lower oscillator shell that is already closer to (or inside) the bulk of the nucleus when it is spherical.

In the opposite limit of large deformation, the $\mathbf{l} \cdot \mathbf{s}$ and $\mathbf{l}^2$ terms in Eqs. 8.2 and 8.5 are negligible and the Hamiltonian simply reduces to an anisotropic harmonic oscillator

whose form shows that the motion clearly separates into *independent* oscillations in the
z direction and in the xy plane. Therefore the number of quanta in these directions,
n_z and $(n_x + n_y)$, separately become good quantum numbers. The eigenvalues of the
one dimensional harmonic oscillator with quanta n_i are simply $\hbar\omega_i(n_i + 1/2)$. This
gives the familiar result for an isotropic three-dimensional harmonic oscillator that $E \approx$
$\hbar\omega(N + 3/2)$ where $N = n_x + n_y + n_z$. Thus, in the present case of *large* δ, the
eigenvalues of the anisotropic harmonic oscillator of Eq. 8.2 go asymptotically to

$$E(n_x, n_y, n_z) = \hbar\omega_x (N - n_z + 1) + \hbar\omega_z \left(n_z + \frac{1}{2}\right) \qquad (8.10)$$

Since H is independent of the angle ϕ around the z axis, the Hamiltonian corresponding to
Eq. 8.10 is invariant with respect to rotations about the z axis. Therefore the z-projection
of both the orbital and spin angular momenta of a particle must be constants of the
motion. As we have stated, these quantum numbers—the eigenvalues of the operators l_z
and s_z—are commonly denoted by Λ and Σ while their sum, the projection of the total
angular momentum on the symmetry axis, is indicated by K. The asymptotic energies
$E(n_x, n_y, n_z)$ can then be more conveniently expressed in terms of the quantum numbers
$K[Nn_z\Lambda]$ of the familiar Nilsson orbit notation.

Asymptotically, these energies are dependent on n_z and independent of Λ. The separation according to n_z, that is, according to the extent of the motion in the z direction
or perpendicular to it, simply reflects the point made at the beginning of this chapter
that, since the nuclear force is attractive, equatorial orbits will be favored and polar orbits unfavored in energy. The independence of Λ occurs for large δ because the terms
in $\mathbf{l} \bullet \mathbf{s}$ and $\mathbf{l}^2$ are negligible. Since $K = \Lambda \pm \Sigma$, this independence of Λ *becomes* an
independence of K for large δ. This is exactly opposite to the small deformation limit.

In general, a given value of n_z will have a number of degenerate states that can be
specified by Λ, taking on the values $(N - n_z)$, $(N - n_z - 2)$, $(N - n_z - 4)$, ... 0 or 1.
For finite deformation where the $\mathbf{l} \bullet \mathbf{s}$ and $\mathbf{l}^2$ terms cannot be ignored, the eigenvalues
will also split according to the value of Λ and, therefore, of K.

The asymptotic separation of the Nilsson diagram for large deformation according
to n_z and the approximate independence of Λ or K are surprisingly little known, but can
easily be seen in the Nilsson diagram for large deformations. As evident in Fig. 8.4 for
large ε, the lowest-lying orbits have $n_z \approx N$, $N - 1$, while in midshell $n_z = 1 - 2$ orbits
predominate and near the end of a shell the $n_z = 0$ and $n_z = 1$ orbits are collected.
The independence of Λ or K is illustrated nicely by the nearly degenerate and parallel
orbits pairs $7/2^-[503]$ and $9/2^-[505]$, $3/2^-[512]$ and $1/2^-[510]$, or $3/2^+[422]$ and
$1/2^+[420]$.

A nice empirical verification of the separability of the motion into components along
and perpendicular to the symmetry axis comes from the properties of certain orbits differing by ± 2 in their principle quantum number N. In principle, the Nilsson Hamiltonian
(specifically the $\mathbf{l}^2$ and $\mathbf{l} \bullet \mathbf{s}$ terms) can couple states with $\Delta N = \pm 2$, although these
couplings are normally neglected since such states are separated by two oscillator shells
(≈ 10 MeV). For large deformations, however, the sensitivity of the energies to n_z leads
to the phenomenon that steeply upsloping orbits from oscillator shell N may eventually

cross steeply downsloping orbits from the $N + 2$ shell. These orbits will have small and large values of n_z, respectively. An example of such $\Delta N = 2$ orbit pairs are the $3/2^+[402]$ and $3/2^+[651]$ orbits. A priori, their mixing might be expected to be large in the near crossing region. However, that mixing has been empirically deduced from single nucleon transfer cross sections. The extracted interaction matrix elements are typically only ≈ 50–100 keV. Such small coupling matrix elements between states with very different distribution of quanta in the z and xy directions points to the approximate validity of the separation of motion in these two perpendicular directions. It is worth noting, however, that the presence of other deformation components, such as hexadecapole (β_4) shapes, can greatly increase $\Delta N = 2$ mixing.

To recapitulate some of the preceding points, we see two limiting situations of the Nilsson scheme. For small deformations δ, the energies are approximately given by Eq. 8.8. They are linear in δ, j remains an approximately good quantum number (the configuration mixing is still small), and the orbits are separated principally by their K quantum numbers. For large deformations, the energies (Eq. 8.10) are again linear in δ (recall that the ω_i are linear in δ). The slopes, however, now depend on n_z and the energies separate according to the distribution of motion along and perpendicular to the z axis. For intermediate deformations, a transition between these two coupling schemes takes place.

8.2 Examples

Having discussed the Nilsson model both physically and formally, we can now turn to its application to odd mass deformed nuclei. Actually, this works in much the same way as the shell model for the single-particle excitations of spherical nuclei. The principal difference lies in the degeneracy of the orbits. In the shell model, an orbit j can contain $2j + 1$ nucleons. In the Nilsson model, the degeneracy is broken according to the orbit orientation, or K value, and each Nilsson orbit can contain only two nucleons, corresponding to the two ways ($\pm K$) in which the nucleon can orbit the nucleus (clockwise or counterclockwise). Neglecting pairing for a moment, in a deformed region the Nilsson orbits are sequentially filled, two protons and neutrons to each, until the last odd nucleon is placed. This defines the ground state. Excited single-particle excitations can be obtained two ways, either by raising the last odd nucleon to a higher orbit, thereby changing its Nilsson quantum numbers, or by lifting a nucleon from one of the filled orbits to the last orbit, completing a pair of nucleons in the latter and leaving a hole below the Fermi surface. One therefore expects to have a sequence of intrinsic excitations whose energies and quantum numbers can be simply read off from the Nilsson diagram once the deformation is specified.

Here, in effect, is the major difference between the spherical shell model and the Nilsson model: $N = 105$ corresponds to 21 holes relative to the magic number 126. A typical shell model calculation would diagonalize some residual interaction among 21 neutrons in the 82–126 shell and the complexity would be enormous. By switching to a deformed basis, the Nilsson model regains a "single-particle" picture, but with *deformed* single-particles orbits, each a relatively simple mixture of spherical j orbits. Multiparticle (or quasi-particle) excitations (the equivalent of seniority $\nu \geq 3$ in the spherical shell model) only begin to appear near the pairing gap at 1.5–2.0 MeV. The deformed ansatz gives a remarkable simplification.

We recall and emphasize here that the Nilsson wave function is only a specification of the orbital motion of the nucleons in a body fixed coordinate system: the full specification of the wave function requires a consideration of the rotational behavior. This is absolutely crucial for an understanding of the structure of odd mass deformed nuclei and, indeed, for a deeper understanding of the Nilsson model itself, as well as its testing and application to real nuclei. We shall turn to the rotational motion shortly.

It is nevertheless useful at this point to indicate how well and simply the Nilsson model can be applied to deformed nuclei by way of a brief example or two. Consider the nucleus ^{177}Hf with 72 protons and 105 neutrons. All the protons will be paired off to total angular momentum zero and, at least for the low-lying single-particle excitations, can be ignored. The same applies to the first 104 neutrons. Simple counting in the Nilsson scheme for $\varepsilon \approx 0.3$ shows that the 105th neutron will enter the $7/2^-[514]$ orbit. We therefore expect that the ground state of ^{177}Hf will be $7/2^-$. (Actually, this is not so trivial: we have implicitly assumed that the lowest angular momentum will be equal to the K value for a given orbit. While this is generally true, it is not always the case, especially when strong Coriolis effects are present. This is a question that must be dealt with when we consider the rotational motion of an odd nucleus in more detail. For the moment we accept this assumption.) A low-lying excited intrinsic (i.e., not just a rotational) state can clearly be formed by lifting the last neutron to the $9/2^+[624]$ orbit, giving a $9/2^+$ state. Similarly, one of the two nucleons in the $5/2^-[512]$ orbit may be raised into the $7/2^-[514]$ orbit leaving a hole with spin $5/2^-$. Other low-lying excitations should correspond to the $1/2^-[521]$, $7/2^+[633]$, and $7/2^-[503]$ orbits at appropriate energies. A partial empirical level scheme for ^{177}Hf is given in Fig. 8.5, showing the bandhead levels corresponding to each intrinsic Nilsson excitation. It corresponds rather well with our predictions. If we now go to ^{179}Hf, we would expect the ground state to be $9/2^+[624]$ with the $7/2^-[514]$ an excited (hole) state. Moreover, all the excitations that were *above* the Fermi surface in ^{177}Hf will now *decrease* in energy while those that were *below* the Fermi surface will *increase* in excitation energy. Comparisons of the two-level schemes in Fig. 8.5, which uses the convention that particle excitations are shown on the right and hole excitations on left, reveals exactly this behavior. In general, as one sequences through a series of isotopes (or isotones, if one is dealing with odd proton nuclei), the energy of a given Nilsson orbit should descend along the right-hand side of the "V." At some point it should become the ground state, or at least occur very low in the spectrum, then increase in energy along the left arm of the "V." At least approximately, the behavior exemplified by the systematics in Fig. 8.6 (here hole energies are shown as negative values) is typically observed. Deviations from it can be due to changes in deformation across such a sequence (we have implicitly assumed a constant deformation), or to shifts in the relative positions of the Nilsson orbits from effects such as higher order deformation components (hexadecapole deformations), or to Coriolis effects.

A nearby nucleus that shows one such case is ^{183}W, whose level scheme will be discussed at great length in the next chapter and is illustrated in Fig. 9.1. Simple count-

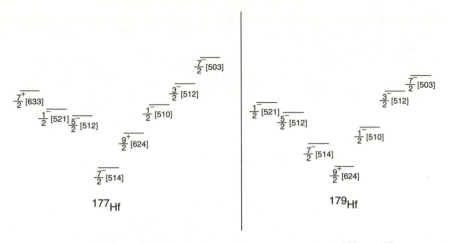

FIG. 8.5. Empirical bandheads of intrinsic Nilsson excitations in ^{177}Hf, ^{179}Hf. Particle (hole) states are on the right (left).

ing would suggest that the ground state is $7/2^-[503]$, with low-lying $1/2^-[510]$ and $3/2^-[512]$ particle excitations. Yet, the empirical level scheme shows that the latter two orbits are near the ground state and the $7/2^-[503]$ occurs at a few hundred keV excitation energy. An explanation of this will be given in Chapter 9.

We have seen that it is as easy in the Nilsson model as in the shell model to determine the expected order of single-particle excitations and their energies and to deduce, virtually by inspection, an anticipated level scheme. Though this seems a trivial exercise, one should not lose sight of the fact that by considering a deformed shell model potential, one is able to account instantly for the low-lying levels of literally hundreds of deformed odd mass nuclei, ranging from $A \approx 20$ to the actinides. There have been innumerable tests of this model over the last three decades and it has proved capable of correlating a vast amount of data, particularly when some rather simple refinements (primarily Coriolis mixing and hexadecapole deformations) are incorporated.

8.3 Prolate and oblate shapes

It is interesting to break the discussion at this point to discuss an extremely basic question that is seldom alluded to but is now easy to answer. It is an empirical fact that the vast majority of deformed nuclei are prolate rather than oblate in their ground states. The only candidates for oblate nuclei are those in which either N or Z is near the very end of a major shell (e.g., Hg). However, the Nilsson diagram can be applied equally on the oblate side.

Although we will not discuss it explicitly, the derivation of the model for oblate shapes should be self-evident by now. Here the "core" nucleus is disc-shaped, and the lowest energy orbits will be polar with high K values. The sequence of levels will be more or less inverted relative to the prolate case. For example (see the sketch in Fig. 8.3), the $i_{13/2}$ shell model state will again split into seven orbits in order of increasing energy

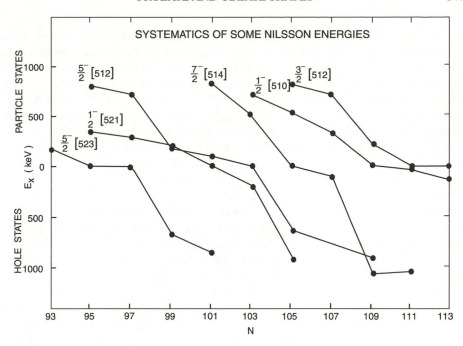

FIG. 8.6. Systematics of some Nilsson orbit excitation energies in the rare earth region. Hole states are given negative energies (extracted from Bunker, 1971).

$K = 13/2, 11/2, 9/2, \ldots, 1/2$. For small $|\beta|$, the energies are again linear in K^2, and hence the spacings between different K levels are proportional to K.

Although the empirical preference for prolate shapes is well known, it is much less understood *why* it should be so. Years ago, this issue was discussed with some frequency, but since has been ignored without having received a satisfactory or at least well-known explanation. However, a simple understanding actually involves only two considerations, namely the specific *sequence* of single-particle spherical j shells and the relative *angular orientations* of different K orbits in the Nilsson model.*

A casual inspection of the Nilsson diagram seems to show no preference for oblate or prolate shapes. Energy-favored downward sloping orbits appear on both sides. Furthermore, if we imagine a spherical closed shell nucleus with a single valence nucleon, the resulting disc-like orbital "ring" would appear to be oblate. There is no elongation along a symmetry *axis* but an orbital *plane* superimposed on the spherical core.

Why, therefore, are most nuclei, and in particular those at the beginning of major shells, prolate? Consider a single j shell such as $i_{13/2}$. We recall from an earlier discussion (see Table 8.1) that the orientations of the orbital planes change very little for low K values, but increase rapidly for higher K. Thus, while on the oblate side a single $K = 13/2$ orbit may descend as rapidly in energy as its $K = 1/2$ partner on the

*I am grateful to C. J. Lister for illuminating discussions of these points.

prolate side, there are *several* prolate orbits that have comparable, strongly downsloping energies. The accumulation of downward sloping orbits on the prolate side is not, in itself, however, enough to energetically favor prolate deformations, as calculation with Eq. 8.8 shows. Another ingredient is needed, namely the presence of several j shells. Each j shell has low K orbits ($K = 1/2, 3/2, \ldots$) but only one has the highest K value in a major shell. Therefore, in general, there are fewer high K orbits than low K orbits. Hence, while on the oblate side the downsloping orbits are isolated, on the prolate side orbits with the same K value from *different* j shells will be close together (see Fig. 8.4): They will interact (this is the mixing that leads to the spreading of the Nilsson wave functions in j) and the lowest orbits will be pushed still lower in energy—*below* the trajectories that would be given by the straight line dependence of Eq. 8.8. For close lying low K orbits from relatively high j spherical single particle states, this mixing can be large since, as we just noted, the orbital planes are similar, the particles are therefore close together, and they therefore interact strongly. This effect is seen in Fig. 8.4 as the downward sloping curvatures of the lowest orbits above the 50 and 82 magic numbers, and it gives an energetic advantage to prolate deformations, as seen experimentally.

There is one important element in this scenario that is worth highlighting. Above, we invoked the presence of several close lying high j spherical levels, as is present at the beginning of major shells in heavy nuclei (see again, Fig. 8.4). However, suppose that, instead of this order, the lowest orbit had been $p_{1/2}$, $p_{3/2}$, or even $f_{5/2}$ instead of a higher j orbit. The preceding argument would then have had little weight. It is because the lowest orbits after a shell closure have relatively *high* j, with many K values, that a distinction between the oblate and prolate behavior can be made and a preference for prolate deformations can develop. Thus, the second key feature is the modification of the shell model potential to include components that favor lower energy for higher l and j orbits. The nearly universal preference for prolate shapes in nuclei therefore stems from the specific radial shape of the shell model central potential that is intermediate between harmonic oscillator and square well and which favors large l values, from the attractive nature of the nuclear force, and from the properties of the sine (the orbit inclinations as a function of K). Once again, we have an example of how a very simple but physically intuitive appreciation leads to important results even without detailed calculation.

8.4 Interplay of Nilsson structure and rotational motion

We have discussed the structure of the Nilsson wave functions as linear combinations of single j shell model wave functions with expansion coefficients C_j and have seen a number of examples of such wave functions. It cannot be emphasized too strongly that these are wave functions in the body-fixed system, that is, the nonrotating nucleus. In this system, j is clearly not a good quantum number. However, the nucleus exists in space and the total angular momentum J must be a constant of the motion. The projection of this angular momentum K, (and the expectation value $\langle j \rangle$ of the single-particle angular momentum) on the nuclear symmetry axis are also good quantum numbers. The Nilsson wave function is known as an *intrinsic* state or a state of excitation of an isolated body, which in this case is the deformed nucleus. The real nuclear states are combinations of this intrinsic motion and a superimposed rotational motion of the core. Phrased another

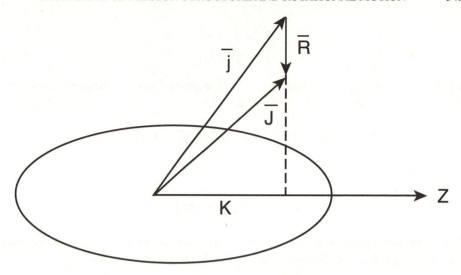

FIG. 8.7. Angular momentum diagram for an odd-mass deformed nucleus. Note: this figure can be very misleading—see text and Fig. 8.9.

way, the Nilsson wave function does not possess a fixed angular momentum **J**; rather, it can be projected onto states of many different angular momenta. It is because of this seemingly abstract idea that a particle in a given Nilsson orbit actually gives rise not only to a single state in a deformed nucleus (as would be the case for a single particle in a given orbit j), but rather to a set of states comprising a *rotational band*.

A proper understanding of the interplay of this rotational motion and the intrinsic motion leads to a much deeper understanding of the wave functions for odd mass deformed nuclei, of the actual nature of the rotational motion involved, of the reasons why single nucleon transfer reactions are such powerful probes of Nilsson model wave functions, and of the effects of the Coriolis interaction.

The same approach used to obtain the first order rotational energy expression in even–even nuclei can also be used for odd mass nuclei. If the odd nucleon, considered for a moment to be in a single j orbit, does not polarize the even–even core, then the total angular momentum results from the vector sum of the core rotation and the odd particle angular momentum. This is illustrated in Fig. 8.7. We now start with the same rotational Hamiltonian as before and obtain, using the notation for the different angular momenta given in Fig. 8.7,

$$\mathbf{H} = \frac{\hbar^2}{2I}\mathbf{R}^2 = \frac{\hbar^2}{2I}(\mathbf{J} - \mathbf{j})^2 = \frac{\hbar^2}{2I}(\mathbf{J}^2 + \mathbf{j}^2 - 2\mathbf{J} \bullet \mathbf{j}) \qquad (8.11)$$

We can convert this to a more useful form by defining the familiar raising and lowering operators

$$\mathbf{J}_\pm = \mathbf{J}_1 \pm i\mathbf{J}_2$$
$$\mathbf{j}_\pm = \mathbf{j}_1 \pm i\mathbf{j}_2 \qquad\qquad (8.12)$$

where we use subscripts 1, 2, 3 for x, y, z. Simple multiplication of these operators gives $\mathbf{J}_+\mathbf{j}_- + \mathbf{J}_-\mathbf{j}_+ = 2(\mathbf{J}_1\mathbf{j}_1 + \mathbf{J}_2\mathbf{j}_2)$. Therefore

$$\mathbf{J} \bullet \mathbf{j} = \mathbf{J}_1\mathbf{j}_1 + \mathbf{J}_2\mathbf{j}_2 + \mathbf{J}_3\mathbf{j}_3 = \frac{1}{2}(\mathbf{J}_+\mathbf{j}_- + \mathbf{J}_-\mathbf{j}_+) + \mathbf{J}_3\mathbf{j}_3$$

and hence

$$H = \frac{\hbar^2}{2I}\left[\mathbf{J}^2 + \mathbf{j}^2 - 2\mathbf{J}_3\mathbf{j}_3 - (\mathbf{J}_+\mathbf{j}_- + \mathbf{J}_-\mathbf{j}_+)\right]$$

Replacing these operators with their eigenvalues where possible and using the fact that, for low-lying states, both J_3 and j_3 have the same projection K, gives

$$E(J) = \frac{\hbar^2}{2I}\left[J(J+1) - 2K^2 + \langle\mathbf{j}^2\rangle - (\mathbf{J}_+\mathbf{j}_- + \mathbf{J}_-\mathbf{j}_+)\right] \qquad (8.13)$$

or

$$E(J) = \frac{\hbar^2}{2I}\left[J(J+1) - 2K^2 + \langle\mathbf{j}^2\rangle\right] + V_{\text{Coriolis}} \qquad (8.14)$$

where

$$V_{\text{Coriolis}} = -\frac{\hbar^2}{2I}(\mathbf{J}_+\mathbf{j}_- + \mathbf{J}_-\mathbf{j}_+)$$

The first term on the right in Eq. 8.14 is identical to the rotational energy expression for an even–even nucleus. It is clear from Fig. 8.7 that this must be the case, since this figure would collapse to that of an even mass rotor if there were no single-particle angular momentum j. The other terms in Eq. 8.14 arise specifically from the presence of the odd particle and are intimately connected with the coupling between rotational and particle degrees of freedom. Even the resemblance of the first term to the symmetric top formula is fundamentally misleading. We shall see momentarily that the simple picture illustrated in Fig. 8.7 conceals some important physical effects, and that the rotational motion is not as simple as commonly believed, but an alternate picture that is nearly as simple will allow us to retrieve Eq. 8.14 in a transparent, elegant way that will disclose a much different understanding of rotational motion in odd-mass nuclei.

The third and fourth terms in Eq. 8.14 involve the j structure of the deformed single-particle wave function (Nilsson wave function). Whatever value $\langle\mathbf{j}^2\rangle$ takes, both it and K should be constant within a rotational band (neglecting bandmixing). The last term is called the *Coriolis interaction* because its effects are very similar to the classical Coriolis force acting on any rotating macroscopic body. The Coriolis interaction has important consequences in both even and odd deformed nuclei and will be extensively discussed later. We will show then that, in first order, its effects on energies simply correspond to a change in the value of $\hbar^2/2I$.

To this order then, the energy levels of a given rotational band in an odd-mass nucleus should behave as $\hbar^2/2I \, J(J+1)$. In Fig. 8.8, we illustrate a number of examples of rotational bands in heavy nuclei and show that this simple formula works remarkably well. Also included in the figure are two examples where it clearly fails to provide even a reasonable first-order estimate. One of these involves a $K = 1/2$ band ($1/2^-$ [521] in ^{169}Er) that we shall later see incorporates a special (diagonal) Coriolis interaction. The other is a "band" ("$h_{11/2}$" in ^{133}La) that appears to be partly "upside down" (e.g. $E(11/2^-) << E(5/2^-), E(7/2^-)$) and unrelated to the kind of structure that we have been examining. It too involves especially strong Coriolis effects and will be discussed later. Here, we wish to raise, and resolve, an apparent paradox that arises from Fig. 8.7 (and that is intimately related to this type of inverted structure).

Equation 8.11 states that the total angular momentum results from the vector combination of the rotational angular momentum R and the particle angular momentum j, that is $\mathbf{J} = \mathbf{R} + \mathbf{j}$. The situation was sketched in Fig. 8.7, which is a simplification since the Nilsson wave function contains, in general, a linear combination of functions of different j. That is not the point. Consider, for simplicity, the one case in which a single j value does nearly characterize the Nilsson wave functions—the unique parity orbits. To be specific, let us take a Nilsson wave function such as that for the $1/2^+$ [660] orbit from the $i_{13/2}$ neutron shell in the rare earth region. (For convenience we neglect Coriolis mixing.) Now, we have seen examples of rotational bands with spins $J = K, K+1, K+2, \ldots$ whose energies vary approximately as $J(J+1)$, or in this case, a sequence with $J = 1/2, 3/2, 5/2, 7/2, \ldots$. The common view (Fig. 8.7) of this band as consisting of a single particle in the $1/2^+$ [660] orbit coupled to a sequence of successively faster core rotations is seriously in error. To see this, recall that we have taken a simple case where the Nilsson wave function consists of only one j value, $j = 13/2$. This is therefore the *only* single-particle angular momentum in the system. This in turn implies that any *total* angular momentum *other* than $J = 13/2$ *must* incorporate angular momentum from another source. That source can only be the rotational motion. Thus, as in Fig. 8.7, one can imagine a $J = 17/2$ state obtained by coupling a $j = 13/2$ single-particle angular momentum to a core rotational angular momentum $R = 2$. Similarly, $J = 9/2, 5/2$ states could be formed by the antiparallel coupling of $j = 13/2$ and $R = 2$, $R = 4$, respectively. The only energies in the system are the Nilsson energy, which is constant (independent of J), and the rotational energy, $\hbar^2/2I \, \mathbf{R}^2$. Hence, the $J = 9/2$ state (with $R = 2$) should have higher energy than the $J = 13/2$ state (with $R = 0$). The $J = 5/2$ and $1/2$ states should be expected still higher. Moreover, the energy difference $E(9/2) - E(13/2)$ should equal $E_{R=2} - E_{R=0}$, or in other words, $E_{2_1^+}$ of the neighboring even–even core nucleus; $E(5/2) - E(13/2)$ should equal $E_{4_1^+}$, and so on. This picture leads to "upside down" rotational bands with the lower spin states lying higher than the state with $J = j$. Such band structures do indeed exist (as we noted in discussing the right-most band in Fig. 8.8). They have become highly interesting as a particular manifestation of important Coriolis effects in high-spin states. However, they

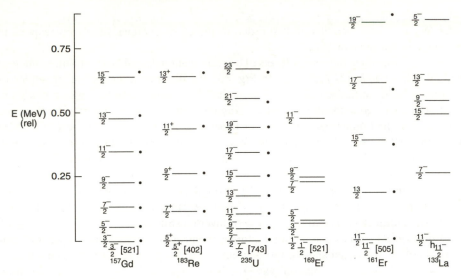

FIG. 8.8. Rotational bands in some deformed odd-mass nuclei. The dots give the rotational energy predictions from Eq. 8.14 after normalization to the first two levels. Dots for the energies of the $K = 1/2$, $1/2^-$[521] band in ^{169}Er are omitted since they require a "decoupling parameter" term (to be discussed in Chapter 9). As discussed later in the text, the rightmost "band" is so Coriolis mixed that no single Nilsson label is possible.

are not the normally observed situation and are certainly not consistent with most of the empirical rotational bands shown in Fig. 8.8. Clearly, there is something wrong with this picture.

A clue to a more accurate understanding of the rotational motion begins by recalling that we are dealing with an axially symmetric deformed nucleus. This means that *any* orientation of the angular momentum vector **j** with respect to the symmetry axis z that maintains a projection K is indistinguishable from any other orientation and therefore is equally likely: the angular momentum vector **j** is free to *precess* around the z axis. Figure 8.7 showed only one *particular* orientation of this angular momentum vector— that corresponding to the smallest possible value of $|R|$. In contrast, imagine that the angular momentum **j** were rotated $180°$ to that shown in the figure so that it lay in the plane of the page but pointed downward, below the z axis. This situation is depicted on the left in Fig. 8.9. The amount of core rotation required to produce a final total angular momentum **J** would clearly be much larger. If we extend this idea to other angles of the angular momentum vector **j**, then, as **j** precesses, **R** will point in a continually varying direction and $|R|$ will take on a constantly changing series of values.

Consider again our example of a $j = 13/2$ particle in a $K = 1/2$ orbit. For $J = 1/2$, J points nearly along the z axis, and $|R|$ remains roughly constant in magnitude, although not in direction, as **j** precesses, at a value $|R| \approx 6$. On the other hand, for $J = 13/2$, $|R|$ takes on values ranging from 0 to 12 as **j** precesses. Since the rotational Hamiltonian,

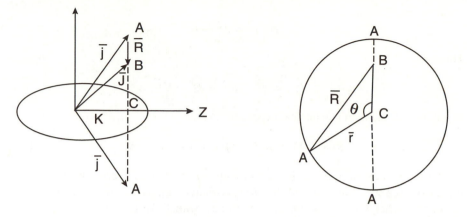

ROTATIONAL MOTION
IN ODD A NUCLEI

FIG. 8.9. (Left) Angular momentum diagram for an odd-mass deformed nucleus. This is a more rigorous version of Fig. 8.7 that incorporates important refinements. (Right) Path (circle) followed by the tip of the **j** vector with time as it precesses about the z axis (point C). This is an "end view" of the time dependence of the diagram on the left. I am grateful to D. D. Warner for this figure and the discussion relating to it.

$\hbar^2/2I\mathbf{R}^2$, is quadratic in $\mathbf{R}$, large $|R|$ values carry more weight than small ones. Therefore, a situation in which $|R|$ varies smoothly from 0 to 12 will have, on average, higher energy than one in which $|R|$ is approximately constant at a value $|R| = 6$, and the $J = 13/2$ state will require more (rotational) energy than the $J = 1/2$ level.

This analysis can be made more quantitative by viewing the nucleus end on. This situation is depicted in the side and end views in Fig. 8.9. Point C is along the symmetry axis at a distance K from the origin. Point B is the terminus of the fixed total angular momentum vector $\mathbf{J}$, and point A marks that of the single-particle angular momentum $\mathbf{j}$. The circle in the end view shows the path followed by the angular momentum vector $\mathbf{j}$ as it precesses about the symmetry axis z. We can now calculate the expectation value of $\mathbf{R}^2$ by allowing $\mathbf{j}$ to rotate about the point C at a radius r. From simple trigonometry

$$\mathbf{R}^2 = r^2 + (BC)^2 + 2r(BC)\cos\theta \qquad (8.15)$$

Integrating this expression over θ gives

$$R_{\text{ave}}^2 = r^2 + (BC)^2$$

but

$$r^2 = (AC)^2 = j^2 - K^2$$

and

$$(BC)^2 = J^2 - K^2$$

Thus

$$R_{\text{ave}}^2 = j^2 - K^2 + J^2 - K^2 \tag{8.16}$$

or, taking the quantum mechanical expectation values

$$R(R+1) = J(J+1) + j(j+1) - 2K^2 \tag{8.17}$$

This is *exactly the same* as Eq. 8.14 except for the Coriolis term that has not been included since we have assumed that K is constant and, as we shall see, the Coriolis interaction inherently mixes different K values. (Since we assumed a single-j wave function, $\langle j^2 \rangle$ has become $j(j+1)$.) Note that the assumption of constant K is equivalent to the assumption that the particle angular momentum vector $\mathbf{j}$ precesses exactly about the z axis. We shall shortly encounter a special, though not uncommon, situation in which precession is about an axis perpendicular to the z axis: clearly in such a situation it will be K, not R, that changes continuously. This then will bring us back to the picture in Fig. 8.7, which leads to "inverted" rotational sequences.

We therefore see that the derivation of Eq. 8.14 for the eigenvalues in a rotational band built on a given Nilsson orbit was correct, even though the simple picture that is commonly used to illustrate this situation, Fig. 8.7, is too simplistic. The more accurate view shown in Fig. 8.9 gives the same formula in a trivial manner. It has been a constant emphasis in this book that most results in nuclear structure physics can be derived, at least semi-quantitatively, by very simple, often intuitive, analyses. This example warns us that such an approach cannot be careless handwaving, but must accurately reflect the correct underlying physics.

The most important conclusion from the present analysis is the recognition that the rotational motion in a deformed odd-mass nucleus is far from simple. Not only does the *magnitude* of the rotational angular momentum $|R|$ vary with time but its *direction* in space also changes. In the $J = 1/2$ state, for example, the nucleus at times rotates clockwise about the y axis, at other times counterclockwise about this axis, and at still other times about a continuously varying axis in the xy plane. The rate of rotation is relatively constant corresponding to $|R| = 6$. In the $J = 13/2$ state, on the other hand, the nucleus is at times stationary ($R = 0$) while at others it rotates at frequencies varying from values corresponding to $R = 0$ to those for $|R| \approx 12$! We can see that the *rotational* motion is really a complex, time dependent, variation that includes not only a true rotational component but also a kind of tumbling motion. The fact that the principal term in Eq. 8.14 (i.e., the term that depends on J, the others being constant) has the same form as in the symmetric top or in the rotational energy expression for an even–even nucleus, is really almost an accidental result of the particular combination of the precessing single-particle motion and the varying core rotational motion needed to produce a constant total momentum J^*.

---

*The author is grateful to D. D. Warner, with whom this analysis was worked out.

For typical (nonunique parity) Nilsson wave functions, several j values commonly appear and the "rotational" motion is even more complex. From the formal standpoint, one need not worry about this since, as we saw in deriving Eq. 8.14, as long as the proper vector character of the angular momenta are taken into account the correct results always emerge.

There is one other important point implicit to this discussion. It should be clear by now that while many values of $|R|$ contribute to the wave function for a state of spin J in a rotational band built on a given Nilsson orbit, the special value $R = 0$ can *only* occur if the Nilsson wave function ψ_{Nils} contains an amplitude ϕ_J—that is, *only* if it contains an amplitude for the single nucleon in the orbit with $j = J$. For any other j value, the state with total angular momentum J can be constructed only by incorporating some rotational motion. As we shall see, this has important consequences for single nucleon transfer reactions. We recall that for a given oscillator shell N, $j_{max} = N + 1/2$ (e.g., $j_{max} = 11/2$ for $N = 5$). It is thus clear that *all* states with $J > N + 1/2$ *must* have $R \neq 0$.

Having discussed the basic Nilsson model and the basic rotational motion in odd mass deformed nuclei, we turn in the next chapter to a detailed discussion of specific tests of and refinements to the model, with emphasis on the crucial and pervasive effects of Coriolis coupling.

9

NILSSON MODEL: APPLICATIONS AND REFINEMENTS

9.1 Single nucleon transfer reactions

The description of odd mass deformed nuclei in terms of Nilsson orbits and their configuration-mixed nonspherical wave functions has been an extremely successful model for over four decades. One of its most appealing features is that it is extremely easy to test empirically and to measure the detailed shell model (j) composition of individual Nilsson wave functions. As we shall see momentarily, single nucleon transfer reactions provide a direct and specific measure of each successive component in the Nilsson wave functions.

This remarkable property stems from a particularly simple feature of the interplay of rotational and single-particle motion in producing final states of given total angular momentum J. The underlying reason relates to a point about the rotational motion brought out at the end of the previous chapter. Since $\mathbf{J} = \mathbf{R} + \mathbf{j}$, a component with $R = 0$ in a state of spin J can *only* occur if the Nilsson wave function $\psi = \Sigma C_j \phi_j$ contains an amplitude for the single-particle angular momentum $j = J$. If we could somehow sample the $R = 0$ components of a sequence of states in a rotational band, we would be sampling the successive C_j amplitudes in the Nilsson wave function for the underlying intrinsic state. Single nucleon transfer reactions do just that.

We will pursue this point in a moment, but first it is interesting to consider how and under what conditions such reactions occur, as well as some of the experimental considerations applicable to them. We shall do this for the (d, p) reaction, but the same reasoning applies to other light-ion single nucleon transfer reactions such as (d, t), (^{3}He, α), (α, t), and so on. First, consider a schematic diagram of such a reaction as was shown in Fig. 3.7. An incident nucleon, in this example a deuteron, passes near a target nucleus. As it experiences the nuclear force (we neglect Coulomb effects) several things may occur. The deuteron may simply scatter from the nucleus elastically or inelastically (producing excited states). It could be absorbed by the target nucleus producing a nucleus $A' = A + 2$ in a highly excited "compound" nuclear state. It could attract one nucleon, such as a neutron, from the target nucleus and emerge from the collision as a triton, leaving behind a residual nucleus with $A' = A - 1$. This is a (d, t) reaction. Or, a neutron could be stripped off and enter an orbit around the target nucleus, producing a final nucleus with mass $A' = A + 1$, a (d, p) reaction. These reactions can be experimentally selected by detecting the outgoing particle and identifying it. This can be done with a number of different techniques we will not discuss here. Some utilize the different magnetic rigidities of the outgoing particles. Others exploit the dependence on the mass and charge of the ratio of the energy loss ΔE in a thin detector to the total energy.

In any case, we assume we have identified an outgoing proton, thereby "tagging" a (d, p) reaction event. This reaction is not necessarily a single step process. It could be accompanied by inelastic scattering or Coulomb excitation. Or it could be the result of a compound nuclear reaction in which the deuteron was first fully absorbed. To select appropriate experimental conditions to favor a simple single step process, we consider some of the parameters describing the reaction process of Fig. 3.3. If the closest distance of the incident projectile from the target nucleus is large, the interaction is weak, and stripping occurs with low probability. When scattering occurs it will not be at large angles. Two-step processes involve a product of such single-step amplitudes, and will be negligible. For a close collision (distance of closest approach), the nuclear interaction and hence the scattering angle are much greater. The probability of a single event occurring is much larger but so is that for multistep processes. The optimum situation of large, direct, reaction cross sections but small multistep amplitudes occurs for an intermediate angle, typically $40°$–$125°$ degrees, corresponding to a "grazing" collision. Of course, this discussion is highly qualitative since quantum mechanical interference effects lead to significant oscillations of $\sigma(d, p)$ with θ. Similar considerations apply for heavier projectiles, except that, for a given bombarding energy per nucleon, the heavier projectile brings in more angular momentum and is more likely to excite final states involving larger angular momentum transfer.

We have gone through this discussion, not because such work is common these days, but because it helps illuminate the nature of the Nilsson-rotor wave functions and because such reactions will take on new importance with the advent of beams of exotic nuclei (see Chapter 11). We now consider the population with (d, p) of members of a rotational band built on some Nilsson orbit in a deformed odd mass final nucleus. Since the reaction is single step, sequential processes, such as transfer followed by inelastic scattering, are eliminated by the choice of experimental conditions. Indeed, by definition—the *only* thing that can occur is that a single nucleon can be transferred to a given, quantized, empty valence orbit (Nilsson wave function ψ_{Nils}). In particular, the process *cannot* induce any *rotation* of the target nucleus. Since no rotational notion can be imparted, it follows that the probability of populating a state with a given total angular momentum J must be proportional to the probability, $C_{j=J}^2$, for a shell model single-particle wave function $\phi_{j=J}$ in the Nilsson wave function. Thus, even though the *intrinsic* wave function for *each* state in the rotational band is identical, the (d, p) cross section for populating each successive state samples a specific component of the Nilsson wave function. The cross sections for populating the $J = K, K + 1, K + 2, K + 3, \ldots$ states directly give $C_K^2, C_{K+1}^2, C_{K+2}^2, C_{K+3}^2$ in ψ_{Nils}. Since the sequence of C_j values is characteristic of each specific Nilsson wave function, the pattern of cross sections is an identifying signature of a particular orbit and is commonly called a fingerprint pattern. Much of the study of Nilsson orbits in heavy nuclei has historically been grounded in this basic property of single nucleon transfer cross sections.

Note the interesting point brought out in the last chapter that since $j_{\max} = N + 1/2$ for a given shell, single-step single-nucleon transfer reactions can *never* populate states with $J > N+1/2$. All components in the wave functions of such states must have $R \neq 0$. If such states are populated, it is immediate evidence either for multistep processes or for wave function admixtures from higher major shells (e.g., $\Delta N = 2$ mixing that can be induced by large hexadecapole deformation components).

The formal expression for the single nucleon transfer cross section to a specific state of spin j of a rotational band in a deformed nucleus built on a Nilsson orbit i, is given by

$$\frac{d\sigma_i}{d\Omega}(j) = 2\mathcal{N}\phi_l^{DW}(\theta)\, C_j^{i^2} P_i^2 \tag{9.1}$$

where we have included a pairing factor denoted by P_i^2, which we shall discuss in a moment. (The quantity $(C_j P)^2$ is analogous to the spectroscopic factor (see Chapters 2 and 3) for spherical nuclei.) This formula is well known and has been extensively used in probing the structure of Nilsson wave functions; strangely, its simple origins described earlier are often only vaguely understoood, and the power of (d, p), (d, t), and other single-particle reactions in elucidating Nilsson structure often seems almost magical.

Before discussing this equation in relation to the empirically deduced structure of various Nilsson orbits in typical deformed nuclei, we add a few more comments on the other factors appearing in it. In a formal derivation of Eq. 9.1, there must occur an integral linking the initial and final states

$$\langle \phi_p \psi_{\text{Nils}}(A+1) | \phi_d \psi_{e-e}(A) \rangle$$

giving the overlap of the initial deuteron–A–nucleon even–even target system with the final proton–$(A + 1)$–nucleon odd-mass system. This matrix element involves the internal structure of the incoming and outgoing projectiles and the degree to which the final nucleus looks like the target plus a neutron in a specific orbit. It is usually assumed for simplicity that the latter point is satisfied: $\langle \psi_{\text{Nils}}(A+1) | \phi_n \psi_{e-e}(A) \rangle^2$ is unity. This is really just the single-step process assumption. The former aspect concerning the projectile/ejectile structure is absorbed into the arbitrary normalization constant $\mathcal{N}$ in Eq. 9.1.

The reaction process also depends on kinematic effects (E_d, θ, etc.). These kinematic factors are included in the function $\phi_l(\theta)$, which can be calculated by standard DWBA techniques. Typically, $\phi_l(\theta)$ has a diffractive oscillatory pattern that is a function of θ. The specific extent and locations of maxima and minima are functions of the transferred angular momentum l. Starting from an even–even nucleus, the final angular momentum $J = j = l \pm 1/2$ and the final state parity is $\pi = (-1)^l$. In principle, a measurement of the angular distribution of the outgoing particles can provide information on the J^π values of various final states. In practice, this information is somewhat unreliable in deformed nuclei, and measurements are typically made at only two or three angles: the ratio of the cross sections at these angles provides at least a qualitative guide to the transferred orbital angular momentum l. On account of the centrifugal barrier, it should not be surprising that, for low-energy, light projectiles, the population of higher l values is inhibited: the cross sections decrease with increasing l (and therefore J). Generally, it turns out that the angular distributions for small l values are somewhat forward peaked, while those for large l transfers are backward peaked. Therefore, a ratio such as $\sigma(125°)/\sigma(60°)$

increases with transferred orbital angular momentum l. The cross sections $\phi_l(\theta)$ also have a dependence on the reaction Q value (the difference in incoming and outgoing projectile energies). The Q value is easily deduced from the known nucleon separation energies. For example, for a (d, p) reaction

$$Q(\text{d, p}) = E_p - E_d = S(n) - E_x - \text{B.E.(d)}$$

where B.E.(d) is the deuteron binding energy 2.23 MeV. Since $S(n) \approx 5$–8 MeV in heavy nuclei, $Q(\text{d, p})$ is typically positive for low E_x and decreases as E_x increases.

As noted, the use of reactions that carry more momentum into the system such as (^{3}He, α), favors high l transfers. Therefore, the ratio of populations of a given state in (α, ^{3}He) and (d, p), $\sigma(\alpha, ^3\text{He})/\sigma(\text{d, p})$, can also serve as a "meter" for the transferred angular momentum l. Indeed, at back angles, it singles out the highest jl values accessible (e.g., $j = 13/2$ in the odd neutron rare earth nuclei: see the following discussion (Fig. 9.9)).

The factor P^2 is U^2 for a stripping reaction such as (d, p), and V^2 for a pickup reaction such as (d, t). It represents the probability that the single nucleon orbit involved is initially either empty or filled, respectively. It is reasonable that this factor is present. In a (d, p) or (d, t) reaction, a given orbit can be populated only to the extent that it is initially empty or full, respectively. Thus (d, p) tends to populate orbits above the Fermi surface, while (d, t) populates orbits below the Fermi surface most intensely.

Now we can turn to the extraction of specific nuclear structure information from these reactions. It is easiest to show this using a specific example. Consider the final nucleus ^{183}W whose level scheme is shown in Fig. 9.1 with the states arranged according to Nilsson assignment and rotational band in the same format as in Fig. 8.5 for ^{177,179}Hf. Figure 9.2 shows (d, p) and (d, t) spectra leading to ^{183}W, while Table 9.1 summarizes the measured cross sections (at 90°) for those negative parity states that were assigned to specific Nilsson states. Table 9.2 gives similar information for ^{185}W. We assume that the "kinematic" factors $\mathcal{N}\phi_l(\theta)$ are known so that the cross sections may be used to extract empirical values of $C_j^i P_i$ for the ith band.

Look at Fig. 9.2, bearing in mind the strong l dependence of the DWBA cross sections. Typically $\sigma(\text{d, p})$ drops by an order of magnitude as l changes from $l = 1$ to $l = 5$. Thus, large cross sections may not imply large $C_j^2 P$ values and vice versa.

Extracting the C_j^i coefficients and the structure of each band now only requires an estimate of the pairing factors U_i or V_i. The simplest way to do this is to assume a Fermi energy, and the simplest assumption here is to assume that it coincides with the energy of the Nilsson orbit that forms the ground state in the odd-mass nucleus. For a reasonable choice of the gap parameter Δ (typically 0.75–1 MeV), it is easy to solve the quasi-particle Eq. 5.23, to obtain U and V as a function of excitation energy. However, the Fermi energy need not coincide exactly with any specific Nilsson orbit. For example, in ^{183}W the ground state is the $1/2^-$[510] orbit (see Fig. 8.4) and the $3/2^-$[512] orbit occurs at an excitation energy of approximately 200 keV, while in ^{185}W the order is reversed but the two bands occur within $\approx$20 keV each other. If the deformation has not changed, one cannot account for this asymmetric situation by placing the Fermi surface

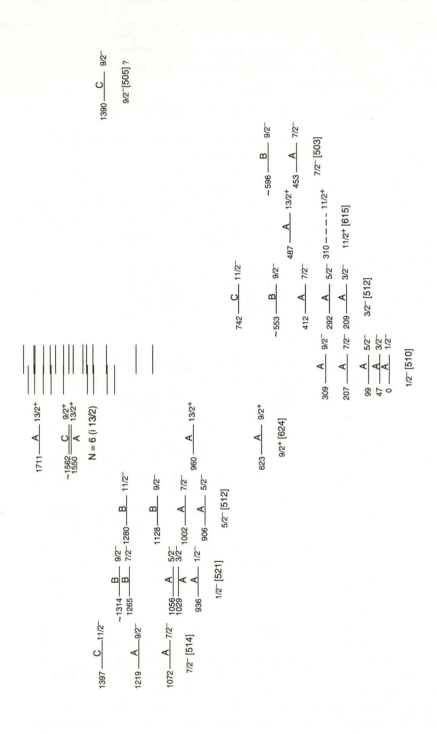

FIG. 9.1. Level scheme for ^{183}W. Levels without Nilsson assignments are given at top center (Casten, 1972). Particle (hole) excitations are on the right (left). Letters on levels (A, B, C) give confidence estimates.

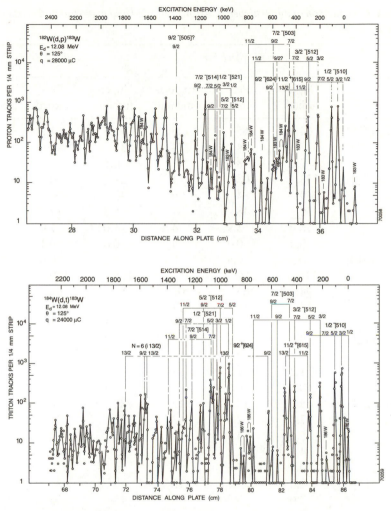

FIG. 9.2. (d, p) and (d, t) spectra leading to ^{183}W. The peaks are labeled by the Nilsson assignments (Casten, 1972).

at, say, the position of the 1/2$^-$[510] orbit in ^{183}W and at the 3/2$^-$[512] orbit in ^{185}W. Given the separation of these two orbits in the Nilsson diagram, ≈150 keV, it is clear that in ^{185}W, the Fermi surface must be approximately centered between these two orbits, producing (see Fig. 5.7) low excitation energies for both. Since the U and V factors change rapidly near the Fermi surface, such fine details of the Fermi surface location can have significant effects on single nucleon transfer cross sections for low-lying orbits.

A more empirical way to extract U and V factors is from the ratio of (d, p) and (d, t) cross sections to the same state in an odd-mass nucleus of mass A. This procedure is

Table 9.1 *Comparison of unperturbed, Coriolis coupled and experimental cross sections in* ^{183}W $(\theta = 90°)^*$

State	$\sigma(d, p)\mu b/sr$			$\sigma(d, t)\mu b/sr$		
	Unper-turbed	Per-turbed	Experi-mental	Unper-turbed	Per-turbed	Experi-mental
9/2 9/2⁻[505]	64	65	25	1.8	1.2	
7/2 7/2⁻[503]	357	364	284	38	48	72
9/2	7.0	8.0	≈3	0.6	1.3	≈1
3/2 3/2⁻[512]	138	208	131	29	60	58
5/2	249	184	96	48	34	35
7/2	30	48	87	5.8	17	44
9/2	6.0	3.9	4	1.0	1.0	<1
11/2	0.2	0.4	4	0.0	0.1	1
1/2 1/2⁻[510]	0.3	0.4	8	0.2	0.6	5
3/2	334	269	264	195	179	150
5/2	98	163	202	53	65	103
7/2	33	18	11	18	15	14
9/2	3.3	5.5	11	1.6	1.6	7
5/2 5/2–[512]	0.0	0.0	6	0.2	0.2	19
7/2	17	7.9	29	269	217	237
9/2	1.2	1.2	3	14	25	18
11/2	0.1	0.0		1.3	0.8	4
1/2 1/2–[521]	15	15	15	280	280	364
3/2	7.0	2.8	3	128	114	54
5/2	5.1	4.3		86	87	66
7/2	4.3	1.8		71	64	69
9/2	0.7	0.6		10	10	8
7/2 7/2–[514]	2.0	3.8	16	40	81	91
9/2	2.6	3.0	3	45	35	29
11/2	0.0	0.1	<4	0.8	1.3	15

* Casten, 1972.

slightly inconsistent, since the U^2 factors refer to the emptiness of Nilsson orbits in the target nucleus A in the (d, p) reaction while the V^2 factors relevant to (d, t) refer to the orbit occupancies in its $A + 2$ target nucleus. Nevertheless, this technique is widely used and is adequate for essentially all cases of practical importance.

We could now extract *empirical* C_j values from the measured (d, p) and (d, t) cross sections. For technical reasons, it is easiest and most common to use *theoretical* sets of C_j coefficients to calculate theoretical cross sections and to compare these with the measurements.

Table 9.2 *Comparison of unperturbed, Coriolis coupled and experimental cross sections in ^{185}W ($\theta = 90°$)**

State	$\sigma(d, p)\mu b/sr$			$\sigma(d, t)\mu b/sr$		
	Unper-turbed	Per-turbed	Experi-mental	Unper-turbed	Per-turbed	Experi-mental
9/2 9/2⁻[505]	66	62	25	3.3	3.9	10
7/2 7/2⁻[503]	334	341	316	58	72	154
9/2	6.4	10		1.0	2.1	≈ 3
11/2	0.5	0.5	≈ 6	0.1	0.2	1
3/2 3/2⁻[512]	107	30	5	59	1.7	1
5/2	195	247	301	97	173	207
7/2	23	7.2	11	12	1.0	1
9/2	4.7	6.5	11	2.1	4.5	11
11/2	0.2	0.1		0.1	0.0	≈ 0.4
1/2 1/2⁻[510]	0.2	0.2	4	0.3	0.9	3
3/3	180	262	357	343	423	308
5/2	53	0.3	≈ 4	92	14	11
7/2	18	37	104	32	55	99
9/2	1.8	0.0		2.7	0.1	≈ 1
11/2	0.1	0.2	5	0.2	0.4	4
5/2 5/2⁻[521]	0.0	0.0	< 3	0.2	0.2	14
7/2	17	7.7	32	268	208	206
9/2	1.2	1.1		14.1	24	≈ 20
1/2 1/2⁻[512]	14	14	< 62	281	280	≈ 266
3/2	6.4	2.8		129	107	≈ 43
5/2	4.7	3.8		86	89	≈ 20
7/2	3.9	1.6		72	60	≈ 25
9/2	0.5	0.6		10	10	11
7/2 7/2⁻[514]	2.1	4.1	15	40	86	80
9/2	2.7	1.2		45	34	19

* Casten, 1972.

By identifying appropriate sequences of cross sections, the states in an odd mass nucleus may be sorted into rotational bands whose Nilsson wave functions can be identified. Such assignments are made in Tables 9.1 and 9.2 for ^{183,185}W, and the theoretical and experimental cross sections are compared. In Fig. 9.2, the deduced J values and Nilsson quantum numbers are indicated above the corresponding peaks.

Careful inspection shows very different patterns for the different rotational bands, justifying the term *fingerprint patterns*. For example, the 1/2⁻[510] band has a very small cross section to the 1/2⁻ state, large cross sections to the 3/2⁻ and 5/2⁻ levels, and smaller cross sections thereafter. The $7/2^-[503]$ band has a large cross section only for the $7/2^-$ state. For the positive parity levels, essentially only the $13/2^+$ states are populated. (Note that we have not attached specific Nilsson quantum numbers to some of the latter levels. The reason will be clear after we have discussed the strong Coriolis mixing between these bands.) The U^2 and V^2 dependence in Eq. 9.1 is also evident in Fig. 9.2. For the low-lying bands where U^2 and V^2 are roughly comparable, the same states were populated in both (d, p) and (d, t), whereas at higher energies, the levels separate according to whether they are populated in stripping (U^2) or pick up (V^2). For example, the 1/2 $1/2^-[521]$ and 7/2 $5/2^-[512]$ hole states are stronger than the ground band states in (d, t), but much weaker in (d, p).

Tables 9.1 and 9.2 are well worth careful inspection. Although there are small differences in detail, especially for weaker states, the characteristic fingerprint patterns are often observed experimentally. Examples are the $5/2^-[512]$ and $1/2^-[510]$ bands in ^{183}W and the $7/2^-[503]$ band in ^{185}W. Indeed, these fingerprint patterns are often the technique used to identify the specific Nilsson orbits in the first place. The reader should not minimize the impressive successes of such a simple model, many of whose predictions can be anticipated without calculation despite the presence of perhaps dozens of valence nucleons.

Nevertheless, while qualitative *patterns* emulate the data, the detailed predictions often disagree substantially with the experimental results. Examples are the 5/2 $3/2^-[512]$, 5/2 $1/2^-[510]$, 3/2 $1/2^-[521]$, and 7/2 $7/2^-[514]$ states in both nuclei. There are differences of nearly an order of magnitude in the cross sections for populating certain corresponding states in the two nuclei. For example, in (d, p) in ^{183}W, the 5/2 $1/2^-[510]$ state is strongly populated while the 5/2 $3/2^-[512]$ state is weak; in ^{185}W, it is just the opposite.

Both of these phenomena are striking manifestations of the importance of the Coriolis interaction in odd mass nuclei. As we shall see, we can greatly improve the predicted cross sections if we take this residual interaction into account. The importance of the Coriolis interaction goes far beyond the question of sorting out difficulties with single nucleon transfer reactions. It is especially important for high-spin states in both odd and even mass nuclei, and has been shown to lead to a new coupling scheme—the so-called rotation aligned coupling scheme characterized by "decoupled" bands in many nuclei, and by the backbending phenomenon. It is in fact difficult to overestimate its significance in understanding odd mass deformed nuclei. We turn now to a systematic treatment of the Coriolis interaction with emphasis on its physical origin, its principal effects, and a simplified discussion of some easy ways to estimate its effects by inspection.

9.2   The Coriolis interaction in deformed nuclei

The origin of the Coriolis interaction has already been seen in Eq. 8.13, where the rotational energy expression for a single-particle coupled to a deformed rotor contains a term $-\hbar^2/2I\,(\mathbf{J}_+\mathbf{j}_- + \mathbf{J}_+\mathbf{j}_-)$ where $\mathbf{J}_\pm$ acts on the total angular momentum and $\mathbf{j}_\pm$ on the particle angular momentum. This term is an interaction between the rotational and single-particle motion and has physical effects similar to that of the Coriolis force on a classic rotating body. The simple properties of operators such as $\mathbf{J}_\pm, \mathbf{j}_\pm$ are discussed in any standard quantum mechanics text. They serve as raising and lowering operators for the z projections, K and Ω, of the total and single-particle angular momenta. Their matrix elements are (equating K and Ω and calling both K):

$$\langle K \,|\mathbf{J}_{\pm}| \, K \pm 1 \rangle = \sqrt{(J \mp K)(J + K + 1)} \tag{9.2}$$

$$\langle K \,|\mathbf{j}_{\mp}| \, K \pm 1 \rangle = \sqrt{(j \mp K)(j + K + 1)} \tag{9.3}$$

Note that both $\mathbf{J}_+$ and $\mathbf{j}_-$ *decrease* K while both $\mathbf{J}_-$ and $\mathbf{j}_+$ *increase* K.

The physical nature of the Coriolis interaction is easy to see. It is analogous to the classic effect that occurs in any rotating body. Consider the analogy of a projectile traveling northeast from the equator on the Earth's surface, illustrated in Fig. 9.3. A projectile launched at the equator initially travels eastward at a rate given by the rotational speed of the Earth at the equator. Since the Earth's circumference at higher latitudes is less, an observer at a northern latitude has a smaller rotational velocity. To this observer, the projectile will appear to be deflected further toward the east. This is the Coriolis effect, and while it is sometimes called a fictitious or apparent force, it has real physical effects. (It accounts for the fact that river banks tend to be eroded more on the right (facing downstream) side in the northern hemisphere. It also accounts for the fact that, following continental drift, South America has not merely separated from Africa but rotated away in the southern latitudes as well.) We see that the Coriolis interaction effectively tilts the orbit relative to the equator. If we now picture the orientation of the angular momentum vector perpendicular to the orbit instead of the orbit itself, the Coriolis effect is equivalent to a change in its projection onto the equator. Thus it is understandable that, in the nuclear case, the Coriolis interaction alters the projection of the angular momentum K on the symmetry axis, *admixing* different K values. Another way of looking at this that will be useful later is to recall that K is only a good quantum number if the nuclear potential is axially symmetric. Therefore, the Coriolis interaction effectively introduces small amounts of axial asymmetry as it mixes K values.

We now evaluate the Coriolis matrix element between Nilsson states explicitly. We consider two intrinsic Nilsson states characterized by K and $K + 1$. Since J is a good quantum number, we can replace $\mathbf{J}_{\pm}$ with its eigenvalue. This cannot be done with $\mathbf{j}_-$ because of the configuration (j) mixing in Nilsson wave functions, although it can be done approximately for the unique parity orbits where one j value dominates. Noting that only one of the two terms in $\mathbf{J}_+\mathbf{j}_- + \mathbf{J}_-\mathbf{j}_+$ gives a nonvanishing result, we get

$$\langle K \,|\mathbf{V}_{\mathrm{Cor}}| \, K + 1 \rangle = \frac{-\hbar^2}{2I}\sqrt{(J - K)(J + K + 1)}\,\langle K \,|\mathbf{j}_-| \, K + 1 \rangle \,(U_1 U_2 + V_1 V_2) \tag{9.4}$$

where the symbols K, $K + 1$ in the $\mathbf{j}_-$ matrix element are a shorthand for the two Nilsson wave functions and the effects of pairing are included in the factor $(U_1 U_2 + V_1 V_2)$. The pairing factor has the general effect of *reducing* the Coriolis matrix elements since its maximum value is unity. The reduction is least for orbits in similar positions relative to

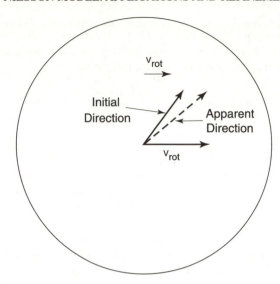

FIG. 9.3. Coriolis effect in a rotating system resulting from the dependence of rotational velocities on "latitude."

the Fermi surface (then $U_1 \approx U_2$ and $V_1 \approx V_2$, hence $(U_1 U_2 + V_1 V_2) \approx U_1^2 + V_1^2 = 1$). It is least for orbits laying on far opposite sides (then $U_1 \approx V_2$ and $U_2 \approx V_1$, so $(U_1 U_2 + V_1 V_2) \approx (U_1 V_1 + U_2 V_2)$ and one factor in each term is small). For diagonal Coriolis matrix elements, the factor is obviously unity.

For a single j shell, the $\mathbf{j}_-$ matrix element is

$$\langle jK \,|\mathbf{j}_-|\, jK + 1 \rangle = \sqrt{(j - K)(j + K + 1)} \qquad (9.5)$$

This is approximately correct for unique parity orbits. For an arbitrary Nilsson wave function, terms like this occur for each j, so that, in general,

$$\langle K \,|\mathbf{j}_-|\, K + 1 \rangle = \sum_j c_j^K c_j^{K+1} \sqrt{(j - K)(j + K + 1)} \qquad (9.6)$$

where the C_j^K and C_j^{K+1} coefficients are those describing the specific Nilsson wave functions. An interesting limiting case occurs for $J, j \gg K$, if j is approximately a good quantum number. This applies for low K orbits from the high-spin unique parity states or for high spin, low K states generally if β is small (little configuration mixing). Then Eqs. 9.4 and 9.6 give

$$V_{\text{Cor}} = \frac{-\hbar^2}{2I} J j \qquad (9.7)$$

This is a general upper limit on the strength of the Coriolis interaction. For typical inertial parameters (say 15 keV for rare earth nuclei) and, say, $J, j \approx 11/2$, this attains ≈ 400

keV! Typical spacings between Nilsson orbits are $\approx$150 keV. Coriolis mixing is not necessarily a minor perturbation!

Since the Coriolis matrix elements change K by $\Delta K = \pm 1$, it is generally a nondiagonal interaction. However, it has an important *diagonal* matrix element that contributes to certain rotational energies. For $K = 1/2$, the symmetrization of the wave function gives rise to terms with $K = \pm 1/2$, allowing a diagonal $\Delta K = 1$ contribution to the energies from the cross terms. Substituting Eqs 9.4 and 9.6 in Eq. 8.14, we get for the rotational energies for $K = 1/2$ bands including Coriolis mixing

$$E(J) = \frac{\hbar^2}{2I}\left[J(J+1) + \delta_{K\frac{1}{2}}a(-1)^{J+\frac{1}{2}}\left(J+\frac{1}{2}\right)\right] \tag{9.8}$$

where a is the well-known decoupling parameter given by substituting $K = -1/2$ in Eq. 9.6

$$a = \sum_j (-1)^{j-\frac{1}{2}}\left(j+\frac{1}{2}\right)C_j^2 \tag{9.9}$$

The two phase factors in Eqs. 9.8 and 9.9 come from the symmetrization of the wave functions (e.g., Eq. 6.15). The phase factor in Eq. 9.8, $(-1)^{J+1/2}$, means that the contribution to the rotational energies from the Coriolis interaction *alternates* in sign with J.

It is clear from Eq. 9.9 that a can be either positive or negative. The behavior of $E(J)$ in Eq. 9.8 as a function of a is shown in Fig. 9.4. If $a < 0$, states with spins $3/2, 7/2, 11/2 \ldots$ are lowered in energy while the alternate spin states are raised. For $a > 0$, the opposite situation occurs. On account of the factor $(J + 1/2)$ the effect grows with spin. On account of the factor $(j + 1/2)$ the effect, on average, increases for heavy nuclei ($j_{\text{ave}} \propto N$). For $|a| = 1$, Eq. 9.8 shows that the levels occur in degenerate pairs. If $a = -1$, the $J = 3/2$ state coincides with the $J = 1/2$ level. Similarly, the $(5/2, 7/2)$ and $(9/2, 11/2)$ pairs are degenerate. If $a = +1$, the degenerate pairs are $(3/2, 5/2), (7/2, 9/2), \ldots$. For $|a| > 1$, the level order within a rotational band is no longer monotonic in spin. Clearly, the typical rotational spacings can be so severely perturbed as to obscure the normal $J(J + 1)$ spacings and even the ordering of different spin states. As we shall see, these effects propagate via nondiagonal Coriolis mixing, and affect many bands with $K \neq 1/2$. We can now understand one of the anomalous rotational spacings and sequences in Fig. 8.8, specifically that for the $1/2^-[521]$ band. This is a $K = 1/2$ band with its decoupling parameter close to unity.

For an arbitrary Nilsson wave function, many terms can appear in Eq. 9.9 for a. Frequently these terms (each carrying a phase) largely cancel and the resultant a values are rather small, typically less than unity. However, very large a values can be obtained if the wave function is dominated by few terms with high j values. The classic example of this is the unique parity orbits for which $\langle j \rangle \approx N + 1/2$ (e.g., $j = 13/2$ in the $N = 6$ shell), and the wave functions are nearly pure in j. For these special orbits, Eq. 9.9 gives $a \approx (-1)^N(N + 1)$. For example, $a \approx 7$ for the $1/2^+[660]$ orbit. This enormous value so perturbs the normal spacing that the $13/2^+$ level is among the lowest-lying levels in the rotational band. For $\hbar^2/2I = 15\,\text{keV}$, the decoupling term is $\approx -750\,\text{keV}$! In general, the sign of a is always such that the $J = N + 1/2$ level (e.g., $J = 13/2$ for the $i_{13/2}$ shell) is lowered.

The reader may recall an apparent paradox in the order and spacing of rotational energies in odd-mass nuclei that was discussed in Chapter 8. The simplest view led to the notion that rotational bands should be "upside down." We showed that the "normal" order was regained when the precession of j around the symmetry axis was considered. We also pointed out that in some cases an "upside down" pattern does in fact occur. We have just encountered that case where large Coriolis effects in unique parity orbits upset the monotonic order of rotational energies with J. Having gone to great lengths to explain away this paradox in Chapter 8, why does it now appear in the data? In other words, what happened to the precession argument? As we shall see later, the physical difference here is that K is no longer a good quantum number ($\pm 1/2$ values are admixed), so the precession need not be about the symmetry axis: the Coriolis force, by mixing K values, forces the angular momentum vector to switch back and forth (precess) about the *rotation* axis instead. Thus, **j** and **R** are nearly parallel or antiparallel, so now $J = 13/2$ does correspond to $|R| \approx 0$, $J = 9$ to $|R| \approx 2$, and so on.

One often reads that a is called the *decoupling parameter* because it represents a decoupling of the rotational and single-particle motion. It is now easy to see why this name is appropriate. For $a = +1$, the energy differences of the $1/2, 5/2, 9/2, \ldots$ states are exactly the same as those between the $0^+, 2^+, 4^+, \ldots$ states of the even–even core: the odd particle acts like a spectator to the rotation. In fact, Fig. 9.4 and Eq. 9.8, in comparison to Eq. 6.17, show that degenerate pairs of levels ($J = 3/2 - 5/2, 7/2 - 9/2, \ldots$) occur at the same energies as the $2^+, 4^+, \ldots$ states of the even–even core nucleus. For $a = -1$, the alternate level pairs $5/2 - 7/2, 9/2 - 11/2, \ldots$ occur at energies (above the $1/2 - 3/2$ pairs) equal to the rotational spacings of the core nucleus. These facts have been associated with the identical band phenomenon discussed earlier (Chapter 6) although such an explanation still does not, in itself, account for the absence of an $A^{-5/3}$ scaling of energies that one would expect from Eq. 6.10. Moreover, it is clear from the comment above that changes in core rotation have little effect on the orientation of $\mathbf{J} = (\mathbf{R} + j)$ when a is large (i.e., when $|j|$ is large, K/j is small, so $\sin\theta = K/j$ is small, and **J** is nearly aligned along the rotation axis). Extrapolating to very large a values in Fig. 9.4 shows that the $J = 13/2$ state will lie lowest, followed by the $9/2$ level ($13/2 - 2$), and then by the $5/2$ level ($13/2 - 4$). The alternate spin levels are much higher. Thus the rotational energies of alternate J values (with $|J - j|$ even) are nearly parabolic in $|J - j|$ where j is the dominant j of the unique parity orbit. Sequential states differ mainly in R: the rotational motion is effectively decoupled from that of the odd particle. We shall encounter a related but even more dramatic effect later as a consequence of the nondiagonal Coriolis interaction.

Turning now to these nondiagonal Coriolis effects, there are two significant observable effects. One concerns rotational energies and stems from a propagation of the highly perturbed level order in the $K = 1/2$ band to higher K bands via successive $\Delta K = 1$ Coriolis mixings. Precisely because of the large decoupling parameters, this is most important for unique parity states and, as we shall see, accounts for their importance

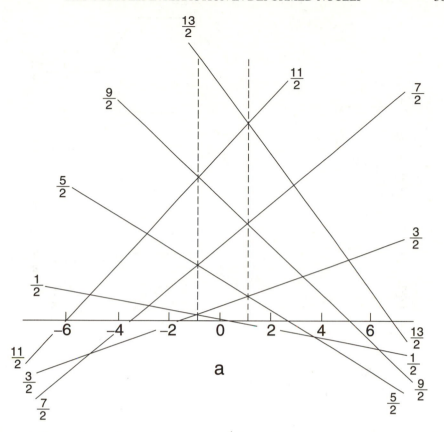

FIG. 9.4. Dependence of rotational energies (J) on the decoupling parameter a. The dashed lines are for $|a| = 1$ (based on Preston, 1975).

in high-spin studies where the $(J + 1/2)$ factor in Eq. 9.8 becomes crucial. The other effect occurs in "normal" rotational bands (especially in their impact on single nucleon transfer cross sections). We shall discuss this first.

Consider the admixture of two bands as shown in Fig. 9.5. Recalling our discussion in Chapter 1 of two-state mixing and the fact that the Coriolis mixing increases with J, the perturbed energies will behave as illustrated. It is easy to show that, to first order, the effect of the Coriolis interaction is to decrease the effective inertial parameter, $\hbar^2/2I$, for the lower band and to increase it for the higher band.

Equation 1.6 gives the energy shifts of the two interacting states relative to the unperturbed spacing for a given spin J

$$\frac{\Delta E_s}{\Delta E_u} = \frac{1}{2}\left[\sqrt{1 + \frac{4V^2}{\Delta E_u^2}} - 1\right] \approx \frac{V^2}{\Delta E_u^2}$$

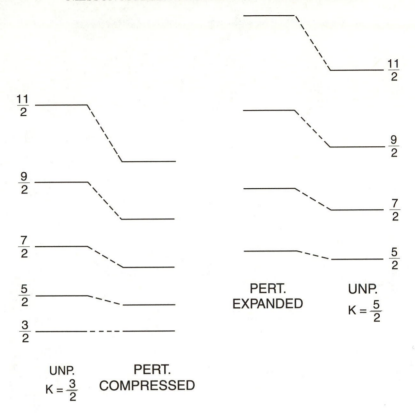

CORIOLIS MIXING OF 2 BANDS

FIG. 9.5. Illustration of the changes in effective rotational parameters, $\hbar^2/2I$, resulting from two-state Coriolis mixing.

where the last step assumes small mixing (Eq. 1.12). Thus $\Delta E_s \propto V_{\text{Cor}}^2$. Isolating the spin dependence Eq. 9.4 gives

$$\Delta E_s \propto (J - K)(J + K + 1)$$

or

$$\Delta E_s \propto [J(J+1) - K(K+1)] \tag{9.10}$$

which is just the rotational energy expression (J, K dependent parts). The Coriolis interaction merely alters the effective rotational spacings. The lower band is compressed, the upper one expanded. This simple result breaks down for very large or multistate mixing, but even in these cases gives a useful framework.

In general, there can be many low-lying Nilsson bands with assorted K values; thus a realistic Coriolis mixing calculation will be a multistate diagonalization. The admixed wave functions can be written

$$\psi(J) = \sum_i \alpha_i \phi^i_{\text{Nils}}(J) \qquad (9.11)$$

where the α_i are the mixing amplitudes that depend on J.

In this way, for example, a predominantly $K = 3/2$ band may contain admixtures of $K = 7/2$ through the intermediary of $K = 5/2$. We shall see later that this is especially important for the unique parity case. Although such mixing can be rather complicated, one can almost always estimate the effects of a full multistate diagonalization rather accurately by carrying out a sequence of two-state mixing calculations. We shall see examples of this shortly.

At this stage it is important to discuss the actual magnitudes of the Coriolis mixing matrix elements. The quantities $\hbar^2/2I$, $(J - K)(J + K + 1)$, and the U, V factors are easy to estimate, as we have done. They are dependent on the orbits involved but not on their detailed structure (Nilsson wave functions). However, $\langle \psi_{\text{Nils}}(K) |\mathbf{j}_-| \psi_{\text{Nils}}(K + 1)\rangle$ depends explicitly on the structure of the states included, as evident in Eq. 9.6.

As with the decoupling parameter, the nondiagonal $\mathbf{j}_-$ matrix elements for the unique parity orbits are both large and particularly simple to calculate, since one j term in Eq. 9.6 dominates. For example, for $\beta \approx 0.23$ and typical Nilsson parameters κ and μ, the Coriolis matrix element of $\mathbf{j}_-$ connecting the $1/2^+[660]$ and $3/2^+[651]$ Nilsson orbits is ≈ 6.6. If we had assumed the wave functions consisted only of the $j = 13/2$ component, Eq. 9.6 gives $\langle K |\mathbf{j}_-| K + 1\rangle \approx N+1$, which is the maximum possible value for any matrix element of j_- in a given shell. Thus, to a good approximation, the unique parity Coriolis matrix elements can be estimated using the single j approximation, and moreover, they have very nearly the maximum possible values.

Table 9.3 provides a number of other examples of off-diagonal $\mathbf{j}_-$ matrix elements in the $N = 5$ shell. The reader may easily derive these numbers by applying Eq. 9.6, using the wave functions given earlier in Table 8.2. Inspection of Table 9.3 reveals two global features. First, Coriolis matrix elements for nonunique parity orbits are considerably smaller. Second, their relative values can differ by more than an order of magnitude.

In considering a given level scheme and attempting to determine whether Coriolis mixing effects will be important, it is often useful to be able to estimate, without calculation, the approximate magnitude for Coriolis matrix elements. A very simple rule allows one to do this. For nonunique parity orbits, Coriolis matrix elements divide roughly into two classes, allowed and nonallowed. The allowed matrix elements are those in which both n_z and Λ change by one unit, but their sum remains constant ($\Delta n_z = -\Delta\Lambda$). (The unique parity case is, of course, in this class.) Examples from the table are the matrix elements between the $3/2^-[512]$ and $1/2^-[521]$ orbits or between the $7/2^-[514]$ and $9/2^-[505]$ orbits. Such allowed $\mathbf{j}_-$ matrix elements are typically on the order of $N/2 - N/3$. All of the others are nonallowed matrix elements and are typically < 1. Examples of these are the matrix elements between $7/2^-[514]$ and $5/2^-[512]$ or $1/2^-[510]$ and $3/2^-[512]$.

Table 9.3 *Theoretical values* $\langle K|j_-|K+1\rangle^*$

	9/2⁻[505]	7/2⁻[503]	3/2⁻[512]	1/2⁻[510]	5/2⁻[512]	1/2⁻[521]	7/2⁻[514]
9/2⁻[505]		−0.973					2.847
7/2⁻[503]	−0.973				2.858		
3/2⁻[512]				0.951	0.045	2.546	
1/2⁻[510]			0.951			−2.541	
5/2⁻[512]		2.858	0.045				−1.151
1/2⁻[521]			2.546	−2.541			
7/2⁻[514]	2.847				−1.151		

* The Nilsson model parameters are $\delta = 0, 2, \kappa = 0.0637, \mu = 0.42$.

It is important to note that Coriolis mixing between orbits in the latter class is not always negligible, especially if they have the same n_z values. As we have seen, the Nilsson diagram separates approximately according to n_z values for large deformations. Therefore, it can often occur that two orbits with identical n_z values differing in K by $\Delta K = \pm 1$ lay very close to each other. Their Coriolis mixing can be large even with a small matrix element. A classic example of this occurs for the 1/2⁻[510] and 3/2⁻[512] orbits in [183,185]W. We shall discuss this case in some detail momentarily.

Having dealt with typical values of the $\mathbf{j}_\pm$ matrix elements, it is useful to develop a feeling for the absolute magnitudes of the full Coriolis matrix elements in Eq. 9.4. For well-deformed nuclei, $\hbar^2/2I \approx E_{2_1^+}/6$ where $E_{2_1^+}$ is given by a neighboring even–even nucleus. For rare earth nuclei, $\hbar^2/2I \approx 14\text{–}18$ keV, while for the actinides, $\hbar^2/2I \approx 7$ keV. The matrix elements of $\mathbf{J}_\pm$ given by the square root factor are typically 2–3 for moderate spin states, although they can become very large for high spins. Finally, as noted, the pairing factor becomes very small for high-lying orbits on opposite sides of the Fermi surface, while for orbits near the Fermi surface, or for those on the same side, this factor is typically between 0.7 and 1.0. Thus, typical nonunique parity *allowed* Coriolis matrix elements in the rare earth region are roughly $V_{\text{Cor}} \approx (16)(3)(1)(0.8) \approx 40$ keV. This estimate is only accurate to a factor of 2–3. *Non*allowed Coriolis matrix elements will, of course, be less.

For unique parity orbits, the $\mathbf{j}_-$ matrix elements are $\sim N$. In addition, the *observed* states are typically of rather high J, since single nucleon transfer reactions preferentially populate the $J = N + 1/2$ states for which $C_j \approx 1$ and heavy ion reactions tend to feed the high-spin unique parity levels. Taking $\sqrt{(J-K)(J+K+1)} \approx N$, the Coriolis matrix elements linking unique parity orbits can be extremely large, typically reaching V_{Cor} (unique parity) $\approx 16(6)(6)(0.8) \approx 400$ keV. Such matrix elements mixing states often only a couple of hundred keV apart have enormous structural effects.

From extensive experience with Coriolis mixing calculations, it has been found that the *actual empirical* matrix elements are generally about 20–50% lower than these theoretical estimates. This conclusion emerges from comparisons of extensive data on level energies and single nucleon transfer cross sections in many deformed nuclei. We will not detail this evidence here, but one example of it is trivially evident in Fig. 9.1, which shows that the $7/2^-$ states of the $7/2^-[514]$ and $5/2^-[512]$ bands in ^{183}W are separated by only 70 keV. If we consider this an isolated two-state system (an approximation good enough for the present purposes although not for detailed calculations) and recall that such states can never be closer than twice their mixing matrix element, then the Coriolis matrix element between these two states must be ≤ 35 keV. Since both bands are hole excitations, the pairing factor is near unity. Using $\hbar^2/2I = 18$ keV $[E_{2^+_1}(^{184}$W$) = 111$ keV$]$ and the $\mathbf{j}_-$ matrix element from Table 9.3 gives a predicted Coriolis matrix element of 55 keV. The maximum matrix element allowed empirically is only 65% of this. Other similar examples abound.

Despite this attenuation, Coriolis mixing effects, especially among unique parity orbits, represent a substantial perturbation to the rotational picture and can seldom be ignored. One final point to emphasize before considering some actual calculations is that in weakly deformed and transitional nuclei, Coriolis matrix elements are far larger than in well-deformed nuclei because of the smaller moments of inertia. Matrix elements between unique parity orbits may reach an MeV or more, and under certain circumstances may even lead to a new coupling scheme, the so-called rotation aligned scheme we shall discuss later.

As an example of multistate Coriolis mixing, let us consider the level scheme of ^{183}W shown in Fig. 9.1. In principle, a full calculation cannot neglect the unseen bands that occur at higher energies, but in practice, one assumes that their effects are small (at least for nonunique parity orbits). We will see one way to estimate whether such an assumption is grossly violated. Under this assumption, the Coriolis mixing among the negative parity bands involves diagonalizing matrices of varying size, 2×2 for $J = 1/2$, up to 6×6 for $J = 11/2$. For the positive (unique) parity states, the strength of the Coriolis mixing precludes safely ignoring unseen bands and therefore one usually carries out a full 7×7 diagonalization.

We consider first the negative parity states. We note the empirical result (from ratios of (d, p) to (d, t) cross sections) that the Fermi surface is slightly below the $1/2^-[510]$ orbit. For simplicity we ignore the $9/2[505]$ band. (In any case, it can only affect $J \geq 9/2$ states.) From our earlier discussion, we anticipate that the principal mixing effects will occur between the $1/2^-[521] - 3/2^-[512]$ and $5/2^-[512] - 7/2^-[503]$ pairs. That this is not quite true highlights the other factors that must be taken into account in practical situations. Although the matrix element connecting the $1/2^-[510]$ and $3/2^-[512]$ orbits is rather small (≈ 1), they lay so close to each other that the mixing is substantial. Likewise, the "forbidden" matrix element between the $5/2^-[512]$ and the $7/2^-[514]$ orbits (≈ 1.1), strongly admixes these close-lying bands. In contrast, despite the large $\mathbf{j}_-$ matrix element between the $5/2^-[512]$ hole orbit and the particle excitation $7/2^-[503]$, the pairing factor substantially reduces the overall matrix element. The one exception to the simple rule given above for estimating Coriolis matrix elements among these bands occurs for the $1/2^-[510]$ and $1/2^-[521]$ pair: the supposedly forbidden $\mathbf{j}_-$ matrix element has a value ≈ 2.5. Although these bands are nearly an MeV apart, the coupling between them is nonnegligible. Thus, the principal Coriolis admixtures will be between the $1/2^-[510]$ and $3/2^-[512]$ bands, the $5/2^-[512]$ and $7/2^-[514]$ bands, and the

Table 9.4 *Calculated mixing amplitudes* $(\alpha_i \times 100)$ *for the* $7/2^-$ *states in* ^{183,185}W

$K\pi[Nn_z\Lambda]$	A	E_x	$7/2^-$[503]	$3/2^-$[512]	$1/2^-$[510]	$5/2^-$[512]	$1/2^-$[521]	$7/2^-$[514]
$7/2^-$[503]	185	244	+100	—	—	+5	—	—
	183	453	+100	—	—	+5	—	—
$3/2 - $[512]	185	174	—	+91	+42	—	+2	—
	183	412	—	−96	+26	—	−8	—
$1/2 - $[510]	185	334	—	−41	+90	—	−11	—
	183	207	—	+27	+96	—	−5	—
$5/2 - $[512]	185	986	−5	—	—	+98	—	−19
	183	1002	−5	—	—	+98	—	−17
$1/2 - $[521]	185	1335	—	+6	−9	—	−99	—
	183	1265	—	−6	−7	—	−100	—
$7/2^-$[514]	185	1058	+1	—	—	−19	—	−98
	183	1072	+1	—	—	−17	—	−99

* Casten, 1972.

$1/2^-$[510] and $1/2^-$[521] bands. Second-order mixtures of, say, the $1/2^-$[510] into the $5/2^-$[512] band, will be very small. Thus, a rather good simulation of the full diagonalization should be obtainable by considering sequential two-state mixing of the preceding three pairs. As an example, consider the $7/2^-$ states with a Coriolis attenuation factor 0.7. With $\hbar^2/2I = 18$ keV and a pairing factor of 0.9, we obtain ≈ 41 keV for the full $1/2^-$[510]–$3/2^-$[512] Coriolis matrix element. The final spacing of the $7/2$ $1/2^-$[510] and $7/2$ $3/2^-$[512] states is 205 keV. Working backwards in Fig. 1.7, we see that R must be rather large and therefore the energy shift induced in each state by the mixing is a small fraction (≈ 0.05) of their unperturbed spacing. We can therefore estimate that $R \approx 3.8$. Another application of Fig. 1.7 or Eq. 1.8 gives the admixed wave functions as $\psi("7/2\ 1/2^-[510]") = (0.97)\ 7/2\ 1/2^-[510] + (0.24)7/2\ 3/2^-[512]$ and the orthogonal combination. For the $7/2\ 1/2^-$[510]–$7/2\ 1/2^-$[521] mixing we take a pairing factor of 0.6, giving the full Coriolis matrix element of ≈ 88 keV. $\Delta E_{\text{final}} = 1058$ keV, so we can use the final spacings to obtain $R \approx 12$. The admixed wave functions are $\psi("7/2\ 1/2^-[510]") = (0.99)7/2\ 1/2^-[510] + (0.08)\ 7/2\ 1/2^-[512]$ and the orthogonal combination. Finally, for $7/2\ 3/2^-$[512] and $7/2\ 1/2^-$[521], the pairing factor is ≈ 0.5. Calculations again give $R \approx 12$, and final wave functions of $\psi("7/2\ 3/2^-[512]") = (0.99)\ 7/2\ 3/2^-[512] + (0.08)\ 7/2\ 1/2^-[521]$, and the orthogonal combination. Note that, instead of working backwards in Fig. 1.7, we could have directly used Eq. 1.5a instead. In all three cases the signs of the amplitudes are arbitrary.

We can test these estimates by reference to the detailed wave functions resulting from a full diagonalization given in Table 9.4. The three admixtures just calculated are $1/2^-$[510]–$3/2^-$[512] $= 0.24$, $1/2^-$[510]–$1/2^-$[521] $= 0.08$, and $3/2^-$[512]–$1/2^-$[521] $= 0.08$. The exact calculations give almost the same results.

We can also estimate the energy shifts. Using the same R values we get the following results (in keV):

$$7/2 \; 1/2^-[510] : -12.0(3/2^-[512]) - 7.2(1/2^-[521]) = -19.2 \, \text{keV}$$

$$7/2 \; 3/2^-[512] : +12.0(1/2^-[510]) - 5.8(1/2^-[521]) = +6.2 \, \text{keV}$$

$$7/2 \; 1/2^-[521] : +7.2(1/2^-[510]) + 5.8(3/2^-[512]) = +13.0 \, \text{keV}$$

where the orbits in parenthesis give the mixing partner that induced each shift. Again, these estimates are close to the results of an exact calculation.

With these shifts, and similar ones for other J values, the $1/2^-[510]$ band is compressed and the $1/2^-[521]$ and $3/2^-[512]$ bands are expanded, reflecting the derivation in Eq. 9.10 that, to first order, Coriolis induced energy shifts can be absorbed into changes in $\hbar^2/2I$. Indeed, one clue to the presence of Coriolis effects in empirical level schemes is unequal *empirical* $\hbar^2/2I$ values (i.e., after mixing), with larger magnitudes for the higher-lying (expanded) bands and smaller values for the lower (compressed) states. A measure of the adequacy of a calculation is whether the *input* (unperturbed) $\hbar^2/2I$ values are substantially closer: they should be if the deformation is the same for all excited states and small microscopic "blocking" effects are neglected.

9.3   Coriolis mixing and single nucleon transfer cross sections

Strong Coriolis effects are at the heart of most current research in high-spin states, and thus occupy a crucial role in modern nuclear structure physics. They were first studied extensively in single nucleon transfer reactions, however, and, although such work is not so common nowadays, it is an appropriate starting point for our discussion since the effects of Coriolis mixing are so dramatic and also easy to understand physically in this context.

While the energy shifts previously discussed may seem rather small and perhaps easily negligible, such an impression is misleading because even small admixtures can have large effects on single nucleon transfer cross sections. The expression for the cross section to a state of spin J in a given band in the presence of Coriolis mixing is an obvious generalization of Eq. 9.1 given by

$$\frac{d\sigma}{d\Omega}(J) = 2\mathcal{N}\phi_l(\theta)\left[\sum_i \alpha_i C_j^i P_i\right]^2 \tag{9.12}$$

where the α_i's are the Coriolis mixing amplitudes and the sum is over the admixed bands. Note that the sum is coherent, thus magnifying the effects. Simple manipulations also show that the total cross section is conserved for each spin J: that which is lost by some states must be gained by others.

Before considering the example of ^{183}W in detail, it is useful to emphasize how small mixing amplitudes can have significant effects. For simplicity, assume a two-state mixing of bands with identical C_j coefficients and pairing factors for some spin J. Then, if the mixing amplitude of each band in the other is 0.22 (meaning that the amplitude for the "parent" state is still 0.975), this gives a 50% increase in the cross section of one state and a 50% decrease in the other [$\sigma_1 \propto (1.22)^2$, $\sigma_2 \propto (0.78)^2$]. The two cross sections that would have been equal without mixing now differ by a factor of three!

Another feature is evident from Eq. 9.12. If two admixed states have very *different* unperturbed C_i values for some J, the state with the larger C_j value will be relatively unaffected while that with the smaller may be drastically altered. Indeed, much of the resultant cross section may easily come from the small admixture rather than from the parent orbit itself. To be specific, suppose the two bands have equal pairing factors, that $C_j^1 = 0.2$ and $C_j^2 = 0.8$, and that the mutual mixing amplitudes are ± 0.22. Then, assuming that the phases are such that the cross section for the state of spin J in band 1 is increased, the ratio of perturbed to unperturbed cross sections is ≈ 3.5 for band 1 $[(0.2 + 0.22(0.8)]^2$ and ≈ 0.9 for band 2 $[(0.8 - 0.22(0.2)]^2$. This is another example (bandmixing in even nuclei was the first) of how relatively small mixing interactions and amplitudes can lead to drastic effects on certain observables, especially when one of the unperturbed transition rates is small or forbidden. If the phases were reversed (which would happen if the *unperturbed* positions of the two bands were *exchanged*), the same analysis shows that despite the small mixing, the cross section for band 1 would essentially vanish while that for band 2 would increase only by about 6%. Finally, if one C_j coefficient is nearly zero, the cross section will come only from the mixing. It will therefore be independent of the signs of the mixing amplitudes and will always be increased by the mixing.

Simple application of Eq. 9.12 to the mixing amplitudes such as those given in Table 9.4 for the $7/2^-$ states of the negative parity bands in ^{183}W gives the cross sections labeled "perturbed" in Table 9.1. The point of this section is highlighted by the enormous differences between perturbed and unperturbed cross sections even in cases where the mixing amplitudes of Table 9.4 are small. For example, the cross sections for the $3/2^-$ levels of the $1/2^-[510]$ and $3/2^-[512]$ bands are significantly shifted by the mixing. The same applies to the $J = 5/2$ and $7/2$ states of these bands and to the $7/2$ states of the $5/2^-[512]$ and $7/2^-[514]$ bands. Table 9.2 shows similar Coriolis mixing results in ^{185}W.

Figure 9.6 shows the systematics of some empirical and Coriolis calculated cross sections across the odd-mass W isotopes. It includes a comparison with the unmixed cross sections. The latter are constant except for small, smooth changes in the pairing factor P_i^2 in Eq. 9.1. The figure highlights the changes in single nucleon transfer cross sections brought about by the Coriolis interaction as well as the dramatic shifts that can occur from one isotope to another. This is particularly evident for the $1/2^-[510]$ and $3/2^-[512]$ bands in ^{183}W and ^{185}W. An understanding of this is obvious from our discussion of two-state mixing in Chapter 1. As we have stated before, these two bands have interchanged positions (energies) in these two nuclei. In ^{183}W, the $1/2^-[510]$ band is the lower, ground state orbital, while the $3/2^-[512]$ band is the first excited intrinsic excitation. In ^{185}W, the $3/2^-[512]$ orbital forms the ground state and the $1/2^-[510]$ band is close, lying just above it. Therefore, the *signs* of the mixing amplitudes are inverted between ^{185}W and ^{183}W. In the familiar terminology of first-order perturbation theory, the sign of the energy denominator has changed from one nucleus to the other. Therefore, for those states where the two unperturbed Nilsson wave functions have comparable C_j coefficients, cross sections that were increased in ^{183}W will be decreased in ^{185}W and vice versa. For cases where the C_j coefficient in one is negligible, the cross section to

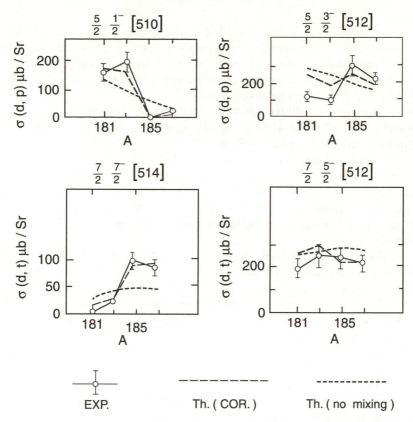

FIG. 9.6. Systematics of experimental, unmixed, and Coriolis coupled (d, p) and (d, t) cross sections in W isotopes.

that state will increase relative to the unmixed case in both nuclei. Another example of inversion concerns the $7/2^-$[514] and $5/2^-$[512] bands, in which the $7/2^-$ levels interchange positions between ^{181}W and ^{183}W. Figure 9.6 shows the dramatic effect on the weaker cross section.

Before turning to the positive parity levels, it is worth re-emphasizing the extremely large effects involved here. Empirical fingerprint patterns automatically incorporate the effects of Coriolis mixing and can differ from those predicted by the Nilsson model by sufficiently large quantities as to completely obscure the identification of the bands if Coriolis mixing is not taken into account. Moreover, the mistakes that one would make would not even necessarily be the same in neighboring nuclei, and the systematics of the Nilsson orbits deduced could be completely wrong.

9.3.1 Unique parity states

We now consider the unique parity orbits. Here, the Coriolis mixing becomes at once much stronger but also somewhat simpler to interpret. The reason is that, for practical purposes, only the $J = N + 1/2$ ($J = 13/2$ for the odd neutron rare earth nuclei) state is important because of the utter dominance of the $C_{j=N+1/2}$ coefficients in the Nilsson wave functions for these orbits. Moreover, the Coriolis matrix elements are all very similar and only directly link adjacent orbits stemming from the same j shell (although second-order ($\Delta K = 2$) mixing is significant).

Before discussing practical calculations, let us take a schematic model. Assume the Fermi surface lies below the whole group of unique parity orbits in some nucleus. The order of their excitation energies is $K = 1/2, 3/2, \ldots, (N + 1/2)$. Each mixes with the $K + 1$ and $K - 1$ member of the series. We thus have a situation analogous to one discussed in Chapter 1, in which equally spaced states each mix with adjacent levels. One general result is that lowest band will be pushed much lower. Moreover, given the increase of the Coriolis matrix elements with J, it will be severely compressed, and its wave functions will be a complex mixture of several components with all wave function components in *phase*.

The principal difference between this schematic situation and the real one arises because of the large decoupling parameter for the $K = 1/2$ band. For $N = 6$, this has the effect of greatly lowering the $13/2^+, 9/2^+, 5/2^+, 1/2^+$ states and raising the $11/2^+, 7/2^+, 3/2^+, \ldots$ states of that band prior to mixing. Consider now the effect of Coriolis mixing on the nearby $K = 3/2$ band. The situation is illustrated in Fig. 9.7. The reordering of energies in the $K = 1/2$ band because of the large diagonal Coriolis effect (decoupling parameter) causes the unperturbed spacings between the $3/2, 7/2,$ and $11/2$ states of the two bands to be much larger than between the $5/2, 9/2,$ and $13/2$ states. Therefore, in the lower-lying $K = 3/2$ band, the $5/2, 9/2, \ldots$, group is shifted down substantially more than the $3/2, 7/2, \ldots$, group. The perturbed energies of the $K = 3/2$ band take on an alternating pattern as well, relative to a pure $J(J + 1)$ rotational spacing, and appear *as if* the $K = 3/2$ band had a decoupling parameter of the same sign and slightly smaller magnitude than the $K = 1/2$ band. When the $K = 3/2$ band in turn mixes with the $K = 5/2$ band, this "signature" is passed on in a somewhat reduced form. In effect, the Coriolis mixing "propagates" the decoupling parameter throughout the entire sequence of unique parity orbitals.

If the Fermi surface is below the $K = 1/2$ orbit, the effect is reversed. The $K = 1/2$ band lies *below* the $K = 3/2$ band and therefore the closest lying pairs are the $3/2, 7/2,$ and $11/2$ states. Also the propagation is severely damped by the pairing factor as one goes from hole states to particle states. In the W isotopes that we have been considering, the Fermi surface is near the $9/2^+[624]$ and $11/2^+[615]$ orbits, and the effect of the $K = 1/2$ band is negligible. (This is linked with a point we will make shortly, that strong diagonal Coriolis matrix elements are most effective in inducing a rotation aligned coupling scheme when the Fermi surface lies near the low K orbits.) In W, the primary observable mixing among the unique parity orbits should be in the $K = 5/2, 7/2, 9/2,$ and $11/2$ orbits. As contrasted with the normal parity states, here the matrix elements and spacings are comparable and a two-state mixing calculation is hopelessly crude. The results for ^{183}W of an explicit calculation of the single nucleon transfer strengths ($C_{j=13/2}$ coefficients) to $13/2^+$ states is shown in the top two panels of Fig. 9.8. Since the $\mathbf{j}$ matrix elements arise almost solely from the $j = 13/2$ term in Eq. 9.6, and since the

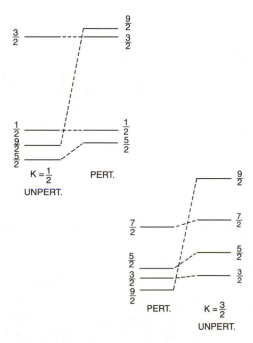

FIG. 9.7. Illustration of how the irregular rotational spacings in strongly decoupled $K = 1/2$ bands can propagate to $K \neq 1/2$ bands via Coriolis mixing.

$C_{j=13/2}$ coefficients all have the same sign, the phases of the resulting wave functions are such that in any two-state mixing of these orbits, the cross section to the *lower* state is increased while that to the higher is decreased. This persists in the multistate extension, and the net effect is to transfer cross section from the higher-lying bands to the lower ones. This is the point alluded to in the schematic model at the beginning of this discussion. We now compare this calculation with the empirical situation.

In ^{183}W, the lowest-lying unique parity orbit is $11/2^+[615]$. The $9/2^+[624]$ and $7/2^+[633]$ orbits are hole excitations. An ideal reaction to study these unique parity orbits is the $(^3\mathrm{He}, \alpha)$ reaction, which preferentially excites higher-spin hole levels. In the rare earth region it can almost be used as a "$J = 13/2^+$ meter." Typical $(^3\mathrm{He}, \alpha)$

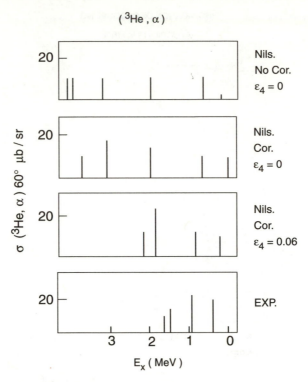

FIG. 9.8. (^{3}He, α) cross sections for ^{183}W (based on Kleinheinz, 1973).

spectra, for $^{181-185}$W, are shown in Fig. 9.9. Comparison with the (d, t) reaction in Fig. 9.2 vividly illustrates the selectivity. In the absence of Coriolis mixing, the (^{3}He, α) reaction to $13/2^+$ states in ^{183}W should look like that shown schematically at the top of Fig. 9.8, in which there are five nearly equally strong peaks. (That for the $K = 11/2^+$ is weaker than the others due to the smaller V^2 factor and the peak from the $13/2^+$[606] orbit is absent since $V \approx 0$.) In contrast, the data show only four peaks but with the same total cross section expected for the six unperturbed states. This illustrates both the shifting and the descent of strength just discussed. (The empirical (^{3}He, α) spectrum (Fig. 9.9) for ^{181}W shows this effect even more; only two peaks consume nearly all the $13/2^+$ strength.) The bottom-most panel of Fig. 9.8 summarizes the empirical $C_{j=13/2}$ coefficients in bar graph form for ^{183}W.

The second panel of Fig. 9.8 includes Coriolis mixing, and is somewhat better than the unmixed calculations shown in the top panel. Further improvement, however, requires, as we shall now see, the introduction of hexadecapole deformations. The study of such shape components offers us an ideal situation to apply the same kind of intuitive approach we used for the Nilsson model itself.

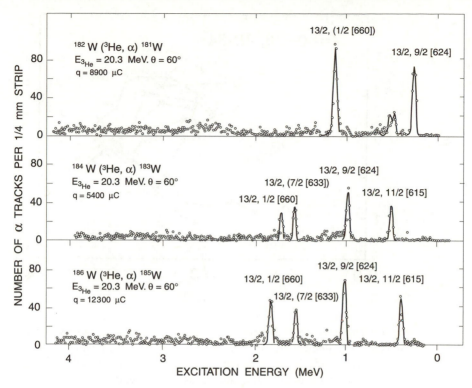

FIG. 9.9. (^{3}He, α) spectra in W isotopes (Kleinheinz, 1973).

9.3.2 Hexadecapole deformations and unique parity states

Although quadrupole distortions dominate heavy nuclei, hexadecapole effects are not at all negligible and, as we shall see, frequently have major impact. It is by now well known that most heavy deformed nuclei have either positive or negative hexadecapole deformations superimposed on their quadrupole distortions. Figure 6.10 illustrated the effect of hexadecapole deformations on the nuclear shape for both signs of ε_4. For $\varepsilon_4 > 0$, the nucleus takes on a so-called pin cushion or barrel shape, while for $\varepsilon_4 < 0$, it is "clover leafed." Remarkably, we can now understand, without calculation, the origin of the shape components, their expected systematics, and their effects on Nilsson energies, on Coriolis mixing, and on single nucleon transfer cross sections. To illustrate the usefulness of this approach, we shall carry out the following discussion without any formal derivations. We will then compare our understanding with the results of actual calculations.

It is obvious that creating a positive ($\varepsilon_4 > 0$) hexadecapole shape requires the occupation of orbits situated at large radii relative to the bulk of the nuclear matter in orientations roughly 45° to the equatorial plane, so that the "corners" of the mass distributions will be filled in. In a given shell, the orbits with the largest radii are the unique parity orbits since they stem from the next higher oscillator shell. From the relation $\sin \theta \sim K/j$, we see that $\theta = 45°$ corresponds to $K/j \sim 0.7$. For the $i_{13/2}$ neutron orbit in

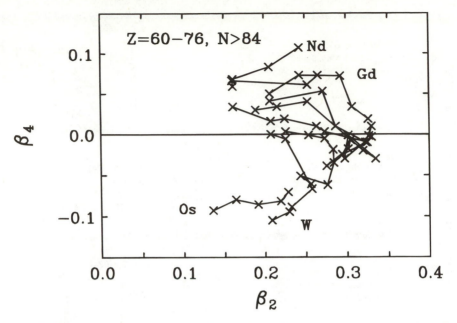

FIG. 9.10. Empirical systematics of β_4 deformations ($\beta_4 \sim -\varepsilon_4$) in the rare earth nuclei, plotted against β_2. The sideways horseshoe pattern is typical: large positive β_4 early in the shell as β_2 increases, β_4 changing sign just after midshell, and then large negative β_4 values as the quadrupole deformation decreases near the end of the shell.

the rare earth region, this gives $K \sim 9/2$. Thus, the largest contributions to an increasing positive hexadecapole deformation in this region occurs when the $9/2^+[624]$ orbit is filling near the Fermi surface. Simple counting in the Nilsson scheme shows that this occurs around $A \sim 180$.

At the other extreme, analogous reasoning shows that negative ε_4 values are favored when very low or very high K $i_{13/2}$ orbits are filling. The former occur near the beginning of the deformed region at neutron number $N \sim 92$. The latter occur near the end of the shell and serve to reduce the positive ε_4 values for $A > 180$. The expected ε_4 systematics should therefore consist of large negative values at the beginning of the deformed region that decrease in absolute value with mass, cross zero, turn positive, and maximize around the W isotopes, followed by a rather rapid decrease towards zero as the ^{208}Pb closed shell is approached. This is *exactly* the systematics observed empirically, as shown in Fig. 9.10.

[A technical point is worth mentioning here. In this discussion, we have used ε_4 as the deformation parameter. In Nilsson's original paper, the principal discussion was carried out in terms of deformations β_2 and β_4. The disadvantage of these parameters when discussing $\Delta N = 2$ mixing and hexadecapole deformations is that, even for $\beta_4 = 0$, there will be finite $\Delta N = 2$ mixing. In contrast, the ε representation discussed in an appendix to Nilsson's original article, was *designed* so that $\Delta N = 2$ mixing vanishes when $\varepsilon_4 = 0$. The relation between ε_2, ε_4 and β_2, β_4 is complex and coupled. Either set of deformation parameters may be converted into the other by the use of Fig. 9 of the article by S.G. Nilsson et al (1969): However, one should note that there is a mistake in this figure and that its proper use requires the reversal in the sign of ε_4. *Very* crudely, for typical quadrupole deformations, $\beta_2 \sim \varepsilon_2$, $\varepsilon_4 \sim -\beta_4$. β_4 is used in Fig. 9.10]

We can go one step further. Since the orbit inclination θ changes slowly for low K, there will be more low-angled orbits ($\theta < 45°$) than orbits near 45°. Therefore, negative ε_4 deformations should predominate and the "crossing point" to positive values should occur past midshell. This is also seen in Fig. 9.10.

The principal effect of hexadecapole deformations on the Nilsson wave functions is to admix components with $\Delta N = \pm 2$. Thus, the $N = 6$ $i_{13/2}$ Nilsson orbits will now contain some components from the $N = 4, 8$ shells, and the $N = 5$ normal parity orbits will contain contributions from $N = 3, 7$. Normally, this $\Delta N = 2$ mixing is minuscule because of the large energy separation of oscillator shells. However, inspection of the characteristic form of the Nilsson diagram (downsloping early, upsloping late), shows that there are a few isolated regions where steeply downsloping unique parity orbits from one shell (e.g., $i_{13/2}$) intersect upsloping orbits from the next lower shell (here, $N = 4$). If the nuclei in such regions have large ε_4, then substantial $\Delta N = 2$ mixing can occur.

Thus far we have discussed the origin of hexadecapole deformations, their systematics, and their relatively minor effects on Nilsson wave functions. It remains to discuss their enormous impact on Coriolis mixing and single nucleon transfer cross sections. This impact arises mostly from the effect of ε_4 on Nilsson energies of orbits that can Coriolis mix. It is easy to see what the main effects will be. Consider, for example, a large positive ε_4 and the $i_{13/2}$ orbits. It is obvious that both equatorial ($K = 1/2, 3/2, 5/2$) and polar ($K = 13/2$) orbits will be, on average, further from the nuclear matter than for $\varepsilon_4 = 0$, and therefore their energies will increase. The mid-K orbits ($K = 7/2, 9/2$) will be closer to the nuclear matter and their energies will decrease. Hence the overall effect will be a *compression* of the energy separations from $K = 1/2$ to $7/2$ or $9/2$. Moreover, this compression will become more extreme as ε_4 increases. In fact one can imagine sufficiently large ε_4 values that some of these K orbits may actually cross and interchange their relative positions. Figure 9.11 shows an explicit calculation of the $i_{13/2}$ energies for fixed ε_2 as a function of ε_4. All these features appear. There is a compression, and even a crossing, near $\varepsilon_4 \sim 0.1$. The envelop of the energies can easily be compressed by a factor of two and, therefore, the already large Coriolis mixing among the unique parity orbits will increase still further (the Coriolis matrix elements themselves will not substantially change).

Recall from our discussion of cross sections to $13/2^+$ states in the W isotopes that Coriolis mixing calculations produce some improvement in the predictions, but that significant discrepancies remain. The mixing casts some cross section from higher lying levels into the lower ones. With the increased Coriolis mixing that now occurs with a large positive ε_4, this effect will be exaggerated, as shown in the third panel in Fig. 9.8, where we see much better agreement with the empirical cross sections.

One of the beauties of the Nilsson model is its easy extendibility. We see here an excellent example where the basic model predictions are in strong disagreement with the data but where simple and physically reasonable refinements easily remove most of the

i $_{13/2}$ ORBITS

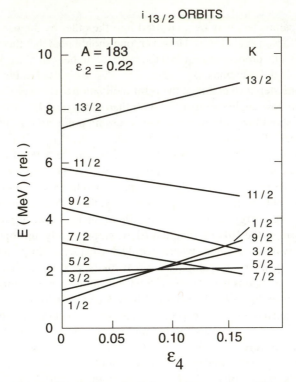

FIG. 9.11. Effect of hexadecapole deformations on i$_{13/2}$ Nilsson energies.

discrepancies, and thereby show both the usefulness of the model as a starting point and also the absolutely crucial need to incorporate certain of these extensions. The particular case we have been considering is historically interesting as well: the large changes in i$_{13/2}$ energies, Coriolis mixing, and (^{3}He, α) cross sections as ε_4 varies from 0 to 0.06 provided the first definitive evidence for large hexadecapole deformations in the odd W isotopes. Another interesting point is that while ε_4 values of zero and 0.06 can be easily distinguished in this way, the approximate constancy of the envelope of $K = 1/2$ to 7/2 orbits from $\varepsilon_4 = 0.06$–0.16 precludes a further specification of the actual ε_4 values.

There is one other consequence of large hexadecapole deformations that should be mentioned. We have been discussing permanent or static hexadecapole shape components. However, it is also possible that the nuclear potential energy surface may be "soft" in ε_4, and that this will lead to hexadecapole vibrations, just as softness in β and γ leads to β and γ vibrations. It is not surprising that the heavy even–even rare earth nuclei, especially the Os isotopes, display rather low-lying (~ 1 MeV) $K = 4$ bands that have been interpreted in terms of hexadecapole vibrations by Baker and co-workers. Of course, such $K = 4$ intrinsic excitations can also be thought of as double γ vibrations: the particular states in the Os isotopes appear to be mixtures of both modes, and interestingly, their γ decay seems to pick up the two-phonon character while their single nucleon transfer cross sections ≈ reveal the hexadecapole aspect. This is a nice illustration of how a complementarity of experimental approaches can highlight different features of nuclear excitations.

9.4 Coriolis effects at higher spins

Our rather brief treatment of the unique parity orbits does not even begin to hint at their importance in modern nuclear structure physics, especially at high spin. Indeed, the unique parity states are the most thoroughly studied of high-spin levels and are nearly always used as the first testing ground for new theoretical ideas. Some of the reasons for this should be obvious from the preceding discussion. Primary is the purity of the unique parity states in j, resulting from the shell model spin orbit interaction that separates them from other levels of the same parity by nearly the distance between major shells. Their properties can be calculated with extremely high reliability and more simply than most other states. Second, since Coriolis effects are crucial to most of the physics at high spins, the fact that these states form an isolated but strongly admixed set enhances the ease with which one may spot the influence of extra degrees of freedom. Colloquially speaking, these states "close under the Coriolis interaction."

Another key feature of the unique parity states is that Coriolis effects are largest amongst the low K orbitals, both because the matrix elements are themselves slightly larger and also because the energy separations are smallest. Noting that these are precisely the orbits whose angular momenta are aligned most nearly parallel to the nuclear rotation axis (their orbital motion is most nearly equatorial), we have three factors contributing to the development of the so-called rotation aligned scheme to which we have alluded and to which we now turn.

9.4.1 Rotation aligned coupling

This coupling scheme, illustrated in Fig. 9.12, was originally introduced to account for apparently anomalous rotational spacings in certain orbits in odd mass nuclei (see extreme right in Fig. 8.8). Figure 9.12 shows angular momentum diagrams for the cases of large and small K. We assume for simplicity that the Nilsson wave functions can be approximated by a single j value. We can write the total Hamiltonian as $\mathbf{H} = \mathbf{H}_{\text{Nils}} + \mathbf{H}_{\text{rot}}$. The Nilsson energy of a single nucleon can be represented by $E_{\text{Nils}} = \varepsilon_j + \Delta E(jK)$ where the ε_j are the spherical single particle energies and the ΔE are the shifts due to the deformed potential. Writing Eq. 8.8 for $\Delta E(jK)$ as a constant plus K-dependent term, we have $E_{\text{Nils}} \approx \varepsilon_j + \text{const} + C\delta K^2$, and hence (neglecting the constant term)

$$E = \varepsilon_j + \frac{\hbar^2}{2I}[J(J+1) - 2K^2 + \langle \mathbf{j}^2 \rangle] + \mathbf{V}_{\text{Cor}} + C\delta K^2$$

$$= \varepsilon_j + \frac{\hbar^2}{2I}[J(J+1) + j(j+1)] + \left(C\delta - \frac{2\hbar^2}{2I}\right)K^2 + \mathbf{V}_{\text{Cor}} \qquad (9.13)$$

This is the Nilsson energy of the odd particle plus the total rotational energy. At the top of Fig. 9.12 K is large and the K^2 term dominates the $\mathbf{V}_{\text{Cor}}$ term. Since the coefficient

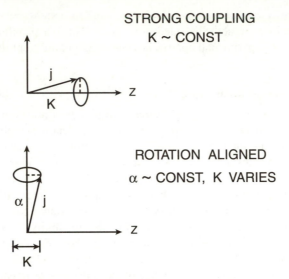

FIG. 9.12. Diagram for strong coupling and rotation aligned or decoupled level schemes (based on Stephens, 1975).

of K^2 is linear in δ and since the inertial parameter, $\hbar^2/2I$, decreases with increasing deformation, this is a situation that is valid for large deformation and/or large K. The dominance of the K^2 term implies that K is a good quantum number (the K mixing terms are relatively small) and one has the so-called deformation aligned or strong coupling scheme we have been discussing. Coriolis mixing effects are a small (but important) perturbation that causes the angle of the vector j to "wobble" slightly as it precesses about the z axis.

However, there are situations in which this coupling scheme does not occur. An obvious one is for high spins for which $\mathbf{V}_{Cor} \propto J$; when this term dominates the K^2 term, the solutions must be approximate eigenfunctions of $\mathbf{V}_{Cor}$ that correspond to a new coupling scheme in which K specifically is *not* a good quantum number. One then has the situation illustrated in the lower part of Fig. 9.12, where the particle angular momentum vector, $\mathbf{j}$, precesses about an axis perpendicular to the symmetry axis (i.e., about the rotation axis). Clearly, K will vary significantly and include negative values. It is now the alignment along the rotation axis, commonly called α, that is the good projection quantum number. This coupling scheme will be realized when

$$\left(C\delta - \frac{2\hbar^2}{2I} \right) K^2 \ll V_{Cor} \tag{9.14}$$

that is, especially for low K values. This is physically plausible since $\mathbf{j}$ then already points nearly along the rotation axis. Clearly, if K is large, an enormous Coriolis interaction (extremely high J values) would be required to enforce precession about the rotation axis, whereas for low K, such precession can occur at relatively low spins. Inspection of Eq. 9.14 shows that the rotation aligned scheme can also be realized for low J if the coefficient of K^2 vanishes. Since $\hbar^2/2I \sim 1/\delta$, it is clearly possible to choose a δ value that satisfies this cancellation requirement. For the $A \approx 130$ region, numerical estimates give $\delta \approx 0.17 (\beta \approx 0.18)$. This is a rather moderate deformation and accounts for the fact that the rotation aligned scheme often manifests itself in transitional, moderately deformed prolate nuclei early in a shell (where the low K orbits are filling). For larger deformations, the inertial parameter $\hbar^2/2I$ drops rapidly while the Nilsson energies further split and the Coriolis effects decrease below a critical value. In most well-deformed nuclei we see normal (strongly coupled) rotational behavior.

The energies in the rotation aligned scheme are very easy to visualize. The total angular momentum points essentially along the rotation axis and is composed of the particle angular momentum $\mathbf{j}$ plus a core rotation $\mathbf{R}$. Thus, from Eq. 9.13

$$E_{\text{rot}}(J) \approx \frac{\hbar^2}{2I}[J(J+1) + j(j+1) + \mathbf{V}_{\text{Cor}}]$$

For high spin unique parity states, low K values and moderate deformations, $J, j >> K$ and j is nearly a good quantum number. Moreover, in the rotation aligned scheme, $|j| \approx |\alpha|$. Therefore, using Eq. 9.7 and neglecting terms independent of J, we have

$$E_{\text{rot}}(J) \approx \frac{\hbar^2}{2I}[J(J+1) - 2J\alpha] \sim \frac{\hbar^2}{2I}[(J-\alpha)(J-\alpha+1) - \alpha^2] \qquad (9.15)$$

This equation is simply that for a rotor of spin $(J - \alpha)$. But $|(J - \alpha)| \approx |R|$, the *core* rotational angular momentum! So the energies do *not* behave like those of a rotor with spin J, but rather like those of the *rotational core*. Moreover, the lowest energies occur for the highest alignments, α. The reader may recall that when we derived Eq. 8.17, we set the problem up as the solution to why rotational bands were not upside down, and why for example, the core angular momenta R were not $0, 2, 4 \dots$ for states with $J = 13/2$ and $(9/2, 17/2), (5/2, 21/2), \dots$, respectively. The solution involved recognizing the *variation* of $|R|$ values that occurs when the particle angular momentum vector precesses around the symmetry axis. We alluded to the possibility that rotational bands with core rotational spacings did indeed exist in certain circumstances. We now see those circumstances—when the precession is no longer about the symmetry axis but rather about the rotation axis—so that R is nearly a constant of the motion.

To understand the implications of Eq. 9.15, let us take as an example a situation of maximum alignment for the rare earth nuclei where the unique parity orbit has $j = 13/2$. We take $\alpha = 13/2$. The energy difference $E_{17/2} - E_{13/2}$ is given by the energy difference between $R = 2$ and $R = 0$, that is, by the energy of the first 2^+ state in the even–even core nucleus or $6(\hbar^2/2I)$. This is completely different from the strongly coupled case in which $E_{17/2} - E_{13/2} = \hbar^2/2I [17/2(19/2) - 13/2(15/2)] = 32(\hbar^2/2I)$. The difference is enormous, as is the energy saving if the rotation aligned scheme is applicable. It was precisely the observation of such spin and energy sequences in moderately deformed odd mass nuclei, such as the La isotopes, with energy spacings nearly *identical* to those of the adjacent even–even Ba nuclei, that inspired the development of the rotation aligned scheme. The remarkable La–Ba comparison is shown in Fig. 9.13. The similarity of

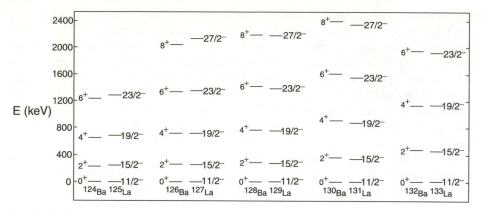

FIG. 9.13. Comparison of rotational spacings in Ba and La nuclei. This is the classic example of decoupled band structure (based on Stephens, 1975).

spacings in the odd and even mass nuclei is striking, extending even to the way they track with neutron number.

The sequence of $R = 0, 2, 4, 6, \ldots$ values leads in turn to low-lying rotational states that differ in spin by two units (the alternate spin states are higher in energy). This is another difference from the strongly coupled case of "normal" rotational sequences $J = K, K + 1, K + 2, K + 3, \ldots$.

Note that it is also possible for $\mathbf{R}$ and $\mathbf{j}$ to be *antialigned*, producing total spin values $J < j$. Thus, the state with $J = 9/2$ also corresponds to a core angular momentum $R = 2$ and will have an identical energy to that for the $J = 17/2$ level. Since the energies follow an $R(R + 1)$ rule, and since states with spins $J = j \pm R$ have identical energies, it follows that a plot of $E(J)$ vs. J will be a parabola whose minimum is at or near $J = j$. It is possible, of course, that the alignment along the rotation axis does not attain its maximum value. For example, if $\alpha = 11/2$, one can still apply Eq. 9.15. In this case the lowest aligned state, with $R = 0$, will have $J = 11/2$. States with $J = 7/2$ and $J = 15/2$ will occur higher with $R = 2$. This sequence of states also forms a parabola, but one lying slightly higher than for the case of maximum alignment. Continuing this process, sequences of parabolas will occur: each succeeding one corresponds to the next lower α value. The states along the lowest parabola are called the *favored* states and may be either *favored aligned* or *favored antialigned*. Those on the higher parabolas are called *unfavored* states and may also be aligned or antialigned.

In a sense, this picture resembles that of a weak coupling model. There is, however, a qualitative difference. If the coupling of a particle with angular momentum $\mathbf{j}$ to a core state $\mathbf{R}$ is weak, all states with $(j - R) \le J \le (j + R)$ will form a nearly degenerate multiplet since $\mathbf{j}$ and $\mathbf{R}$ can have any relative orientations. Here, however, only the energies of the favored states approximate those of a weak coupling model. The other, unfavored states are pushed considerably higher.

Of course the exact solution for the rotation aligned scheme can be obtained by explicit diagonalization of the Nilsson particle-plus-rotor Hamiltonian. An example taken

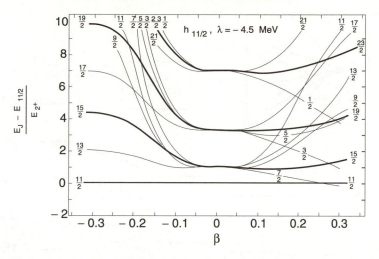

FIG. 9.14. Behavior of particle-rotor level energies with β for unique parity levels including Coriolis mixing. The Fermi surface is below the entire $h_{11/2}$ set of orbits. Note the strongly coupled pattern ($\Delta J = 1$, $J(J+1)$ spacings) on the left and the decoupled pattern ($\Delta J = 2$, E_{min} for $J = 11/2$, compressed (core) rotational spacings) on the right (Stephens, 1975).

from Stephens is shown in Fig. 9.14, for the case where the unique parity orbit is the $h_{11/2}$ and the Fermi surface is below the $K = 1/2$ orbit. (Note that in all of the preceding discussion of the $i_{13/2}$ orbits, the only relevant property of the unique parity orbits was the purity in j, and therefore nearly identical effects result for any other unique parity orbits simply by substituting a different j. For example, if the unique parity orbit is $h_{11/2}$ rather than $i_{13/2}$, the lowest lying spin state will be $J = 11/2$ and the favored aligned states will have spins $J = 11/2, 15/2, 19/2, \ldots$. This makes for a very generally applicable scheme with close correlations from mass region to mass region). In Fig. 9.14, the favored aligned states are given by the thick lines, the others as thin lines. The characteristic feature of the decoupled band emerges clearly on the prolate side, whereas on the oblate side, the Fermi surface is near the high K orbits, so the lowest states form a normal strongly coupled band. For the rotation aligned scheme, the favored aligned energies remain remarkably close to those of the core energies (which can be seen at $\beta = 0$) even out to relatively large deformations.

This brief summary of rotational alignment shows that it can be a rather widely applicable phenomenon, occurring especially in moderately deformed nuclei, where Coriolis effects are strong and deformation effects still rather weak, whenever the Fermi surface is near the low K unique parity Nilsson orbits. The rotation aligned scheme relies on the notion that it is energetically easier to achieve a given spin by combining a particle angular momentum aligned nearly along the rotation axis with a small amount of core rotation than it is to couple a particle angular momentum aligned elsewhere with a large core rotation.

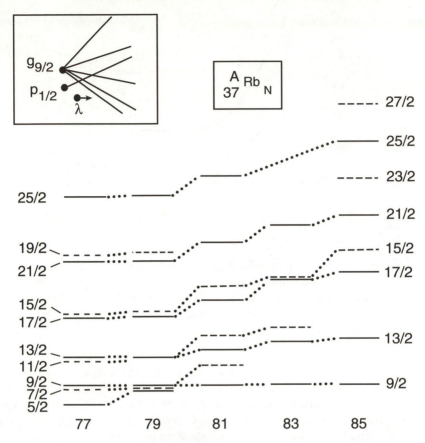

FIG. 9.15. Empirical proton unique parity ($\pi = +$, $g_{9/2}$) levels of odd mass Rb isotopes. The isotopes span a strongly coupled (deformed) toward decoupled (weakly
deformed) transition. The inset indicates the proton orbits and the movement of the
proton Fermi surface (with *decreasing* neutron number). Solid levels are favored states
(dashed are unfavored) in the rotational aligned picture (based on Tabor, 1989).

If we inspect a sequence of nuclei, it is possible to observe smooth transitions from
rotation aligned behavior when the Fermi surface is near the low K unique parity orbits
early in a shell (moderate deformation) to strongly coupled as the Fermi surface rises. The
odd Z Rb isotopes, shown in Fig. 9.15, provide a nice example. The low neutron number
isotopes are reasonably well deformed: a transition toward smaller deformations takes
place as neutron number increases. The proton Fermi surface is just below the entire $g_{9/2}$
unique parity shell. Near ^{85}Rb, a fine decoupled structure exists: the low-lying levels
form a $\Delta J = 2$ sequence, starting at $J = 9/2$; that is, a fully aligned ($\alpha = 9/2$)
configuration. The alternate spins are shifted quite high. This splitting of favored and
unfavored states is known as "signature splitting". As N decreases, the deformation
increases, simultaneously reducing the Coriolis strength among the $g_{9/2}$ proton orbits

and lowering the $K = 1/2, 3/2$ orbit energies so that the Fermi energy moves into this group (see inset): a transition to a strongly coupled scheme ensues. By ^{77}Rb, a nearly monotonic $\Delta J = 1$ sequence appears although a fully developed rotational pattern has yet to emerge.

Another beautiful example of such a transition from a decoupled toward a strongly coupled scheme occurs in the neutron deficient, near-stable Er isotopes. Stephens displays this transition in a way that highlights a couple of important points. Consider a state of spin J in an odd-mass nucleus. If the state is fully aligned ($\alpha = j$), then the core rotational angular momentum $|R| = |J - j|$ and the energy difference in the odd-mass nucleus $E(J) - E(j) = (\hbar^2/2I)|J - j|^2 = E_{ee}(|J - j|)$ of the neighboring even–even nuclei. If the strongly coupled limit applies, these rotational energy differences in the odd-mass nucleus will be much larger than those in the even–even nucleus for states with $J > j: E(J) - E(j) = (\hbar^2/2I)[J(J+1) - j(j+1)]$. Figure 9.16 shows the data for the lowest band based on the $i_{13/2}$ orbital in the odd-mass Er isotopes. The ordinate is the ratio R_{eo} of the energy difference $E^0(J+2) - E^0(j)$ of two states in each odd-mass nucleus divided by the energy difference $E^e(J + 2 - j) - E^e(J - j)$ taken from the data for the neighboring even–even nucleus. If the odd-mass nucleus has full rotation alignment ($\alpha = 13/2$), then $R_{eo} \approx 1$ independent of spin.

If the rotational band in the odd-mass nucleus is strongly coupled, the larger spacings will lead to $R_{eo} >> 1$. Empirically, near ^{157}Er $R_{eo} \approx 1$, but a clear transition toward a strongly coupled limit is observed with increasing neutron number. This is caused both by an increase in deformation, which reduces the strength of the Coriolis interaction ($V_{Cor} \propto \hbar^2/2I$), and by an increase in the Fermi surface from near the low K unique parity orbits toward the mid-K orbits. In addition, we note that the transition proceeds much more slowly for higher spins. The energy spacing $E_{29/2} - E_{25/2}$ remains very close to the rotation aligned limit. This is simply because, as we have noted several times, the Coriolis interaction increases with J for high spin states and therefore the rotation aligned coupling scheme persists longer.

One last point, of some interest in terms of testing this picture of rotation alignment and favored and unfavored states, concerns the relative role of high-and low-spin levels. The entire picture described so far assumes a simple axially symmetric rotational core and its coupling to the odd particle motion. It entirely neglects any effects due to rotation–vibration coupling, axially asymmetry, mixing with quasi-particle states, and so on. If one considers the rather general situation of a fixed number of valence nucleons spanning a specific set of single-particle states it is clear that, while there are many ways of constructing low and intermediate spin states from different combinations of the single-particle angular momenta, there is only one way of constructing the highest spin level—by aligning all these individual particle angular momenta along the same direction in space. Therefore, *any* model for this highest spin state has the same structure, independent of the assumptions of the model. For other high-spin (but not the highest-spin) states there will, in general, be only a few ways of constructing them and different models may present somewhat different, but mostly likely not very different, predictions. For low-spin states, however, different models with different interactions may have entirely different effects on specific subsets of states. Their energies and structure may differ markedly from one

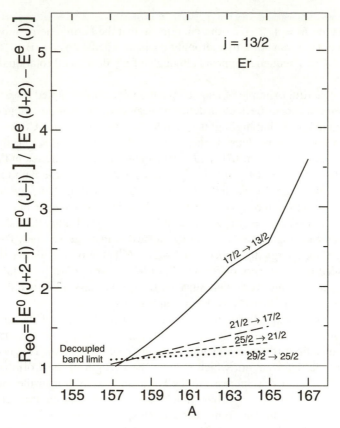

FIG. 9.16. Plot of R_{eo} (see text) against A for several energy differences in the Er nuclei (based on Stephens, 1975).

model to another. Although the beauty of and evidence for rotation aligned behavior is most dramatic in the high-spin states, perhaps the most sensitive tests of such models occur in the low-spin, unfavored, antialigned levels. Study of such levels may provide evidence for other degrees of freedom of some importance.

We should briefly apply our understanding of Coriolis effects to even–even nuclei. This will provide a simple understanding of the widespread phenomena of backbending. Consider a specific two-quasi-particle state with both particles in a low K unique parity orbit paired to $J^\pi = 0^+$ and with the Fermi surface below all the unique parity orbits. Neglect for a moment the interaction between the two particles. Since the particles are in the lowest unique parity orbits, their energies are greatly lowered due to the strong Coriolis interaction with particles in higher K orbits. This is analogous to the situation discussed in Chapter 1 of a set of equally spaced states with equal matrix elements connecting adjacent levels. The lowest level is pushed down and becomes a collective (strongly admixed) combination of amplitudes. From the size of the unique parity Coriolis mixing matrix elements, we have seen that this energy lowering can be substantial. The tendency will be for each particle to align its angular momentum with the rotation axis. Since the Coriolis interaction grows with spin, it may well be that at some J value it becomes energetically favorable to form a state, not from a core angular momentum $R = J$ superimposed on a spin 0^+ pairing condensate, but rather by breaking the pair of particles in a high j orbit so that their spins couple to $(2j - 1)$ ($2j$ is forbidden by the Pauli principle) and coupling this angular momentum nearly parallel to a much lower core rotational angular momentum of $R \approx J - (2j - 1)$. The lower energy required because of the lower expenditure of core rotational energy more than compensates for the energy lost in breaking the pair of particles.

In such a case, a plot of the yrast energies against J will increase parabolically until this transition or critical angular momentum (J_{crit}), where the sudden drop in required R values leads to smaller energy jump to the next higher angular momentum. This is illustrated in Fig. 9.17 (lower), where $J_{crit} = 14$. The figure also shows the continuation of the "decoupled pair" band below the critical angular momentum. That is, the config-uration in which the angular momenta of the two particles are aligned nearly along the rotation axis also exists below spin J: it is just not yrast. The "crossing" spin is simply that point where this rotation aligned band begins to be favored energetically. This spin depends on the deformation and the location of the Fermi surface relative to the low K unique parity orbits. It therefore varies with N or Z.

We saw in Chapter 6 that a band crossing shows up vividly in a "backbending" plot of $2I/\hbar^2$ against rotational frequency $\hbar\omega = E_\gamma/2$. The rotational alignment process is an excellent example of this phenomenon.

The rotation aligned band has a larger moment of inertia than the fully paired band because the pairing correlations are reduced due to one fewer $J = 0$-coupled pairs in the pairing condensate. Hence, the pairing gap Δ is reduced, and so therefore is the denominator in Eq. 6.12a.

We commented in Chapter 6 that backbending plots, in the early years of the phe-nomenon, were often in the form of plots against $(\hbar\omega)^2$ instead of $\hbar\omega$. So that the reader encounters both forms we show an $(\hbar\omega)^2$ plot in the upper part of Fig. 9.17.

As we have discussed in Chapter 6, both ordinate and abscissa in such a plot are trivially obtained (without ever knowing the actual excitation energies) simply from the observed γ-ray *transition* energies between adjacent rotational levels. The smaller energy spacing between the $J = 12$ and $J = 14$ states in the lower part becomes the sharp backbend in the upper part. The concept of crossing frequency now replaces that of J_{crit}.

One can view the reason for the upbend or backbend in two equivalent ways, both based on the rotational energy expression $(\hbar^2/2I)R(R + 1)$. In one, the R values are effectively lower, as we have described. This is the view in terms of a new aligned coupling scheme in which R suddenly decreases to take advantage of the angular momentum gained from the aligned particle. The other view is phrased directly in terms of Coriolis coupling in that the coupling causes a substantial lowering of the states of the two-quasi-particle aligned band. This lowering is larger for higher spins so that, at some spin, they cross the "normal" levels. (Of course, there can be other origins for backbending

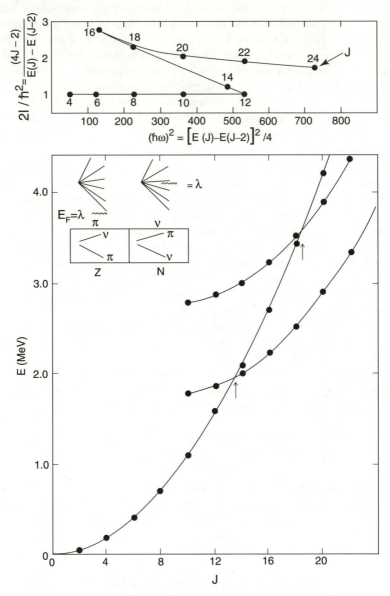

FIG. 9.17. Illustration of the idea of crossing bands and frequencies in a plot of $E(J)$ against J (lower part) leading to the backbending phenomenon (top). The insets give an example, appropriate to the neutron deficient Ce region, of how one can understand the systematics of proton (π) and neutron (ν) crossing frequencies (sketched in the small boxes) against Z and N (based on Wyss, 1989).

behavior, such as shape changes of nucleus as a whole: we do not discuss these here.)

In a realistic situation, there will be an interaction between the ground band and the decoupled band near the critical or crossing frequency. Just below J_{crit}, the decoupled band is higher. Its interaction with a ground band will lower the energies of the latter, effectively increasing $2I/\hbar^2$, while simultaneously reducing the transition energies. This causes a slight upbend for $J < J_{\text{crit}}$: mixing smooths out the sharply angled ideal pattern. The net effect is to lead to an S-shaped curve where the sharpness of the backbending or upbending depends on the strength of the interaction between the two bands.

The data for many nuclei have now been accumulated and backbending is a widespread phenomenon. As expected, it is typically observed when the Fermi surface is near the low K orbits of the unique parity orbit. It tends to disappear with increasing mass for a given sequence of isotopes as the Fermi surface rises toward the higher K orbits.

How this works in practice can be illustrated by the example in Fig. 9.17. The lower part, discussed earlier, shows a normal plot of $E(J)$ vs J and depicts two crossings occurring at different E, J and crossing frequencies ($\hbar\omega$). The insets show typical Nilsson diagrams for proton and neutron unique parity orbits and indicate where the Fermi energies (E_F) are in this example. We can now understand the expected systematics of backbending in this region. As Z increases, E_F^π (proton Fermi energy) increases. Moreover, the deformation β increases as deformation driving orbits are filled. For both reasons, the energy required to occupy the low K unique parity orbits decreases. Hence, the proton crossing (labeled π in the inset) occurs at lower energy and angular momentum and hence lower $\hbar\omega$. As N increases, the deformation decreases as polar orbits are encountered. Hence the *proton* crossing frequency increases with N as the energies of the proton unique parity orbits rise. The *neutron* crossing frequencies have an opposite behavior. They decrease with increasing N (decreasing β) since the energy separation to the low K orbits decreases, but increase with Z because the deformation increases. These ideas are sketched in the figure as crossing frequencies for protons and neutrons. While the arguments are qualitative, they are borne out both experimentally and theoretically in a cranked shell model calculation in a region such as the neutron deficient Xe–Ba–Ce nuclei, to which this example applies.

There is no reason why there should not be backbending in odd nuclei, and this has indeed been observed. Again, the study of such states has become a major "industry" in itself and cannot be described in any detail here. There is one particular point deserving mention, however: the use of backbending studies in odd mass nuclei to identify the nature of the *specific* particles producing the backbend in the neighboring even–even nuclei. So far, we have made little mention of whether or not the unique parity orbits involved are the proton or neutron orbits. In the rare earth region, backbending can be caused, for example, either by aligning two protons in an $h_{11/2}$ orbit or two neutrons in an $i_{13/2}$ orbit. It is difficult to distinguish these two from the backbending plot in the even nucleus itself. However, the odd nucleus can be used to resolve this ambiguity with the technique known as *odd particle blocking*. The basic idea is extremely simple. Consider an even–even nucleus in which the backbending is caused by alignment of two-*neutron* quasi-particles in a specific low K unique parity orbit. Now consider the rotational states in the neighboring odd proton nucleus where the odd *proton* is in a low

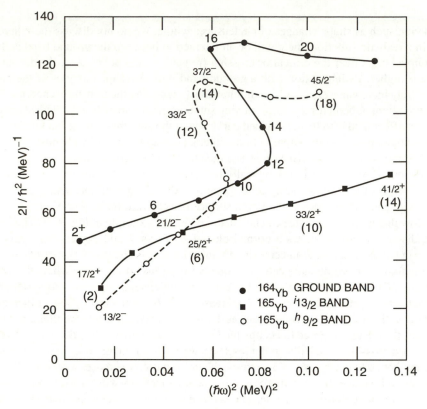

FIG. 9.18. Example of the odd particle blocking technique for intrinsic excitations in $^{164-165}$Yb (Riedinger, 1974).

K unique parity orbit. As the spin increases, there is no reason why two low K neutron quasi-particles cannot align. Therefore backbending should ensue, and at approximately the same frequency as in the neighboring even–even nucleus. On the other hand, if one considers a rotational band built on the same unique parity low K orbit in the neighboring odd *neutron* nucleus, the alignment of two neutrons in that orbit will be blocked by the prior occupation by the last odd neutron. There should be no backbend, or at least a greatly reduced effect. If this is observed, one deduces that the backbend in the even–even nucleus results from neutron alignment. This technique has been exploited in many nuclei and an example is given in Fig. 9.18. Here we see a case where backbending occurs in the even nucleus ^{164}Yb. In the neighboring odd nucleus ^{165}Yb, the band based on the $h_{9/2}$ orbit also backbends, but not that based on the $i_{13/2}$ orbit. Thus one deduces that the backbending in ^{164}Yb nuclei is caused by the $i_{13/2}$ neutrons and not $h_{9/2}$ neutrons, at least at this transition frequency.

The study of high-spin states, backbending, and related phenomena in heavy nuclei has become a major area of activity in the last decade and an enormous literature has developed. It is not our purpose here to summarize either it or the many refinements and improvements that have been made to this simple picture: Our aim is rather to provide some insight into the "traditional" background. Suffice it to say that much more sophisticated calculations are now standard, involving a "cranked" or rotated shell model that quantitatively accounts for the transformation from the body-fixed Nilsson scheme to the actual rotating system in the laboratory. Some of these calculations include the effects of axial asymmetry or hexadecapole deformations as well.

Empirically, double and even triple backbending has also been observed at higher transition frequencies. This has been interpreted in terms of the successive alignment of first one pair of particles along the rotation axis and then a second pair and sometimes even a third. Note that this reduces the overall pairing strength. At very high rotation, the alignment of all pairs can lead to a situation of complete pairing collapse. This is essentially the situation that arises in superdeformed bands at high spin. The collapse of pairing correlations also increases the moment of inertia towards the rigid body value, Eq. 6.12, leading to very small rotational spacings.

Clearly, as ever higher spins are sought and attained, increasing sophistication in data acquisition and analysis is required and a remarkable richness of phenomena unfolds. From all this has developed a deeper understanding of rotational-particle coupling, nuclear shapes, and potential energy surfaces, the phenomena of superdeformation and pairing collapse, and so on. The reader is encouraged to scan the fascinating literature on these topics.

10

MICROSCOPIC TREATMENT OF COLLECTIVE VIBRATIONS

Although there are many microscopic approaches to deriving the structure, energies and systematics of collective states from the shell model, there is one that has been widely used for nearly three decades, is still commonly encountered, and is easily adaptable to higher-order treatments. Moreover, it clearly shows the basic microscopic physics that must be at the heart of any approach. The technique referred to appears in two forms, the *Tamm–Dancoff Approximation* (TDA) and the *Random Phase Approximation* (RPA). Doubtless, the reader who is at all versed in nuclear structure physics has encountered calculations carried out in the RPA or references to such techniques. Except to the practitioners of this art, unfortunately, the phrases TDA and RPA tend to elicit fear and mystery. Those sections of theoretical papers describing such calculations are frequently glossed over, and the reader quickly leaps to tables or figures of the results. This is unfortunate for two reasons. First, these sections often contain important information on the input physics (e.g., single-particle energies, interaction strengths, etc.). Second, these techniques are actually rather easy to understand if they are presented in a simplified form that illustrates the essential physics rather than in an abstract formalism designed to cover every generalization. In the next few pages we shall present a simple derivation and discussion of the basic ideas involved, and then illustrate the techniques with a particular calculation for rare earth nuclei. As the reader will see, the end result will not only be a set of predictions for comparison with experiment but a deeper understanding of the microscopic nature of many aspects of collective behavior as well as of a very useful, but often obscure, tool. It will then be easy to make predictions of the basic structure of particular collective states without detailed or complex calculations. Simply by visual inspection of a Nilsson diagram, we will be able to predict the energy behavior of collective vibrations (e.g., γ or octupole), the systematics of their collective properties (e.g., B(E2) or B(E3) values), and even such details as whether they should be seen in single nucleon transfer reactions. In short, a simple understanding of the basic ideas behind the microscopic structure of collective states will give us the ability to anticipate, without calculation, many results of detailed RPA or TDA calculations.

The following derivation, inspired by Lane, may seem rather formal and abstract, but it is in fact easy to follow and leads to some very simple, powerful, and useful results.

10.1 Structure of collective vibrations

To start, we denote the ground state wave function of a many-body system by Ψ_0. Ψ_0 is an eigenfunction of a Hamiltonian, $\mathbf{H}$, such that

$$\mathbf{H}\Psi_0 = E_0\Psi_0 \tag{10.1}$$

We call Ψ_0 the vacuum state and will see exactly what this means later.

Consider now an arbitrary operator $\mathbf{O}_\alpha^\dagger$ that acts on Ψ_0, giving a new wave function Ψ_α

$$\Psi_\alpha = \mathbf{O}_\alpha^\dagger \Psi_0 \qquad (10.2)$$

Now, *suppose*, and this is the key point on which all else depends, that $\mathbf{O}_\alpha^\dagger$ happens to satisfy the operator relation

$$\left[\mathbf{H}, \mathbf{O}_\alpha^\dagger\right] \equiv \left(\mathbf{H}\mathbf{O}_\alpha^\dagger - \mathbf{O}_\alpha^\dagger \mathbf{H}\right) = \omega_\alpha \mathbf{O}_\alpha^\dagger \qquad (10.3)$$

Although we have yet to specify what form $\mathbf{O}_\alpha^\dagger$ must have for Eq. 10.3 to be obeyed, or what ω_α is, let us center our attention for the moment on the implications of this equation. Writing out Eq. 10.3 explicitly and acting on Ψ_0, we have

$$\mathbf{H}\mathbf{O}_\alpha^\dagger \Psi_0 - \mathbf{O}_\alpha^\dagger \mathbf{H}\Psi_0 = \omega_\alpha \mathbf{O}_\alpha^\dagger \Psi_0$$

Using Eq. 10.2, this is

$$\mathbf{H}\Psi_\alpha - \mathbf{O}_\alpha^\dagger \mathbf{H}\Psi_0 = \omega_\alpha \Psi_\alpha$$

Using Eq. 10.1, we have

$$\mathbf{H}\Psi_\alpha - \mathbf{O}_\alpha^\dagger E_0 \Psi_0 = \omega_\alpha \Psi_\alpha$$

or, since E_0 is a number, not an operator,

$$\mathbf{H}\Psi_\alpha - E_0 \mathbf{O}_\alpha^\dagger \Psi_0 = \omega_\alpha \Psi_\alpha$$

Using Eq. 10.2 again, we obtain

$$\mathbf{H}\Psi_\alpha - E_0 \Psi_\alpha = \omega_\alpha \Psi_\alpha$$

or, finally

$$\mathbf{H}\Psi_\alpha = (\omega_\alpha + E_0)\Psi_\alpha \qquad (10.4)$$

Thus, Ψ_α, the result of acting on Ψ_0 with $\mathbf{O}_\alpha^\dagger$

a) is *also* an eigenfunction of H, and

b) has an excitation energy ω_α with respect to the ground state E_0.

Thus, $\mathbf{O}_\alpha^\dagger$ *is a creation operator for the excitation* α. This basic idea, plus some simplifying assumptions about the structure of $\mathbf{O}_\alpha^\dagger$, is the basis for the TDA and RPA.

To proceed, we obviously need to find a set of linearly independent operators $\mathbf{O}_\alpha^\dagger$ that satisfy Eq. 10.3. Alternatively, we can ask what the *condition* is such that Eq. 10.3

is satisfied. To see this, let us expand the $\mathbf{O}_\alpha^\dagger$ in another set of arbitrary operators, $\mathbf{O}_r$, that form a complete set:

$$\mathbf{O}_\alpha^\dagger = \sum_r X_{\alpha r} \mathbf{O}_r \tag{10.5}$$

For the moment, we specify nothing about the $\mathbf{O}_r$, even whether they consist of creation or destruction operators. The wave function for excitation α is then

$$\Psi_\alpha = \left(\sum_r X_{\alpha r} \mathbf{O}_r \right) \Psi_0 \tag{10.6}$$

This is basically a definition of the expansion coefficients $X_{\alpha r}$. Using Eq. 10.5 in Eq. 10.3 gives

$$\left[\mathbf{H}, \sum_s X_{\alpha s} \mathbf{O}_s \right] = \omega_\alpha \sum_r X_{\alpha r} \mathbf{O}_r$$

where, to avoid confusion, we use different subscripts in the independent summations on the two sides. Since the X coefficients are just numbers, we have

$$\sum_s X_{\alpha s} [\mathbf{H}, \mathbf{O}_s] = \omega_\alpha \sum_r X_{\alpha r} \mathbf{O}_r \tag{10.7}$$

Now, we *define* a set of coefficients, M_{rs}, by the relation

$$[\mathbf{H}, \mathbf{O}_s] = \sum_r M_{rs} \mathbf{O}_r \tag{10.8}$$

and substitute this into Eq. 10.7:

$$\sum_s X_{\alpha s} \sum_r M_{rs} \mathbf{O}_r = \omega_\alpha \sum_r X_{\alpha r} \mathbf{O}_r$$

or,

$$\sum_{s,r} M_{rs} X_{\alpha s} \mathbf{O}_r = \omega_\alpha \sum_r X_{\alpha r} \mathbf{O}_r \tag{10.9}$$

This, however, is just a matrix equation that must be satisfied for each r and can be written

$$\mathbf{M} \mathbf{X}_\alpha = \omega_\alpha \mathbf{X}_\alpha \tag{10.10}$$

where $\mathbf{M}$ is the matrix whose elements are $\mathbf{M}_{rs}$ and $\mathbf{X}_\alpha$ is a column vector with elements $X_{\alpha r}$.

Thus, Eq. 10.4 and the condition on the $\mathbf{O}_\alpha^\dagger$, Eq. 10.3, follow if Eq. 10.9 holds. In other words, if Eq. 10.9 is true, then defining the $\mathbf{O}_\alpha^\dagger$ by Eq. 10.5, we find that Eqs. 10.3 and 10.4 are obeyed.

At this point, it is probably not clear why we have indulged in this process of piling definition upon definition. The aim was to produce Eq. 10.9. The reason, and the *practical* use of all this, is as follows.

The basic idea of the TDA and RPA (or higher-order approximations) is

a) to make assumptions about the operators $\mathbf{O}_r$ defined in Eq. 10.5
b) then use the definition given in Eq. 10.8 to solve for M (that is for the elements M_{rs})
c) to then use these M_{rs} in Eq. 10.9 to solve for the $X_{\alpha r}$.

By Eq. 10.5 we then know the *detailed structure* of the operators $\mathbf{O}_\alpha^\dagger$ that create the excitation Ψ_α! If this seems amazing and magical, good. If it seems abstract and artificial, be patient.

To remove a little of the magic (but hopefully not the awe), let us consider two simple assumptions for the $\mathbf{O}_r$ and see what results. Suppose the $\mathbf{O}_r$ have the schematic form:

$$\text{i)} \quad \mathbf{O}_r^\dagger \approx \mathbf{A}_i^\dagger \tag{10.11}$$

This operator creates a particle–hole excitation by raising one nucleon to orbit i.

$$\text{ii)} \quad \mathbf{O}_r^\dagger \approx \mathbf{A}_i^\dagger + \mathbf{A}_j \tag{10.12}$$

This operator creates and destroys particle–hole excitations. For the moment, we label these excitations by a single subscript. Later, when it becomes necessary to specify both orbits involved, we will expand the notation.

Approximation or assumption (i) is called the TDA and (ii) is called the RPA. Before proceeding, we use the general forms of (i) and (ii) to understand the meaning of the TDA and RPA.

We need only to recall that Ψ_0 is the *ground state*. That is, any destruction operator acting on Ψ_0 must give zero:

$$\mathbf{O}_\alpha \Psi_0 \equiv 0 \tag{10.13}$$

For the TDA (assumption (i)), this is equivalent to

$$\mathbf{O}_\alpha \Psi_0 = \sum_r X_{\alpha r} \mathbf{A}_r \Psi_0 = 0$$

This will hold if $\mathbf{A}_r \Psi_0 = 0$ for all r. Therefore, this Ψ_0 has no particle–hole excitations r. If it had any, $\mathbf{A}_r$ could destroy one, giving a nonzero wave function that did not contain a particle in that orbit. Thus, Ψ_0 must be a closed shell nucleus. Alternatively phrased, Ψ_0 has no built-in correlations. The TDA therefore corresponds to creating excitations from a very simple, uncorrelated, uncollective ground state.

Assumption (ii), the RPA, on the other hand, means that we define $\mathbf{O}_\alpha^\dagger$ in terms of Eq. 10.5 by

$$\mathbf{O}_\alpha^\dagger = \sum_r \left(X_{\alpha r} \mathbf{A}_r^\dagger + Y_{\alpha r} \mathbf{A}_r \right)$$

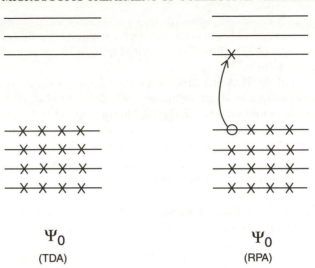

$$\Psi_0 \qquad\qquad\qquad \Psi_0$$

$$\text{(TDA)} \qquad\qquad\qquad \text{(RPA)}$$

FIG. 10.1. Highly schematic illustration of the difference between the TDA and RPA approximations to the microscopic structure of collective excitations.

where, for convenience, we have separated those arbitrary operators "O_r" that are particle–hole creation operators (denoted $A_r^\dagger$) from those that are particle–hole destruction operators (denoted A_r). Thus, we have

$$O_\alpha \Psi_0 = \left(\sum_r X_{\alpha r} A_r + Y_{\alpha r} A_r^\dagger \right) \Psi_0 = 0$$

This implies that

$$A_r^\dagger \Psi_0 = 0 \qquad\qquad (10.14)$$

for those r for which $Y_{\alpha r} \neq 0$. But $A_r^\dagger$ *creates* a particle–hole excitation r. Equation 10.14 holds then if Ψ_0 already contains such a particle–hole excitation so that $A_r^\dagger$ is "blocked". Equation 10.14 can also be satisfied by cancellations of several terms ($X_{\alpha r}$ and $Y_{\alpha r}$ terms), which again implies the existence of particle–hole excitations in Ψ_0. Thus, although Ψ_0 is itself the ground state wave function, it already has some built in excitations (or correlations, or collectivity). We illustrate the TDA and RPA in a very schematic way in Fig. 10.1. We expect that the RPA will be a somewhat more realistic or better approximation, at least for nonclosed shell (or subshell) nuclei.

Now, let us use the approximations (i) and (ii) to see how they allow us to solve for **M**, and hence for $O_\alpha^\dagger$, Ψ_α, and ω_α. If we do this, we will then know both the *microscopic structure* of the excitation α directly in terms of the occupation of the single-particle shell model orbits and the excitation energy of this excitation!

For simplicity, we consider the TDA, approximation (i). Then, Eq. 10.8 is

$$\left[\mathbf{H}, \mathbf{A}_s^\dagger\right] = \sum_r M_{rs} \mathbf{A}_r^\dagger \qquad (10.15)$$

Multiplying by $\mathbf{A}_i$ on the left, writing out the commutator, and taking the matrix element of Eq. 10.15 for state Ψ_0, gives

$$\left\langle \Psi_0 \left| \mathbf{A}_i \mathbf{H} \mathbf{A}_s^\dagger \right| \Psi_0 \right\rangle - \left\langle \Psi_0 \left| \mathbf{A}_i \mathbf{A}_s^\dagger \mathbf{H} \right| \Psi_0 \right\rangle = \left\langle \Psi_0 \left| \mathbf{A}_i \sum_r M_{rs} \mathbf{A}_r^\dagger \right| \Psi_0 \right\rangle$$

Recalling that $\mathbf{A}_s^\dagger$ and $\mathbf{A}_s$ simply create and destroy particle–hole excitation s, and abbreviating wave functions $\Psi_s \equiv \mathbf{A}_s^\dagger \Psi_0$ by s, (and similarly for i), we have

$$\langle i \left| \mathbf{H} \right| s \rangle - \left\langle \Psi_0 \left| \mathbf{A}_i \mathbf{A}_s^\dagger \mathbf{H} \right| \Psi_0 \right\rangle = \sum_r M_{rs} \left\langle \Psi_0 \left| \mathbf{A}_i \mathbf{A}_r^\dagger \right| \Psi_0 \right\rangle \qquad (10.16)$$

where we also used the fact that the quantities M_{rs} are just numbers.

Now, the right side of Eq. 10.16 vanishes unless $r = i$ (since for $r \neq i$ it contains the factor $\langle \Psi_0 | \mathbf{A}_i \mathbf{A}_r^\dagger | \Psi \rangle = \langle \Psi_0 | \mathbf{A}_i | \Psi_r \rangle$, which vanishes by orthogonality since Ψ_r has no particle–hole excitation i to be destroyed). For $r = i$, we have on the right side, $\langle \Psi_0 | M_{is} \mathbf{A}_r \mathbf{A}_r^\dagger | \Psi_0 \rangle = M_{is}$. So

$$\langle i \left| \mathbf{H} \right| s \rangle - \left\langle \Psi_0 \left| \mathbf{A}_i \mathbf{A}_s^\dagger \mathbf{H} \right| \Psi_0 \right\rangle = M_{is}$$

But the second term on the left, $\langle \Psi_0 | \mathbf{A}_i \mathbf{A}_s^\dagger \mathbf{H} | \Psi_0 \rangle = \langle \Psi_0 | \mathbf{A}_i \mathbf{A}_s^\dagger E_0 | \Psi_0 \rangle$, also vanishes unless $i = s$, in which case it is given by the ground state energy E_0. Thus, we have the simple result

$$M_{is} = \langle i \left| \mathbf{H} \right| s \rangle - E_0 \delta_{is} = \langle i \left| \mathbf{H} \right| s \rangle \qquad (10.17)$$

since we can choose to set $E_0 = 0$. In the RPA, a similar derivation yields a slightly more complicated result that the reader can easily work out using analogous arguments.

This result, Eq. 10.17, simply states that the matrix elements M_{is} of the matrix $\mathbf{M}$ that, via Eq. 10.9, give the coefficients (vectors) X_α and the energy ω_α defining the excitation α, are equal to matrix elements of the Hamiltonian between states i and s. In order to calculate this explicitly, we need only to make some choices for $\mathbf{H} = \Delta \varepsilon + \mathbf{V}$. The particle–hole (or quasi-particle) energies contained in $\mathbf{H}$ are energy differences between empty and filled orbits. Starting from a shell model or Nilsson model that gives the particle–hole energies $\Delta \varepsilon_i$, we need only specify the interaction $\mathbf{V}$. Suppose we make the reasonable assumption that $\mathbf{V}$ is a quadrupole interaction. Then,

$$\langle i \left| \mathbf{V} \right| j \rangle \equiv C Q_i Q_j \qquad (10.18)$$

where the Q are proportional to the transition quadrupole matrix element from the ground state to a particle–hole state with the particle in orbit i or j. That is, Q_i means $\langle i \| \mathbf{Q}_i \| \Psi_0 \rangle = \langle i \| \mathbf{r}^2 \mathbf{Y}_2 \| \Psi_0 \rangle$. In practice (see below) we will specify the μ in $\mathbf{Y}_{2\mu}$ according to the character of the vibrational mode we are studying. C is the strength of the interaction. For an attractive interaction, $C < 0$. To keep the notation simple, recall that we have specified each particle–hole excitation by a single subscript. Technically this is incomplete; each such excitation involves elevating a particle to some empty orbit and vacating a filled one. As long as no confusion results, we shall keep to this notation but the reader should keep in mind that each $\Delta\varepsilon_i$ involves a pair of orbits and the energy difference between them.

Substituting Eq. 10.18 in Eq. 10.9 and using Eq. 10.17, we have

$$\sum_{s,r} \langle r|\mathbf{H}|s\rangle X_{\alpha s}\mathbf{O}_r = \omega_\alpha \sum_r X_{\alpha r}\mathbf{O}_r$$

or

$$\sum_{s,r} [\Delta\varepsilon_r \delta_{rs} + \langle r|\mathbf{V}|s\rangle] X_{\alpha s}\mathbf{O}_r = \omega_\alpha \sum_r X_{\alpha r}\mathbf{O}_r$$

or

$$\sum_r \Delta\varepsilon_r X_{\alpha r}\mathbf{O}_r + C \sum_{s,r} Q_s Q_r X_{\alpha s}\mathbf{O}_r = \omega_\alpha \sum_r X_{\alpha r}\mathbf{O}_r$$

This must be satisfied for all r, so

$$\Delta\varepsilon_r X_{\alpha r} + C \sum_s Q_s Q_r X_{\alpha s} = \omega_\alpha X_{\alpha r}$$

or

$$(\Delta\varepsilon_r - \omega_\alpha) X_{\alpha r} = -C \sum_s Q_s Q_r X_{\alpha s}$$

or, finally, we get the expansion coefficients or amplitudes for the wave function Ψ_α of Eq. 10.6

$$X_{\alpha r} = \frac{C Q_r \sum_s Q_s X_{\alpha s}}{\omega_\alpha - \Delta\varepsilon_r} \qquad (10.19)_{\text{TDA}}$$

To find the eigenvalues ω_α, we multiply Eq. 10.19 by Q_r and sum over r, obtaining

$$\sum_r X_{\alpha r} Q_r = C \sum_r \left[\frac{Q_r^2 \sum_s Q_s X_{\alpha s}}{\omega_\alpha - \Delta\varepsilon_r} \right]$$

But $\sum_r X_{\alpha r} Q_r$ on the left is identical to $\sum_s X_{\alpha s} Q_s$ on the right, so cancelling these, we have

$$\frac{1}{C} = \sum_r \frac{Q_r^2}{\omega_\alpha - \Delta\varepsilon_r} \qquad (10.20)_{\text{TDA}}$$

which is the desired result. Note that the wave functions specified by the $X_{\alpha r}$ and the energy ω_α are now written in terms of completely known quantities: single-particle energies and single-particle transition quadrupole moments. Given a single-particle model

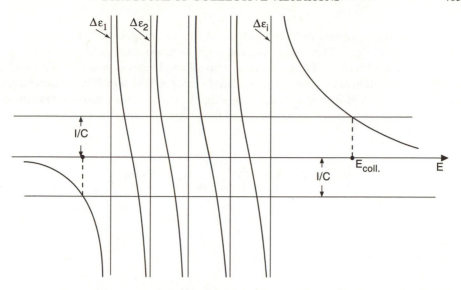

FIG. 10.2. Illustration of the solution to Eq. (10.20)$_{\text{TDA}}$. The curves give values of the right side of Eq. (10.20)$_{\text{TDA}}$ as a function of energy. The solutions to Eq. (10.20)$_{\text{TDA}}$ correspond to those ω_α where the vertical curves cross the "1/C" lines. The lowest (rightmost) solution is the collective one (Ring, 1980).

(shell, Nilsson, etc.), they are easily calculable. Indeed, we shall see shortly that, using these results, we can for example virtually "derive" γ vibrational wave functions by inspection of the Nilsson diagram without detailed calculations.

We emphasize that, in Eq. 10.19, the quantity $\Sigma Q_s X_{\alpha s}$ is just a number, so that the coefficients of the vibrational wave functions for the excitation α are simply proportional to $Q_r/(\Delta\varepsilon_r - \omega_\alpha)$. This fundamental result is of great importance.

Recall that the Q_r are the quadrupole transition matrix elements connecting the ground state with a particular (rth) particle–hole excitation. Thus the amplitude for particle–hole excitation r in the vibrational wave function is proportional to the matrix element for a quadrupole transition to this particle–hole excitation and is inversely proportional to the difference between the particle–hole energy, $\Delta\varepsilon_r$, and the vibrational energy, ω_α. Particle–hole excitations that require energies close to ω_α and have large quadrupole matrix elements are strongly favored.

The eigenvalue Eq. 10.20 has several possible solutions that are labeled α. These correspond to those energies ω_α such that the right-hand side takes on the value $1/C$. These solutions are illustrated in Fig. 10.2. One solution always appears significantly lowered in energy. This is the *collective vibration*.

It is easy to see this if we take a highly simplified example. Suppose there are N identical particle–hole energies $\Delta\varepsilon$ and that the Q_r values are also all equal. Then the lowest solution has the wave function amplitudes

$$X_{\alpha r} = aQ_r$$

where a is just a constant. By normalization of this N-component wave function, we have $a = 1/(\sqrt{N}Q_r)$. This means that the wave function consists of a sum of *equal* amplitudes for all particle–hole excitations, with all amplitudes *in phase*. This is exactly the formal definition of a *coherent* or *correlated wave function*, which has many comparable amplitudes contributing with the same sign. To see how this collectivity manifests itself, let us calculate an E2 (or quadrupole) transition rate from the vibration α to the ground state:

$$\left\langle \Psi_\alpha \left\| r^2 Y_2 \right\| \Psi_0 \right\rangle = \sum_{r=1}^{N} \left\langle \frac{1}{\sqrt{N}} \Psi_r \left\| r^2 Y_2 \right\| \Psi_0 \right\rangle$$

$$= \frac{N Q_r}{\sqrt{N}} = \sqrt{N} Q_r$$

Hence,

$$\mathrm{B}(\mathrm{E2} : \Psi_{\mathrm{vib}\,\alpha} \to \Psi_0) \propto \left\langle \Psi_\alpha \left\| r^2 Y_2 \right\| \Psi_0 \right\rangle^2 = N Q_r^2$$

which is N times the single-particle B(E2) value. Since this exhausts the total strength, it follows that no other solution has any strength. Of course, this is an extreme simplification, but it does demonstrate what is meant by collectivity and coherence and how they arise from this microscopic formalism.

Finally, the eigenvalue for the collective solution in this degenerate case is given from Eq. 10.20 by setting all $(\omega_\alpha - \Delta\varepsilon_r)$ constant, giving

$$E_{\mathrm{vib}\,\alpha} = \omega_\alpha = \Delta\varepsilon_r + C \sum_{r=1}^{N} Q_r^2 = \Delta\varepsilon_r + NCQ_r^2 \tag{10.21}$$

The vibrational energy is lowered (recall that $C < 0$ for an attractive residual interaction), relative to the (common) particle–hole energy by the large amount NCQ_r^2. This leads to an alternate viewpoint on the structure and origin of the vibrational wave functions since they are identical to those we obtained in Chapter 1 for the mixing of a set of degenerate levels. The lowest wave function (see Eq. 1.13) was a sum of equal amplitudes for all basis states: this state was lowered by an amount proportional to the number of admixed states while all of the others were raised in energy.

Before closing this discussion and looking at applications of these ideas, it is worth recapitulating the key ingredients that led to the results obtained:

- We discussed a simple but abstract operator formalism leading to the essential Eqs. 10.3, 10.4, 10.5, and 10.8. Equations 10.3 and 10.4 show the properties of the excitation α whose structure we are interested in. Equation 10.5 defines the creation operators $\mathbf{O}_\alpha^\dagger$ and Eq. 10.9, with the definitions of Eq. 10.8, gives a means for solving for the coefficients $X_{\alpha r}$ that define the wave function Ψ_α of Eq. 10.6, and for the energies ω_α.

- We made simple choices for the *form* of the various operators $\mathbf{O}_\alpha^\dagger$. These correspond to the TDA ($\mathbf{O}_\alpha^\dagger$ contains only particle–hole creation operators) and the RPA ($\mathbf{O}_\alpha^\dagger$ contains both particle–hole creation and destruction operators).
- We made a simple choice for the residual interaction V. This step is often called a *schematic model*. Doing this allowed us to calculate the X_α, hence the $\mathbf{O}_\alpha^\dagger$, and hence the structure and energy (ω_α) of the vibrational state Ψ_α.

The TDA represents the gross approximation that the ground state is a pure, undisturbed, closed shell nucleus. This is generally not realistic and the use of the RPA to incorporate ground state correlations is more common. We therefore present (without derivation) the eigenvalue results analogous to Eq. 10.20$_{\mathrm{TDA}}$, for the RPA:

$$\frac{1}{C} = \sum_r Q_r^2 \left[\frac{1}{\omega_\alpha - \Delta\varepsilon_r} - \frac{1}{\omega_\alpha + \Delta\varepsilon_r} \right]$$

or

$$\frac{1}{C} = 2 \sum_r Q_r^2 \frac{\Delta\varepsilon_r}{\omega_\alpha^2 - \Delta\varepsilon_r^2} \tag{10.20$_{\mathrm{RPA}}$}$$

This is similar to the TDA expression except for the more complicated factor multiplying the Q_r^2. Note that, in the degenerate case (all $\Delta\varepsilon_r$ equal) this gives

$$\omega_\alpha^2 - \Delta\varepsilon_r^2 = 2C\Delta\varepsilon_r \sum_r Q_r^2$$

or

$$\omega_\alpha = \left[\Delta\varepsilon_r^2 + 2C\Delta\varepsilon_r \sum_r Q_r^2 \right]^{\frac{1}{2}}$$

or, if all the Q_r are equal,

$$\omega_\alpha = \left(\Delta\varepsilon_r^2 + 2C\Delta\varepsilon_r N Q_r^2 \right)^{\frac{1}{2}}$$

Using Eq. 10.20$_{\mathrm{TDA}}$ and labeling the energies ω_α as $\omega_\alpha^{\mathrm{TDA}}$ or $\omega_\alpha^{\mathrm{RPA}}$, we have

$$\omega_\alpha^{\mathrm{RPA}} = \left[\left(\omega_\alpha^{\mathrm{TDA}} \right)^2 - C^2 \left(\sum_r Q_r^2 \right)^2 \right]^{\frac{1}{2}}$$

$$= \left[\left(\omega_\alpha^{\mathrm{TDA}} \right)^2 - \left(CN Q_r^2 \right)^2 \right]^{\frac{1}{2}}$$

if all the Q_r are equal. This shows that $\omega_\alpha^{\mathrm{RPA}}$ is always less than $\omega_\alpha^{\mathrm{TDA}}$. The RPA leads to greater collectivity.

We close this section by noting that the discussion has been phrased in terms of spherical single-particle–hole energies ε_i and single-particle quadrupole moments (or other moments for other choices of V, such as octupole moments). However, as hinted at a couple of times, the single-particle energies can be Nilsson energies equally well, and the $\mathbf{A}_i^\dagger$ and $\mathbf{A}_i$ can be creation and destruction operators for Nilsson orbits. Then Ψ_0 is the ground state of the deformed nucleus. A related point concerns our choice for V in our example of a schematic model in Eq. 10.18. It seems to ignore the short-range parts of the residual (nonsingle particle) interaction such as the pairing interaction. However, the whole formalism is identical if the pairing interaction is incorporated into the definition of the single-particle or Nilsson energies by defining these as quasi-particle energies instead of single-particle energies. Then the TDA or RPA can give the structure of vibrations in deformed nuclei in terms of amplitudes for *specific Nilsson quasi-particle excitations*. We thus see how a rather formidable looking formalism leads simply and elegantly to an easy way of deducing the particle (or quasi-particle) composition of specific collective vibrations. This is one answer to the apparent dichotomy that we noted earlier between the independent particle picture of the nucleus and the existence of collective excitations and correlations. The key element, of course, is in a sense inserted a priori by defining the *operators* $\mathbf{O}_\alpha^\dagger$ as linear combinations of single-particle (or two-quasi-particle) operators and by simplifying the interaction V. Ultimately, the method is tested by its agreement with experiment. This test has been passed many times, making the result a useful, elegant, and powerful approach to the microscopic structure of collective vibrations.

10.2 Examples: vibrations in deformed nuclei

The formalism just developed is ideally suited for understanding the microscopic structure of vibrations in spherical nuclei. Indeed, calculations for nuclei such as Pb or Sn were among its first applications. Since these nuclei are singly magic, the simpler TDA is a reasonable approximation and is frequently used. Pioneering applications of this formalism to such cases were made by Kisslinger and Sorenson and are described in many texts. We will turn instead to the application of these ideas to deformed nuclei.

Departing from singly magic nuclei, the p–n interaction rapidly induces configuration mixing that obscures shell structure. The onset of deformation affects the single-particle energies, as we have seen in the Nilsson scheme. So, for $\beta = 0.3$, any remnants of the 50, 82, or 126 shell gaps almost totally disappear. Excitations involving orbits from the shells below or above the valence shell must be incorporated in realistic calculations, and the RPA becomes a necessary refinement.

Even higher-order forms of the operator $\mathbf{O}_\alpha^\dagger$ are sometimes applied and are often called *higher* RPA (HRPA) for obvious reasons. Other approaches such as that of Kumar and Baranger or its refinement, the dynamic deformation theory, may be used. The full sweep of many-body theory encompasses many varied and complementary approaches.

One of the most interesting, informative, and physically transparent applications of the RPA is to the γ bands in deformed nuclei. The relevant calculations were carried out by Bès and co-workers in the early 1960s and remain the standard for the microscopic structure of these modes.

The basis states here are naturally the Nilsson orbits. Pairing must be included via a BCS calculation, so that, for the ground state, there is a distribution of occupation amplitudes over several orbits near the Fermi energy λ. This can be seen quantitatively in a modification to Eq. 10.20_{RPA}. For this discussion, we must specifically label both orbits involved in the quasi-particle excitation as we must specify occupation amplitudes for each. Replacing the particle–hole energy $\Delta\varepsilon_r$ with $E_i + E_j$, where the E are quasi-particle energies defined by

$$E_i = \sqrt{(\varepsilon_i - \lambda)^2 + \Delta^2}$$

we have, for the energies ω_α, the equation

$$\frac{1}{C} = \frac{\sum\limits_{ij} Q_{ij}^2 (E_i + E_j)(U_i V_j + U_j V_i)^2}{\omega_\alpha^2 - (E_i + E_j)^2} \tag{10.22}$$

where Q_{ij} now means $\langle i \,|\mathbf{r}^2 Y_{2\pm2}|\, j\rangle$. The wave functions are now linear combinations of two-quasi-particle excitations rather than single-particle–hole excitations, but the physical idea is identical. Note that in Eq. 10.22, E_i changes more slowly than ε_i when $(\varepsilon_i - \lambda) < \Delta$. Therefore, a wider range of ε_i values and orbits can contribute. Also, the pairing factor favors pairs of quasi-particle excitations on opposite sides of the Fermi surface; as an analogue to particle–hole excitations, this is not surprising.

The matrix elements of $\mathbf{r}^2 Y_{2\pm2}$ that determine the important γ-vibrational amplitudes are easily deduced by writing $\mathbf{r}^2 Y_{2\pm2}$ in Cartesian coordinates as

$$\mathbf{r}^2 Y_{2\pm2} \sim (\mathbf{x} \pm i\mathbf{y})^2 \tag{10.23}$$

This is a field (operator) that does not change n_z (there is no effect in the z-direction) but that changes Λ by ±2. (For a table giving the structure of all the low multipole operators in Cartesian form, see Mottelson and Nilsson, 1959). Another practical selection rule is that the sum $n_z + n_x + n_y = N$ should be conserved; otherwise the matrix element would involve Nilsson wave functions $[Nn_z\Lambda]$ and $[N'n_z'\Lambda']$ with $N' = N \pm 2$. Such states are far apart (about 10 MeV). Thus, the important components of the γ-vibration will be two-quasi-particle excitations involving a Nilsson orbit within about 2 MeV of the Fermi surface differing by $\Delta K = \pm2$, $\Delta\Lambda = \pm2$ (N, n_z unchanged) from another near the Fermi surface.

Since both K and Λ change by the same amount, the projection Σ of the intrinsic nucleon angular momentum will not change ($\Delta\Sigma = 0$). In considering which orbits satisfy this rule, recall that the ground state consists of a correlated wave function $\Psi_0 = X_1\phi_1 + X_2\phi_2 + \dots$, each of whose components describes two particles in the same Nilsson orbit but with opposite K values so that the resultant $K = 0$. Since $K = \Lambda \pm \Sigma$, these two nucleons also have equal and opposite Λ values ($\pm\Lambda$). Hence, $\mathbf{r}^2 Y_{2\pm2}$ will also have matrix elements satisfying the above rule that connect orbits such as [1/2 521] and 3/2[521], since the $\Lambda = \pm1$, $K = \pm1/2$, and $K = \mp3/2$ components differ by $\Delta\Lambda$ and $\Delta K = 2$, respectively.

Table 10.1 *Principle neutron two-quasi-particle amplitudes* ($\times 100$) *for the γ-vibration in several rare earth nuclei*[*]

Two-quasi-particle states	^{154}Gd	^{160}Dy	^{164}Dy	^{170}Er	^{174}Yb	^{178}Hf	^{184}W	^{186}W
$11/2^-[505] \otimes 7/2^-[514]$	—	—	—	7	7	—	—	—
$11/2^-[505] \otimes 7/2^-[503]$	11	12	15	25	26	21	11	8
$9/2^-[514] \otimes 5/2^-[523]$	9	7	—	—	—	—	—	—
$9/2^-[514] \otimes 5/2^-[512]$	25	23	21	20	14	—	—	—
$9/2^-[505] \otimes 5/2^-[503]$	—	—	—	—	—	—	11	16
$7/2^-[523] \otimes 3/2^-[521]$	19	12	—	—	—	—	—	—
$7/2^-[514] \otimes 3/2^-[512]$	—	—	—	17	35	48	34	26
$7/2^-[503] \otimes 3/2^-[501]$	—	—	—	—	—	7	20	26
$5/2^-[523] \otimes 1/2^-[541]$	9	—	—	—	—	—	—	—
$5/2^-[523] \otimes 1/2^-[521]$	20	36	51	29	14	8	—	—
$5/2^-[512] \otimes 3/2^-[521]$	—	—	—	9	7	—	—	—
$5/2^-[512] \otimes 1/2^-[510]$	—	—	9	31	56	56	32	24
$3/2^-[532] \otimes 1/2^-[530]$	24	8	—	—	—	—	—	—
$3/2^-[521] \otimes 1/2^-[541]$	9	—	—	—	—	—	—	—
$3/2^-[521] \otimes 1/2^-[521]$	27	45	46	25	13	8	—	—
$3/2^-[512] \otimes 1/2^-[510]$	—	—	—	8	13	22	66	70

[*]Bès, 1965. The table only includes amplitudes from the $N = 5$ shell.

Let us now apply these ideas to a couple of examples. Consider ^{184}W. The ground states of ^{183,185}W are $1/2^-[510]$ and $3/2^-[512]$. This approximately locates the neutron Fermi energy in the Nilsson diagram (see Figs. 9.1 and 8.4). Then we deduce by inspection that the important neutron components of the γ-vibration in ^{184}W are two-quasi-particle states of the form

$$1/2^-[510] \otimes 3/2^-[512], \quad 1/2^-[510] \otimes 5/2^-[512]$$

and

$$3/2^-[512] \otimes 7/2^-[514] \tag{10.24}$$

The analogous proton amplitudes can be similarly deduced. To see how well this estimate works, we show the detailed results of RPA calculations in Table 10.1, from which it is evident that we have indeed identified the most important two-quasi-particle components.

As a second case, consider $^{170}_{68}\text{Er}_{102}$. Here the principal neutron two-quasi-particle components of the γ-vibration would be expected to be

$$1/2^-[521] \otimes 5/2^-[523]$$
$$1/2^-[521] \otimes 3/2^-[521]$$
$$1/2^-[510] \otimes 5/2^-[512]$$
$$3/2^-[512] \otimes 7/2^-[514]$$
$$7/2^-[503] \otimes 11/2^-[505] \tag{10.25}$$

Comparison with Table 10.1 again shows that these comprise most of the main neutron components of the γ-vibration in ^{170}Er. Perusal of this table shows how various two-quasi-particle amplitudes systematically grow and decay across the region as the Fermi surface rises, and how different amplitudes are favored by the energy denominator and pairing factors in Eqs. 10.20$_{RPA}$ and 10.22.

It is sometimes possible to test such predictions using single nucleon transfer reactions. Since the ^{183}W target ground state is the $1/2^-[510]$ orbit, the (d, p) reaction can only populate two-quasi-particle states in ^{184}W of the specific form $1/2^-[510] \otimes \psi_{Nils}$. Moreover, they will contribute proportionally to U^2, the emptiness of ψ_{Nils} in ^{183}W. In ^{183}W, therefore, the hole state $5/2^-[512]$ component should not contribute significantly to σ(d, p), for the γ-vibration, while neutron transfer involving the $3/2^-[512]$ orbit should. (Of course, in ^{182}W the γ-vibration must have similar structure. Hence, $5/2^-[512]$ transfer will be important for the ^{183}W(d, t)^{182}W cross section.) In any case, the (d, p) cross section to the various spin states of the γ rotational band in ^{184}W will be determined primarily by the C_j coefficients for the $3/2^-[512]$ orbit.

The explicit expression for σ(d, p) into an even–even nucleus is not quite as simple as given in Chapter 9 for an odd-mass final nucleus, since each state of spin J can be constructed by coupling a component (j_1) of the $1/2^-[510]$ orbit with a j_2 from the $3/2^-[512]$ orbit. In general, several j_1, j_2 pairs can lead to the same J. Their relative contributions are given by Clebsch–Gordan coefficients. The population of a two-quasi-particle state in an even–even nucleus in (d, p) is thus given by a generalization of Eq. 9.1

$$\sigma(\text{d, p})_{e-e} = 2N \left[\sum_{l_2, j_2} \phi_l \langle j_1 K_1 j_2 K_2 | J K \rangle C_{j_2} \right]^2 U_2^2 \tag{10.26}$$

where the ϕ_l are DWBA cross sections ($l = j \pm 1/2$). It can easily happen that several l values contribute to this expression for a given final state. In the case of ^{184}W, for example, the $J^\pi = 2_\gamma^+$ state can arise by coupling the target $j = 1/2^-$ state with $j = 3/2^-$ or $j = 5/2^-$ components of the $3/2^-[512]$ wave function; these proceed by $l = 1$ and $l = 3$ transfer, respectively. For a higher spin target such as ^{167}Er ($7/2^+$), there are even more possibilities. A 4^- level can be formed by coupling the $7/2^+$ ground state with $j_2 = 1/2, 3/2, 5/2, 7/2, 9/2, 11/2$ (the upper limit here is provided not by angular momentum conservation but by the available orbits in the $N = 5$ shell where $j_{max} = 11/2$). This involves six contributions, with $l = 1, 1, 3, 3, 5, 5$, respectively.

Thus, generally the "fingerprint patterns" in even–even nuclei will be somewhat "washed out" and less orbit sensitive than in odd nuclei. Nevertheless, the (d, p) data for population of the γ-band in ^{184}W does have a pattern very close to that predicted for the transfer of a $3/2^-$[512] neutron Nilsson orbit, and the corresponding (d, t) data in ^{182}W confirms the presence of the $5/2^-$[512] orbit in the γ-vibration there. Note that not only can the important orbits be *identified* in this way, but their *amplitudes* in the γ-vibration may be deduced (approximately) by comparing, for example, the cross sections into the even–even nucleus to those for transfer of the same orbit into the neighboring odd A nucleus. In this way, even the detailed structure of RPA calculations can be confirmed, at least for a couple of important amplitudes. Other amplitudes, such as $3/2^-$[512] $\otimes$ $7/2^-$[514] cannot be directly tested.

In a few special cases it is possible to test for amplitudes involving a non-target-ground state orbit. As an example, ^{185}W has a ground state $3/2^-$[512] and a low-lying $1/2^-$[510] excited state. If ^{185}W were stable, permitting the relevant (d, t) reaction to ^{184}W, it would be possible to study the $1/2^-$[510] $\otimes$ $5/2^-$[512] component in the ^{184}W γ-vibration, because the $(3/2^-)$ ground state of ^{185}W has a strong Coriolis admixture of the $3/2$ $1/2^-$[510] state. (The case starting with ^{183}W is not the same, even though here too the $1/2$[510] and $3/2^-$[512] orbits are admixed, since the $1/2^-$ ^{183}W *ground state* cannot contain a $3/2$ $3/2^-$[512] admixture.) In the ^{185}W(d, t)^{184}W case, the γ band would be primarily populated by a coherent combination of the components $3/2^-$[512]$\otimes$ $7/2^-$[514] and $1/2^-$[510] $\otimes$ $5/2^-$[512].

To return to our discussion of the structure of collective excitations, we see how an understanding of the basic formalism and philosophy of the RPA allows us to understand and even anticipate the detailed microscopic structure of excitations such as the γ vibrations of deformed nuclei. A little further thought reveals five additional basic features of these excitations.

First, the unique parity orbit is generally unimportant in the γ vibration, since this vibration has positive parity and therefore *both* quasi-particles have to be in a unique parity orbit. But, from the sequence of asymptotic Nilsson quantum numbers, we see that any pair with $\Delta K = 2$ also has $\Delta n_z \neq 0$, and therefore will not be connected by a $r^2 Y_{2\pm 2}$ operator.

Second, since the important orbits are those within a certain distance of the Fermi surface, the γ vibrational wave functions will not change radically from one nucleus to the next or even over a small region. This is almost a requirement of a collective mode. We can go even further and predict how the structure will vary. From our discussion of the relevant (large) matrix elements and of the role of the energy denominator, it is clear that a given Nilsson orbit will, for some mass A, be high above the Fermi surface and will contribute little. As A (N or Z really) increases, this Nilsson orbit will drop closer to the Fermi surface and gain importance. Later, it will become a hole state and begin decreasing in amplitude as it recedes further from the Fermi energy. Thus, for most two-quasi-particle excitations, a plot of their contributions against N or Z (whichever is relevant) will be bell-shaped. Inspection of Table 10.1 shows that, for most orbits, this is precisely the observed behavior. Amplitudes for two-quasi-particle states such as $9/2^-$[514] $\otimes$ $5/2^-$[512], containing an excitation from the $h_{11/2}$ orbit from the next lower major shell, peak early in the deformed region. The $5/2^-$[523] $\otimes$ $1/2^-$[521] component increases into the deformed region and attains its maximum amplitude near

$N = 98$ (^{164}Dy), but contributes little for $N > 104$. Finally, the $7/2^-[514] \otimes 3/2^-[512]$ combination is negligible until rather late in the shell, but becomes strong near $N = 108$ and again drops off in importance near $N = 112$.

Third, we note that for any given nucleus, the number of significant components is surprisingly low, typically three to five for neutrons and slightly fewer for protons (since the protons are filling the lower shell with fewer orbits). Thus the word "collective" must be taken in context. One should not imagine 50 to 100 nucleons or amplitudes significantly involved in this or other collective excitations.

Fourth, one should address the question of how the collectivity (as measured, for example, by large B(E2: $\gamma \to g$) values) arises if only a few orbits contribute strongly. This is especially so since "single-particle" B(E2) values in odd-mass deformed nuclei (i.e., B(E2) values where one particle changes its Nilsson orbit) are much smaller than in "shell model" nuclei. The former are typically 10^{-3} to 10^{-4} s.p.u., since the shell model strength is fragmented by configuration mixing in deformed nuclei. Yet B(E2: $\gamma \to g$) values are typically ≈ 10 s.p.u. This can only arise then as a specific effect of coherence: the dominant contributions to these E2 matrix elements must add in phase.

This can be understood (at least qualitatively) by expressing the structure of the vibration in an equivalent but, in a sense, inverted picture that we alluded to briefly earlier. Instead of conceiving the vibration as built up of components that arise by operating with $r^2 Y_{2\pm2}$ on the orbits near the Fermi surface, imagine a set of closely spaced two-quasi-particle states at some excitation energy (e.g., 2 to 3 MeV) that mix by an appropriate interaction (that need not be specified). Then, by the arguments concerning two- and multistate mixing in Chapter 1, one level will be pushed down and its wave function will have the most coherent admixture of amplitudes. More specifically, if one starts with the idealized case of a set of degenerate levels with equal mixing matrix elements, the lowest state after mixing is described by a thoroughly mixed wave function (see Eq. 1.13) with all amplitudes identical and with the *same phase*. For the still schematic case of nondegenerate but equally spaced initial states that mix with equal matrix elements, a similar result is obtained. For the more realistic case of more or less random initial spacings but roughly comparable $r^2 Y_{2\pm2}$ matrix elements (since otherwise the orbits in question would not be important in the γ vibration), the same qualitative feature persists. Sample diagonalizations show this. In particular, the lowest state always has its various wave function amplitudes in phase. Thus the B(E2: $\gamma \to g$) values will have the maximum possible collectivity consistent with this set of orbits and matrix elements.

This leads to the fifth feature—the energy systematics of the γ vibration—which we can again deduce qualitatively without explicit calculation. We need only refer once again to the aforementioned mixing calculations and recall that (taking for simplicity the case of N degenerate levels mixing with each other with equal matrix elements) *one state* is pushed down well below all others and that its energy lowering is $(N - 1)V$, where V is the mixing matrix element. Thus the γ vibration will be lower when there are more contributing (mixing) states. This occurs precisely when there is an abundance of Nilsson orbits near the Fermi surface with identical n_z values so that many large $r^2 Y_{2\pm2}$ matrix elements with $\Delta n_z = 0$, $\Delta \Lambda = \pm 2$ contribute.

It was shown in our earlier discussion of the Nilsson model that at the beginning of a shell, the steepest downsloping orbits have high n_z (for prolate nuclei). Since these orbit energies are so strongly correlated with the extent of the wave function in the z direction, even neighboring orbits will have different n_z values. This makes it difficult to find nearby orbit pairs with $\Delta\Lambda = \pm 2$ and $\Delta n_z = 0$. Near the top of a shell, the large changes in orbit angles for different K values means that there will be fewer orbits and that they will be further apart on average. Thus, both cases only permit a few important γ-vibrational amplitudes. Just before midshell, however, one encounters a region where orbits with various n_z and Λ values congregate so that there will be more contributing $\Delta n_z = 0$, $\Delta\Lambda = \pm 2$ orbit pairs, hence greater collectivity and a lower γ-band energy. This qualitative behavior is sketched in Fig. 10.3 (which also includes actual values calculated by Bès). These may be compared with the observed systematics shown earlier in Fig. 2.17. The similarity of both observed and calculated patterns to our simplified estimates is remarkable. (The sharp drop empirically observed in E_γ near $A = 190$ in Fig. 2.17 is beyond this approach since it involves the onset of γ-softness and large γ_{rms} values even in the ground state.)

It is worth noting that this systematics stems from the *specific detailed microscopic structure* of the γ vibration. It is not simply a handwaving statement that vibrations will be lowest, and be most collective, at midshell where there are the most valence nucleons. Few of the valence nucleons actually participate. Moreover, other excitations, such as the β vibration (see the following), have radically different systematics.

Similar arguments can be applied to other vibrational modes such as octupole (or hexadecapole) vibrations. For example, for $K = 0^-$ octupole vibrations, the relevant operator is Y_{30} and hence $\Delta n_z = \pm 3$ $\Delta\Lambda = \pm 0$ amplitudes are critical. Being of negative parity, octupole vibrations need two orbits of opposite parity. Thus, they *must* involve the unique parity orbit, and therefore, there will not generally be as many available two-quasi-particle states that can contribute. Octupole vibrations thus tend to be rather high-lying and not very collective. They should lie lowest early in the deformed region where there are a number of $\Delta n_z = \pm 3$, $\Delta K = 0$ combinations involving excitations from the high j normal parity orbits into the unique parity orbits.

For neutrons in the rare earth region, for example, amplitudes such as $1/2^-[530] \otimes 1/2^+[660]$, $3/2^-[521] \otimes 3/2^+[651]$, and $5/2^-[512] \otimes 5/2^+[642]$ can contribute. As the shell fills, the low K unique parity orbits, whose participation is essential, begin to fill in the ground state, effectively blocking the collectivity. Thus $K = 0^-$ excitations should rise in energy. Late in the shell, the normal parity orbits are all low K, while the nearest unique parity ones are high K. Hence $K = 0^-$ octupole excitations are inhibited. Now, however, $K = 3^-$ octupole vibrations, with amplitudes satisfying $\Delta n_z = 0$, $\Delta\Lambda = \pm 3$ such as $11/2^+[615] \otimes 5/2^-[512]$ or $9/2^+[624] \otimes 3/2^-[521]$, begin to increase in collectivity and drop in energy. Thus one expects the qualitative systematics shown in Fig. 10.3, where the K sequence of successive octupole excitations in a given nucleus is seen to change from $K = 0^- \rightarrow 3^-$ to $K = 3^- \rightarrow 0^-$ as a major shell fills. This inversion of K ordering across a shell is verified in detailed calculations by Neergaard and Vogel and confirmed by experimental systematics. It is interesting that this inversion arises from the interplay of the quadrupole (Nilsson diagram structure) and octupole

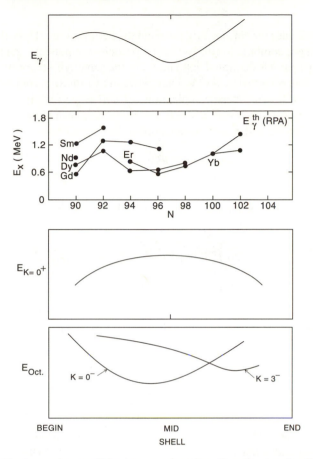

FIG. 10.3. Qualitative estimate of the systematics of γ, β, and octupole vibrations across a shell. The second panel shows the actual calculations for a γ vibration from Bès (1965).

($\mathbf{Y}_{3\pm K}$ operators) modes. As a consequence, the locus of the lowest $J^\pi = 3^-$ excitation (regardless of its K value) will have an undulating character. First the $K = 0^-$ excitation drops. Then it rises, and the $K = 3^-$ excitation drops in energy.

It is instructive to apply similar arguments to the β vibration. Here one encounters a different situation because of the $K = 0$ requirement, for which the relevant operator is $\mathbf{r}^2 \mathbf{Y}_{20}$. This has large matrix elements only between Nilsson orbits with $\Delta N = \pm 2$, $\Delta n_z = +2$, $\Delta \Lambda = 0$. However, any $\Delta N = 2$ matrix elements will naturally be small because of the large energy denominators. It is not surprising then that β vibrations are expected to be less collective than γ vibrations.

Collectivity in $K = 0^+$ states must arise from a completely different origin. It occurs as a consequence of pairing correlations according to an idea outlined in Chapter 1. The pairing interaction smooths out the Fermi surface, leading to a gradual falloff in occupation amplitudes for Nilsson orbits around the Fermi energy. Thus, the ground state consists of a linear combination of two-quasi-particle components, each with $K = 0$ and with the two particles coupled antiparallel in the same Nilsson orbit. We then need to evaluate the matrix elements of $\mathbf{r}^2\mathbf{Y}_{20}$ not between two-quasi-particle (Nilsson) states but between two different linear combinations of *pair* wave functions. For simplicity, consider a case of just two orbits, a and b, mixed by the pairing interaction, and let the ground state and an excited $K = 0$ state be orthogonal admixtures:

$$\psi_{g.s.} = a\phi_a + b\phi_b$$
$$\psi_{K=0} = -b\phi_a + a\phi_b$$

The calculation of $\langle \psi_{K=0}|\mathbf{r}^2\mathbf{Y}_{20}|\psi_{g.s.}\rangle$ is just a special case of the general result discussed in Chapter 1. We have now

$$\left\langle \psi_{K=0}\|\mathbf{r}^2\mathbf{Y}_{20}\|\phi_{g.s.}\right\rangle = ab\left[\left\langle \phi_b\|\mathbf{r}^2\mathbf{Y}_{20}\|\phi_b\right\rangle - \left\langle \phi_a\|\mathbf{r}^2\mathbf{Y}_{20}\|\phi_a\right\rangle\right]$$
$$+ \left(a^2 - b^2\right)\left\langle \phi_a\|\mathbf{r}^2\mathbf{Y}_{20}\|\phi_b\right\rangle \tag{10.27}$$

The last term vanishes since $\mathbf{Y}_{20}$ cannot change the orbits of both particles. The first two matrix elements are just the quadrupole moments of two particles in orbits b and a, respectively. Thus,

$$\left\langle \psi_{K=0}\|\mathbf{r}^2\mathbf{Y}_{20}\|\phi_{g.s.}\right\rangle = ab\left[Q_b - Q_a\right] \tag{10.28}$$

and this is small unless the orbits a and b have very different quadrupole moments.

As we have discussed, the orientation of a Nilsson orbit (and hence its quadrupole moment) is closely linked to its up- or downsloping character. For prolate nuclei, downsloping orbits are equatorial and have prolate quadrupole moments, while upsloping orbits are oblate. So $K = 0^+$ vibrations should be relatively collective and low-lying only in regions of the Nilsson diagram where orbits with very different slopes lie near each other. Inspection of Fig 8.4 shows that at the beginning and end of major shells, regions of strongly up- and downsloping orbits approach each other from different shells. (The fact that they have different N values is inconsequential, since the allowed $\mathbf{r}^2\mathbf{Y}_{20}$ matrix elements in Eqs. 10.27 and 10.28 are diagonal: they do not connect the two orbits involved.) Near midshell, there are virtually no such orbit combinations. Thus, $K = 0^+$ vibrations should be low in energy at the start and end of a shell and should peak (and have the lowest collectivity) near midshell. This is illustrated in Fig. 10.3 and can be compared to the empirical situation in Fig. 2.17.

The agreement for $K = 0$ excitations is at best qualitative. The data show little real trend and are quite erratic-as opposed to the smooth systematics for the γ vibration. This again points to the ambiguity in the nature of these $K = 0^+$ modes-β vibration, $\gamma\gamma$ mode, pairing mode, 2 quasi-particle excitations, mixtures of all four, or with some other structure entirely.

The brief discussion of the TDA and RPA formalism in this chapter, and its application to some of the most important low-lying vibrations in medium and heavy nuclei is meant only as an introduction to the field. In practical calculations there are numerous subtleties (e.g., detailed choices of single-particle energies) and sophistications (e.g., higher-order operators, O_r, than in Eqs. 10.11 and 10.12, or other related approaches involving self-consistent many-body theory). Nevertheless, the underlying physics is always similar to that outlined here. It is hoped that the present discussion will have removed some of the mystery from such calculations and indicated how simple arguments, based on the *form* (operator) for each type of vibration and the available single-particle orbits, can lead to reasonable deductions of the principal wave function components, their mass dependence, collectivity, and energy systematics.

11

EXOTIC NUCLEI AND RADIOACTIVE BEAMS

This chapter, which replaces Chapter 10 in the first edition, focuses on a new field of study in nuclear structure (indeed, in many areas of nuclear physics and astrophysics), namely the study of exotic nuclei with radioactive nuclear beams (RNBs)*. Exotic nuclei refer to β-unstable nuclei with extreme ratios of proton to neutron number—on both the proton and neutron rich sides of stability. Exotic can also refer to extremes of mass number, A, namely the heaviest (superheavy) nuclei that can be bound by the nuclear force. RNBs refers to beams of unstable nuclei produced in new generations of accelerators that are opening up these new horizons of the nuclear chart to study.

We first review the extent of existing knowledge of nuclei, then present a brief description of the main production methods for RNBs, and end with both generic and specific discussions of the physics opportunities presented by the study of exotic nuclei.

Most experimental studies of nuclei to date have focused on nuclei either in the valley of stability or near it. This, of course, is natural. Beams and targets used in accelerators are virtually always composed of stable nuclei and the nuclei produced in reactions are therefore in or near stability.

There are two major exceptions to this. As we discussed in Chapter 6, the fusion of two nuclei in the collision of a heavy ion with a target nucleus invariably produces a nucleus relatively more neutron deficient than either beam or target, simply because of the curvature of the valley of stability towards the neutron rich side. Usually these reactions form compound nuclei at relatively high excitation energies, well above the binding energy for nucleon emission, and at high angular momenta. For such states, nucleon evaporation is a more efficient means of lowering the system energy and angular momentum than γ ray emission. Hence, such reactions normally lead to the evaporation of several nucleons, until the nucleus has cooled so that γ decay competes successfully with nucleon emission. Heavy ion reactions in which 3 or 4 neutrons are evaporated are very common: fusion evaporation reactions therefore often lead to nuclei that are even more neutron deficient than the initial compound nucleus. From countless studies over the last quarter century, many neutron deficient (proton rich) nuclei, even out to the proton drip line, have been studied.

---

*Much of the material to be presented below has been developed in conjunction with the writing of several "White Papers" on this field and in numerous discussions with other members of the Writing Panels for those reports. Specifically, the White Paper of the IsoSpin Laboratory (ISL) Steering Committee and that following the 1997 "Columbus" Workshop provided key material for this section. Indeed, several figures in this Chapter are taken from the "Columbus" Report. [ISL is a generic, laboratory-independent, concept for an advanced ISOL facility; the White Paper from the Steering Committee discusses both the Physics Opportunities and technical issues.] These source references are listed in the Bibliography under "White Papers".

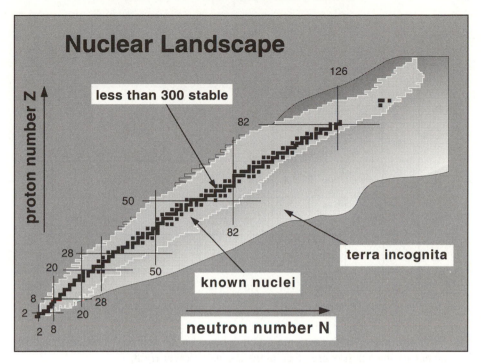

FIG. 11.1. The nuclear landscape showing the valley of stability, the limits of anticipated
nucleon binding, and the unknown regions that are being made accessible with RNBs.

The other means by which large numbers of unstable nuclei have been studied is
through the spectroscopy of fission products. Since fission is the breakup of a neutron
rich nucleus (e.g., ^{235}U), the products are neutron rich. Spontaneous or low energy
induced fission produces a mass asymmetric distribution of daughter nuclei, typically
centered around masses $A \sim 80$–110 and $A \sim 130$–150. Neutron rich nuclei in these
two regions therefore form pockets of well-known nuclei far from stability.

However, most of our knowledge of nuclei, and most of the empirical underpinning
for nuclear models, stems from nuclei near the valley of stability. The situation is illustrated in Fig. 11.1 which shows the landscape of the nuclear chart, with the stable
nuclei, those that have been studied to some extent, and the approximate position of the
proton and neutron drip lines—the limits of nuclear existence with respect to the strong
interaction. There is a vast, uncharted "terra incognita," especially on the neutron rich
side of stability, where virtually nothing is known. Even the location of the neutron drip
line is highly uncertain. Access to this region is highly important since there are reasons
to expect new nuclear physics in nuclei with highly asymmetric (relative to stability)
N/Z ratios, and in weakly bound nuclei near the drip lines.

We should note at the outset of this discussion, however, that it is not necessary to go
to the drip lines to obtain new information, especially on the neutron rich side. Indeed in
many mass regions, even the first unstable neutron rich nuclei are little, if at all, studied.

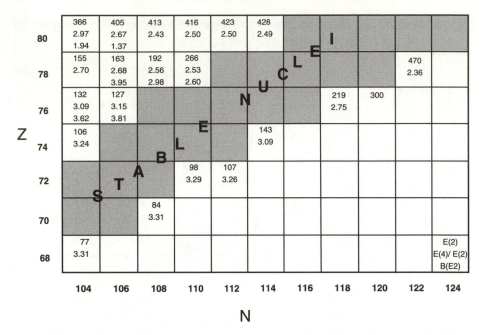

FIG. 11.2. Illustration of the extreme paucity of data on the neutron rich side of stability. The entries in each box are given in the legend at lower right.

This is illustrated in Fig. 11.2 which shows the nuclear chart (even–even nuclei) from Er to Hg. In each box, the experimental values for three key structure observables $E(2^+_1)$ (in keV), $R_{4/2}$, and the B(E2: $0^+_1 \to 2^+_1$) value (in e^2b^2) are given, where known. These quantities are invariably known for stable nuclei and the figure also shows abundant information on the proton rich side. On the neutron rich side, however, there is virtually nothing known. There is not even a single nucleus past stability on this side in this region where all three quantities have been measured. Going just two nucleons past stability is already truly terra incognita. Hence, even modest forays into the neutron rich side of stability would produce critical new information on the evolution of structure.

For both nuclear structure and nuclear astrophysics reasons (as well as to enhance certain tests of the Standard Model of elementary particles and of fundamental conservation laws, and for a myriad of practical applications in medicine, electronics, condensed matter, and industrial processes), the opportunity to study the terra incognita of unstable nuclei is of immense importance. We will delve into these opportunities in more detail in Sections 11.2 and 11.3.

Access to these new horizons is not an idle hope. If it were, this chapter would not exist. We now possess the technology to produce and accelerate beams of unstable nuclei and to use these nuclei for nuclear physics experiments. We can study the unstable nuclei in the beam itself, for example by using them in Coulomb excitation, elastic and inelastic scattering experiments. We can study the daughter nuclei into which they decay. And, we can use them as beams to induce reactions to make other (sometimes even more exotic) nuclei.

The field of RNB physics has burst onto the scene in the last decade and created an immense surge of interest and excitement. It has led to a renaissance in nuclear structure. Indeed, in many respects, it represents the future of the field. The excitement offered by these new opportunities is matched by the suddenness of their emergence. Readers of the first edition of this book (written in the late 1980s) will find no mention of RNBs at all!

The technological breakthroughs are twofold: Advances in accelerators and ion sources on the one hand, and in detectors for experiments on the other. The first advance allows production of the RNBs themselves. But, these beams are and will be, usually, several (or many) orders of magnitude weaker in intensity than typical beams of stable nuclei. Therefore, to exploit the opportunities they present, we need comparable advances in the efficiency and resolution of the instruments used in experiments.

Here, two major advances stand out. One is in the high resolution electromagnetic separation of heavy ions near to each other in mass. These devices are necessary both to obtain highly pure beams of a given type (N, Z) of exotic nucleus and to separate, with high efficiency, the reaction products they lead to when used for secondary reactions. These devices are called, respectively, isobar separators (they distinguish nuclei of the same mass number A but slightly different actual masses), and fragment mass analyzers or recoil mass separators.

Secondly, advances in detection of γ-rays following nuclear reactions gives us the capability to extract useful information from experiments with low intensity beams. These devices generally consist of arrays of semiconductor Ge detectors, often supplemented by other (charged particles, x-ray or neutron) detectors that help select a particular reaction channel [e.g., to distinguish a (H.I., $4n$) from (H.I., $3n$), (H.I., $p2n$) or (H.I., fission) reactions] and thereby to select a particular final nucleus. The best existing arrays of this type are Gammasphere in the US and Euroball in Europe. A novel approach, already used in RNB Coulomb excitation studies at MSU, involves an array of parallel position-sensitive Ge detectors that allows one to correct for angle-dependent Doppler effects and permits productive studies with beams as low as 100 particles/sec or even less!

Next generation arrays of substantially higher efficiency and "resolving power" are being developed. One such, known as GRETA, would track each γ-ray through the detector. Even the best of the existing arrays, however, have only a few percent total efficiency. Therefore a parallel, complementary approach utilizes low resolution (e.g., NaI scintillator) detectors that cover nearly 4π solid angle and have, perhaps, 50% efficiency. As with the acceleration methods for RNBs, the detector technology is advancing rapidly.

The recent development of this field, and its promise for the future, are the reasons we have included this chapter in the second edition (and indeed replaced the earlier, somewhat scatter-shot sampling of experimental techniques formerly in Chapter 10). The aim of the pages that follow is to discuss methods of producing RNBs, and to outline briefly some of the physics opportunities for the study of exotic nuclei with these beams. Given the low RNB intensities, we will need to be able to extract more nuclear structure information from less data. One way to do this, which we will discuss, is to exploit the new, highly efficient signatures of structure based on VCSs, and on the correlations of collective observables, discussed in Chapter 7.

11.1 Methods of producing RNBs

A variety of techniques, both specialized and general, are used to produce RNBs. An example of one of these specialized techniques is single nucleon transfer reactions with ^{6}Li and ^{7}Li beams used at Notre Dame to yield beams of selected light nuclei, for studies of halo nuclei (see below) and reactions of astrophysics interest. Such early initiatives have played a seminal role in spurring interest in exotic nuclei.

The two most important general techniques for producing RNBs, however, are Projectile Fragmentation (PF) and the Isotope Separator-on-line (ISOL) method. The production mechanism for exotic nuclei is similar in both cases but the techniques are, in a sense, inverses of each other. They are also complementary, in approach and in physics goals, and each has its advantages and disadvantages.

The PF and ISOL methods are illustrated schematically in Fig. 11.3. Both have many variants. We will present a simple, typical scenario for each. In the PF approach a high energy heavy nucleus (typically 50–500 MeV per nucleon) impinges on a thin target. Nuclear reactions produce a large variety of nuclei which exit the target at essentially the incoming beam velocity and pass through a mass separator that chooses a specific beam of interest. This beam of exotic nuclei then proceeds directly to a target area where it induces a secondary reaction on a target in the experimental area. Appropriate instruments at this point detect the resulting γ rays or charged particles.

In the classic ISOL technique, a light beam (two examples are a 1 GeV proton beam, or a couple hundred MeV neutron beam) impinges on a thick, hot target of heavy nuclei. Similar reactions occur (in the center of mass) as in PF. However, in this case, the exotic nuclei produced are thermalized in the target (or within a closely linked secondary "catcher" target). They diffuse and desorb within this material, eventually reaching the surface where they are ionized, pre-accelerated after mass separation, and introduced into a second accelerator which brings them to the desired energy to induce a nuclear reaction or scattering process on a downstream target. Alternatively, instead of re-acceleration, the exotic, unstable species of interest can be collected in order to study their decay, or introduced into an ion or atom trap for specialized experiments. In recent versions of the ISOL approach, the primary beam can itself be a heavy ion (e.g., C) but then its range in the target is short and the target is usually thinner.

For obvious reasons the ISOL technique is sometimes called the 2-accelerator or re-acceleration method and PF is sometimes called in-flight production.

The two methods are almost ideally complementary. The advantages of PF are that the exotic nuclei are produced and made available instantly (that is, within microseconds, which is instantaneous on the time scale of most β decays), and with no chemical selectivity—all species produced in the target are available for isotope separation and secondary reactions.

The main disadvantage of PF relates to the beam energy and quality. The beam energies of ~ 100 MeV per nucleon are well suited for nuclear reaction studies and selected aspects of nuclear structure studies, but are generally too high in energy for the majority of nuclear structure work. To reiterate a point we have made before, there is a

Projectile Fragmentation

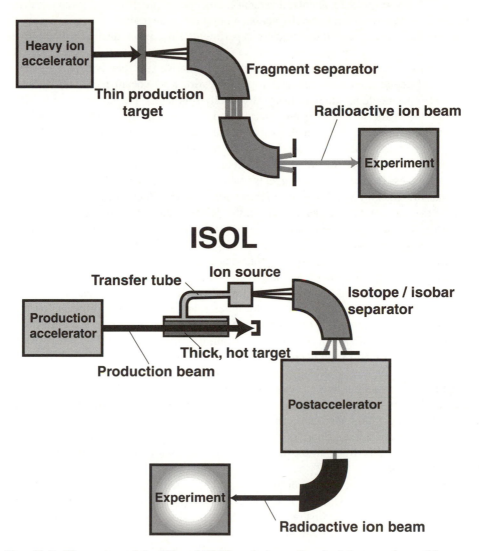

FIG. 11.3. Illustration of the PF and ISOL techniques for obtaining accelerated beams of exotic nuclei (RNBs). See text for explanation.

relation between a probe, the system probed, and the physics extracted. It is often not optimal to destroy the object of study. Therefore much nuclear structure work proceeds at or below the Coulomb barrier or in a modest energy region above it that permits transfer or fusion reactions, or non-destructive scattering experiments (i.e., 0–15 MeV per nucleon). Due to the production method in PF, beam quality–that is, beam purity (in N and Z), beam spot size, angular divergence, and energy precision and spread–is often modest. Up to now, it has generally not been practical to degrade the beam in energy to the Coulomb barrier region for structure experiments since beam quality is then further compromised.

The ISOL technique therefore provides an alternative specifically useful for many nuclear structure studies, as well as astrophysics work where beam energies at or below 1.5 MeV per nucleon simulate many stellar environments. The advantages of ISOL are that the beam of interest is pre-selected and accelerated to the desired energy in a secondary accelerator. Beam quality (in all the above senses) can therefore be as high as obtained with any stable beam accelerator. Beam energy can be varied at will.

However, there are two major disadvantages of ISOL, which are interrelated. The release process from the primary target is often slow (perhaps 100s of milliseconds up to seconds or longer). Therefore, many of the exotic nuclei produced will decay before they can be extracted and re-accelerated. Given that RNB studies are a continual struggle for intensity, this can be a significant hardship in certain cases (or none at all in others) which raises the second issue. Secondary beam production with ISOL is element and lifetime dependent. Diffusion and desorption processes and time scales are element dependent, target temperature dependent, and target composition dependent. A third issue is the power density deposited in the target, the target's ability to withstand this heating, and the attainment of optimal temperatures for quick release of the ions produced.

Much work has gone into (and continues) on ISOL target optimization. For example, some targets are composed of thousands of thin discs so that the "local" thickness of the target seen by an exotic nucleus before it reaches the vacuum (or a gas carrier) for transfer to an ion source is orders of magnitude less than the composite target. Other targets are fibrous on a micron scale so that diffusion is again enhanced. Some targets are specifically porous to certain elements. The chemical and physical dependence of extraction efficiencies and times can, of course, be an advantage in selecting a specific element to re-accelerate. Enormous progress in target-ion sources is continually being made but it remains true that the development of RNBs for each element is a research project in itself.

There are a number of important advanced PF facilities in operation worldwide, notably at MSU in the US, GSI in Europe, and RIKEN in Japan. The ISOL technique is more in its infancy. The first operational ISOL facility, which pioneered this method, is at Louvain-la-Neuve in Belgium. Here, a high current, low energy cyclotron is used as the primary accelerator while the secondary RNBs are accelerated with a second cyclotron. In the US, HRIBF at Oak Ridge has now been in operation for a couple of years. Louvain-la-Neuve and the HRIBF are first generation facilities. More advanced ISOL laboratories are currently under construction at TRIUMF in Vancouver, Canada and GANIL, France.

In the end, the pluses and minuses of PF and ISOL balance and both techniques are valuable, needed, and as we have stated, complementary. And, in fact, hybrid methods incorporating features of both have been developed, in which the exotic nuclei are produced in a thinner target, fly out, are separated magnetically, stopped, ionized, and then re-accelerated. An advantage of this approach is that the stopper material can be different from the production target material and can be chosen for optimal and fast release. With this technique, the ISOL method might be used for shorter half-lives than with traditional in-target stopping, and hence nuclei further from stability should become accessible. The US is planning to build an advanced next-generation facility, tentatively named RIA for rare isotope accelerator, that will use this hybrid technique. Such a facility has the additional advantage that it can also produce direct fast in-flight beams.

Both PF and ISOL techniques can produce an enormous variety of exotic nuclei as beams, with intensities ranging from less than one per second to 10^{10}/sec which is comparable to the intensities of heavy ion beams of stable nuclei. These beams are used in a variety of experiments employing many different techniques and requiring beam intensities spanning the full range just cited. With RNBs, one is often studying essentially unknown nuclei. Therefore, the goal is to get the initial basic information on their structure, not to study their spectroscopy at the exhaustive level currently possible for nuclei produced with stable beams and targets. The *techniques* that will be used (Coulomb excitation, transfer reactions, fusion, decay studies and so on) will be the familiar tested techniques used for decades (and, in fact, often, primarily, decades *ago*) to glean information on "normal" nuclei. Indeed, we will exploit our often rather sophisticated understanding of these techniques and reaction processes to gain the first data in new regions. Rather than the experimental techniques per sec, it is the nuclei, their structure, and the physics phenomena and insights they will yield, that are new. Typical ranges of cross sections for different techniques and the beam intensities needed are shown in Fig. 11.4.

Together, PF and ISOL represent the technological basis for the birth of a new field of nuclear physics, with expanded horizons, and every expectation of producing new nuclear and astrophysics results, and exotic nuclear structure phenomena. They have already disabused us of the parochial view of nuclear structure resulting from the limited vistas available within the valley of stability.

11.2 Nuclear structure physics opportunities in exotic nuclei

There is every reason to believe that nuclear structure in exotic nuclei will yield new phenomena and new insights into the nuclear many-body problem. There are many reasons for this expectation. We briefly discuss six of them.

The first is history itself. We have already had access to some exotic nuclei and they have usually revealed new, and important, facets of structure. Early studies of fission products near $A = 100$ revealed the most dramatic onset of deformation known-between neutron numbers $N = 58$ and 60. The relevant data for the Zr isotopes are shown in Fig. 11.5: $E(2_1^+)$ drops from 1223 keV in ^{98}Zr to 212 keV in ^{100}Zr. These data (see Chapter 7) revealed the dissolution of the $Z = 40$ *proton* magic gap as a function of *neutron* number, which in turn, via the seminal calculations of Federman and Pittel, confirmed the importance of the residual $T = 0$ p–n interaction in the development of collectivity and deformation in nuclei.

These studies were later complemented by similar investigations of the evolution of shell structure in other mass regions and led to an interpretation of the onset of

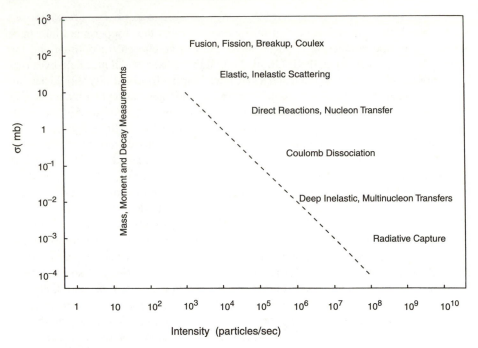

FIG. 11.4. Cross section ranges and required beam intensities for various techniques with radioactive beams (based on the Columbus ISOL White Paper, 1997). Recent advances in experimental techniques have greatly lowered some of the required intensities (e.g., Coulomb excitation) since this figure was developed.

deformation near $A = 150$ (see Chapter 7) in terms of the obliteration of a subshell gap at $Z = 64$. Still more recent work has shown that ^{32}Mg ($N = 20$) is also deformed, not spherical, and there are suggestions for the same in ^{44}S ($N = 28$). As we shall see, one of the pervasive themes in extremely neutron rich exotic nuclei will be just this idea (generalized and with new ingredients) of the fragility of magicity. Indeed, the benchmark magic numbers (2, 8, 20, 28, 40, 50, 82, 126) and the shell structure they reflect, which have been the underlying microscopic basis for nuclear structure for half a century, are now being recognized as but one reflection of a much more general concept of shell structure (and even of shell "melting")–the particular embodiment of shell structure that happens to occur near stability.

Further evidence of the importance of the $T = 0$ interaction comes from studies of unstable nuclei on the proton rich side. Access to successively heavier nuclei with $N \sim Z$ by heavy ion reactions has revealed the nearly singular dominance of the $T = 0$ p–n interaction in these nuclei and, again, the fragility of shell structure: Proton rich ^{80}Zr, with $N = Z = 40$, is deformed, not doubly magic, although ^{90}Zr (which is stable) with $Z = 40$, $N = 50$ is indeed doubly magic.

In the heaviest nuclei, advances (with stable beams) in the producton of heavier and heavier nuclides are revealing new insights into shell effects and the stability of

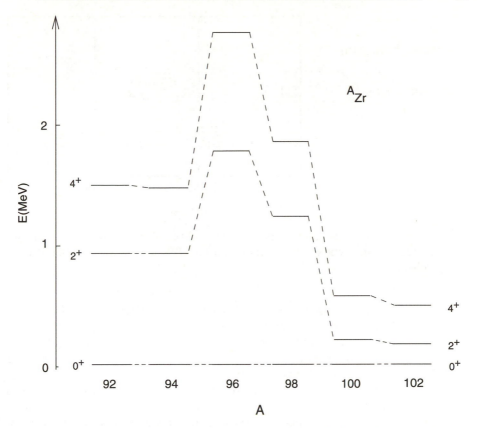

FIG. 11.5. Systematics of Zr isotopes illustrating the rapid phase/shape transition at $A = 100$.

superheavy nuclei. The race toward heavier and heavier nuclei may now have reached $Z = 118$. Further progress with both stable and radioactive beams will push this frontier into heavier regions. Near $Z = 126$, a region of special stability and long lifetimes is predicted.

Finally the study of very light near-drip line neutron rich nuclei, such as ^{11}Li, has revealed the existence of "halo" nuclei which, in effect, comprise a new form of matter (see below).

It is therefore evident that history itself suggests an on-going stream of new phenomena and insights if we can access increasingly more exotic nuclei with RNBs.

Secondly, the vast number of possible beams made available in RNB facilities allows greater flexibility in choosing the optimum ones either to study a specific effect or to gain access to a new nucleus of specific interest (see Fig. 11.6). In studying nuclei to date various probes have been used—a few "elementary" particles (protons, neutrons, electrons, muons, and photons to name some), and a variety of light and heavy stable nuclei (ranging from the deuteron to some 50–100 heavy ions). The potential number of

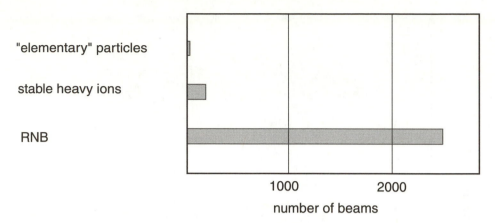

FIG. 11.6. Schematic illustration of the potential number of beams available from simple projectiles, heavy ions, and RNBs (see text). (Based on a figure obtained from I. Tanihata.)

stable beams is limited to the number of stable nuclei, which is less than 300. In contrast, 1000s of RNB species are, in principle, available (albeit at vastly varying intensities). The point is not at all the sheer numbers (we will never be able to, want to, or have the time to, use all of them), but the *flexibility of choice* and *selectivity of physics* that the availability of this 'gene pool' of nuclei presents.

Thirdly, a point we have mentioned above. One way to study any physical system is to stress it and study its response. Historically, we have made great strides in our understanding by stressing nuclei by pouring large amounts of energy or angular momentum into them–in both cases, with exciting results. With RNBs, we have the *isospin* degree of freedom available to stress nuclei to extremes in N/Z composition, extreme limits of binding (i.e., to the drip lines) and to extreme bounds of proton–neutron asymmetry in Fermi energy, density, spatial extent, and diffuseness.

Stressing nuclei to extremes in their N/Z composition will amplify certain interactions between nucleons (and decrease the importance of others) so that, in these nuclei, we can sample a different balance of terms in the nuclear Hamiltonian. By understanding better these amplified interactions, we can then return to stable nuclei and improve our descriptions.

The fourth and fifth reasons relate specifically to the basic nature of the nuclear many-body fermionic system interacting under the strong force (see Fig. 11.7). There are three ingredients in the argument: the fact that nucleons are fermions and obey the Pauli principle, the fact that shell model orbits typically have low j, and the inherent *weakness* of the strong force in the nuclear medium. The first two imply that only a finite (and relatively small) number of nucleons can occupy a given orbit. Therefore, the addition of new nucleons must lead to the occupation of new orbits, with new quantum numbers n, l, j. The weakness of the nuclear force refers to the fact that, for example, it is only strong enough to barely bind the deuteron. Combining this idea with the effects of the Pauli principle leads to a second point which is reflected in the very validity of the independent particle model itself. Nucleons comprise about 60% of the nuclear volume, yet undergo $\sim 10^{20}$ orbits per second without collision. The reason is that nucleons in inner, occupied orbits cannot collide (perhaps more accurately, the collision is ineffectual) unless they can be scattered to other orbits. But, inner orbits are already filled and the nuclear force is generally too weak to excite inner nucleons to the valence or higher shells. Hence most nucleon interactions involve the valence orbits and it is the *valence space* that determines the nature and evolution of structure.

Combining now these two results we see that if new orbits, with different quantum numbers, are occupied, and if such new valence orbits determine structure, then new types of structure should occur. It is nuclei featuring the occupation of new orbits, or new combinations of proton and neutron orbits, that RNBs can provide.

The fifth reason goes to the heart of the nuclear structure of loosely bound nuclear systems. Nucleons in the normal orbits of stable nuclei are bound (see Fig. 1.1) typically by 5–8 MeV. The wave function of a nucleon that is bound in a potential well by such an amount (imagine for simplicity a finite square well a few fermis wide) is almost completely contained within the well. On the other hand, near the drip line, the last nucleons will be bound by only (typically) 100s of keV. Their wave functions dissipate exponentially with $[2m(E - V)]^{1/2}$ within the potential barrier. Since $(E - V)$ is so small the spatial distribution extends to large radii, especially if the nucleons occupy low angular momentum orbitals (smaller angular momentum barrier) and especially for neutrons (no Coulomb barrier): loosely bound neutron wave functions are spatially extended, with diffuse, low density probability distributions.

We will discuss below an extreme example of this in the form of a new nuclear topology, halo nuclei. But the effect is not limited to such exotica. We have said many times in this text that the nuclear radius $R \sim r_0 A^{1/3}$. Actually, from data on light unstable nuclei that we already have, we now recognize that this is a good approximation *only* near the valley of stability. Near the drip lines, as shown in Fig. 11.8, strong deviations occur. In fact, the radii are often not even monotonic in A, as exemplified by the light C nuclei ($Z = 6$) or the $N = 8$ nuclei in Fig. 11.8. Figure 11.8 shows that, along an "iso-chain" (either isotopic or isotonic or even a diagonal isobaric sequence) the radii tend to follow a parabolic evolution, large values in weakly bound near-drip line nuclei and smaller for the normally bound nuclei near stability.

The low binding of the outermost nucleons also means that their interactions will occur not only with bound nucleons but with nucleons in the continuum (positive energy) where wave functions go, not as e^{-kr} as in the barrier region, but as e^{ikr}. Also, the asymmetry in the numbers of protons and neutrons (on the neutron rich side) means that protons and neutrons occupy very different orbits.

The consequences of all these arguments are (as we shall see in more detail below) that the shell model potential itself may be altered, as may the spin-orbit force, and so may residual interactions between valence nucleons. This surely would lead to altered structure and altered structural evolution. In a moment we will turn to one specific scenario illustrating this.

Finally, the sixth reason covers astrophysics. The energy sources that power stellar

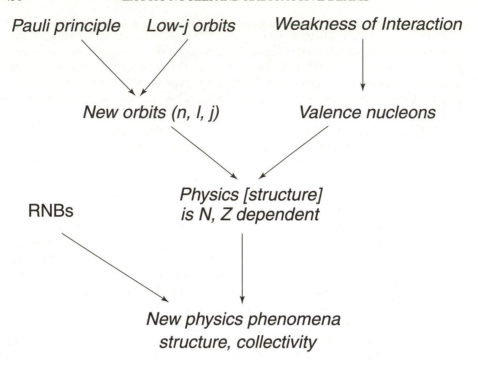

FIG. 11.7. Generic illustration of why access to exotic nuclei with RNBs is almost certain to lead to new phenomena and new manifestations of nuclear structure (see text).

objects are primarily nuclear reactions. The fusion of hydrogen into He and of He nuclei into ^{12}C, and then the sequences of (p, γ) and (p, α) reactions and β decay processes known as the CNO cycle describe the evolutionary history of most normal stars. To understand stellar energy generation and the life-history of stars we need to know the nuclear reaction rates involved, as a function of stellar temperature. Many of the key reactions, and nearly all of those that have not been studied, involve unstable nuclei. Some indirect information on these rates has been obtained from reactions with stable beams and targets by exploiting the concepts of detailed balance (i.e., looking at the inverse reaction) and isobaric analogue states. Nevertheless, far too little is known to claim a quantitative understanding of, for example, the CNO cycle, or the breakout from this cycle to still heavier nuclei (rp-process region). Studies with RNBs, for the first time, are giving access to exactly the reactions we need to study.

Heavy element nucleosynthesis also occurs primarily in stellar objects. Indeed, it has validly been said that we, and the world around us, are all merely stellar debris, the result of cataclysmic stellar explosions. Most of the nuclei beyond Fe, for example, have probably been created in robust stellar explosions known as supernova. These explosive events last (at their peak power) only a few seconds yet, during that time, generate more energy than all the known galaxies combined in the same time period. The primary nuleosynthetic mechanism is the r-process of rapid neutron capture. The

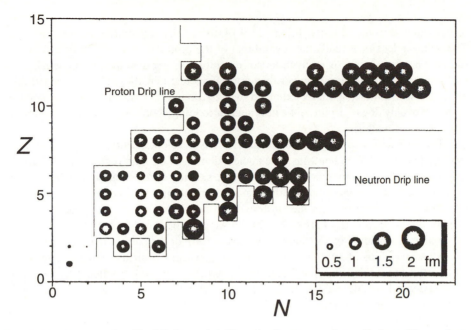

FIG. 11.8. Measured radii of light nuclei. Note the "suppressed zero"–the radii are given relative to that for ^{4}He, that is, the radii are shown in the units $R - R_\alpha$ using the size scale at lower right. (Based on a figure obtained from I. Tanihata.)

resulting nuclei formed are determined by the balance between (n, γ) and (γ, n) rates and the values of several key β decay lifetimes. To understand nucleosynthesis therefore requires knowledge of $T_{1/2}(\beta)$, masses and basic properties of nuclei along the very neutron rich trajectory followed by the r-process in the $N - Z$ plane. Access to these nuclei is only available by exploiting the capabilities of RNB facilities such as RIA.

Having argued from general grounds (historical, nuclear physics, and astrophysics) that exotic nuclei are likely to yield new nuclear physics and astrophysics we now discuss a few specific nuclear structure issues of interest in this new frontier.

11.3 Facets of structure in exotic nuclei

One of the most dramatic examples of new structure and of new nuclear topologies far off stability is that of "halo" nuclei. In these nuclei the last nucleon(s) is (are) so weakly bound that the wave functions extend to many fermis. The best known and studied example is ^{11}Li, which consists of a normal ^{9}Li core and two halo neutrons. It has one bound state, bound by only ~ 300 keV. The radius (rms radius) of ^{11}Li is about the same as ^{208}Pb: the deviation from $r_0 A^{1/3}$ with $r_0 \sim 1.2$ fm is huge.

The binding of ^{11}Li is a specific consequence of the residual interaction of the two last neutrons since ^{10}Li is unbound. Such a system, namely one in which a 3-body system (^{9}Li and 2 neutrons) becomes unstable if *any one* of its constituents is removed, is called a Borromean system after the three interlocking rings of the family crest of the Italian

Renaissance Boromeo family. Figure 11.9 illustrates the remarkable features of ^{11}Li. The neutron density extends far beyond that of the proton, out to 10–20 fms. At these radii the neutron density is many orders of magnitude larger than the proton density, yet both are orders of magnitude lower than normal nuclear densities. This region exhibits a low density, diffuse, spatially extended, weakly bound region of nearly pure neutron matter. Not only does ^{11}Li present a new nuclear topology but its outer reaches comprise a new form of matter.

Heavier nuclei with large excesses of neutrons develop an outer layer, a skin, consisting of perhaps 10–30 neutrons. The nuclear density in the skin region falls off slowly (albeit not as slowly as for halos since most of the skin neutrons remain bound by an MeV or more) so that one again has a diffuse neutron rich outer surface. This phenomenon of a low density neutron skin has profound consequences on structure. Such a diffuse density distribution is unlikely to be able to support the normal shell model potential with rather sharp outer edges (the Woods–Saxon form or the harmonic oscillator plus l^2 form). One can think of this in terms of a Fourier transform. A diffuse density distribution will not lead to a self-consistent mean field potential with high frequency components (sharp edges). Instead, one expects a more rounded potential to develop. To date, calculations using different microscopic methods such as Hartree–Fock or relativistic mean field approaches, with various forces (e.g., Skyrme or Gogny or any of the others on the market) are producing quite different scenarios for the resulting single particle levels. Only experiment will likely decide the issue.

For now, we illustrate one scenario, that of Nazarewicz and colleagues (Dobaczewski, 1996), for the near-neutron drip line Sn nuclei. We caution that these results are speculative at present. Even more so are the qualitative conclusions we draw from them. Moreover, it is unlikely that one will reach the neutron drip line for nuclei as heavy as Sn, at least in the foreseeable future. We therefore present the discussion below not as a prediction to be tested but to illustrate the *types* of phenomena one might encounter.

Figure 11.10 illustrates the single particle level sequence of a normal shell model potential (left) along with one inspired by the drip line calculations of Nazarewicz and colleagues (middle). (We will discuss the scenario on the right in a moment). One notes several interesting changes in shell structure. First, the highest l orbits increase in energy. This is especially true for the unique parity orbits ($h_{11/2}$ and $i_{13/2}$ in the figure) which revert back to their parent shells. The reason is easy to see. The original motivation for a deeper potential at large radii (i.e., for the squaring off of the potential) was to make it more attractive for higher angular momentum states whose wave functions are shifted outwards by the centrifugal force (see Fig. 3.1b). A rounding off of the potential at large radii for drip line nuclei will reverse this effect, namely, increasing the energy of high l, high j orbits. Analogous arguments account for the decrease in energy of the lowest l orbits.

Secondly, the magic numbers and shell gaps disappear or change. In the former case, the phenomenon has been called "shell quenching" or a melting of shell structure. Note that such a change or disappearance in magic numbers has a somewhat different origin than we have discussed in the context of the origin of deformation in the $A = 100$ and 150 regions and for the presence of "intruder" states in nuclei such as Cd and Hg. In

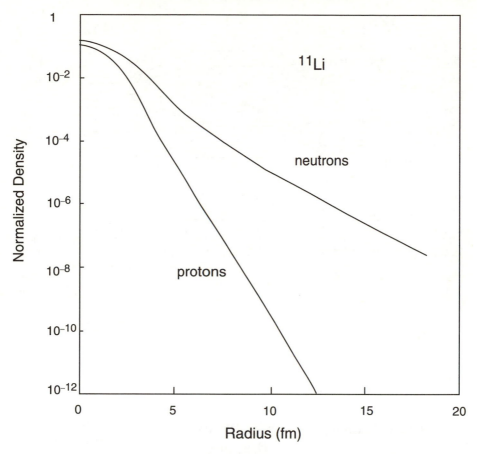

FIG. 11.9. Calculated proton and neutron density distributions in ^{11}Li (based on B. Sherrill and A. Sustich, 1994).

these cases the effect was due largely to the residual p–n interaction (specifically, the monopole component) acting on proton and neutron orbits of similar character [similar (n, l, j) values]. Here, it is an effect of changes in the central potential itself.

The consequences of such single particle changes, as illustrated in Fig. 11.10, would be enormous if they are realized in practice. If the traditional spherical magic numbers are replaced by new ones (for example 82 and 126 would be replaced by 70 and 112 if the $h_{11/2}$ and $i_{13/2}$ orbits shift up to their original major shells) then clearly the locus of collectivity and deformation as a function of neutron number will be far different near the drip line compared to near stability. If the magic numbers disappear, then what becomes of the meaning of a valence space? Indeed, how would one be justified in using shell model calculations that focus on the valence nucleons, or using other valence space models such as the IBA?

Thirdly, the sequence of single particle levels within a shell is significantly altered.

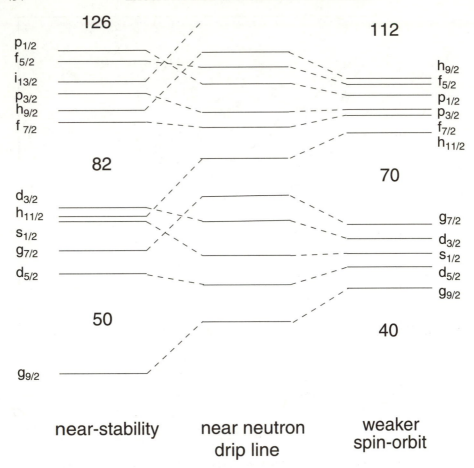

FIG. 11.10. Single particle energy levels: (left) normal levels for the shell model; (middle) possible scenario that could occur in a more rounded potential in exotic near-drip line nuclei; (right) same as middle panel but with a spin-orbit interaction strength half as large. (Modified from a figure by W. Nazarewicz.)

Let us compare the specific sequences of single particle energies on the left and middle of Fig. 11.10. In the normal shell potential the sequence through a shell (for the normal parity orbits) is generally one of decreasing j values, often decreasing by $\Delta j < 1$. In the drip line scenario, there is a "nested" pattern with the highest j values surrounding middle ones, which, in turn, surround the lowest orbits. This corresponds, in fact, to a $\Delta j = \Delta l = \pm 2$ sequence (e.g. $h_{11/2} - f_{7/2} - p_{3/2} - p_{1/2} - f_{5/2} - h_{9/2}$: note that the change from $p_{3/2}$ to $p_{1/2}$ is also $\Delta j = \Delta l = 2$ since the angular momentum vectors can be anti-parallel).

This new sequencing would have significant structural effects (ignoring complementary and probably offsetting changes in residual interactions for the moment). Since the quadrupole interaction ($\propto Y_{20}$) preferentially couples orbits with just these changes in l and j, there could be a tendency for a more rapid onset of collectivity and deformation outside doubly magic numbers (if they exist at all).

Moreover, recall that in a normal shell, the high degeneracy orbits (those that can accommodate the largest number of particles) lie lowest. This was one of the ingredients that led (see Chapter 8) to the predominance of prolate deformations in nuclei. In the nested sequence the highest j orbits are at the beginning and end of the shell. Hence the preference for prolate deformations early in a shell could be muted. On the other hand, the tendency towards reduced prolate deformations (and even a few oblate nuclei) near shell end might be replaced by a renewed push towards larger prolate deformations at one or more mid-shell points. Noting the new groupings of the levels in the middle of Fig. 11.10—a somewhat isolated $h_{11/2}$ orbit, a pair of orbits $f_{7/2}$ and $p_{3/2}$, and a grouping $p_{1/2}$, $f_{5/2}$, and $h_{9/2}$—one might expect a tendency towards an oscillatory pattern of deformations as a shell fills.

Of course, this whole discussion ignores a number of important ingredients. One is the fact that the protons are largely immune to these effects and will tend to modulate them. A second is that residual interactions will also likely be very different. The latter idea is illustrated in Fig. 11.11.

The p–n interaction could, in fact, be substantially reduced since the outermost proton and neutron orbits are quite different and will have less overlap than in normal nuclei. Secondly, the neutron pairing interaction, which is related to the number and density of orbits into which zero-coupled pairs can scatter, could be greatly increased by virtual scattering to the large reservoir of continuum single particle states nearby in energy at the neutron drip line. Some of these states (with $l \neq 0$) can be quasi-bound due to the centrifugal barrier. Also, when pairs of particles are scattered to positive energy states, the strong attractive residual interactions can effectively bind the $J = 0^+$ state. Thus, the presence of a neutron skin, and of skin excitations, along with increased pairing interaction, may even invalidate the concept of single particle orbits in the outer reaches of neutron drip line nuclei. Both of these effects on residual interactions, a decrease in the p–n interaction and an increase in the pairing interaction, would exert a restraint on deforming tendencies. So, the effects discussed above in the context of Fig. 11.10 could be significantly reduced in magnitude.

We illustrate (somewhat reluctantly since it is so speculative and qualitative) a possible outcome in Fig. 11.12. Compared with the simple evolution of deformation across a shell, one might expect a quicker onset of deformation, albeit with smaller magnitude, and oblate nuclei at several points. This figure illustrates a possible scenario but should be taken with tons of salt. Nevertheless, it shows the potential for radically new manifestations of structural evolution.

There are still other effects of the altered single particle order in Fig. 11.10. The return of the unique parity orbit to its parent shell will impact nearly all manifestations of high spin phenomena such as backbending, superdeformation and the like. Furthermore, the simplifications in interpretation resulting from the purity of the unique parity orbit in a normal shell sequence will disappear and one will have to treat high spin phenomena in a multi-j, configuration-mixed space. A second consequence will be that negative

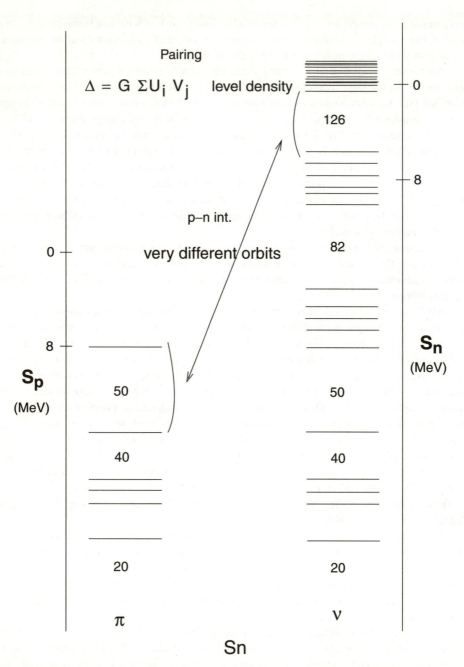

FIG. 11.11. Residual interactions (p–n and pairing) in neutron rich near-drip line nuclei
(see text).

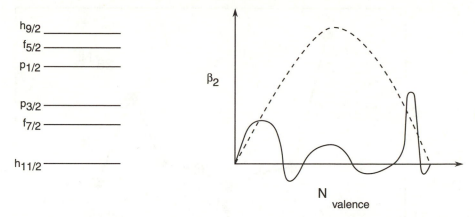

FIG. 11.12. A possible scenario (taken from the middle panel of Fig. 11.10) for single particle levels in neutron rich near-drip line nuclei and, on the right, a speculation (solid curve) as to how the quadrupole deformation might vary as this shell fills, based on the single particle levels on the left and the likely changes in p–n and pairing interactions illustrated in Fig. 11.11. The dashed curve shows roughly how deformation varies across a shell in normal nuclei.

parity excitations will now no longer be low-lying in nuclear spectra since a given shell would consist solely of orbits of a given parity. Collective octupole correlations or stable octupole deformation might be absent at low energies.

There is yet one further potential consequence of a diffuse, shell model potential at large radii that could have an important impact. The spin-orbit interaction (see Chapter 3) has a radial dependence given by the derivative of the shell model potential: it is surface peaked. With a more gradual fall-off in the shell model potential the spin-orbit force may be reduced in strength. Some calculations directly suggest this. If the spin-orbit splittings in Fig. 11.10 (middle) decrease by a factor of two one obtains the spectrum on the right. Again, this is very speculative. Given such a scenario, though, one would have compact, full major shells. "Full" here means containing all levels from $j = 1/2$ to $j = l_{max} + 1/2$. This approximates the conditions for fermionic SU(3) structure as proposed by Elliott in the late 1950s for light nuclei. For heavy nuclei such a scheme has not been applicable since the disappearance of the unique parity implies that the j-sequence is incomplete. With the sequence of states on the right in Fig. 11.10, an Elliott SU(3) scheme could be more widespread along the neutron drip line.

An excellent opportunity to study the evolution of structure, and to perhaps glimpse some of these effects, is provided by the Ni isotopes. These are unique in the nuclear chart in that they span four magic numbers, from ^{48}Ni ($N = 20$) to ^{56}Ni ($N = 28$) to ^{68}Ni ($N = 40$) and finally to ^{78}Ni ($N = 50$). The Ni isotopes thus involve five major neutron shells: $N \leq 20$, $N = 21 - 28$, $N = 29 - 40$, $N = 41 - 50$, and $N > 50$. The existing data for 2_1^+ energies in Ni are shown in Fig. 11.13, illustrating the $N = 28$ and 40 shell closures. Strenuous and ingenious efforts are currently underway in several

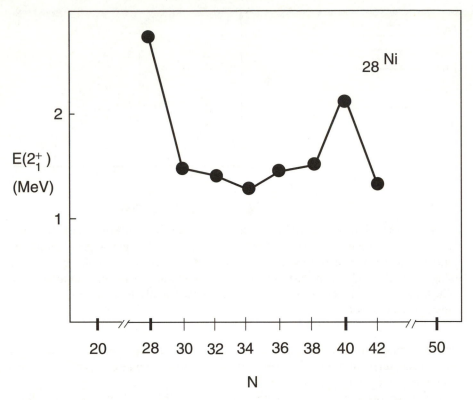

FIG. 11.13. 2_1^+ energies in Ni isotopes showing evidence for magic numbers at $N = 28$ and 40 and the constancy of $E(2_1^+)$ that is predicted by the seniority scheme in between. [Based on a figure obtained from P. Mantica and using recent data from Grzywacz, 1998.]

laboratories to study ^{78}Ni and success is likely within a couple of years.

Ni is also an important stage in nucleosynthesis between the rp-process and the r-process. Its study is important in understanding the mechanisms of supernova explosions and the role of neutrino emission in the early stages of such violent stellar events.

Turning to the proton rich side of stability there is also interesting new physics with RNBs. Here, the territory is not nearly as unknown since we have already mapped much of the proton drip line up to Bi. However, there are special nuclei on the proton rich side that are well known goals of RNB physics. One is ^{100}Sn ($N = Z = 50$). It is the heaviest particle stable $N = Z$ nucleus. It is also expected to be doubly magic. ^{100}Sn has been produced and identified but, as of this date, only a handful of nuclei have been produced. Little is known with precision about its lifetime or mass (although such data will surely come soon, perhaps before this edition reaches print). Less is known about its structure. The question of its magicity is important for all the reasons mentioned above and also because one can then compare the single particle structure near ^{132}Sn and ^{100}Sn. These nuclei differ by 32 neutrons (the 50–82 shell) and the effects of both the like nucleon and the p–n interactions of these 32 neutrons will tell us much about the residual interactions at work. We have already seen (see Fig. 3.11) the strong changes in single particle energies across this shell due to the p–n interaction. Another appeal of ^{100}Sn is that doubly magic nuclei above Ni are rare indeed. Only ^{132}Sn and ^{208}Pb are well studied. (Some would count ^{90}Zr as doubly magic as well.) We need as many examples of such potential benchmarks as we can study.

$N = Z$ nuclei present singular features, due to the effects of the $T = 0$ p–n interaction. We recall from Chapters 3, 4 that an isolated 2-nucleon p–n system can exist in two states, called $T = 1$ (spins anti-parallel) and $T = 0$ (spins parallel)—see Fig. 11.14 which is similar to Fig. 3.12. The $T = 0$ interaction strength (see Chapters 3, 4) is about twice as strong as the $T = 1$ (and accounts for the binding of the deuteron). In $N = Z$ nuclei protons and neutrons fill identical orbits and, excluding the effects of the Coulomb interaction, their wave functions should have nearly complete spatial overlap. Hence, the $T = 0$ interaction rises to a sudden dominance in $N = Z$ nuclei. Effects of this probably account for deformation in ^{80}Zr. RNB-based studies of $N = Z$ nuclei beyond ^{84}Mo (^{88}Ru is the last for which even a rudimentary level scheme exists) will help understand the locus of deformation in these unique nuclei and the role of the $T = 0$ interaction.

The effects of a suddenly enhanced $T = 0$ interaction in $N \sim Z$ nuclei could have many repercussions. One that has been much discussed is p–n pairing. In an odd–odd $N = Z$ nucleus the last proton and neutron should be (assuming that changes to the proton single particle energies due to the Coulomb force have not altered the orbit sequence) in the same shell model orbit, j. For the $T = 0$ state of this p–n pair, the residual interaction will be very strong in the $J = 2j$ state (see discussion of $T = 0$ interactions in Chapter 4). Such a state could form the ground state and there could even be a "pairing gap" despite the fact that this is an odd–odd nucleus. Also, other phenomena could be altered. If the nucleus is non-spherical and rotates, the Coriolis force would not affect the two last nucleons differently, in contrast to even–even nuclei where this force tends to break $J = 0$ - coupled like nucleon pairs. Hence, any backbending might be delayed to higher spins.

Other dramatic manifestations of the $T = 0$ interaction show up in nuclear binding energies. Here, it is useful to introduce the concept of "interaction filters". These are specially chosen functions of binding energies, usually of nearby nuclei, that are designed to isolate specific interactions. One of the most important was mentioned in Chapter 7 in providing an empirical justification for the $N_p N_n$ scheme. Called δV_{pn}, it is a double difference of binding energies that, at least in first and second order, cancels out all residual interactions except that of the last proton with the last neutron. It is defined by

$$\delta V_{pn}(Z + 1, N + 1) = 1/4\{[B(Z + 2, N + 2) - B(Z + 2, N)]$$
$$- [B(Z, N + 2) - B(Z, N)]\} \tag{11.1}$$

Empirical values for δV_{pn} are shown in Fig. 11.15, where lower values correspond to stronger interactions.

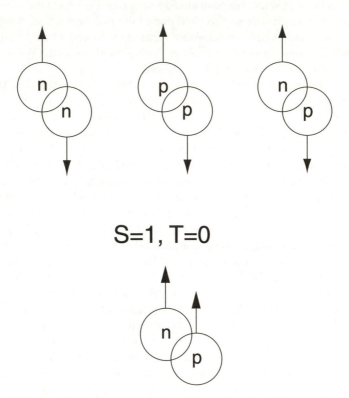

FIG. 11.14. Spin couplings of two nucleons, illustrating the difference between $S = 0$, $T = 1$ and $S = 1$, $T = 0$ systems.

We see that the general trend is a smooth decrease in $-\delta V_{pn}$ towards heavier nuclei. This is easily understood. In heavier nuclei the protons and neutrons are filling orbits with less and less overlap and the wave functions are also spread out over larger volumes. Thus, short range interactions will be weaker. Superposed on this very smooth trend, however, are a series of spikes (singularities) for $N = Z$ nuclei, where, suddenly, $|\delta V_{pn}|$ increases by a factor of 2–4. This huge increase occurs only at $N = Z$. It is understood in the sense that realistic shell model calculations including the $T = 0$ interaction reproduce the spikes almost exactly. It also results as a consequence of basic symmetry arguments.

With increasing Coulomb effects on proton energy levels in heavier $N = Z$ nuclei, a key question is whether the effect in Fig. 11.15, and similar ones for odd-A nuclei, persist. The apparent decrease in δV_{pn} for $N = Z$ as a function of A in Fig. 11.15 could either reflect the weakening of this effect or a natural evolution of its strength across each shell. [One sees (and calculations reproduce) a similar evolution within the s–d shell.]

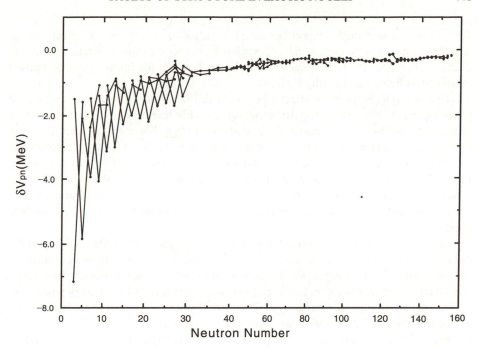

FIG. 11.15. Empirical δV_{pn} values for all even–even nuclei. Note scale change on abscissa at $N = 40$. Note the hugely attractive (downward-going) singularities which occur precisely for $N = Z$ nuclei (based on Brenner, 1990).

With exotic nuclei, we can evaluate δV_{pn} for several more $N = Z$ nuclei, eventually closing in on ^{100}Sn.

Finally, for proton rich nuclei, we raise a most intriguing subject, that of proton radioactivity. At, and even beyond the proton drip line, it is still possible to study nuclei that, in a sense, should not exist. Although the last proton is unbound by the nuclear potential beyond the proton drip line, proton emission is delayed by both the angular momentum and Coulomb barriers (especially the latter).

This effect of the Coulomb force might seem counter-intuitive, at first. How can a repulsive force bind nucleons. The simple answer is that it acts in both inward and outward directions. To the extent that the Coulomb barrier is primarily *outside* the probability distribution for a given proton, it will help to keep the protons in. Over most of its distance from the nuclear center to its edge, as illustrated in Fig. 3.3, the repulsive Coulomb potential reduces the strength of the attractive proton potential relative to that for neutrons. The net potential is still strongly attractive. However, the Coulomb force extends to larger radii than the nuclear and hence, when the latter has vanished, the only net potential is the repulsive Coulomb interaction which now tends to force the protons inward.

For a proton to "drip" out of a proton-drip line nucleus it must tunnel through the Coulomb barrier. The effect (delay) is sufficient that proton decay lifetimes are not the

10^{-18} seconds one might expect but range from μs to ms. When the effects of the angular momentum barrier are added, protons in higher l orbits may take seconds to decay. Hence, the structure of nuclei at, and even beyond, the limits of their supposed existence, is being successfully studied.

This is a purely quantum effect which is both interesting in itself and useful as a probe of structure as well since the tunneling is highly sensitive to the proton angular momentum and to the 3-dimensional shape of the nucleus. For example, near $A \sim 160$, a proton unbound by 1 MeV takes 10s of ms to decay if it is in an s state, but seconds if the orbital angular momentum is $l = 5$. Fine structure in proton decay (decay to different final states) has also been observed and can be used to probe both the structure of the parent and that of the daughter nucleus. With RNBs, we can push the frontiers of our knowledge past the proton drip line over much of the proton rich edge of the nuclear chart.

There are, of course, many other fascinating areas of study with RNBs, but, hopefully, our discussion of some of the main ones has been sufficient to illustrate the opportunities present. The 1997 "Columbus" White Paper on the scientific opportunities with radioactive beams (specifically for an ISOL facility) cited a number of key areas of study with exotic nuclei—in nuclear structure, nuclear astrophysics, tests of the Standard Model, and of fundamental conservation laws. These are schematically illustrated in Fig. 11.16. As the field of study progresses, other opportunities, currently unimagined, will doubtless present themselves.

11.4   Some simple signatures of structure

While RNBs provide new horizons for nuclear physics, we have also noted that beam intensities will often be orders of magnitude less than we are accustomed to. Therefore, we need to improve the efficiency with which we extract information from experiments on exotic nuclei. There are two generic ways of doing this, as indicated schematically in Fig. 11.17. In order to extract the same amount of physics as incident beam (or source decay) rate falls, we need to increase either the efficiency of our detectors or of our signatures of structure or, of course, of both. The complementarity of these two routes to new physics in exotic nuclei is highlighted by the contours of "constant physics" in Fig 11.17.

We have already briefly mentioned the first route, namely new generations of instruments for the study of exotic nuclei. Truly remarkable advances have been made in this area. As just one example, fifteen years ago Coulomb excitation experiments generally required beams of roughly 10^9-10^{10} particles/sec. In stark contrast, recent detector developments, coupled with the use of inverse kinematics in which the traditional roles of target and projectile are reversed, now enable such experiments to be done with a few particles/sec.

Nevertheless, it is equally crucial to develop the second route, better and more efficient signatures of structure so that more physics can be extracted from less data. There have been enormous advances in this area as well in the last years, and it is the purpose of this section to discuss a few of these. Much of the discussion will relate back to ideas previously developed, in particular in Chapter 7.

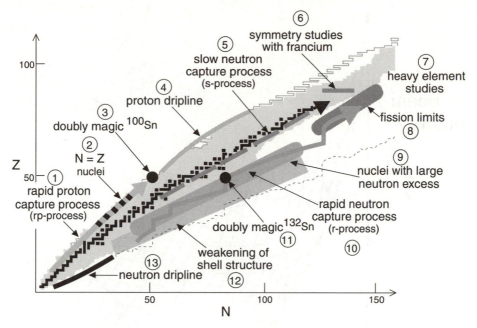

FIG. 11.16. Summary of some of the scientific opportunities with a next generation
ISOL-based RNB facility (based on the "Columbus" White Paper, 1997).

To start, we note that even with the weakest beams, it will still be possible to measure
nuclear masses, that is, binding energies and nucleon separation energies. We introduced
just above the concept of mass or interaction filters, and discussed one of the most
important of these, δV_{pn}. Clearly, further δV_{pn} values will be obtainable, on both sides
of stability, with advanced exotic beam facilities. Here, we want to briefly mention
another mass filter which we will use in a somewhat unconventional manner. This filter
is given by

$$\delta V_{nn}(N, N+1) = 1/2\{[B(Z, N+1) - B(Z, N-1)]$$
$$- [B(Z, N) - B(Z, N-2)]\} \qquad (11.2)$$

It was originally designed to isolate the non-pairing part of the like-nucleon interaction.
However, if the structure of the even–even core nucleus changes in going from a nucleus
with $N-1$ neutrons to one with $N+1$, then the filter in Eq. 11.2 actually does *not* represent
the non-pairing n–n interaction (or, really, any identifiable interaction). Nevertheless, one
can still evaluate the double difference of binding energies empirically even though it
loses its simple physical interpretation.

The reason this is worthwhile is seen from the empirical values for δV_{nn} shown
in Fig. 11.18. We see that, as expected, δV_{nn} is indeed generally positive (repulsive).
(The reader will recall that, though the pairing part of the like-nucleon interaction is
strongly attractive, the non-pairing ($J > 0$) part is, on average, slightly repulsive.) More

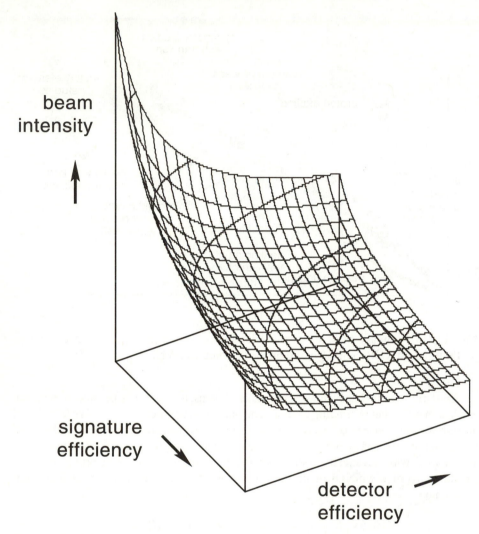

beam
intensity

signature
efficiency

detector
efficiency

FIG. 11.17. Schematic illustration of the need for better and better detectors and signatures of structure as beam (or source) intensities of exotic nuclei decrease far from
stability. The curved arcs symbolize contours of "constant physics" for a given beam
intensity. (I am grateful to N. Pietralla for help with this figure.)

interesting, though, are the spikes. The positive spikes occur precisely at the known magic
numbers, the negative spikes at spherical-deformed transition regions. Though we do not
even know what δV_{nn} represents physically at these core-changing points, it does not
matter. What is useful is that it provides an observable signature either for magicity or
for shape changes. Since masses are relatively easy to measure, simple measurements
with RNBs can provide clues to the underlying structure, even in the absence of any

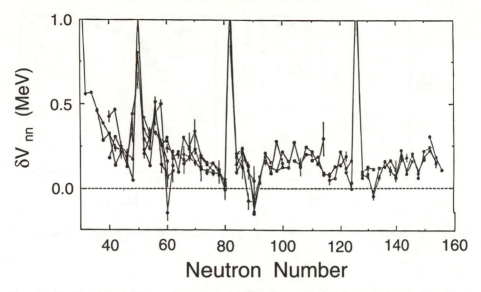

FIG. 11.18. Empirical δV_{nn} values. See text for discussion of the meaning and implications of these data (based on Zamfir, 1991).

spectroscopic data.

With at least rudimentary spectroscopic data [$E(2_1^+)$, $E(4_1^+)$, B(E2 : $2_1^+ \rightarrow 0_1^+$)] we can exploit the correlation schemes of Chapter 7. For example, plots such as Fig. 7.17 can help identify phase transitional behavior and possible phase coexistence in new mass regions. Moreover, if we plot $E(4_1^+) - 2E(2_1^+)$ we get (see Fig. 7.19) the anharmonicity ε_4 of the AHV formula. We saw, in our discussion of Fig. 7.19, that anomalous ε_4 values (even at the level of a few 10s of keV) can give clues either to shape transition regions (low ε_4), or 2-hole nuclei (high ε_4) and, therefore, in the latter case, to the location of new magic numbers.

The $N_p N_n$ scheme likewise provides a very powerful tool in magnifying anomalies in deviant nuclei. We illustrate this for known nuclei in Fig. 11.19 but the application to exotic nuclei is analogous. In Fig. 11.19 we show the ratio of ground band 4^+ and 2^+ energies in the $A \sim 190$ region. The normal plot on the left reveals a complex distribution of points but nothing striking stands out. The values range from near 2.0 (vibrator) towards the rotor value of 3.33. There is large scatter. On the right, however, the $N_p N_n$ scheme gives a remarkably smooth trajectory.

As we have noted many times, a paradigm gives a new standard of normal behavior and highlights anomalous behavior. In this case the $R_{4/2}$ value for ^{184}Hg stands out far from the smooth trajectory in the $N_p N_n$ plot on the right but is invisible on the left. In this case, the explanation is known. There are coexisting spherical and deformed states in ^{184}Hg that mix with each other. As a consequence, the ground band 4^+ is repulsed to higher energies.

Interestingly, by assuming that this data point would otherwise have been on the

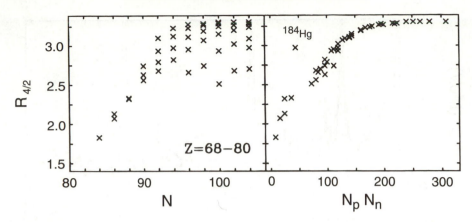

FIG. 11.19. Normal and $N_p N_n$ plots of $R_{4/2}$ for $Z = 68$–80 (based on the "Columbus" White Paper, 1997).

curve, we can deduce the energy shift due to mixing (the 2^+ spherical and deformed states are much further apart and mix very little). Hence, by 2-state mixing, a mixing matrix element of about 85 keV is trivially obtained, which is a value typical of coexisting states, and about equal to the value previously extracted by a far more complex analysis. In new nuclear regions, analogous plots may also magnify deviant behavior and lead to interesting physical insights.

The $N_p N_n$ scheme can also be used to point to new magic numbers in exotic nuclei (recall our discussion of the fragility of magicity). The idea is that one needs to choose magic numbers in order to obtain N_p and N_n themselves so as to construct the $N_p N_n$ plot. We have seen in Chapter 7 how this kind of analysis helped reveal the dissolution of subshell gaps at $Z = 40$ and 64 with increasing neutron number in the $A = 100$ and 150 mass regions.

Here we illustrate the approach by an artificial example. Figure 11.20 shows, on the left, an $N_p N_n$ scheme plot of measured B(E2: $2_1^+ \rightarrow 0_1^+$) values in the $A = 130$ region using the well known magic numbers for this region $Z = 50$, $N = 82$. As expected, a smooth, nearly linear, trajectory results. Suppose, however, that we *intentionally* used incorrect magic numbers. We do this on the right by assuming that $Z = 50$ is not magic. A chaotic plot results. This technique can be used in new mass regions by testing which choices of N_p and N_n give smooth compact trajectories. Note that one can assess the robustness of various magic numbers even if nuclei with these numbers of nucleons are not themselves accessible.

We stress that any exploitations of correlation schemes, such as those discussed here and in Chapter 7, and the behavioral paradigms they reveal, *assume* that the correlation schemes remain simple in exotic regions of nuclei. This situation leads to a key and profound question concerning the evolution of nuclear structure that can only be addressed by studies of exotic nuclei far from stability. If the correlations are truly universal they must result from very general features of nuclear interactions. They can therefore be ex-

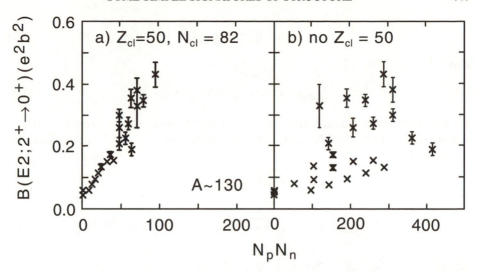

FIG. 11.20. Empirical B(E2) values in the $A \sim 130$ region plotted against $N_p N_n$ using a) normal definitions of the magic numbers and b) purposely (for the sake of illustration) incorrect values by assuming that $Z = 50$ is not a magic number.

ploited as indicated above but, more importantly, they point to a remarkable, and not yet understood, generality in nuclear structural evolution that bridges extreme ranges of N/Z composition, shell structure, and residual interactions, and that challenges theoretical descriptions. On the other hand, if the correlations break down in new mass regions, the intriguing question is then what is special about nuclei near the valley of stability—what aspects of shell structure and interactions are responsible for their remarkably simple correlations. On the theoretical side, recent studies with random interactions are beginning to address such questions. It appears that the basis space, rather than the interactions, may be the key ingredient (Johnson, 1998, Bijker, 2000, Kusnezov, 2000).

As a final example of new and more efficient signatures of structure, we turn to the concept of Q-invariants or, more accurately, approximate Q-invariants. These provide enormously powerful and extremely simple signatures of both basic nuclear structure information and of some of the most subtle structural insights obtainable. The information they can provide has seldom, if ever, been available for *any* nuclei. With recent developments, it will now be available from simple data even in exotic nuclei.

The idea of Q-invariants, introduced years ago by Kumar (1972) and Cline (1986) relates directly to a perspective on nuclei as geometrical objects. Let us consider a quadrupole operator, Q, acting on the nuclear ground state. Largely following Werner (1999) and references cited therein, one can define various moments of this operator as

$$q_n = C_n \langle 0_1^+ \| [Q^n]^\circ \| 0_1^+ \rangle \tag{11.3}$$

where C_n are useful normalization constants and the products of Q operators are appropriately coupled as scalar or tensor products to spin zero. For example,

$$q_2 = \langle 0_1^+ \| (Q \cdot Q) \| 0_1^+ \rangle \tag{11.4}$$

and

$$q_4 = \langle 0_1^+ \| (Q \cdot Q)(Q \cdot Q) \| 0_1^+ \rangle \tag{11.5}$$

By introducing and summing over intermediate states, the q_n can be written in terms of infinite sums over reduced E2 matrix elements, connecting 0^+ and 2^+ states. Thus, for the above cases:

$$q_2 = e^2 \sum_i \langle 0_1^+ \| Q \| 2_i^+ \rangle \langle 2_i^+ \| Q \| 0_1^+ \rangle \tag{11.6}$$

$$q_4 = e^4 \sum_{i,j,k} \langle 0_1^+ \| Q \| 2_i^+ \rangle \langle 2_i^+ \| Q \| 0_j^+ \rangle \langle 0_j^+ \| Q \| 2_k^+ \rangle \langle 2_k^+ \| Q \| 0_1^+ \rangle \tag{11.7}$$

where e is an E2 effective charge.

The interest in these quantities stems from the fact that normalized dimensionless shape invariants, defined by $K_n = q_n/q_2^{n/2}$, can be directly related to expectation values of the geometric shape variables β and γ as follows

$$q_2 = e^2 \left(\frac{3ZeR^2}{4\pi} \right)^2 \beta_{\text{eff}}^2 \tag{11.8}$$

$$K_3 = \frac{\langle \beta^3 \cos 3\gamma \rangle}{\langle \beta^2 \rangle^{3/2}} = \cos 3\gamma_{\text{eff}} \tag{11.9}$$

$$K_4 = \frac{\langle \beta^4 \rangle}{\langle \beta^2 \rangle^2} \tag{11.10}$$

$$K_5 = \frac{\langle \beta^5 \cos 3\gamma \rangle}{\langle \beta^2 \rangle^{5/2}} \tag{11.11}$$

$$K_6 = \frac{\langle \beta^6 \cos^2 3\gamma \rangle}{\langle \beta^2 \rangle^3} \tag{11.12}$$

Thus q_2 and K_3 can give effective β and γ values. Even more interestingly, the higher K_n shape invariants are related to *fluctuations* in β and γ. Following Werner (1999) we can define quantities σ_β and σ_γ, approximately representing fluctuations in β and $\cos 3\gamma$, as

$$\sigma_\beta = \frac{\langle \beta^4 \rangle - \langle \beta^2 \rangle^2}{\langle \beta^2 \rangle^2} = K_4 - 1 \tag{11.13}$$

$$\sigma_\gamma = \frac{\langle \beta^6 \cos^2 3\gamma \rangle - \langle \beta^3 \cos 3\gamma \rangle^2}{\langle \beta^2 \rangle^3} = K_6 - K_3^2 \tag{11.14}$$

Unfortunately, these invariants have seldom been used since infinite (or even extensive) sets of B(E2: $0_i^+ \rightarrow 2_j^+$) values are never known. However, recent work by Jolos, Brentano, Pietralla, and co-workers [Jolos, 1997; Palchikov, 1998; Pietralla, 1994, 1995,

1998; and Werner, 1999], has shown that a remarkable simplification is possible in most nuclei, making these Q-invariants an extremely valuable potential new tool for understanding nuclei in general, and exotic nuclei in particular.

Brentano and co-workers noted that, if the first 2^+ state exhausts or nearly exhausts the $0\hbar\omega$ E2 strength from the ground state, these Q-invariants collapse to just one or two terms. The idea that B(E2: $0_1^+ \rightarrow 2_1^+) \approx \sum_i B(E2 : 0_1^+ \rightarrow 2_i^+)$ is equivalent to treating the 2_1^+ state as a 1-phonon excitation of the ground state: $|2_1^+\rangle = Q|0_1^+\rangle$. Interestingly, the data on all known medium and heavy mass nuclei ($A > 100$) very strongly supports this assumption. On average, the B(E2: $0_1^+ \rightarrow 2_1^+$) value is 97% of the full sum; only in γ-soft [O(6)-like] regions does this percentage drop to $\sim 92\%$. Also, the quantity B(E2: $0_1^+ \rightarrow 2_1^+$) + B(E2 : $0_1^+ \rightarrow 2_2^+$) exhausts 99% of the sum. Moreover, the $E(4_1^+)$ vs. $E(2_1^+)$ plots we showed in Chapter 7 as examples of the correlations of collective observables, and the discussion there of anharmonic vibrator (AHV) structure, further supports the phonon ansatz.

Adopting this ansatz, Brentano and co-workers then used the IBA model to show that the K_n are given, in that model, by very simple expressions, in terms of just a few B(E2) values and branching ratios. [Recent calculations with the GCM and the Davydov triaxial rotor model both give similar results as the IBA for several Q-invariants, thus suggesting (albeit not proving) that these expressions are, in fact, not very model dependent.] Figure 11.21 illustrates the simple information needed. To write these expressions we first define the convenient ratios

$$R_1 \equiv \frac{B(E2 : 2_2^+ \rightarrow 0_1^+)}{B(E2 : 2_1^+ \rightarrow 0_1^+)} \tag{11.15}$$

$$G \equiv \frac{7}{10} \frac{B(E2 : 4_1^+ \rightarrow 2_1^+)}{B(E2 : 2_1^+ \rightarrow 0_1^+)} \tag{11.16}$$

$$W \equiv \frac{B(E2 : 2_2^+ \rightarrow 2_1^+)}{B(E2 : 4_1^+ \rightarrow 2_1^+)} \tag{11.17}$$

Then, the following expressions result:

$$\frac{Q(2_1^+)}{\sqrt{B(E2 : 2_1^+ \rightarrow 0_1^+)}} \equiv \frac{8}{7}\sqrt{\pi G(1 + R_1 - W)} \sim 2\sqrt{G(1 - W)} \tag{11.18}$$

$$K_3 \approx \sqrt{G}\left[\left(\frac{1 - R_1}{1 + R_1}\right)\sqrt{1 - \frac{W}{1 + R_1}} - \frac{2}{1 + R_1}\sqrt{\frac{R_1 W}{1 + R_1}}\right]$$

$$\sim \sqrt{G}\left[\sqrt{1 - W} - 2\sqrt{R_1 W}\right] \tag{11.19}$$

$$K_4 \approx G \tag{11.20}$$

and hence

$$\sigma_\beta \sim \frac{7}{10}\frac{B(E2 : 4_1^+ \rightarrow 2_1^+)}{B(E2 : 2_1^+ \rightarrow 0_1^+)} - 1 \tag{11.21}$$

Q - Invariants (key matrix elements)

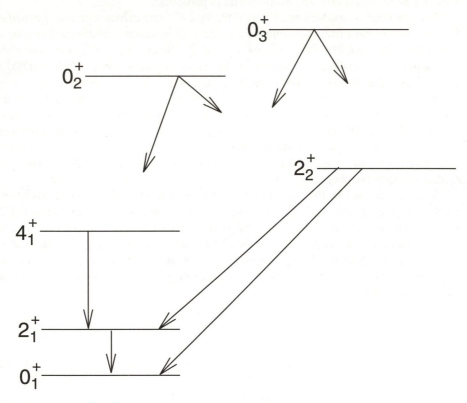

FIG. 11.21. Simple E2 observables needed to evaluate the approximate Q-invariants q_2, $K_3 - K_6$. The transition arrows from the 0^+ states indicate branching ratios, which enter only the expressions for K_5 and K_6.

K_5 and K_6 are more complex, involving the branching ratios from the 0^+ states indicated in Fig. 11.21. On the far right in Eq. (11.18) and again in Eq. (11.19) we have assumed that $R_1 = 0$ since (see Eq. 11.15) it is almost always small, being a ratio of forbidden to allowed transitions in the vibrator and inter- to intraband in the rotor.

Of course, these expressions, being model dependent, need to be verified experimentally. This has recently been done in ^{152}Sm (Klug, 2000) for one of the most important of the Q-invariants, K_4, by comparing the results using extensive and precise B(E2) data for the exact expression (using Eqs. 11.6 and 11.7) with the approximate relation in Eq. 11.20. Almost perfect agreement was obtained.

Note the simplicity of these results. In particular, it is remarkable that one can glean a sense even of the *softness* in the potential in β merely from a low spin yrast B(E2)

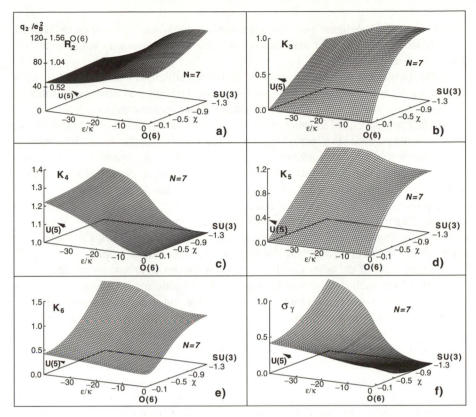

FIG. 11.22. Q-invariants calculated for $N_B = 7$ with the IBA over most of the symmetry triangle of Fig. 6.41. Note that, for space reasons, at lower left in each plot, the calculations do not extend fully to U(5). The scale R_2 in the q_2 plot normalizes q_2 to 1.0 in O(6). (From Werner, 1999)

ratio.

The γ_{eff} values obtained from K_3 are interesting since, traditionally, γ values have been obtained from the Davydov expressions (see Chapter 6). However, the Davydov model invokes rigid triaxiality while we know from the staggering index (see Eq. 6.47) that essentially all equilibrium axial asymmetry in nuclei arises from γ-softness. Therefore, the γ_{Davydov} values should be open to question. With the expression for K_3, however, they can be tested. One finds that the γ_{eff} values obtained from Eq. (11.19) agree with those calculated, using the observables in Fig. 6.25, with the Davydov expressions to within a few degrees.

We end by showing (exact) values of the invariants q_2, K_3, K_4, K_5, K_6, and σ_γ calculated in the IBA (for $N = 7$) over nearly the entire symmetry triangle of Fig. 6.41. The results are shown in Fig. 11.22 (Only the region very near U(5) is omitted for convenience to avoid having the parameter ε/κ go to infinity. The contours go smoothly to the U(5) limit.) This figure gives a guide to the values and variations of these invariants over a broad range of nuclear structural types. (One expects a similar variation in other models such as the GCM.)

Particularly interesting are the results of K_4 (or σ_β) which, as we have noted, gives the β-softness. K_4 is a minimum (1.0) for deformed nuclei (either axially symmetric or γ-soft) and a maximum (1.4) for a spherical vibrator. Thus, measurements of just the simple yrast B(E2) ratio in Eq. 11.20 can sensitively monitor the evolution of structure and the softness of the potential in β across regions of new nuclei. Another facet of K_4 makes it particularly revealing in regions of rapid structural change: K_4 experiences its steepest rate of descent at the phase transition point from vibrator to deformed rotor. Indeed, the phase transition occurs where its second derivative changes sign. Interestingly, this critical point is also characteristic of the correlation in Fig. 7.17, as we have discussed, corresponding to the point where the slope changes from 2.00 to 3.33. Thus, we see that similar and complementary information on phase transitional behavior in finite nuclei can be obtained from two different (but both simple) sets of observables.

The simplicity of the approximate expressions for the Q-invariants and the relative ease with which such data should be obtainable in exotic nuclei, suggests that these approximate shape observables may be very important and useful signatures of structure as the advent of radioactive beam facilities opens up entirely new realms of nuclei for study.

There are a number of other new signatures of structure that we cannot discuss in detail. We only briefly cite two of them. It has been shown that the correlation of B(E2:$2_1^+ \rightarrow 0_1^+$) values with $R_{4/2}$ follows two trajectories, one for nuclei in a vibrator to rotor transition region and another for a γ-soft to rotor region, thus providing a simple clue to the nature of structural change in new regions of the nuclear chart. The second also concerns $R_{4/2}$ values, namely their frequency distribution as a function of $R_{4/2}$. It has been shown recently (Kusnezov, 2000) that such an $R_{4/2}$ abundance distribution is governed by exactly the two key structural issues we discussed above for weakly bound neutron rich nuclei, namely changes in the underlying shell structure, including alterations or quenching of magicity, and in the nature of residual interactions, such as pairing or the p–n interaction amongst the valence nucleons. Distributions of $R_{4/2}$ values in new regions may help elucidate evidence for these two effects and help in unravelling their respective roles.

It should be clear from this discussion that significant advances in increasing "signature efficiency", using the simplest and easiest-to-measure observables, are being made at a rapid rate, complementing advances in instrumentation so that, referring to Fig. 11.17, we are learning to maximize the amount of nuclear structure information that can be extracted from a minimum of data on nuclei in new regions of the nuclear chart.

It is appropriate to close this book, which has continually stressed simple ideas about nuclear structure, on such a note, since it is just such ideas that now offer new hope for future understanding.

REFERENCES

This list includes two categories of references. One consists of those specifically cited in the text (for example, as the source for figures or tables). (The author apologizes for the disproportionate and semi-obscene number of citations of his own work here. This reflects both the fact that the topics covered in the book tend to be those that I am interested in and may have worked on, and a degree of laziness in choosing illustrative figures from my own papers rather than searching the literature for the better ones.) The other category gives a few classic articles and other excellent works, often textbooks, review articles, or compilations, which are suggested for additional study. All of these are recommended. Papers that the author has personally found to be continually and particularly helpful, some indispensable reference works, and a few others are marked with an asterisk. No attempt is made to provide a complete reference list for ideas and data discussed in this book. This list is for the benefit of the reader, not for the recognition of earlier accomplishments. That is the task of appropriate review articles. References are cited in the text by first author only, with date of publication.

Ajzenberg-Selove, F. (1990). *Nucl. Phys. A506*, 1.

Alaga, G., Alder, K., Bohr, A., and Mottelson, B. R. (1955). *Mat. Fys. Medd. Dan. Vid. Selsk. 29*, No. 9.

*Alder, A., Bohr, B., Huus, T., Mottelson, B. R., and Winter, A. (1956). *Rev. Mod. Phys. 28*, 432.

Aprahamian, A., Brenner, D. S., Casten, R. F., Gill, R. L., and Piotrowski, A. (1987). *Phys. Rev. Lett. 59*, 535.

Aprahamian, A., Brenner, D. S., Casten, R. F., Gill, R. L., Piotrowski, A., and Heyde, K. (1984). *Phys. Lett. 140B*, 22

Arenas Peris, G. E. and Federman, P. (1988). *Phys. Rev. C38*, 493.

Arima, A., Harvey, M., and Shimizu, K. (1969). *Phys. Lett. 30B*, 517.

*Arima, A. and Iachello, F. (1976). *Ann. Phys. (N.Y.) 99*, 253.

*Arima, A. and Iachello, F. (1978a). *Ann. Phys. (N.Y.) 111*, 201.

*Arima, A. and Iachello, F. (1979). *Ann. Phys. (N.Y.) 123*, 468.

Arima, A. and Iachello, F. (1978b). *Phys. Rev. Lett. 40*, 385.

Arima, A., Otsuka, T., Iachello, F., and Talmi, I. (1977). *Phys. Lett. 66B*, 205.

Arvieu, R. and Moszkowski, S. A. (1966). *Phys. Rev. 145*, 830.

Aumann, T., Bortingnon, P. F., and Emling, H. (1998). *Annu. Rev. Nucl. Part. 548*, 351.

Baldsiefen, G., *et al.* (1994). *Nucl. Phys. A574*, 551.

*Baranger, M. (1960). *Phys. Rev. 120*, 957.

Barrett, B. R. and Kirson, M. W. (1970). *Nucl. Phys. A148*, 145.

Beausang, C. W. and Simpson, J. (1996). *J. Phys. G22*, 527.

Belyaev, S. T. (1959). *Mat. Fys. Medd. Dan. Vid. Selsk. 31, No. 11*.

Bemis, C. E., Jr., Stelson, P. H., McGowan, F. K., Milner, W. T., Ford, J. L. C., Jr., Robinson, R. L., and Tuttle, W. (1973). *Phys. Rev. C8*, 1934.

Bengtsson, T., Ragnarsson, I., and Aberg, S. (1988). *Phys. Lett. 208B*, 39.

Bertulani, C. A. (1998). *J. Phys. G. 24*, 1165.

* Bès, D. R., Federman, P., Maqueda, E., and Zuker, A. (1965). *Nucl. Phys. 65*, 1.

Bès, D. R. (1961). *Mat. Fys. Medd. Dan. Vid. Selsk. 33, No. 2.*

Bès, D. R. (1963). *Nucl. Phys. 49*, 544.

Bès, D. R. and Sorensen, R. A. (1969). *Adv. in Nucl. Phys. 2*, 129.

Bijker, R. and Frank, A. (2000). *Phys. Rev. Lett. 84*, 420.

Bohle, D., Richter, A., Steffen, W., Dieperink, A. E. L., Lo Iudice, N., Palumbo, F., and Scholten, O. (1984). *Phys. Lett. 137B*, 27.

* Bohr, A. and Mottelson (1969, 1975). *Nuclear Structure*. Vols. I and II, Benjamin, New York; World Scientific, Singapore, 1998.

Bohr, A. and Mottelson (1953). *Mat. Fys. Medd. Dan. Vid. Selsk. 24, No. 16.*

Börner, H. G., Jolie, J., Robinson, S. J., Casten, R. F., and Cizewski, J. A. (1990). *Phys. Rev. C42*, R2271

Börner, H. G., *et al.* (1991). *Phys. Rev. Lett. 66*, 691 and *Phys. Rev. Lett. 66*, 2837.

Bortignon, P. F., Bracco, A., and Broglia, R. A. (1998). *Giant Resonances, Nuclear Structure at Finite Temperature,* Harvard Academic, New York.

Braun-Munzinger, P. (1984). (Editor), *Nuclear Physics with Heavy Ions, Nucl. Sci. Res. Conf. Series*, Vol. 6, Harwood Academic, New York.

Brenner, D. S., *et al.* (1990). *Phys. Lett. B243*, 1.

Brink, D. M., de Toledo Piza, R., and Kerman, A. K. (1965). *Phys. Lett. 19*, 413.

Bromley, D. A., Gove, H. E., Paul, E. B., Litherland, A. E., and Almquist, E. (1957). Can. *J. Phys. 35*, 1042.

Brown, G. E. and Bosterli, M. (1959). *Phys. Rev. Lett. 3*, 472.

*Brussaard, P. J. and Glaudemans, P. W. M. (1977). *Shell Model Applications in Nuclear Spectroscopy*, North Holland, Amsterdam.

Bucurescu, D., *et al.* (1996). *Phys. Rev. B376*, 1.

Bucurescu, D. G., Cata, G., Cutoiu, D., Dragelescu, E., Ivascu, M., Zamfir, N. V., Gizon, A., and Gizon, J. (1989). *Phys. Lett. 229B*, 321.

Bunker, M. and Reich, C. W. (1971). *Rev. Mod. Phys. 43*, 348.

Byrski, T., *et al.* (1990). *Phys. Rev. Lett. 64*, 1650.

*Castaños, O. and Draayer, J. P. (1989). *Nucl. Phys. A491*, 349.

*Casten, R. F. (1985a). *Nucl. Phys. A443*, 1.

Casten, R. F. (1986). *Phys. Rev. C33*, 1819.

Casten, R. F., Aprahamian, A., and Warner, D. D. (1984). *Phys. Rev. C29*, 356.

Casten, R. F., Brenner, D. S., and Haustein, P. E. (1987). *Phys. Rev. Lett. 58*, 658.

Casten, R. F., von Brentano, P., and Haque, A. M. I. (1985b). *Phys. Rev. C31*, 1991.

Casten, R. F. and Cizewski, J. A. (1978). *Nucl. Phys. A309*, 477.

Casten, R. F., Frank, A., Moshinsky, M., and Pittel, S. (1988). (Editors), *Contemporary Topics in Nuclear Structure Physics*, World Scientific, Singapore.

Casten, R. F., Heyde, K., and Wolf, A. (1988b). *Phys. Lett. 208B*, 33.

Casten, R. F., Kleinheinz, P., Daly, P. J., and Elbek, B. (1972). *Mat. Fys. Medd. Dan. Vid. Selsk., 38, No. 13.*

*Casten, R. F., Warner, D. D., Brenner, D. S., and Gill, R. L. (1981). *Phys. Rev. Lett. 47*, 1433.

Casten, R. F., Warner, D. D., and Aprahamian, A. (1983). *Phys. Rev. C28*, 894.

*Casten, R. F. and Warner, D. D. (1988a). *Rev. Mod. Phys. 60*, 389.

Casten, R. F., Warner, D. D., Stelts, M. L., and Davidson, W. F. (1980). *Phys. Rev. Lett. 45*, 1077.

Casten, R. F., Zamfir, N. V., and Brenner, D. S. (1993). *Phys. Rev. Lett. 71*, 227.

Casten, R. F. and Zamfir, N. V. (2000). *Phys. Rev. Lett. 85*, 3584.

Casten, R. F., *et al.* (1998). *Phys. Rev. C57*, R1553.

Casten, R. F., Jolie, J., Börner, H. G., Brenner, D. S., Zamfir, N. V., Chou, W. T., and Aprahamian, A. (1992). *Phys. Lett. B297, 19*, and (1993). *Phys. Lett. B300*, 411.

Cizewski, J. A., Casten, R. F., Smith, G. J., Stelts, M. L., Kane, W. R., Borner, H. G., and Davidson, W. F. (1978). *Phys. Rev. Lett. 40*, 167.

Cizewski, J. A., Casten, R. F., Smith, G. J., MacPhail, M. R., Stelts, M. L., Kane, W. R., Borner, H. G., and Davidson, W. F. (1979). *Nucl. Phys. A323*, 349.

Cline, D. (1986). *Ann. Rev. Nucl. Part. Sci. 36*, 766.

Cohen, B. L. and Price, R. E. (1961). *Phys. Rev. 121*, 1441.

Corminboeuf, F., Jolie, J., Lehmann, H., Fohl, K., Hoyler, F., Börner, H. G., Doll, C., and Garrett, P. E. (1997). *Phys. Rev. C56*, R2101.

Cottle, P. D., and Bromley, D. A. (1987). *Phys. Rev. C35*, 1891.

Davidson, W. F., Warner, D. D., Casten, R. F., Schreckenbach, K., Borner, H. G., Simic, J., Bogdanovic, M., Koicki, S., Gelletly, W., Orr, G. B., and Stelts, M. L. (1981). *J. Phys. G7*, 455.

Davydov, A. S. and Filippov, G. F. (1958). *Nucl. Phys. 8*, 237.

Deleplanque, M. A. and Diamond, R. M. (1988). GAMMASPHERE Proposal.

Délèze, M., *et al.* (1993a). *Nucl. Phys. A551*, 269.

Délèze, M., Drissi, S., Jolie, J., and Vorlet, J. P. (1993b). *Nucl. Phys. A554*, 1.

de Shalit, A. and Goldhaber, M. (1953). *Phys. Rev. 92*, 1211.

*de Shalit, A. and Feshbach, H. (1974). *Theoretical Nuclear Physics*, Vol. I, *Nuclear Structure*, Wiley, New York.

*de Shalit, A. and Talmi, I. (1963). *Nuclear Shell Theory*, Academic Press, New York.

*Dobaczewski, J., Nazarewicz, W., Skalski, J., and Werner, T. (1988). *Phys. Rev. Lett. 60*, 2254.

Dobaczewski, J., Nazarewicz, W., Werner, T. R., Berger, J. F., Chinn, C. R., and Decharge, J. (1996). *Phys. Rev. C53*, 2809.

Eisberg, R. and Resnick, R. (1974). *Quantum Physics*, Wiley, New York.

Eisenberg, J. M., and Greiner, W. (1970). *Nuclear Models*, Vol. 1, *Excitation Mechanisms of the Nucleus, Vol. 2*, and *Microscopic Theory of the Nucleus*, Vol. 3, North Holland, Amsterdam.

Elliott, J. P. (1958). *Proc. Roy. Soc. A245*, 128 and 562.

Elliott, J. P. and Harvey, M. (1963). *Proc. Roy. Soc. A272*, 557.

Endt, P. M. and Van der Leun, C. (1978). *Nucl. Phys. A310*, 1.

Fahlander, C., *et al.* (1996). *Phys. Lett. B388*, 475.

Fallieros, S. and Ferrell, R. (1959). *Phys. Rev. 116*, 660.

*Federman, P. and Pittel, S. (1977, 1978). *Phys. Lett. 69B*, 385 and 77B, 29.

*Gallagher, C. J. (1964). In *Selected Topics in Nuclear Spectroscopy* (ed. B. J. Verhaar). North Holland, Amsterdam.

Garrett, J. D. (1984). *Nucl. Phys. A421*, 313c.

Garrett, J. D., Nyberg, J., Yu, C. H., Espino, J. M., and Godfrey, M. J. (1988). In *Contemporary Topics in Nuclear Structure Physics*, p. 699 (volume referenced elsewhere in this listing).

Garrett, P. E., *et al.* (1997a). *Phys. Lett. B400*, 250.

Garrett, P. E., *et al.* (1997b). *Phys. Rev. Lett. 78*, 4545.

Ginocchio, J. N. and Kirson, M. W. (1980). *Nucl. Phys. A380*, 31.

Gneuss, G. and Greiner, W. (1971). *Nucl. Phys. A171*, 449.

Greenwood, R. C., *et al.* (1978). *Nucl. Phys. A304*, 327.

Grzywacz, R., *et al.* (1998). *Phys. Rev. Lett. 81*, 766.

Gustafson, C., Lamm, I. L., Nilsson, B., and Nilsson, S. G. (1967). *Ark. Fysik 36*, 613.

Hager, R. S. and Seltzer, E. C. (1968). *Nucl. Data A4*, 1.

Haxel, O., Jensen, J. H. D., and Suess, H. E. (1950). *Z. Phys. 128*, 295.

Hecht, K. (1964). In *Selected Topics in Nuclear Spectroscopy* (ed. P. J. Verhaar), North Holland, Amsterdam.

Heisenberg, W. (1930). *The Physical Principles of the Quantum Theory.* Trans. C. Eckart and F. C. Hoyt. University of Chicago Press, Chicago, IL.

Herzberg, R. D., *et al.* (1995). *Nucl. Phys. A592*, 211.

Heyde, K. (1994). *Basic Ideas and Concepts in Nuclear Physics,* Institute of Physics, Bristol, U.K.

*Heyde, K. (1990). *The Nuclear Shell Model. Springer Series in Nuclear and Particle Physics*, Springer-Verlag, Berlin.

Heyde, K. (1989), *Int. J. Mod. Physics A4*, 2063.

Heyde, K., Jolie, J., Moreau, J., Ryckebusch, J., Waroquier, M., Van Duppen, P., Huyse, M., and Wood, J. L. (1987). *Nucl. Phys. A466*, 189.

Heyde, K., Kirchuk, E. D., and Federman, P. (1988). *Phys. Rev. C38*, 984.

Heyde, K. and Sau, J. (1986). *Phys. Rev. C33*, 1050.

*Heyde, K., Van Isacker, P., Casten, R. F., and Wood, J. L. (1985b). *Phys. Lett. 155B*, 303.

*Heyde, K., Van Isacker, P., Waroquier, M., Wood, J. L., and Meyer, R. A. (1983). *Phys. Rep. 102*, 291.

Holmberg, P. and Lipas, P. O. (1968). *Nucl. Phys. A117*, 552.

Honma, M., Misuzaki, T, and Otsuka, T. (1995). *Phys. Rev. Lett. 75*, 1284.

*Hornyak, W. F. (1975). *Nuclear Structure*, Academic Press, New York.

Iachello, F. (2000). *Phys. Rev. Lett. 85*, 3580.

Iachello, F., Casten, R. F., and Zamfir, N. V. (2001), in press.

Iachello, F. and Scholten, O. (1979). *Phys. Rev. Lett. 43*, 679.

Iachello, F., Zamfir, N. V., and Casten, R. F. (1998). *Phys. Rev. Lett. 81*, 1191.

Iachello, F. and Talmi, I. (1987). *Rev. Mod. Phys. 59*, 339.

Inglis, D. (1954). *Phys. Rev. C96*, 1059.

Janssen, D. R., Jolos, R. V., and Donau, F. (1974). *Nucl. Phys. A224*, 93.

Johnson, C. W., Bertsch, G. F., and Dean, D. J. (1998). *Phys. Rev. Lett. 80*, 2749.

Jolos, R. V., *et al.* (1996). *Phys, Rev. C54*, R2146.

Jolos, R. V., *et al*. (1997). *Nucl. Phys. A618*, 126.

Just, M., *et al*. (1972). Conference on Physics and Chemistry of Fission, Jülich, Germany, IAEA, Vienna, p. 71.

Kerman, A. K. (1959). In *Nuclear Reactions*, Vol. 1 (eds. P. M. Endt and M. Demuer), North Holland, Amsterdam.

*Kisslinger, L. S. and Sorensen, R. A. (1960). *Mat. Fys. Medd. Dan. Vid. Selsk 32, No. 9.*

Kleinheinz, P., Casten, R. F., and Nilsson, B. (1973). *Nucl. Phys. A203*, 539.

Kleinheinz, P., *et al*. (1982). *Phys. Rev. C48*, 1457.

Klug, T., *et al*. (2000). *Phys. Lett.*, in press.

Koonin, S., *et al*. (1997). *Phys. Rep. 278*, 1.

Korten, W., *et al*. (1993). *Phys. Lett. B317*, 19.

Krusche, B., Börner, H., Takruri, F., Robinson, S., Michaelson, S., and MacMahon, D. (1989). Private communication.

Krücken, R., *et al*. (1996). *Phys. Rev. C54*, 1182.

Krücken, R., *et al*. (1997). *Phys. Rev. C55*, R1625.

Kumar, K. (1972). *Phys. Rev. Lett. 28*, 249.

Kumar, K. and Baranger, M. (1968). *Nucl. Phys. A122*, 273.

Kumar, K. (1982). In *Contemporary Research Topics in Nuclear Physics* (eds. D. H. Feng, M. Vallieres, M. W. Guidry, and L. L. Riedinger), Plenum, New York.

Kuo, T. T. S. and Brown, G. E. (1966). *Nucl. Phys. 85*, 40.

Kusnezov, D., Zamfir, N.V., and Casten, R. F. (2000). *Phys. Rev. Lett. 85*, 1396.

*Lane, A. M. (1964). *Nuclear Theory*, Benjamin, New York.

Leander, G. A., Dudek, J., Nazarewicz, W., Nix, J. R., and Quentin, Ph. (1984). *Phys. Rev. C30*, 416.

*Lederer, C. M. and Shirley, S. (1977). *Table of Isotopes*, 7th edition, Wiley, New York.

Lehmann, H., *et al*. (1998). *Phys. Rev. C57*, 569.

Lipas, P. O. (1962). *Nucl. Phys. 39*, 468.

Ma, W. C., Quader, M. A., Emling, H., Khoo, T. L., Ahmad, I., Daly, P. J., Dichter, B. K., Drigert, M., Garg, U., Grabowski, Z. M., Holzmann, R., Janssens, R. V. F., Piiparinen, M., Trzaska, W. H., Wang, T. F. (1988). *Phys. Rev. Lett. 61*, 46.

Mampe, W., Schreckenbach, K., Jeuch, P., Maier, B. P. K., Braumandl, F., Larysz, J., and von Egidy, T. (1978). *Nucl. Instr. and Meth. 154*, 127.

Mayer, M. G. (1950). *Phys. Rev. 78*, 16.

Metz, A., Jolie, J., Graw, G., Hertenberger, R., Grögen, J., Günther, Ch., and Warr, N. (1999). *Phys. Rev. Lett. 83*, 1542.

Mobray, A. S., *et al*. (1990). *Phys. Rev. C42*, 1126.

*Molinari, A., Johnson, M. B., Bethe, H. A., and Alberico, W. M. (1975). *Nucl. Phys. A239*, 45.

Mottelson, B. R. and Nilsson, S. G. (1959). *Mat. Fys. Skr. Dan. Vid. Selsk., 1, No. 8.*

Nagai, Y., Styczen, J., Piiparinen, M., Kleinheinz, P., Bazzacco, D., von Brentano, P., Zell, K. O., and Blomqvist, J. (1981). *Phys. Rev. Lett. 57*, 1259.

*Nathan, O., and Nilsson, S. G. (1965). In *Alpha, Beta, and Gamma Ray Spectroscopy*, Vol. 1 (ed. K. Siegbahn), North Holland, Amsterdam.

Neergaard, K. and Vogel, P. (1970). *Nucl. Phys. A145*, 33.

*Nilsson, S. G. (1955). *Dan. Mat. Fys. Medd. 29, No. 16.*

Nilsson, S. G. and Prior, O. (1961). *Mat. Fys. Medd. Dan. Vid. Selsk 32, No. 16.*

Nilsson, S. G., Tsang, C. F., Sobiczewski, A., Szymanski, Z., Wycech, S., Gustafson, C., Lamm, I. L., Moller, P., and Nilsson, B. (1969). *Nucl. Phys. A131*, 1.

Otsuka, T. and Kim, K. -H. (1994). *Phys. Rev. C50*, 1768.

Ower, H., Elze, Th. W., Idzko, J., Stelzer, K., Grosse, E., Emling, H., Fuchs, P., Schwalm, D., Wollersheim, H. J., Kaffrell, N., and Trautmann, N. (1982). *Nucl. Phys. A388*, 421.

Paar, V. (1979). *Nucl. Phys. A331*, 16.

Palchikov, Yu. V. (1998). *Phys. Rev. C57*, 3026.

Pietralla, N., *et al.* (1994). *Phys. Rev. Lett. 73*, 2962.

Pietralla, N., *et al.* (1995). *Phys. Lett. B349*, 1.

Pietralla, N., *et al.* (1998). *Phys. Rev. C57*, 150.

Pietralla, N., *et al.* (1998a). *Phys. Rev. C58*, 796.

Preston, M. A. and Bhaduri, R. K. (1975). *Structure of the Nucleus*, Addison-Wesley, Reading, MA.

Preston, M. A. and Kiang, D. (1963). *Can. J. Phys. 41*, 742.

*Raman, S., Malarkey, C. H., Milner, W. T., Nestor, C. W., Jr., and Stelson, P. H. (1987). *At. Data and Nucl. Data Tables, 36*, 1.

Ratna Raju, R. D., Draayer, J. P., and Hecht, K. T. (1973). *Nucl. Phys. A202*, 433.

*Riedinger, L. L., Johnson, N. R., and Hamilton, J. H. (1969). *Phys. Rev. 179*, 1214.

Riedinger, L. L., Smith, G. J., Stelson, P. H., Eichler, E., Hagemann, G. B., Hensley, D. C., Johnson, N. R., Robinson, R. L., and Sayer, R. O. (1974). *Phys. Rev. Lett. 33*, 1346.

Riley, M. A., Simpson, J., Sharpey-Schafer, J. F., Cresswell, J. R., Cranmer-Gordon, H. W., Forsyth, P. D., Howe, D., Nelson, A. H., Nolan, P. J., Smith, P. J., Ward, N. J., Lisle, J. C., Paul, E., and Walker, P. M. (1988). *Nucl. Phys. A486*, 456.

Ring, P. and Schuck, P. (1980). *The Nuclear Many-Body Problem*, Springer, New York.

Ronningen, R. M., Hamilton, J. H., Ramayya, A. V., Varnell, L., Garcia Bermudez, G., Lange, J., Lourens, W., Riedinger, L. L., Robinson, R. L., Stelson, P. H., and Ford, J. L. C., Jr. (1977). *Phys. Rev. C15*, 1671.

*Sakai, M. (1984). *At. Data and Nucl. Data Tables 31*, 399.

Scharff-Goldhaber, G. and Weneser, J. (1955). *Phys. Rev. 98*, 212.

*Schiffer, J. P. (1971). *Ann. Phys. (N.Y.) 66*, 798.

*Schiffer, J. P. and True, W. W. (1976). *Rev. Mod. Phys. 48*, 191.

Scholten, O., Iachello, F., and Arima, A. (1978). *Ann. Phys. (N.Y.) 115*, 325.

Sheline, R. K. (1960). *Rev. Mod. Phys. 32*, 1.

Sherrill, B. and Sustich, A. (1994). Private communication.

Soloviev, V. G. (1992). *Theory of Atomic Nuclei: Quasiparticles and Phonons*, Institute of Physics, Bristol, U.K.

Streletz, G., *et al.* (1996). *Phys. Rev. C54*, R2815.

*Stephens, F. S. (1975). *Rev. Mod. Phys. 47*, 43.

Stephens, F. S. and Simon, R. S. (1972). *Nucl. Phys. A183*, 257.

Tabor, S. L., Cottle, P. D., Gross, C. J., Huttmeier, U. J., Moore, E. F., and Nazarewicz, W. (1989). *Phys. Rev. C39*, 1359.

*Talmi, I. (1962). *Rev. Mod. Phys. 34*, 704.

Troltenier, D., Maruhn, J. A. and Hess, P. O. (1991). In *Computational Nuclear Physics 1* (eds. K. Langanke, J. A. Maruhn, and S. E. Koonin), Springer-Verlag, Berlin.

Twin, P. J., *et al.* (1986). *Phys. Rev. Lett. 57*, 811.

Twin, P., Nyako, B. M., Nelson, A. H., Simpson, J., Bentley, M. A., Cranmer-Gordon, H. W., Forsyth, P. D., Howe, D., Morhtar, A. R., Morrison, J. D., Sharpey-Schafer, J. F., and Sletten, G. (1986). *Phys. Rev. Lett. 57*, 811.

Van Duppen, P., Coenen, E., Deneffe, K., Huyse, M., and Wood, J. L. (1985). *Phys. Lett. 154B*, 354.

Van Maldeghem, J., Heyde, K., and Sau, J. (1985). *Phys, Rev. C32*, 1067.

Warner, D. D., Casten, R. F., and Davidson, W. F. (1980). *Phys. Rev. Lett. 45*, 1761.

Warner, D. D. (1984). *Phys. Rev. Lett. 52*, 259.

Warner, D. D. and Casten, R. F. (1982). *Phys. Rev. Lett, 48*, 1385.

Warner, D. D., *et al.* (1982a). *Phys. Rev. C26*, 1921.

*Warner, D. D. and Casten, R. F. (1983). *Phys. Rev. C28*, 1798.

Waroquier, M. and Heyde, K. (1971). *Nucl. Phys. A164*, 113.

Werner, V., *et al.* (1999). *Phys. Rev. C61*, 021301.

White Paper. (1991). *The IsoSpin Laboratory–Research Opportunities with Radioactive Nuclear Beams.* The ISL Steering Committee, Casten, R. F., *et al.* LANL 91–51.

White Paper. (1997). *Scientific Opportunities with an Advanced ISOL Facility.* "Columbus" Writing Panel, Casten, R. F., *et al.*

White Paper. (2000). *RIA Physics White Paper.* Casten, R. F., Nazarewicz, W., *et al.*

Wiedenhöver, I., *et al.* (1997). *Phys. Rev. C56*, R2354.

*Wilets, L. and Jean, M. (1956). *Phys. Rev. 102*, 788.

Wilhelm, M., *et al.* (1996). *Phys. Rev. C54*, R449.

*Wilkinson, D. (1983). (Editor), *Collective Bands in Nuclei*, Vol. 9, *Progress in Particle and Nuclear Physics*, Pergamon Press, New York.

Wirowski, R., Yan, J., Dewald, A., Gelberg, A., Lieberz, W., Schmittgen, K. P., von der Werth, A., and von Brentano, P. (1988). *Z. Phys. A239*, 509.

Wolf, A. and Casten, R. F. (1987). *Phys. Rev. C36*, 851.

Wu, C. L., Feng, D. H., Chen, X. G., Chen, J. Q., and Guidry, M. W. (1986). *Phys. Lett. 168B*, 313.

Wyss, R., Granderath, A., Lieberz, W., Bengtsson, R., von Brentano, P., Dewald A., Gelberg, A., Gizon, A., Gizon, J., Harissopulos, S., Johnson, A., Nazarewicz, W., Nyberg, J., and Schiffer, K. (1989). *Nucl. Phys. A505*, 337.

Yeh, M., *et al.* (1998). *Phys. Rev. C57*, R2085.

Zamfir, N. V. and Casten, R. F. (1991). *Phys. Rev. C43*, 2879.

Zamfir, N. V. and Casten, R. F. (1994). *Phys. Lett. B341*, 1.

Zamfir, N. V., *et al.* (1999). *Phys. Rev. C60*, 054312.

Zamfir, N. V. (1999). Private communication.

Zamfir, V., Casten, R. F., and von Brentano, P. (1989). *Phys. Lett. 226B*, 11.

Zhang, J. Y., Casten, R. F., and Zamfir, N. V. (1997). *Phys. Lett., B407*, 201.

Zhang, J. -Y., Casten, R. F., and Brenner, D. S. (1989). *Phys. Lett. 227B*, 1.

INDEX

This index only gives the locations of principal discussions of various ideas and topics. As one of the purposes of this textbook is to show the linkage of ideas, some topics are frequently discussed in a variety of contexts. No attempt is made to cite all such cases. Also, themes running throughout the text, sometimes implicitly, like the attractive nature of the nuclear force, the Pauli principle, two-state mixing, the nuclear potential, configuration mixing, collectivity, deformation, and so on, are only cited where they are explicitly a principal topic.

Alaga rules 218–28, 266–8, 283, 284
algebraic models 3, 252–88
angular range of forces 100–34, 175–7, 332–4
anharmonic vibrator 179–94, 315–30, 444–5,
antisymmetrization 17, 99–100, 103–14, 119, 120,
    132
astrophysics 429–31
axial asymmetry 210, 221–8, 240–50, 267–75,
    285–8, 300

B(E2) values 27–9, 42–6, 153–7, 186, 194, 210–
    30, 244, 245, 256–96, 327–9, 416, 417, 446,
    452
  definition 42
backbending 206–9, 237, 336, 337, 392–7
backbending and Coriolis effects 392–7
band mixing 27–9, 218–28, 281–4
BE/A data 6–11
β decay 31
β softness 447–52
β vibration 28, 210–16, 415, 416
binding energies 6–14, 61, 62, 90–4, 158, 159,
    439–44
Bohr Hamiltonian, Bohr–Mottelson model 197–
    240

Casimir operators 257
centifugal force 63, 64
CFPs 144–6, 183–7
charge independence, charge symmetry 10–13
chiral bands 253
closed shells 66–73, 98, 99, 432–5
coexistence 189–95, 237–9, 302–6, 317, 321–30,
    351, 352, 446
collective models (even–even nuclei) 173–296
configuration mixing 129–31, 150, 162, 173–7,
    332–40
confinement 54–8

consistant Q formalism 276–88
Coriolis coupling 26, 221, 348–55, 364–97
correlations in ground state 401, 402
correlations of collective observables 314–30
Coulomb effects 80
Coulomb excitation 442
Coulomb force 50, 68, 69, 71
cranked shell model 397
critical points 316–30
critical point symmetries 328–30
crossing frequency 237, 206–10, 393–7
cubic terms (in IBA) 288

δ function residual interaction 98–134
Davydov model 240–51, 285, 286, 450, 451
decoupled bands 351, 385–97
decoupling parameter 367
deformation 173–397
deuteron 8, 14–16, 57–9, 86–90
drip lines 429–31
dynamic deformation theory 408
dynamical symmetry 252–88

E(5) 328–30
effective valence nucleon numbers 303–6
electron scattering 7, 8
energy staggering in γ-band 240–250
EUROBALL 216
even tensor interactions 153–7
evolution of structure 30–46, 173–9, 200, 297–
    330, 419–21, 425–42
exotic nuclei (nuclei far from stability) 62, 306–8,
    314, 319, 330, 418–52

Federman–Pittel mechanism 81, 302–4
Fermi gas model 58–62
fermion symmetry models 288
finite boson number effects 253–88
finite range interactions 133
f-spin (and multiplets) 287

g factors 14, 15, 304, 305
γ band energy staggering and γ-softness 248, 249

γ ray intensities, energy dependence 190, 214
γ ray transitions (see also B(E2) values) 184, 185,
    194–7, 208, 209, 233–6, 393–7
γ softness, rigidity 240–50, 267–75, 284–8
γ-vibrations 210–28, 240, 241, 408–14
Gammasphere 216, 236
GAMS 214
gap (pairing) 164–9
g-boson 288
geometric collective model 203, 288–96, 317, 320,
    321, 327
geometric interpretation of residual interactions
    107–9, 116–24, 128–36
geometric models 177–252, 288–96, 331–97
giant resonances 178, 179, 197
GRETA 216, 421
GRID 214
Grodzins rule 298
group theory (IBA) 252–76

halo nuclei 427, 431, 432
harmonic oscillator potential 65–7
harmonic vibrator Hamiltonian 180
Hartree–Fock method 50, 52
heavy ion reactions 233–6
hexadecapole deformation 381–5
high spin states 206–9, 230–40, 385–97

IBM (IBA) 252–88, 317, 320, 321, 327–30
identical bands 240, 367, 368
independent particle model 49–97
Inglis formula 200–2
integer nucleon numbers 315, 323–4
intrinsic excitations in deformed nuclei 202–88,
    331–55
intruder states, coexistence 189–95, 237–9, 302–
    6, 327–30, 446
isobaric multiplets, analogue resonances 86–97
ISOL method 422–5
isospin 10–13, 60, 84–97, 113–16

$j_1 j_2$-configurations 100–22, 125–40
j^2-configurations 100–11, 116–32
j^n-configurations 142–63

K=0, 2 excitations 21, 202–28, 240–4, 408–17
K mixing (see also Coriolis coupling) 221–8, 241–
    4
K quantum number, definition 205, 332, 333

l^2-potential 66–8, 346, 432
Lipas formula 232, 233
loosely bound nuclei 429, 431–7
LS–jj coupling 103, 104

macroscopic-microscopic models 62
magic numbers 7, 32, 33, 50–84, 297–314, 425,
    426, 432–8, 446, 447
magicity 50–84, 297–314, 302–6, 319
magnetic moments 14, 15, 304, 305
magnetic rotation 179, 252
magnetic substate localization 127, 128, 332–5,
    381, 383
mass filters 439–45
mean field 50–3
meson exchange 16
microscopic treatment of vibrations 398–417
Mikhailov plots 223–7, 282, 283
mirror nuclei 6, 10–13
moment of inertia 200–2, 232–40
monopole interaction 129, 302–6
monopole p–n interaction 302–6
Monte Carlo shell model technique 170
morphological classes of transition regions 299,
    300
m-scheme 141–4, 181, 182
multiparticle systems (shell model) 98–170
multi-phonon states 178–197, 213–16
multipole decomposition of residual interactions
    125–36, 302–4
multistate mixing 24–7
    Coriolis coupling 364–85

$N = Z$ nuclei 59–62, 90–2, 96, 438–41
n-hole configurations 138–41
Nilsson model 331–97, 408–17
NN potential 49, 50
normal parity orbits 70
$N_p N_n$ 300–15, 445–7
N-rich nuclei 431–7
nuclear force, short range, attractive 4–16, 71, 72,
    125–134, 241, 428–430
nuclear potential depth 58, 59
nuclear radii 6, 9, 429–434
nuclear size 3, 6, 429–432

O(6) 259, 267–75
octupole vibrations 194, 195, 251, 414, 415
odd particle blocking 395, 396
odd tensor interactions 148–53, 157
off-diagonal matrix elements 100, 129, 150, 162,
    175–7, 334, 338, 339
orbit dependence of nuclear forces 81–4, 302–6
origin of deformation 173–7

p, f bosons 288
pairing 123–5, 132, 133, 163–70, 365, 366, 435
pairing plus quadrupole (PPQ) model 124, 125
parabolic rule 128–36
parity, definition 69, 70

particle versus hole matrix elements 138–40
Pauli principle 4, 16, 17, 71, 72, 84–92, 99–120, 132, 133, 181, 182, 422–30
P-factor 310–13
phase coexistence 316–30
phase transitions 316–30, 451, 452
physical meaning of C-G coefficients 218
π mass and range of nuclear force 16
p–n interaction 10–16, 36, 37, 81–4, 111–24, 173–7, 191–3, 297–314, 435–41
potentials 51–8, 62–73
P-rich nuclei 437–42
probe size 127, 128
projectile fragmentation 422–5
prolate, oblate shapes 197–9, 337, 338, 346–8
proton radioactivity 440–2

Q-invariants 447–52
quadrupole deformation 197–397
quadrupole interactions 124–36
quadrupole moments 8, 15, 28, 153–7, 200, 416
quadrupole vibrations of spherical nuclei 178–97
quasi-particles 163–8, 200–2, 409
quenching of shell structure 301–6, 425–6, 432–8, 446, 447

residual interactions 49–53, 67, 98–177, 398–417, 428–30, 435–7
RIA 425
RNB production 422–5
RNBs 422–8
rotation alignment 385–97
rotational energies 204–10, 230–40, 247, 348–55, 385–97
rotation–vibration coupling 221–30, 240–2
RPA 398–417

s, d bosons *see* IBM, IBA
saturation of collectivity 311–13
scissors mode 178, 179
second quantization 180, 181
selection rules 184–7, 194, 195, 210–21, 245, 246, 259–76
seniority 146–63
separation energies 6–10, 13, 14, 163, 164, 214, 215, 419
separation of rotational and particle motions 331
shape coexistence 189–94, 327–30, 445, 446
shapes of nuclei 197–02
shell gaps, deformed 238
shell model 49–170
shell structure 64–84
signature efficiency 442–4
signature splitting 390
signatures of structure 297–330, 439–52

single-particle energies 64–9
    mass dependence 79, 80
single-particle transfer reactions 77, 78, 356–81
spherical-deformed transition regions 276–330
spin–orbit interaction 66–70, 432–7
square well potential 55–8
structural evolution, *see* evolution of structure
SU(3) 259, 262–7
subshell gaps 237–9, 302–6, 432–5
superdeformation 209, 237–40
superheavy nuclei 426, 427
surface δ-interaction 156
symmetry triangle 276–9, 290–1

$T = 1$, $T = 0$ states, interactions 10–13, 16, 84–97, 100, 111–23, 132, 133, 313, 439, 440
Talmi–Glaubman rule 137
TDA 398–417
transition regions 276–330
tunneling 57, 440–2
two state mixing 4, 18–29

U factors 165–7, 358–62, 365, 367
U(5) 255, 259–62
uncertainty principle 16, 56, 57, 65
unique parity orbit 70, 71, 333–5, 351, 377–97, 435, 436

valence correlation schemes 300–15
validity of single particle motion 435
valley of stability 31, 32, 60, 71, 91, 419, 429
V factors 165–7, 358–62, 365, 366
vibrations of deformed nuclei 210–16
vibrator model 177–97, 327–9
VMI 231

W.u., definition (E2) 42
weak coupling and rotational alignment 388
weakness of nuclear force 14–16, 72, 428, 429
Weizsäcker mass formula 61, 62
Wilets–Jean model 244–50, 267–269

X(5) 328–30

Young tableaux 181
YRAST ball 236
Yrast states 206–9, 234, 235

zero point motion: confinement and quantization 54–8, 65
Z_γ 221–8